KATHY GALANTE

Essentials
of
Cell Biology

Essentials
of
Cell Biology

ROBERT D. DYSON
Oregon State University

Allyn and Bacon, Inc.
BOSTON · LONDON · SYDNEY

Portions of this book first appeared in CELL BIOLOGY: A MOLECULAR APPROACH by Robert D. Dyson, © copyright 1974 by Allyn and Bacon, Inc.

Library of Congress Cataloging in Publication Data

Dyson, Robert D
    Essentials of cell biology.

    Includes bibliographies and index.
    1. Cytology. I. Title. [DNLM: 1. Cells. QH581.2
D998e]
QH581.2.D96   574.8'7   74-23489

ISBN 0-205-04649-5

ISBN 0-205-04664-9 (International)

Second printing . . . July, 1975

To Louise
And to Lisa, Leanna, and Ryan

Contents

Preface   ix

1

The Cellular Basis of Life   1

Cell Theory / The Procaryotic Cell / The Eucaryotic Cell / Viruses / Observing Cells / The Choice of Experimental Systems in Biological Investigations

2

Chemical Bonds and Chemical Equilibrium   65

Gibbs Free Energy and Related Parameters / Bond Energy and Chemical Equilibrium / The Biologically Important Weak Bonds / Energy Changes in Chemical Reactions / Acids and Bases / The State of Intracellular Water

3

Molecular Architecture and Biological Function   93

The Importance of Carbon to Biological Structure / Lipids / Carbohydrates / Nucleic Acids / Proteins

4

The Energetics and Control of Cellular Reactions   141

Enzymes / Enzyme Kinetics / Feedback Regulation / ATP and Coupled Reactions / Glycolysis / Biological Oxidations

5

Mitochondria and Chloroplasts   179

The Mitochondrion / Electron Transport and Oxidative Phosphorylation / The Citric Acid Cycle / The Relationship Between Mitochondrial Structure and Function / The Chloroplast / The Photosynthetic Light Reaction / The Photosynthetic Dark Reaction / Photosynthesis in Procaryotic Cells / Origin of Mitochondria and Chloroplasts

6   Membranes and the Regulation of Transport   230

Membrane Structure / The Membrane as a Passive Barrier / Membrane Transport / Metabolically Coupled Transport / Endocytosis

7   Excitability and Contractility   266

The Neuron / The Nerve Impulse / Propagation of the Nerve Impulse / Muscle Cells and the Mechanism of Contraction / Contraction–Relaxation Control / Neuromuscular Interaction / Contraction in Non-Muscle Systems

8   Genes and Genetic Control   317

The Genetic Concept / The Genetic Message / Protein Synthesis / Gene Regulation in Procaryotes / Gene Regulation in Eucaryotes

9   Cell Division and Its Regulation   382

Cell Division in Procaryotes / Replication of the Eucaryotic Nucleus / Division of the Eucaryotic Cell / The Regulation of Cell Division

10   Cellular Differentiation   427

Gene–Cytoplasm Interaction in Development / Intercellular Communication / Stability of the Differentiated State / Senescence

Index   455

Preface

The goal of this work was to give a unified description of cellular structure and function at the introductory level, presenting the core of our current knowledge in order to build for the student a framework into which new facts will fit as they become available. Unfortunately, however, there seems to be no general agreement as to what topics should be included in such a course. For that reason, I have tried to design a format having a maximum of flexibility. Thus, while the chapters form what to me seems a logical sequence, they are written with a minimum of dependence on the order given here. In addition, chapters are broken down into sections, each of which has its own summary and study guide. Therefore, instructors can rearrange topics with relative ease, or delete material to take advantage of varying needs.

This book is an obvious offspring of its longer (by half) forerunner, "Cell Biology: A Molecular Approach." Some sections have been transferred almost intact from the previous work, others have been deleted entirely. This is not just a simplified abridgement, however, for there has also been extensive rewriting of many subjects, the expansion of others, and the introduction of some new topics. The latter were made at the suggestion of a number of scientists who generously transmitted to me or to the publisher their comments on the longer book after it was published. And of course, every effort was made to bring both the text and references as up to date as possible.

ACKNOWLEDGMENTS

The author is grateful to the following people for their expert counsel and criticism of various portions of this manuscript or the materials from which it was developed: Carl Baker, Richard Barsotti, Anita Bolinger, Janet Cardenas, Ernst Florey, Dean Fraser,

Eric Holtzman, Benjamin Kaminer, Eugene P. Kennedy, John Menninger, Margot Pearson, Hayden Pritchard, and Ralph Quatrano. Special thanks are due Professors Bolinger, Fraser, and Pritchard for reviewing the entire manuscript at a late stage of its development. Whatever inaccuracies appear in the final version were almost certainly introduced by the author too late to be caught by any of the aforementioned.

A great many others contributed in substantial ways to this work, among whom are a number of the author's colleagues at Oregon State University, more than a hundred investigators who provided micrographs, numerous publishers who gave permission to use material copyrighted by them, and Jacqueline Atzet, who prepared most of the drawings.

The staff at Allyn and Bacon have the author's deepest gratitude for their patience and skill, in particular Editor Jay Alexander and Senior Editor William Roberts. It was the latter who both conceived the project and guided it through to its conclusion.

Essentials
of
Cell Biology

CHAPTER 1

The Cellular Basis of Life

1-1 CELL THEORY   2
The Discovery of cells
The Rise of Molecular Biology
The Unity and Diversity of Cells

1-2 THE PROCARYOTIC CELL   10
Procaryotic Diversity
Structure of Procaryotic Cells

1-3 THE EUCARYOTIC CELL   16
Cell Wall
Plasma Membrane
Nucleus
Chromosomes
Nucleolus
Golgi Apparatus
Endoplasmic Reticulum
Peroxisomes
Lysosomes
Vacuoles
Plastids
Mitochondria
Cytoplasmic Filaments
Microtubules
Centrioles and Basal Bodies
Cilia and Flagella
Cell Fractionation

1-4 VIRUSES   34
Discovery
Bacteriophages
Life Cycle of the Bacteriophage
The Structure of Animal and Plant Viruses
Life Cycle of Animal and Plant Viruses
A Comparison of Cells and Viruses

1-5 OBSERVING CELLS   47
Resolution of the Light Microscope
The Electron Microscope
The Scanning Electron Microscope
Advantages and Disadvantages of the Three Types of Microscopy
Sample Preparation

1-6 THE CHOICE OF EXPERIMENTAL SYSTEMS IN
     BIOLOGICAL INVESTIGATIONS   60

SUMMARY   61
STUDY GUIDE   63
REFERENCES   63

*Long ago it became evident that the key to every
biological problem must finally be sought in the
cell; for every living organism is, or at some time
has been, a cell.*

E. B. WILSON, *THE CELL IN DEVELOPMENT AND
HEREDITY.* THE MACMILLAN CO., 1925.

1-1 CELL THEORY

Although the foundations of cell biology were formed in the seventeenth century, it was only in the nineteenth century that the nature and importance of cells became fully recognized. It was during this period that the first serious attacks were made on the problem of determining the molecular nature of life and of living things.

THE DISCOVERY OF CELLS. In 1665, an Englishman by the name of Robert Hooke (1635–1703) published a collection of essays under the title *Micrographia*. One essay described cork as a honeycomb of chambers, or "cells." The chambers are now recognized to be the rigid remains of dried cells (see Fig. 1-1). Hooke thought of the cells he observed as similar to the veins and arteries of animals. He could tell that they were filled with "juices" in living plants, but his microscope did not permit the observation of any intracellular structure, an observation that would have dispelled the notion that cells are merely partitioned channels for the passage of material.

Within a decade after the publication of Hooke's essays, the Dutch inventor and scientist Anton van Leeuwenhoek (1632–1723) had succeeded in greatly improving the art of polishing lenses of short focal length, and had used his lenses to describe a host of "little animals," many of which proved to be single cells (see Fig. 1-2). Leeuwenhoek's microscope also provided the resolution

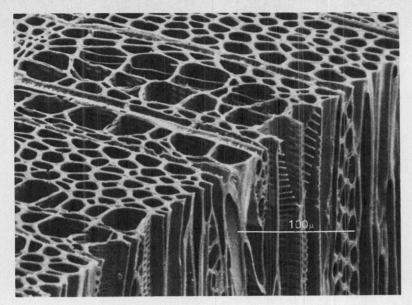

FIGURE 1-1. The Three-Dimensional Structure of Wood. Scanning electron micrograph of magnolia shows the honeycomb of chambers left when wood is dry. A similar pattern in cork was described by Hooke as an array of "cells," although each chamber is actually the rigid remains of a dried "cell," as we now use that word. (Courtesy of B. A. Meylan, Physics and Engineering Lab., Dept. of Scientific and Industrial Research, New Zealand.)

needed to delve into the structure of the cells themselves. However, to take advantage of that capacity, better ways of preparing tissues for examination were needed. By the end of the seventeenth century, considerable progress in tissue preparation had been made, so that microscopists were able to examine slices as thin as 10 microns (10 μ), stained in various ways to bring out different details.[1]

With improved microscopes and preparation techniques, R. J. H. Dutrochet was able to conclude by 1824 that all animal and plant tissues are actually aggregates of cells of various kinds, and that growth results from an increase in either the size or the number of cells, or both. Though Dutrochet recognized that intact cells can be separated from each other, he did not realize that each is capable of its own independent existence including, in most cases, the ability to reproduce itself.

The next few decades saw a rapid increase in the understanding and appreciation of cells. In 1838-1839 the German biologists M. J. Schleiden and Theodor Schwann presented convincing arguments for the position that each cell *is* capable of maintaining an independent existence. Their arguments had a profound influence on the scientific community. Later, in 1858, Rudolf Virchow

---

1. "Micron" (μ) is another way of saying "micrometer" (μm), i.e., 10^{-6} m or 1/25,400 inch. (There is a trend toward the use of μm instead of μ.) It is a convenient unit of size when discussing cells and their components, although its subdivision, 10^{-9} m, called a nanometer (nm) or "millimicron" (mμ), is also used. However, when discussing atoms, molecules, and things of comparable size, we will usually use the traditional angstrom unit (Å), where 10 Å = 1 nm or 1 mμ; 10,000 Å = 1 μ; and 10^{8} Å = 1 cm.

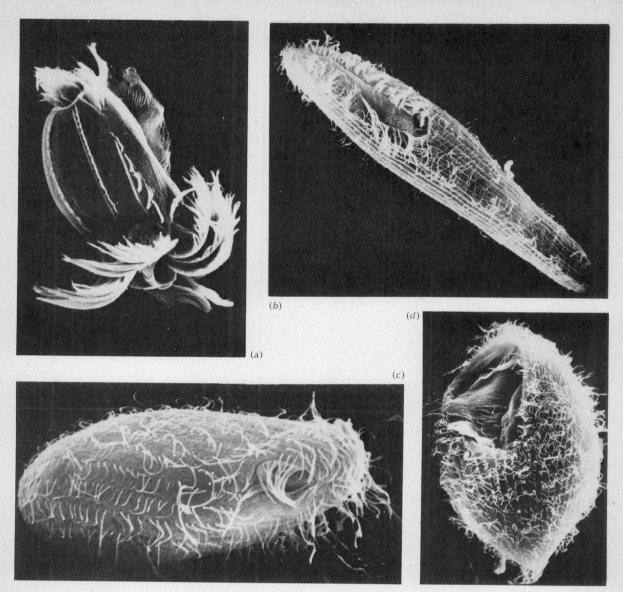

FIGURE 1-2. Leeuwenhoek's "Little Animals"—as he never saw them. These scanning electron micrographs are of protozoa. Each is a single cell. **(a)** *Uronychia* sp., 100 μ long. **(b)** *Blepharisma* sp., about 140 μ. **(c)** *Tetrahymena pyriformis W,* about 25 μ. **(d)** *Turaniella* sp., 250 μ. [Courtesy of E. B. Small and G. Antipa. (a) and (b) from E. Small, D. Marszalek, and G. Antipa, *Trans. Am. Micros. Soc.* **90:**283 (1971).]

published his classic textbook, *Cellular Pathology,* which argued forcefully for the concepts suggested by Schleiden and Schwann and for the extension that attributes every cell to a pre-existing cell (*omnis cellula e cellula*—every cell from a cell). Virchow's ideas completed what has come to be known as the *cell theory* or *cell doctrine,* which holds that: (1) all living things are composed of one or more units called cells, (2) each cell is capable of maintaining its vitality independent of the rest, and (3) cells can arise only from other cells.

THE RISE OF MOLECULAR BIOLOGY. Prior to Virchow, the tendency had been to regard the cell as a structural unit rather than a functional unit of living systems. However, in his book, adapted from a series of lectures given at the Berlin Institute of Pathology, Virchow presented disease as an aberration in normal cellular processes. The properties of a tissue—or of a whole organism—must therefore be due to the properties of its individual cells. Between the 1858 publication of Virchow's book and 1925 when E. B. Wilson wrote the words with which this chapter opened, the cell doctrine, as we now understand it, became firmly entrenched. During that period also, the influence of the physical sciences on biology was growing ever stronger, complementing the older discipline of physiology with the new fields of biochemistry and biophysics.

While the cell doctrine was being formulated, biochemists (though the word itself was not coined until the end of the century) were working hard to dispel the "vitalist" notion that cellular activities are somehow immune to the laws of chemistry and physics. The first blow came with the laboratory synthesis of urea ($H_2N—CO—NH_2$) by Friedrich Wöhler in 1828, proving that organic compounds are not formed by any mysterious process, but by ordinary chemical reactions. Synthetic organic chemistry flourished thereafter, and with it came the first clues to the nature of the metabolic transformations by which a cell turns its foodstuffs into a vast array of quite different molecules.

By the middle of the nineteenth century, considerable controversy had arisen concerning the nature of one commonly observed metabolic transformation, alcoholic fermentation. Alcoholic fermentation is a process by which glucose is changed to ethanol and carbon dioxide. The German chemist Justus von Liebig (1803–1873) maintained that the process is catalyzed by nonliving "ferments," later called *enzymes*, that are naturally present in the juices being fermented. He was hotly opposed by those who would attribute the transformation to a vital process rather than to a chemical process.

In 1871 a Frenchman, Louis Pasteur (1822–1895), demonstrated that a living organism—yeast—is essential to alcoholic fermentation as that process usually occurs. Although vitalists felt that Pasteur's observation proved their point, Eduard Buchner found in 1897 that it is not necessary for the yeast actually to be alive: extracts from yeast cells can also carry out alcoholic fermentation. While the enzymes responsible are constructed of simpler organic materials by the yeast cells, the function of enzymes does not depend on the vitality of the cells that make them. Thus, Liebig's earlier contention that alcoholic fermentation is a straightforward chemical process was vindicated, even though Liebig did not recognize the source of the enzymes responsible for it.

At the turn of the century, then, the identity and importance of individual cells were recognized and serious efforts were being made to understand their function. Buchner's work established that individual components such as enzymes could be isolated and studied in the laboratory. Although the success of his approach led many to view the cell as little more than a "bag of enzymes," the interdependence of structure and function gradually became apparent. This recognition fostered an increasing cooperation between those biologists interested in the structure of individual cells (cytologists), or the cellular structure of tissues and organs (histologists), and those whose primary interest is in function (biochemists, biophysicists, and physiologists). As it became more and more difficult to distinguish one discipline from another, interdisciplinary studies, variously called *molecular biology* and *cell biology*, appeared. This pooling of knowledge and experience permitted the rapid expansion of our understanding of cellular processes, some of the early milestones of which are summarized in Table 1.1

THE UNITY AND DIVERSITY OF CELLS. We now recognize that cells are found in an enormous variety of sizes and shapes, representing their evolutionary adaptation to different environments or to different specialized functions within a multicellular organism. Cells range in size from the smallest bacteria, only a few tenths of a micron in diameter, to certain marine algae and to the yolks of bird eggs with dimensions of centimeters. Between these two extremes, one finds that most bacteria have measurements of a couple of microns, human red blood cells have diameters of 6 to 8 microns, and most other human cells fall in the range of about 5 to 20 microns.[2] (See Fig. 1–3.)

For all their apparent diversity, however, cells have many characteristics in common, the most basic of which is the potential for an independent existence. That is, cells have the ability to continue living in the absence of any other cell, a capacity that requires first, a metabolic machinery capable of obtaining energy from its environment through the capture of light or the degradation of some chemical foodstuff, and second, the ability to use this energy to support essential life processes. These processes include the movement of components from one part of a cell to another, the

---

2. In discussing very small objects, it helps to keep in mind that the human eye has a resolving power of about 0.1 mm, or 100 μ. In other words, when two objects get closer than 0.1 mm, they begin to appear as one. The light microscope stretches the eye's resolving power to about 0.2 μ, while electron microscopes have resolutions of a nanometer (10 Å) or better. Thus, for example, a 20-micron human cell is visible with even a very simple magnifying device, and the better light microscopes are capable of revealing much of its detail.

TABLE 1.1 Some early milestones in cell biology

1665	Hooke publishes *Micrographia*, in which he describes and illustrates the cellular structure of cork.
1675	Leeuwenhoek improves microscope lenses, and with them discovers a variety of single-celled life forms, including (in 1683) bacteria.
1824	Dutrochet correctly concludes that all tissues, animal and plant, are composed of smaller units, the cells.
1828	Wöhler synthesizes urea, discrediting the view that organic compounds can only be made by living things, and paving the way for a systematic investigation of cellular reactions.
1830	Meyen suggests that each plant cell is an independent, isolated unit capable of receiving nourishment and building its own internal structures.
1831	Brown reports the existence of nuclei.
1838–9	Schleiden and Schwann argue convincingly for the cell doctrine, holding that all tissues are composed of cells and that the metabolism and development of tissues are the result of cellular activity.
1840	Liebig proposes that alcoholic fermentation is a purely chemical reaction, independent of living cells but catalyzed by substances (ferments, or enzymes) that are naturally present in juices.
1858	Virchow correctly asserts that cells arise only from other cells and that, as the functional units of life, they are also the primary site of disease.
1862	Pasteur disposes of the spontaneous generation theory of microbial appearance.
1871	Pasteur proves that natural alcoholic fermentation always involves the action of yeast.
1897	Buchner finds that alcoholic fermentation requires only the extract of yeast, not the cells themselves.
1907	Harrison finds a satisfactory way of growing isolated animal cells *fish* in the laboratory, so that future studies of cellular function can be carried out under controlled conditions.

selective transfer of molecules into and out of a cell, and the ability to transform molecules from one chemical configuration to another in order to replace parts as they wear out or to support growth and reproduction. In addition to this metabolic machinery, a cell must have a set of genes that act as blueprints for the synthesis of other components. And finally, a cell must have a physical delimiter—a boundary between it and the rest of the world—called the cell membrane.

Most of our attention in the ensuing chapters will be devoted to the similarities among the various cells rather than to their differences, for once the discussion begins to focus on the molecular basis for cell function, differences among cells appear as minor

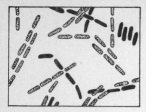

Bacteria: Rods

Flagellated rods

Spirals

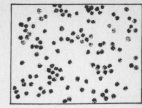

Spheres

Spermatozoon
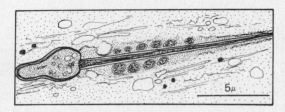

5μ

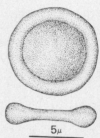

5μ

Mammalian red blood cells
(side and face views)

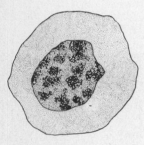

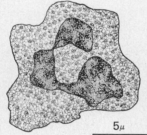

5μ

White blood cells

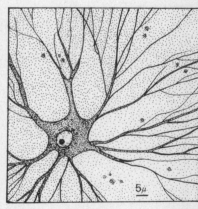

5μ

Neuron

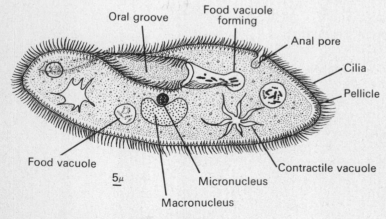

Oral groove

Food vacuole
forming

Anal pore

Cilia

Pellicle

Food vacuole

5μ

Micronucleus

Macronucleus

Contractile vacuole

Paramecium

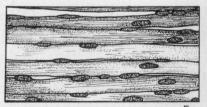

Multinucleate muscle cells

5μ

FIGURE 1-3. The Diversity of Cells. Bacteria are among the smallest of all cells, with typical lengths of a micron or so. Animal and plant cells are more often 10–20 μ in diameter, although some, such as nerve and muscle cells, may be very long with relatively small diameters. Single-celled animals, like the *Paramecium* (a protozoan) shown here, may be incredibly complex and quite large (over 200 μ).

variations on a few central themes. Before attacking the subject at that level, however, we will pause to take a look at some of the more obvious physical features of the typical cell, as those features appear to the microscopist, and to introduce some of the nomenclature used by cell biologists. For these purposes, all cells are divided into two groups, *eucaryotic* and *procaryotic,* according to whether or not their genes are contained in a well-defined nucleus. Only the tiny bacteria and blue-green algae (which are not necessarily either blue or green) belong to the simpler, procaryotic group, to be discussed first (see Fig. 1–4).

FIGURE 1-4. Bacteria Lying on the Surface of Human Forearm Skin. The crevices separate adjacent cells. Note the rod-shaped particle, presumably a bacterium, lying in one of the major crevices. This scanning electron micrograph is presented to emphasize the difference in size between a typical procaryotic and eucaryotic cell. [Courtesy of E. O. Bernstein and C. B. Jones, *Science,* **166**:252 (1969). Copyright by the A.A.A.S.]

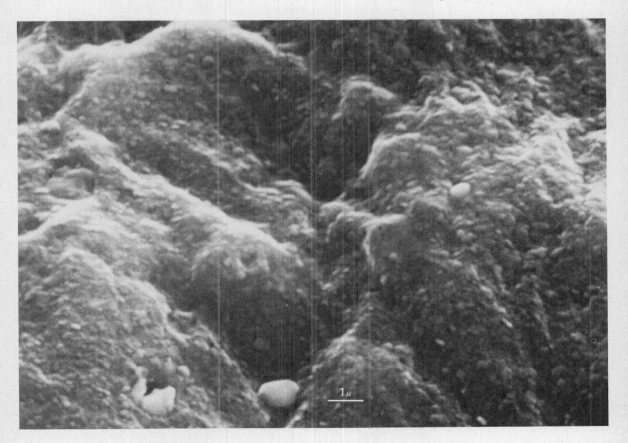

1μ

1–2 THE PROCARYOTIC CELL

Bacteria comprise the largest group of procaryotes. Although they have been identified in fossil remains some two billion (2×10^9) years old, they were unknown until 1683, when Leeuwenhoek first reported seeing them with his improved microscope. He noted the major morphological classes of bacteria—spheres, rods, and spirals—and described them in letters to the Royal Society in London. Leeuwenhoek found that bacteria tend to be associated with decaying organic matter from which, it was generally believed, they arise spontaneously. This concept of bacterial origin persisted until the mid-nineteenth century when Pasteur, using newly developed sterile techniques, proved that bacteria do not arise by spontaneous generation, but by contamination. When Pasteur sealed boiled organic material (e.g., milk) in its container, no bacterial growth could be detected; but when the flask was opened to the air for a time, cells invariably appeared. Pasteur correctly concluded, in his 1862 monograph on spontaneous generation, that airborne microbes were the source of the bacterial growth (see Fig. 1–5). Bacteria, it seems, are no exception to the cell doctrine.

Bacteria (*Schizomycetes*), along with the small group of blue-green algae (*Cyanophyta*), have in common a uniquely simple cell structure that sets them apart from all other forms of life. Because of this simplicity, they are often relegated to a kingdom of their own, *Monera*, rather than to either the plant or animal kingdom. In older schemes, bacteria were either classified as plants or, together with other single-celled life forms, as *Protista*. In spite of their structural simplicity, bacteria and the blue-green algae are true cells. They are bounded by a membrane and contain the genes and metabolic apparatus needed to grow and reproduce, given the proper nutrients and environment.

FIGURE 1–5. Pasteur's Experiment. This experiment did more than any other to convince scientists that life does not arise spontaneously from organic matter. Pasteur found (a) that if he boiled a solution of organic compounds and left the flask undisturbed (b) no contaminating growth would appear. But when a flask was opened directly to the air (c) so that dust could settle into it, a variety of microscopic organisms was soon found flourishing therein (d). The long, curved neck of the boiled flask, though unsealed, was an effective trap for dust particles.

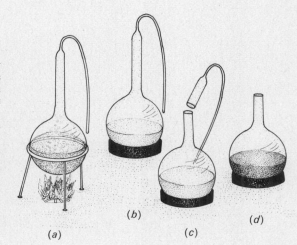

(a)   (b)   (c)   (d)

PROCARYOTIC DIVERSITY. Bacteria can be distinguished from blue-green algae in several ways, one of which is that blue-green algae evolve oxygen whereas bacteria do not. Oxygen is released by blue-green algae and higher plants as a by-product of photosynthesis, which is the process whereby light energy is converted to chemical energy. Although some bacteria are also photosynthetic, they use quite a different mechanism, including a different set of light-capturing pigments. Bacteria, for instance, never have chlorophyll *a*, a green pigment common to all other photosynthetic systems.

Bacteria may be placed into several large groups according to shape: *cocci* (sing., coccus) for spherical cells, *bacilli* for rods, and *spirilla* for helical cells[3] (see Fig. 1–6). The cocci, especially, are prone to hang together after they are formed by binary fission of the parent cells. They are referred to as diplococci when in pairs, streptococci when in chains or filaments, and staphylococci when found in clusters.

Bacteria are also broadly grouped according to their response to a staining procedure developed in 1884 by the Danish bacteriologist Christian Gram. Those cells that are stained by Gram's procedure are called gram-positive, the others are gram-negative. The distinction is important because of a fundamental difference in the structure of the walls of the two classes of cells and a corresponding difference in their sensitivity to various chemical agents other than the stain. For example lysozyme, a bactericidal enzyme found in egg whites, human tears, and elsewhere, is much more effective against gram-positive cells than against gram-negative cells. Penicillin also shows much greater bactericidal action against gram-positive bacteria.

Still another variable among bacteria is nutrition. If, as in all animal cells, complex organic molecules of the kind that would normally be supplied only by the destruction of other cells are required, the organism is called *heterotrophic*. If, however, a cell can utilize carbon dioxide (CO_2) from the air, reducing it to the organic compounds needed, the cell is called an *autotroph* ("self-feeder"). If the energy required for carbon dioxide reduction is obtained from light via photosynthesis, as in ordinary green plants, the cell is *photoautotrophic*. If energy is supplied by the oxidation of inorganic compounds (e.g., H_2S to S, or H_2 to H_2O), the organism is *chemoautotrophic*. Members of this latter category, along with photosynthetic heterotrophs, are uncommon.

Bacteria also differ from one another in their requirement for oxygen. Some are *aerobes*, meaning that they grow only in the pres-

---

3. There is also a genus *Bacillus*, of which the common hay bacterium *Bacillus subtilis* is an example. Although a *Bacillus* is rod-shaped, not all rod-shaped cells belong to that genus. A similar situation exists with the spiral-shaped cells of the genus *Spirillum*.

FIGURE 1–6. Some Representative Bacteria. Parts (a) and (b) are scanning electron micrographs (S.E.M.'s) of *Staphylococcus aureus* and *Streptococcus pyogenes*, respectively. Both bacteria are about 0.7 μ in diameter. (c) S.E.M. of *Proteus mirabilus*. The larger cell, which is about 1.5 μ long, would normally divide to yield two smaller cells such as those next to it. (d) Light micrograph of *Salmonella typhosa*, a flagellated rod. The cell bodies are 1–3.5 μ long. (e) *Spirillum volutans*, a 3–5 μ flagellated spiral. [(a), (b), and (c) courtesy of D. Greenwood, St. Bartholomew's Hospital, London. Part (a) from D. Greenwood and F. O'Grady, *Science*, **163**:1076 (1969), copyright by the American Association for the Advancement of Science (A.A.A.S.). (d) and (e) from the Turtox Collection, courtesy of General Biologicals, Inc.]

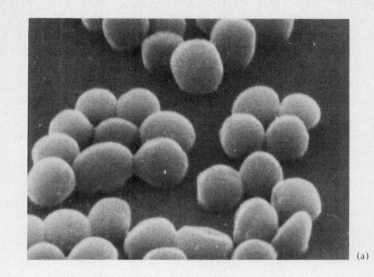

(a)

(b)

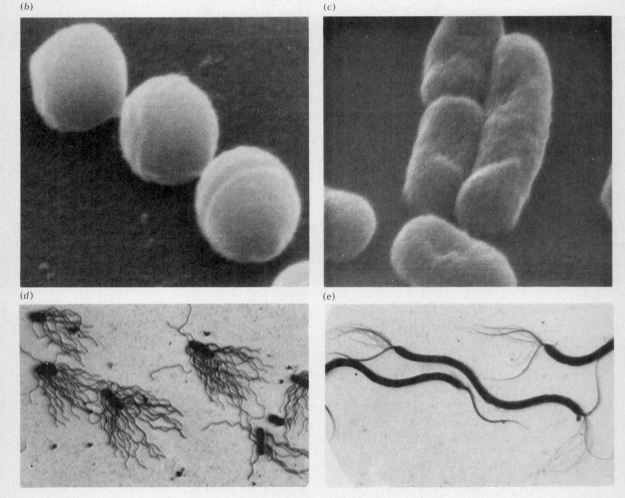

(c)

(d)

(e)

ence of oxygen. *Anaerobes*, on the other hand, grow only without oxygen. Those cells that can grow either in the presence or absence of oxygen are called *facultative* aerobes or facultative anaerobes. Some facultative organisms utilize oxygen when it is present, whereas others ignore it.

Genus and species names are assigned to bacteria on the basis of a combination of these and other physical and chemical differences, but not without occasional ambiguity. In recent years, attempts to bring greater order to the field of bacterial taxonomy have focused largely on measuring the genetic similarity between various strains of cells, using techniques that will be described in later chapters.

STRUCTURE OF PROCARYOTIC CELLS. Bacteria and blue-green algae generally have dimensions of a few microns, although the smallest bacteria (mycoplasmas, rickettsiae, and chlamydiae) measure only a few tenths of a micron. Many bacteria have appendages in the form of *flagella* (see Figs. 1-6 and 1-7), which are 0.01 to 0.02 μ in diameter and up to 10 or 11 μ long. The bacterial flagellum, which is a simple structure consisting of parallel strands of protein wound about each other in ropelike fashion and anchored within the cytoplasm at a *basal body*, can impart motility to the cell with a whiplike flailing movement. In addition to flagella, certain bacteria have numerous finer projections called *pili* ("hair"; sing., pilus) or fimbriae ("fringe"), with diameters of 0.01 μ or less and lengths up to a micron or two.

The surface layer of procaryotes is very often a gelatinous *capsule* or, in the case of the blue-green algae, a *sheath*. Many pathogenic bacteria can no longer cause an active infection when deprived of their capsule, apparently because of the protection it affords against host defenses.

Inside the sheath or capsule, when one is present, is a rigid *cell wall* (see Fig. 1-8). Cellulose is a common component of this structure in the blue-green algae, just as it is in the higher plants;

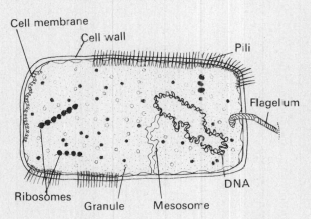

Cell membrane

Cell wall

Pili

Flagellum

DNA

Ribosomes

Granule

Mesosome

FIGURE 1-7. The Generalized Bacterium. Almost all bacteria are surrounded by a cell wall, which is attached to the cell membrane at relatively few points. Either flagella or pili or both may extend from the cytoplasm through the wall. Ribosomes, granules of various kinds, and DNA are found in the cytoplasm, which frequently also includes an inward extension of the membrane known as a mesosome. In some cases mushroomlike stalks have been reported on the inner surface of the plasma membrane (left side of drawing).

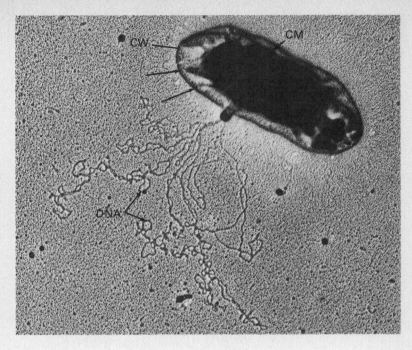

FIGURE 1–8. A Lysed Bacterium. Because of a loss of some cytoplasm, the plasma membrane has collapsed inward, clearly showing that the cell membrane (CM) and cell wall (CW) are separate structures, connected at only a few points (arrows). The extruded fiber is a molecule of deoxyribonucleic acid (DNA), which forms the genetic material of the cell. (Courtesy of M. M. K. Nass, from *Biological Ultrastructure: The Origin of Cell Organelles*, P. J. Harris, ed., Corvallis: Oregon State Univ. Press, 1971.)

sometimes protoplast cells can not divide

however, bacteria have walls composed of materials (e.g., teichoic acids) not found anywhere else. Gram-positive bacteria have walls varying in thickness from about 0.015 to 0.08 μ, comprising some 10–25% of the dry weight of the cell. The gram-negative cell wall, though thinner (generally about 0.01 μ), is a more complex structure, accounting for the different response to the gram stain and to certain antibiotics, as mentioned above.

The bacterial capsule, sheath, and cell wall are constructed from materials secreted by the cells. That these structures are not essential to viability under favorable conditions can be demonstrated by chemically removing them, producing a *protoplast* if removal of the wall is complete, or a *spheroplast* if removal of the wall material is incomplete. Such cells are still capable of growth and replication. In fact, one genus of bacterium, the tiny *Mycoplasma* (formerly called pleuro-pneumonia-like organisms, or PPLO), are always found without walls.

Although a cell might survive very well without flagella, pili, capsule, or walls, it cannot survive without its membrane. This structure encloses the cytoplasm ("cell plasma") or protoplasm ("chief plasma") of the cell, and hence is also called the *cytoplasmic membrane, plasma membrane,* or *plasmalemma*. The membrane is an essential barrier through which nutrients must pass on their way into the cell, and through which secretions such as waste products are passed out of the cell. It exercises considerable selectivity in this traffic, a selectivity that is vital to the continuing life of the cell.

The plasma membrane is a deceptively simple structure, only about 75 to 100 Å thick. Nevertheless, a variety of different molecules are known to be associated with it, some of which act as ferries for the transport of specific materials into and out of the cell (see Chap. 6). In addition, the membranes of some aerobic bacteria appear to have mushroomlike structures pointing toward the interior of the cell (see Fig. 1-7), which are thought to be the major site of oxygen utilization.

While cells from higher organisms have numerous inclusions that are surrounded by their own individual membranes, procaryotic cells are devoid of such inclusions. The small size of procaryotes drastically limits the opportunity for internal structure, since their longest dimension may be the equivalent of only a few thousand hydrogen atom diameters. However, the photosynthetic procaryotes (blue-green algae and the green and purple bacteria) have layers of membranes (*lamellae*) within their cytoplasm that contain light-capturing pigments and certain other parts of the photosynthetic apparatus. In addition, *mesosomes*, which are inward folds of the cytoplasmic membrane, can sometimes be identified in bacteria (see Fig. 1-9).

The bacterial cytoplasm, though it is an impressively complicated chemical factory, has little structure. Its most notable feature is a great number of *ribosomes*, which are roughly spherical particles about 0.02 μ in diameter. Some are attached to the plasma membrane, while others are free in the cytoplasm. As we shall see later, ribosomes are the sites of protein synthesis. Hence, cells that are growing more rapidly, and which therefore require faster protein synthesis, also have more ribosomes—up to 40% of the cell's total dry weight. Aside from ribosomes, granular inclusions found in the cytoplasm of procaryotes merely represent a way to store important materials, and thus they come and go as the nutrition and environment of the cell change.

The information needed to guide the construction and continuing functioning of a cell (any cell) is found in its genes. Genes are segments of long molecules of *deoxyribonucleic acid*, or *DNA*. A procaryotic cell requires only one long double-stranded DNA molecule to hold its entire set of genes, although more than one copy of the DNA molecule may be present. When stretched out, DNA molecules (in both procaryotes and higher cells) are often a thousand times longer than the cell itself (see Fig. 1-8).

Many of the procaryotic structures just described have names adapted from somewhat similar structures found earlier in the cells of higher organisms. Often, however, there is a fundamental difference in organization or function. The nucleus is a case in point,

FIGURE 1-9. Bacterial Mesosome. The mesosome represents an inward-folded, convoluted extension of the plasma membrane. As in this micrograph of *Bacillus subtilis*, it is often found at the site of cell division (just beginning at the points marked by outside arrows) and may have DNA attached (inside arrow). [Courtesy of N. Nanninga, *J. Cell Biol.*, **48**:219 (1971).]

for the nuclear genes of higher cells are enclosed (most of the time) in their own membranous structure, called the nuclear envelope. This is a "true nucleus," and cells that have one are called *eucaryotes*, which is what that name means. Procaryotes, as that name implies, have a more primitive nuclear arrangement consisting of unenclosed DNA.

1–3 THE EUCARYOTIC CELL

All the cells from higher animals, and those of many microscopic organisms as well, are eucaryotic. There is, of course, tremendous diversity among eucaryotic cells, but most features are common to nearly all. It is the unifying elements that will occupy most of our attention, starting here with a brief survey of the more obvious structural elements, most of which will be discussed in much more detail in later chapters. The shape and position within a "typical" plant and animal cell of each of the *organelles* to be discussed can be appreciated by referring to Fig. 1–10.

CELL WALL. Many plant cells have a rigid cell wall outside their plasma membrane. Composed chiefly of cellulose and (especially in woody tissues) lignin, the cell wall is what Hooke found when he examined cork under the microscope. The hardness of plant tissues, as opposed to most animal tissues, is due to the presence of such walls.

PLASMA MEMBRANE. Cell membranes have much the same structure wherever we find them, although the mushroomlike projections that are sometimes present on the inner surface of bacterial membranes are never present in eucaryotes. The plasma mem-

FIGURE 1–10. The Generalized Eucaryotic Cell. **(a)** A complete cell. **(b)** A portion of the nucleus, showing the two layers of the nuclear envelope and two centrioles nearby. **(c)** A cilium (or flagellum) extending from its basal body. (Note that the basal body and centriole have the same construction.) An anchor, in the form of a rootlet, is seen. **(d)** Detail showing a mitochondrion. On one side is a well-developed Golgi body, on the other side rough endoplasmic reticulum. Note vesicle formation from the endoplasmic reticulum and the filaments just under the cell membrane. **(e)** A plant cell. It differs from the animal cell primarily in having a rigid cell wall with penetrating plasmodesmata (pores), in the size of the vacuoles, and in having chloroplasts, here depicted as being about the size of mitochondria.

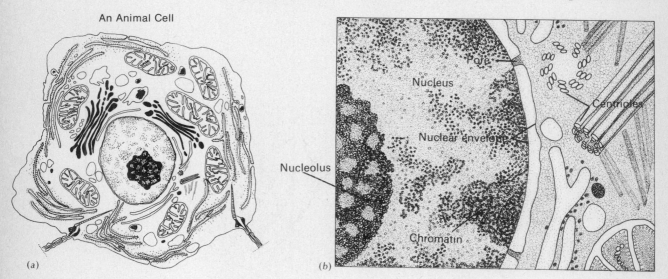

An Animal Cell

(a)

(b)

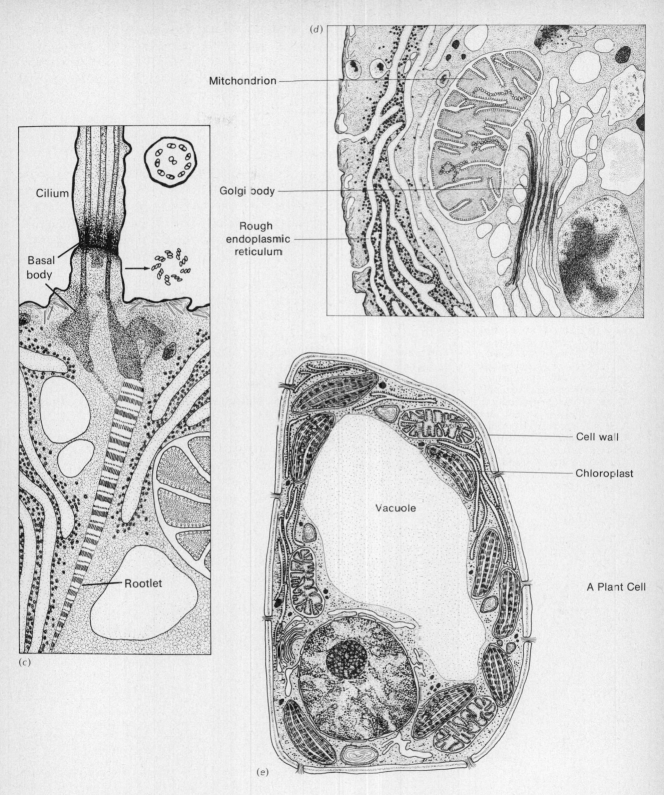

(d)

Mitchondrion

Golgi body

Rough
endoplasmic
reticulum

Cilium

Basal
body

Rootlet

(c)

Cell wall

Chloroplast

Vacuole

A Plant Cell

(e)

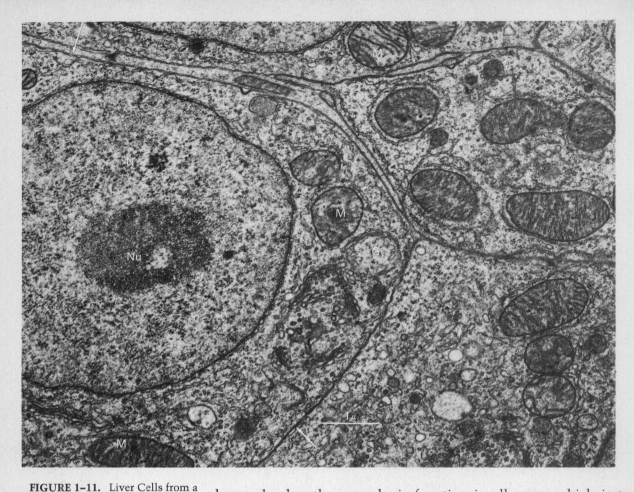

FIGURE 1–11. Liver Cells from a Chick Embryo. This electron micrograph shows portions of several cells. The cell at the lower left (bounded by the plasma membrane running between the two arrows) has a nucleus (N) with a prominent nucleolus (Nu). The nuclear envelope is composed of two membranes, the outer one of which is wavy. Also visible are the Golgi apparatus (G), mitochondria (M), and numerous vesicles within the cytoplasm. [Courtesy of C. A. Benzo and A. M. Nemeth, *J. Cell Biol.*, **48:**235 (1971).]

other genetic
information found
in cytoplasm and
mitochondria

brane also has the same basic function in all cases, which is to maintain the integrity of the cell contents by controlling what gets in and out.

NUCLEUS. The most prominent internal feature of most cells is the nucleus (Fig. 1–10b), discovered by Robert Brown, who also gave us "Brownian motion." The nucleus, which is the repository of nearly all of a cell's genetic information, is bounded by the *nuclear envelope.* (See Figs. 1–11 and 1–12.) The nuclear enve-

FIGURE 1–12a. The nuclear envelope. Seen from its cytoplasmic side after a procedure called freeze fracture and etching (see section 1–5). The outer membrane is broken away in patches, revealing portions of the inner membrane. Note pores (arrows). Large arrowhead indicates direction of metal shadowing. [Courtesy of G. G. Maul, J. W. Price, and M. W. Lieberman, *J. Cell Biol.*, **51:**405 (1971).]

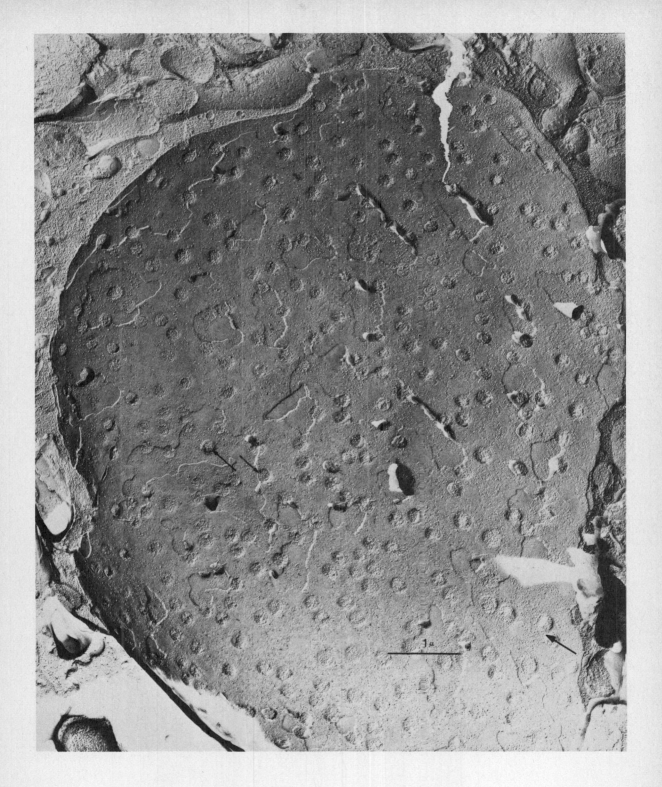

1 μ

19

lope is composed of two similar membranes, each about 100 Å thick, separated by a space of some 100–150 Å. The overall width is about 350 Å. Each of the individual membranes is similar in structure to the plasma membrane; however, the two membranes of the nuclear envelope appear to be fused together periodically to form circular windows called *annuli* or *pores*. The size and spacing of the pores vary somewhat, but a diameter of 0.1 μ (1000 Å) is quite common. Annuli often occupy about a third of the total surface area of the nuclear envelope, and are thought to be responsible for the selective passage of materials into and out of the nucleus.

The prominence and importance of the nucleus are recognized in the nomenclature used to describe the cell contents: everything inside the plasma membrane is collectively called the *protoplasm*, but only that part of the protoplasm outside the nucleus is called the *cytoplasm*.

CHROMOSOMES. The genes of the cell are almost all found in the nucleus, associated with strands of *chromatin*. Chromatin consists mostly of DNA and protein, plus a little RNA (ribonucleic acid). Chromatin may be either dispersed throughout the nucleus or gathered together in discrete, compact bodies called chromosomes ("colored bodies"), a name refering to the fact that they are

FIGURE 1–12b. The Nuclear Envelope. Highly magnified cross-section of the nuclear envelope of a rat kidney cell. Chromatin (arrows) is visible on the nuclear side. The nuclear pore (NP) is plugged with a fibrous material that resembles chromatin. The adjacent indentation may be an immature pore. [Courtesy of G. G. Maul, J. W. Price, and M. W. Lieberman, *J. Cell Biol.*, **51:**405(1971).]

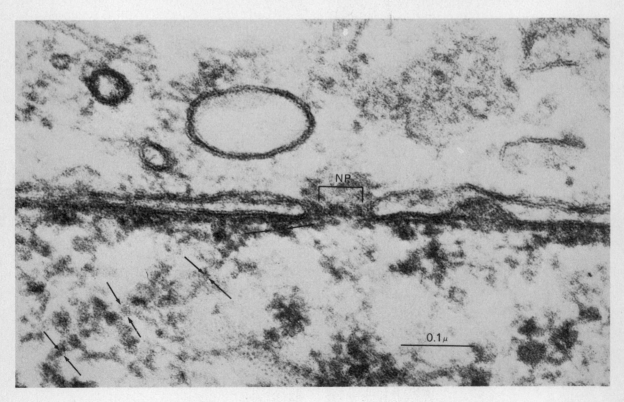

dense enough to stain readily. The genes of a typical human cell, for example, are distributed among its 46 chromosomes. The individual strands of chromatin vary in thickness, from 100 to 400 Å, giving them a "bumpy" appearance in the electron microscope but making them too small to be seen with the light microscope.

Recent evidence indicates that dispersed chromatin fibers are anchored to the inside of the nuclear envelope at the edges of annuli (Fig. 1-12b). During replication of most cells, these anchors are lost and the chromatin condenses into chromosomes; as it does so, the nuclear envelope breaks up. At that point each chromosome consists of two identical strands, called *chromatids*, that are joined at their *centromere* (Fig. 1-13). The two "sister" chromatids contain the same set of genes, a result of chromatin replication while in the dispersed state. In most cell divisions each daughter cell will get one chromatid from every pair, so that both cells receive the same set of genes, packaged into the same number of chromosomes. When the separation of chromatids is complete, a new nuclear envelope forms around each set individually, whereupon the chromatin disperses.

NUCLEOLUS. The most clearly defined feature within the nucleus is often the nucleolus (nu-clē'-o-lus, the Latin diminutive of nucleus—see Fig. 1-11). Nucleolar composition is very much like that of the chromatin itself, except for the presence of large numbers of granules rich in RNA. These granules are the precursors of ribosomes. Since ribosomal components are made there, the size of the nucleolus, and sometimes the number of nucleoli per cell, vary with the requirements for ribosome synthesis. The nucleolus has no membrane of its own, and would not be visible in the light microscope were it not for the relatively high packing density of its fibers and granules. During cellular replication, when the rest of the chromatin is condensed into discrete chromosomes, the material of the nucleoli usually disperses, only to re-form again in the new daughter cells.

GOLGI APPARATUS. Almost all eucaryotes have a complex of vesicles[4] and membranes known as the *Golgi apparatus*, or *Golgi body*, after Camillo Golgi, who first described it at the turn of the century. (He called it an "internal reticular apparatus.") The name *dictyosome* is also used to identify this structure, particularly in plants, where it consists of a stack of flattened sacs, often 200–300 Å apart and flanked by vesicles (see Fig. 1-14). Secretion (e.g., of cell wall material in plants or digestive enzymes from pancreatic cells) frequently seems to involve the freeing of

---

4. A vesicle is a small, membrane-enclosed sphere. A vacuole is a large vesicle.

FIGURE 1–13. Chromosome from a Human White Blood Cell. Note that it is comprised of fibers of chromatin. Chromatin is most often seen in electron micrographs as a granular field, the grains corresponding to the cut ends of the fibers that wind in and out of the plane of section as in Figs. 1–10 and 1–11. [Courtesy of F. Lampert, Nature New Biology, **234**:187 (1971).]

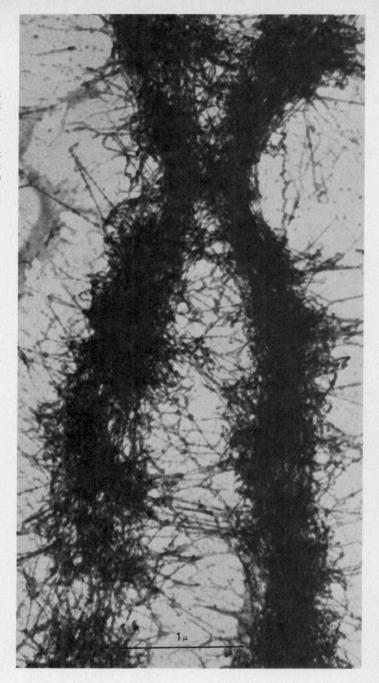

1μ

vesicles from the Golgi stacks and the union of these vesicles with other vesicles pinched off from endoplasmic reticulum to form a *secretory vacuole* or *secretory granule*. The secretory vacuole discharges its contents by fusing with the plasma membrane, opening itself to the exterior in an event called *exocytosis*.

ENDOPLASMIC RETICULUM. Most eucaryotic cells have a complicated network of cytoplasmic membranes that appears to be continuous with the plasma membrane, and probably also with the nuclear membrane. This network is called the *endoplasmic reticulum*, a name derived from the fact that in the light microscope it looks like a "net in the cytoplasm." (Eighteenth-century ladies carried purses of netting called reticules.) When a cell is disrupted and its components separated, a *microsomal fraction* is identified that is composed largely of fragments of the endoplasmic reticulum.

The reticulum ordinarily forms enclosed or semi-enclosed areas in the form of tubules or cisternae. (A cisterna is a cavity.) These channels and reservoirs function in intracellular storage and transport and often in protein synthesis, which is carried out on the ribosomes. Many of the ribosomes are found attached to endoplasmic reticulum, giving the latter a granular appearance in the microscope. In this form the reticulum is known as rough endoplasmic reticulum, or *rough ER* (Fig. 1–15a); without ribosomes, it is *smooth ER*. Materials destined for secretion are transported through the cytoplasm in vesicles that have budded from the endoplasmic reticulum. These vesicles fuse with vesicles from the Golgi apparatus to form secretory vacuoles, as explained above.

PEROXISOMES. Peroxisomes (Fig. 1–15a) are found in almost all cells. When first identified, they were named *microbodies*. The term peroxisome has evolved because this vesicle contains many of the enzymes responsible for producing and degrading peroxides. The most prevalent enzyme within them is *catalase*, whose func-

FIGURE 1–14. The Golgi Body. (a) The Golgi body of an ameba, with vesicles being pinched off from the ends of its cisternae. (b) A dictyosome from maize root cap actively engaged in the formation of secretory vesicles. Note the sloughing of cisternae. New ones are formed on the opposite side. [(a) Courtesy of C. J. Flickinger, *J. Cell Biol.,* **49:**221 (1971); (b) courtesy of H. H. Mollenhauer, *J. Cell Biol.,* **49:**212 (1971).]

(a)

(b)

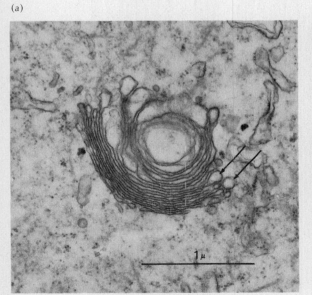

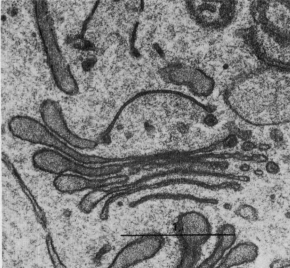

FIGURE 1–15a. Endoplasmic Reticulum and Peroxisomes. TOP: Part of an insect fat body (*Calpodes ethlius*), showing peroxisomes (P) similar to those found in vertebrates. Note the abundance of ribosomes, both free and bound to endoplasmic reticulum (RER). There is a small Golgi complex (G) and numerous mitochondria (M). BOTTOM: A plant microbody from tobacco leaf, containing a crystalline inclusion. The crystals give an intense stain for the enzyme catalase, identifying the body as a peroxisome. [*Top photo,* courtesy of M. Locke and J. T. McMahon, *J. Cell Biol.,* **48:**61 (1971); *bottom photo,* courtesy of S. E. Frederick and E. H. Newcomb, U. of Wisconsin—see *Science,* **163:**1353 (1969) and *J. Cell Biol.,* **43:**343 (1969).]

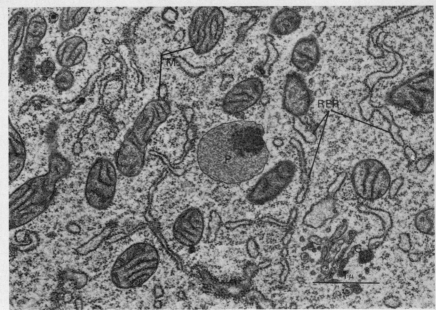

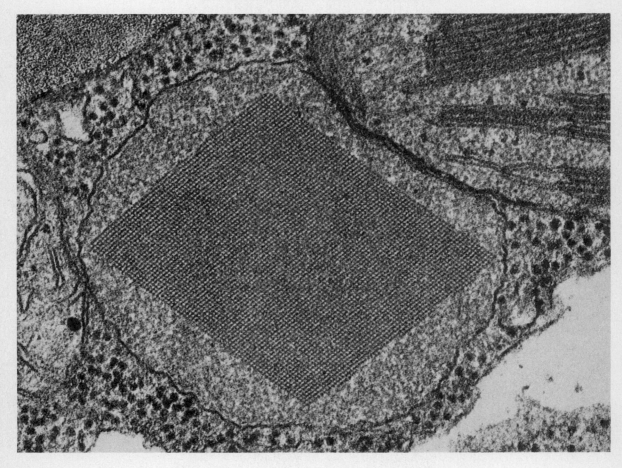

tion is to degrade hydrogen peroxide (H_2O_2) to water and oxygen. In microorganisms and higher plants, peroxisomes may also contain enzymes of the "glyoxylate cycle," an essential part of the pathway by which fats and oils are converted to sugars. They are then known as *glyoxysomes*.

LYSOSOMES. There are numerous other membrane-bound sacs within the cytoplasm for which special functions have been identified. Among them are the lysosomes, discovered by Christian de Duve in 1952 (see Fig. 1–15b). Lysosomes, which appear to originate from the Golgi apparatus, serve to isolate digestive enzymes that are capable of destroying a wide variety of substances. Under periods of prolonged fasting, for instance, protein is readily dissolved and reused from our own muscle cells with the help of lysosomes. Material to be digested may also be brought in from outside the cell by enclosing it in a vacuole (a process called *endocytosis*), that later fuses with a lysosome to form a digestive vacuole. This fusion exposes ingested material to lysosomal enzymes without exposing the rest of the cytoplasm. Breakdown products apparently leave the lysosome via transport across its membrane. Indigestible material normally is eliminated from the

FIGURE 1–15b. Lysosomes. Lysosomes (L, dense interiors) being formed in the Golgi region of an immature white blood cell (rabbit). Numerous smaller vesicles are also seen. Nuclear envelope (NE). Nuclear pore (NP). Chromatin (CH). [Courtesy of B. Nichols, D. Bainton, and M. Farquhar, *J. Cell Biol.*, **50**:498 (1971).]

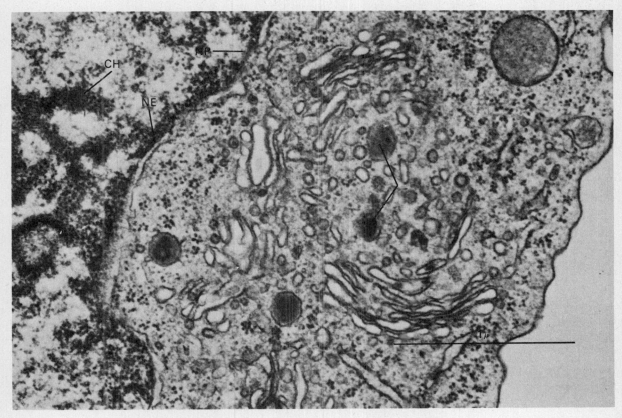

digestive vacuole by exocytosis, in which a vacuole fuses with the plasma membrane and so discharges its contents outside the cell.

VACUOLES. Many kinds of membrane-enclosed sacs or vacuoles other than those already described are found in different cell types, especially in plants. These are generally storage depots. In fact, many plant cells have a single large vacuole occupying as much as 80–90% of the total cell volume, the contents of which may be called the *cell sap* (see Fig. 1–16).

PLASTIDS. Plants contain a family of organelles called plastids, which are more complicated than the simple vacuoles discussed above. Like the nucleus, they are bounded by two membranes. A plastid of a particular type may arise by division and differentiation of a smaller precursor, or *proplastid*, or by changes that take place in another mature plastid. Some plastids, the *leucoplasts*, are colorless and are generally concerned with the storage and metabolism of starches and oils. In contrast, *chromoplasts* contain the pigments that give plants their brilliant colors.

The most impressive chromoplast is the chloroplast (Fig. 1–16), which is often banana-shaped and 4–10 μ long. Chloroplasts are responsible for the bright green color of leaves; although they contain pigments of other colors as well, there is generally so much chlorophyll in them that only the green of the chlorophyll is obvious. When leaves die in the fall or when a tomato ripens, chlorophyll is destroyed first, unmasking the other pigments. Chloroplasts are the site of photosynthesis, using light energy to capture carbon dioxide from the air, reducing it to sugars and other carbohydrates. In recent years much attention has been paid to the semiautonomous nature of chloroplasts, which have been found to contain ribosomes and DNA. The DNA contains a portion of the genetic information needed for the synthesis of chloroplasts, making them only partially dependent on nuclear genes.

MITOCHONDRIA. Most eucaryotic cells, both animal and plant, contain mitochondria (see Figs. 1–10 and 1–11). These organelles are somewhat smaller than chloroplasts, often oval-shaped and

FIGURE 1–16. A Plant Cell. Taken from timothy grass (*Phleum pratense* L.), it has a large central vacuole (V) and several small ones. Five mature chloroplasts can be seen, with numerous grana (G), and interconnecting lamellae (L) traversing the stroma (Sr). The chloroplasts contain starch granules (S) and DNA (light areas). A small dictyosome (D) is present. A plasmodesma (Pd) penetrates the cell wall (CW). Gas-filled cavities (O_2, CO_2, water vapor, etc.) such as the one at lower right account for a major portion of the volume of many leaves. [Courtesy of M. C. Ledbetter and K. R. Porter, from *Introduction to the Fine Structure of Plant Cells*, New York: Springer-Verlag, 1970.]

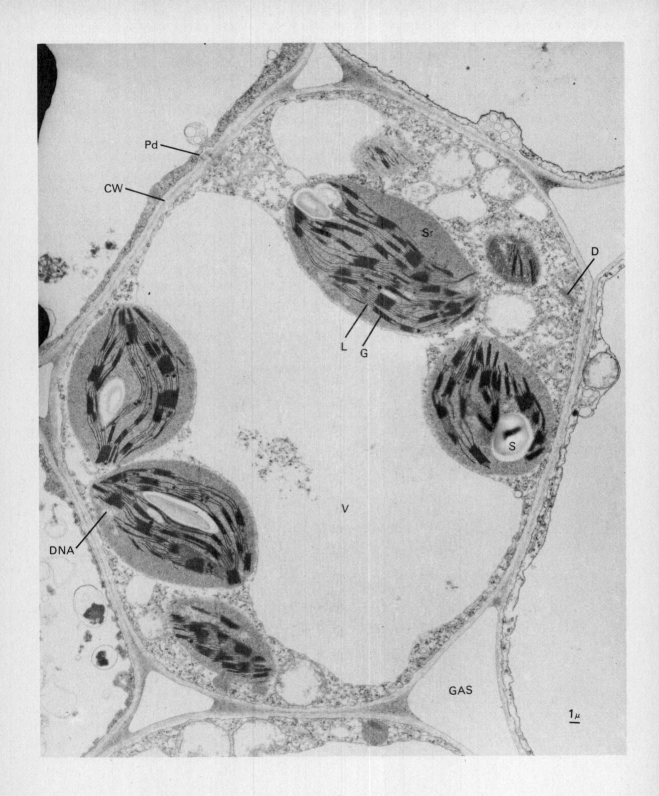

about 2 μ long, although round and Y-shaped mitochondria are also seen. Like chloroplasts, mitochondria are enclosed by a double membrane, contain ribosomes and some of their own genes, and grow and reproduce by binary fission almost as if they were autonomous organisms instead of an important part of the cell's metabolic apparatus. Mitochondria supply most of the energy needed in a typical nonphotosynthetic cell by oxidizing (a slow burning) certain selected food molecules, reducing molecular oxygen to water in the process. Mitochondria are the only major site of oxygen consumption in the cell; hence, eucaryotic cells that do not have mitochondria (mature mammalian red blood cells, for instance) cannot use oxygen.

Mitochondria are readily identified by membranes, called *cristae*, that divide their interior into almost isolated sections. The cristae are actually folds of the inner mitochondrial membrane, and hence are two membrane layers thick, with a space (the intracristal space) between the layers. The inner surface of the cristae is studded with mushroomlike projections similar to those found on the plasma membrane of many aerobic bacteria. These projections are integral parts of the oxidative apparatus.

CYTOPLASMIC FILAMENTS. The cytoplasm is laced with filaments and tubules of various sizes, whose structure and function have only recently started to become clear. There are two basic classes of filaments: the rope-like *microfilaments*, some 40–60 Å in diameter, and the larger, tube-like *100 Å filaments*. (See Fig. 1–17.) Microfilaments are frequently found in dense bundles or networks just under the plasma membrane in a region of the cytoplasm known as the *ectoplasm*. Concentrations of microfilaments are also found at cleavage furrows during cell division, and in pseudopods ("false feet"), which are the cytoplasmic extensions by which ameboid-type locomotion is believed to be achieved. Movement of the cleavage furrow and pseudopod is thought by many to be due to changes in microfilaments, as will be discussed later. This association between microfilaments and movement is reinforced by the observation that one of the two main types of filaments in muscle (the thin, actin filament) closely resembles the more universal microfilaments. Since contraction of muscle is caused by the sliding of filaments past each other, a similar kind of movement has been suggested to explain the contractions associated with microfilaments.

The 100 Å filaments are more complex. Whereas microfilaments appear as single strands, ultrastructural analysis of the larger filaments reveals a tube-like configuration. Though found in many cell types, they have been most extensively studied in nerve cells, where they are called *neurofilaments*. At least in these cells, they consist of a helical arrangement of four strands, each about 30 Å, leaving a hollow 30 Å core. They are clearly different from the much more common microfilaments in chemical composition as well, but relatively little is known of their functions.

MICROTUBULES. Microtubules, as their name implies, are tubular structures, but they are much larger and more complex than the tubular 100 Å filaments described above (see Fig. 1–18). Microtubules have a diameter of about 250 Å, with a hollow core about

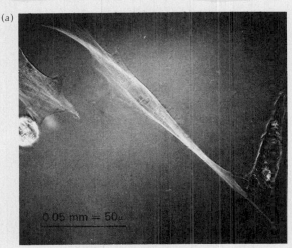

(a)

0.05 mm = 50μ

FIGURE 1–17. Cytoplasmic Filaments and Microtubules. **(a)** Light micrograph of a cultured hamster kidney fibroblast (BHK-21 strain). Fibroblasts establish collagenous connective tissue, including scars. Note the long processes on these spindle-shaped cells. **(b)** An electron micrograph of a longitudinal section through one process, showing the 40–60 Å microfilaments (MF) just inside the plasma membrane, oriented longitudinally (top) and at an angle to the cut (bottom). The 100 Å filaments (F) and 250 Å microtubules (MT), both of which are hollow structures, are at deeper positions and are usually longitudinally oriented. [Courtesy of R. D. Goldman, *J. Cell Biol.*, **51**:752 (1971).]

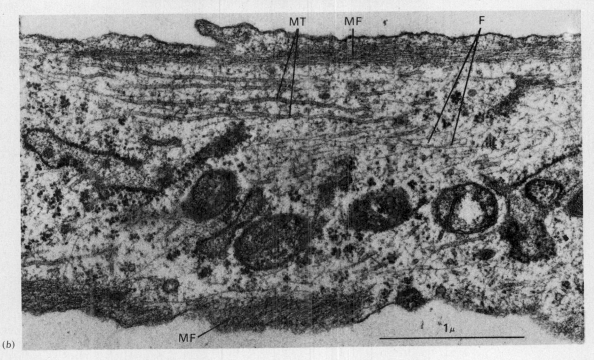

(b)

MT   MF   F

MF

1μ

FIGURE 1–18a. Microtubules. Microtubule cross-section from a plant cell, *Juniperus chinensis* L., a common evergreen shrub. Note that the wall of the tubule is composed of 13 parallel *protofilaments*. The 100 Å thick trilaminate structure nearby (arrows) is the plasma membrane. [Courtesy of M. C. Ledbetter, *J. Agr. Food Chem.*, **13**:405 (1965). Copyright by the American Chemical Society.]

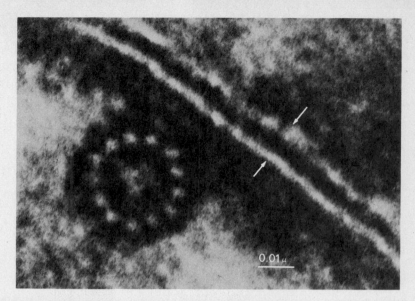

FIGURE 1–18b. Microtubules from the Tail of Sea Urchin Sperm. Partially disassembled to show the continuity of its 13 individual filaments. [Courtesy of P. J. Harris, Oregon State Univ.]

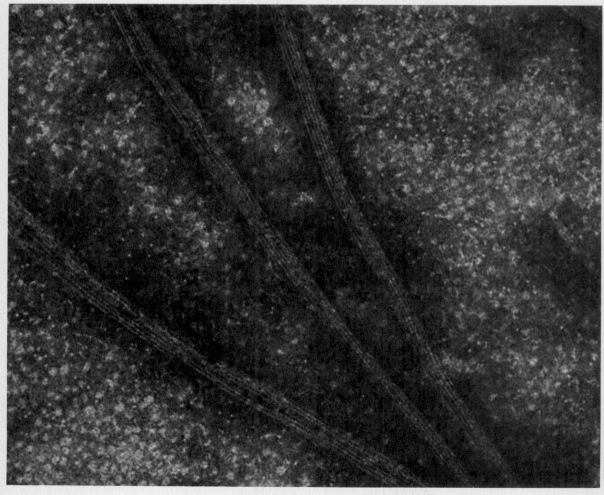

150 Å and walls comprised of generally thirteen (in some lower animals probably twelve) individual protofilaments, each about 40 Å in diameter. Their chemical composition is different from the 100 Å filaments and from the smaller microfilaments.

Where a cell forms definite shapes in the absence of outside restraints (for example, the biconcave shape of the human red blood cell), microtubules are often found to be responsible. They may also be responsible, at least in part, for *cytoplasmic streaming*, which is the constant movement of cytoplasm within the cell. In addition to these functions, microtubules are the basic building blocks of centrioles, basal bodies, cilia, and flagella.

CENTRIOLES AND BASAL BODIES. Animal cells (but not plant cells except for a few primitive algae) regularly contain near the nucleus at least two hollow, cylindrical bodies called centrioles (see Fig. 1–19). Centrioles are commonly about $0.4\ \mu$ long by $0.15\ \mu$ in diameter, and are frequently found in pairs lying at right angles to one another just outside the nuclear envelope. Their position defines the *centrosome* of the cell. An identical structure, called a *basal body* or *basal corpuscle*, gives rise to the motile appendages, cilia and flagella, in both animals and plants. The walls of centrioles and basal bodies are comprised of nine sets of microtubules, with each set generally consisting of three tubules lying in the same plane and embedded in a dense, granular substance (see Fig. 1–20). All microtubules and microtubular structures are thought by some to be derived from centrioles or basal bodies. The centrioles and basal bodies themselves seem to be self-reproducing.

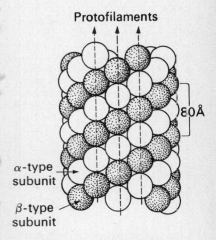

Protofilaments

α-type subunit

β-type subunit

FIGURE 1–18c. Model of Microtubular Structure. Note the helical arrangement of subunits making up the protofilaments. [Model described by J. Bryan, *Federation Proceedings*, **33**:152 (1974).]

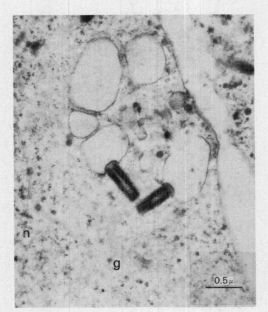

0.5μ

FIGURE 1–19. Centrioles. A pair of centrioles lying at right angles to each other near the nucleus (n). A faint Golgi apparatus (g) is also seen. The centrioles in this organism (*Myrmecaelurus*, an ant) are sometimes very large (up to 8 μ), and are unusual also in that their biosynthesis seems to involve the attached vesicles seen in the micrograph. [Courtesy of M. Friedländer and J. Wahrman, *I. Cell Science*, **1**:129 (1966).]

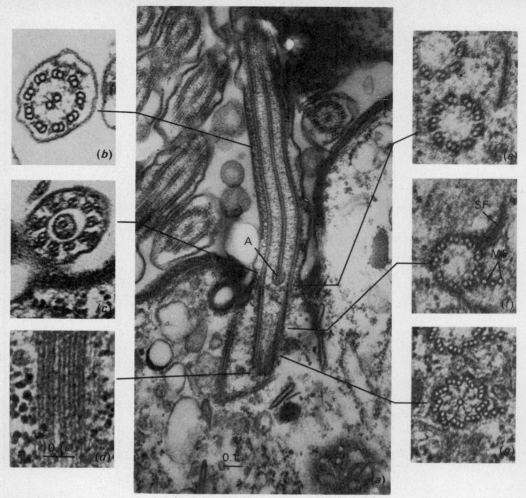

FIGURE 1–20. Structure of a Basal Body and Cilium. **(a)** Longitudinal section of a cilium and associated basal body from the protozoan *Tokophyra infusionum.* **(b)** Cross-section of the cilium. **(c)** Cross-section at the axosome (A), or region where the basal body and cilium join. **(d)** Cross-section taken below the basal body, showing basal microtubules extending from the basal body. **(e)-(f)-(g)** Three levels of the basal body in cross-section. In (e) only the basic structure is present. In (f) a striated fiber (SF) and microtubular fiber (MF) (probably anchors) arc from the basal body toward the cell membrane. Note the cartwheel structure at level (g). [Courtesy of Lyndell Millecchia. In part from L. Millecchia and M. A. Rudzinska, *J. Cell Biol.,* **46**:553 (1970).]

CILIA AND FLAGELLA. Cilia are typically 3–10 μ long by about 0.5 μ in diameter, and are numerous on those cells that have any at all. Flagella are longer (100–200 μ) and have the same diameter and the same structure, but usually there are no more than one or two per cell. Both structures are comprised of microtubule-like cylinders enclosed by an extension of the plasma membrane (Fig. 1–20). The tubules are generally found in nine pairs arranged about the circumference, with two single tubules running down the center. This is the so-called "9 + 2" construction, distinct from the "9 + 0" construction of the basal body or centriole.

Cilia and flagella are motile appendages. They appear to move by sliding their microtubules past one another. Cells lining the air passages in lungs, for example, have cilia that beat in a synchronized, wavelike motion to sweep particles out of the lungs. A similar synchrony is used for locomotion in unicellular animals (see

Fig. 1–21). While cilia may be used for other purposes, as noted above, flagella are ordinarily used only to propel the cell itself—a sperm, for instance, has a single long flagellum for a tail, surrounded near its base with mitochondria that provide the energy for flagellar motion (Fig. 1–3). Note that flagella and basal bodies of eucaryotic cells are very different from the structures of procaryotes that carry the same names. In fact, an entire procaryotic cell would ordinarily fit quite nicely inside the membrane of the usual eucaryotic cilium.

CELL FRACTIONATION. Although the study of the structure of cell organelles is usually best carried out by microscopic examination of intact or sectioned cells, chemical analysis of the various organelles requires that they be isolated. Similarly, experiments involving the function of an organelle may require a pure preparation, free of other cellular constituents. Cilia and flagella can be sheared from the cell; for other organelles, the first step is to break up the cells and fractionate the suspension by centrifugation. The

(a)

(b)

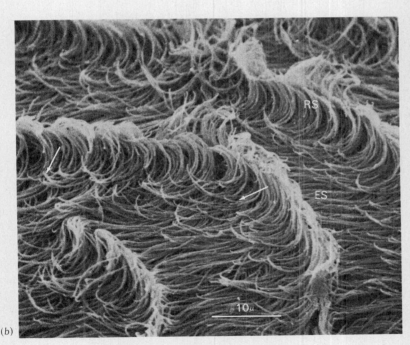

FIGURE 1–21. Synchrony in Ciliary Motion. Scanning electron micrographs of the protozoan *Opalina ranarum*. (a) Whole animal, anterior at top. (b) Different animal, posterior at left. Some cilia are caught in their effective stroke (ES) and some in their recovery stroke (RS). Arrows indicate direction of wave transmission. [Courtesy of G. A. Horridge and S. L. Tamm, *Science*, **163**:817 (1969). Copyright by the A.A.A.S.]

different components vary from one another in their mass and density, and hence in how fast they sediment to the bottom of a tube in a spinning centrifuge. A common protocol is diagrammed in Fig. 1–22. Though none of the fractions contains only the items listed, they are greatly enriched for the indicated organelles and may often be used without further purification.

1–4 VIRUSES

Although viruses are not cells, their study has provided a great deal of information about cells. To understand why, we need to know how viruses, which are cellular parasites, manage to survive.

DISCOVERY. In 1798 an English physician, Edward Jenner, reported that infecting a young boy with a mild disease called cowpox made the child immune to smallpox. Even deliberate attempts to infect him with the much more serious disease failed. This was not as daring an experiment as it sounds, since deliberate exposure to smallpox was then often used in an attempt to produce a mild infection, in the hope of protecting the patient from later, more threatening, infections. Jenner coined the term "vaccination,"

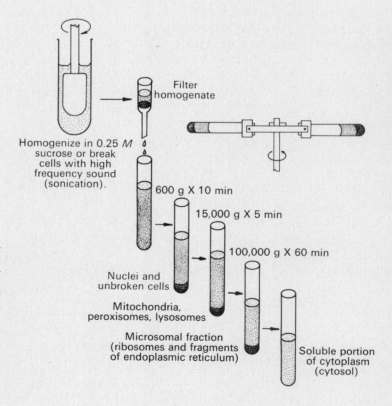

FIGURE 1–22. Cell Fractionation by Centrifugation. The g represents gravitational acceleration. In other words, the centrifugal force on a particle at 600 g is 600 times the force of gravity. This value is calculated as $(rpm)^2 r/9 \times 10^4$, where rpm is revolutions of the rotor per minute and r is the distance in cm from the center of the rotor to the particle. For example, at 2450 rpm, 600 g is obtained at 9 cm from the center of the rotor.

from the Latin name for cowpox (*variolae vaccinae,* meaning small pocks of the cow), to describe this procedure. However, the nature of the disease-causing agent, or smallpox *pathogen,* remained unknown for another century.

By the 1880s, considerable progress had been made in microbiology, with Pasteur's demolition of the concept of spontaneous generation and the development of culture procedures for growing microorganisms. Still, some microorganisms refused to be cultured. The pathogens causing smallpox and rabies fell into that category. They were assumed to be particularly fastidious and very small bacteria that could not be grown in the laboratory because of an inadequate knowledge of their nutritional requirements. Their true nature as subcellular life forms was not appreciated until a similarly mysterious agent was discovered in plants.

In 1892 a Russian scientist, D. Ivanovsky, studied an infectious agent that causes a destructive mottling of the leaves of tobacco plants. Ivanovsky demonstrated that the agent, now called *tobacco mosaic virus* (TMV), is small enough to pass through a filter of unglazed procelain (a porcelain candle) that Pasteur and C. Chamberlain had earlier found capable of stopping all known forms of cellular life, including bacteria. A few years later, M. W. Beijerinck confirmed Ivanovsky's observations, and in 1899 published a paper suggesting that tobacco mosaic virus is not just a small cell, but something more fundamental: a subcellular form of life that can reproduce only as a parasite in living cells.

The concept of the virus as an obligate cellular parasite was extended to the animal world through experiments performed by F. Löffler and P. Frosch, who in 1898 demonstrated that foot-and-mouth disease in cattle is also caused by an agent that passes a porcelain filter—i.e., the agent is a *filterable virus.*[5] It has since been found that a host of human ailments—including smallpox, rabies, measles, mumps, polio, and warts—are virus-caused diseases. In addition, it has been known since the second decade of this century that certain viruses cause cancers in birds, and more recently viruses have also been implicated in some mammalian leukemias and solid cancers.

Viruses may be broadly grouped according to their usual hosts. We have mentioned *animal viruses* and *plant viruses,* but it was the discovery of a third group, *bacterial viruses,* that made possible detailed investigations of viral reproduction. These investigations also revealed much of what is known about the molecular aspects of gene function, as we shall see.

---

5. The word "virus," meaning "poison," was first applied to all pathogens. Later, pathogens were divided into filterable and nonfilterable viruses, using the porcelain filter as a test. As the distinction between cells and what we now simply call "viruses" became clear, however, references to filterability were generally dropped.

BACTERIOPHAGE. An Englishman, F. W. Twort, published a report in 1915 in which he described a remarkable phenomenon. He watched some bacterial colonies growing on the surface of agar go through a "glassy transformation" in which they became watery and transparent. Samples from such colonies could not be used to start new colonies. In fact, when a sample was introduced to a healthy colony, it too went through the same transformation. He diagnosed the situation as a contagious bacterial disease.

Twort observed that even very dilute samples of a "sick" colony could introduce the sickness to healthy cells, and that this process could be repeated as often as he liked, passing material from one colony to the next. Thus, the agent must cause its own reproduction. It proved to be filterable, hence much smaller than the bacteria themselves, and heat sensitive. Twort reasoned that the agent could be an enzyme (also heat sensitive and very small) or a virus, but his investigations were interrupted by World War I and he published no more on the subject.

Only two years later, however, a Canadian, Felix d'Herelle, published a report on a quite independent discovery of a similar nature. He had found a filterable agent capable of causing *lysis* (a clearing of the culture due to a destruction of cells) in cultures of a dysentery bacterium. In addition to observations such as those made by Twort, d'Herelle noticed that a very dilute sample from a diseased colony, when mixed with healthy cells and spread on the surface of agar, produced the expected lawn of colonies. But, in addition, a number of round clear *plaques*, or holes also formed (see Fig. 1–23). When sampled the plaques proved to contain the lytic agent, settling the question of its nature: according to d'Herelle, since "a chemical substance cannot concentrate itself over definite points . . . the antidysentery microbe is an obligatory bacteriophage." There was no doubt in his mind that this bacteriophage (or "bacteria eater") was a virus capable of reproducing only as a parasite of bacterial cells.[6]

Since d'Herelle's bacteriophage was so effective in destroying the dysentery bacillus *in vitro*, it was natural for him to ask whether it was as efficacious *in vivo*.[7] Although he noted that the recovery of patients suffering from dysentery often coincided with the appearance of phage in their feces, neither he nor an army of other workers was completely successful in developing it into a real weapon in man's fight against bacterial disease. There are viruses parasitic to a wide range (perhaps all) pathogenic bacteria, but bacteria can mutate to virus resistance with a frequency that ensures

---

6. Although other "bacteria eaters" are known, the word bacteriophage, or just "phage" for short, is used synonymously only with bacterial virus.

7. *In vitro* means, literally, "in glass"—i.e., in a laboratory vessel. Its counterpart, *in vivo*, means in the living organism itself.

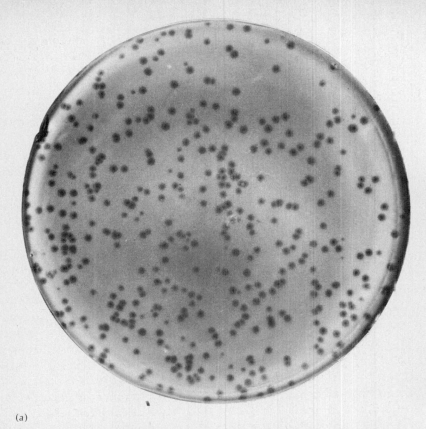

(a)

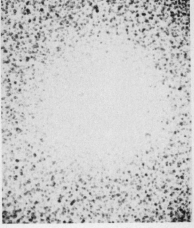

(b)

FIGURE 1-23. Bacteriophage Plaques. (a) A "lawn" of bacterial colonies with bacterial virus plaques. (b) A single plaque, greatly enlarged. Note the spherical colonies of bacteria and how their destruction produces the plaque. (Phage T4, growing on *Escherichia coli*.)

their survival. Combined with the natural defenses of the body against foreign material, including bacterial viruses, the best efforts of medical microbiologists to use phages to cure human disease were frustrated. Finally, the advent of antibiotics around 1940 put an end to the study of phages as a medical tool. However, at about the same time, the easily studied phages were recognized as a useful tool in the investigation of gene activity.

Because of the ease with which they are grown, much of the early work with bacterial viruses was carried out with *coliphages*, which are viruses parasitic to a common intestinal bacterium, *Escherichia coli* (*E. coli* for short). On several occasions reference will be made to the members of this group that are designated as "T" phages (T for "type"), particularly the closely related T-even phages, T2, T4, and T6. (See Fig. 1-24 and Fig. 1-25.) With concentration on just a few virus strains, workers learned a great deal that has proved to be generally true not only for bacteriophages, but for other kinds of viruses as well.

LIFE CYCLE OF THE BACTERIOPHAGE. The life cycle of most bacteriophages is similar to that described by d'Herelle for his antidysentery virus. It consists of (1) adsorption to the host cell, (2)

FIGURE 1–24. Bacteriophage T4. The so-called "T-even" bacteriophages are probably the most complex of all viruses. Each contains about 200 genes. In contrast, simple viruses contain only three or four genes. (Micrograph courtesy of Lee Simon, *New Scientist and Science Journal*, 25 March 1971, p. 670.)

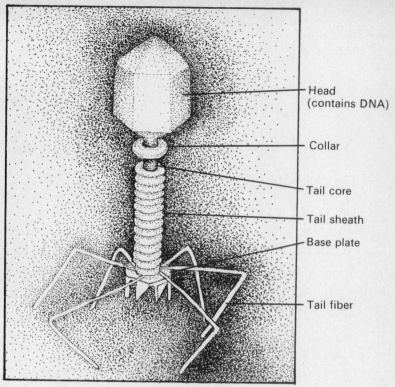

Head (contains DNA)

Collar

Tail core

Tail sheath

Base plate

Tail fiber

viral DNA made in nucleus, viral RNA made in cytoplasm

injection of phage genetic material (DNA or RNA), (3) intracellular production of new particles, and finally (4) lysis of the host cell to release the progeny phage, usually 100 or more.

The *adsorption* of the phage to its host is made possible by the proper juxtaposition of chemical groups on the two during a random collision. Many phages are shaped a little like tadpoles, with a distinct "head" region and a cylindrical "tail." (See Fig. 1–24.) Reactive groups at the end of the tail can join with a complementary set of chemical groups (a *receptor site*) in the cell wall of the bacterium. The T-even phages, and a few others, also have long fibers extending from the tail which, because of their size, are apt to be the first to contact and attach to the cell. The fibers help to position the phage's tail perpendicularly to the cell wall. Probably because of them, the T-even phages are unusually efficient at attacking their hosts—calculation of collision frequencies between virus and cell indicate that under some conditions it may take no more than one viral collision to cause adsorption, remarkable when one considers that the collision occurs with a random orientation. Phages without tail fibers seem less efficient in their attacks, though one cannot be absolutely certain that this is the only reason.

Once the phage is attached to its prospective host, *injection* can take place, involving a movement of the genetic material (DNA or

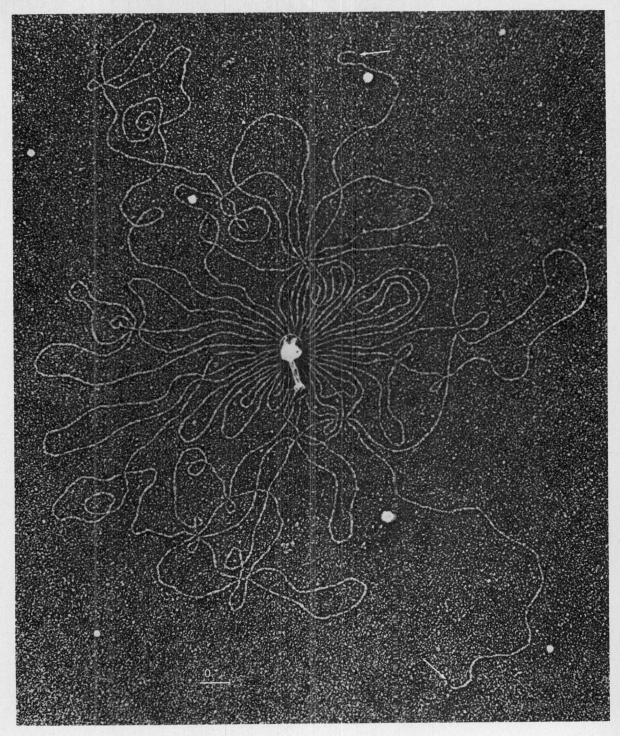

FIGURE 1–25. Phage T2 and Its DNA. The head region, which is about 0.1 μ long, contains a single molecule of DNA with a total length of about 50 μ. (Note the two ends, marked by arrows.) The preparation has been coated with platinum to improve contrast; hence, the diameter of the extruded DNA has been greatly increased. [Courtesy of A. K. Kleinschmidt *et al.*, *Biochim. Biophys. Acta.* **61**:857 (1962).]

RNA) from its position inside the head of the phage through the hollow core of the tail and into the bacterium. Entry is made possible when the bacteriophage penetrates the cell wall, either by a contraction of the outer sheath of the tail or by the action of enzymes carried by the phage tail, or both (see Fig. 1–26a).

Once inside a host cell, phage genes take over the metabolic machinery of the cell and direct it to produce replicas of the infecting virus (Fig. 1–26b). In other words, although the cell continues to procure raw material and energy from its environment, the virus genes allow only virus components to be built. Not only is the normal ability of the host cell's DNA to control the cell lost, but the host's DNA may be destroyed by early products of the viral genes.

Usually the host cell's fate is death by lysis (i.e., the cell splits open). After anywhere from a few score to several thousand new *virions*, or virus particles, have been manufactured the cell bursts, permitting the new particles to diffuse away in search of additional hosts.

In the late 1940s, a type of phage was discovered that occasionally adopts a life cycle startlingly different from that just described (see Fig. 1–27). The genetic material of the phage may be inserted into the chromosome (DNA) of the host. In this condition the host cell may continue to function in the usual way, with no obvious evidence of the viral attack. In other words, the viral genes may remain dormant for many generations, being replicated along with the rest of the host's chromosome. The original infected cell may produce many descendants, each of which carries a copy of the genes of the virus that invaded its ancestor. In its dormant state, the viral genes are said to exist as a *prophage* or *provirus*. The infected cell is said to be *lysogenic* because a prophage may, at any time, become detached from the chromosome of its host and begin directing the synthesis of new virus particles that cause lysis just as if the cell had been invaded recently.

The chances of spontaneously activating a provirus may be as low as one in a million per generation; however, if the lysogenic culture is exposed to ultraviolet light, or to any one of a number of other physical or chemical agents, it may be induced to begin turning out virus particles *en masse*. This phenomenon, in which an apparently healthy culture of cells can all at once begin producing viruses without any new infection, is called *induction*.

We shall have more to say about phages later, because the study of their interaction with hosts has taught us a great deal, not only about viruses but about how genes behave. This is true because a virus is little more than a set of genes wrapped in a protective coat.

THE STRUCTURE OF ANIMAL AND PLANT VIRUSES. Viruses are not cells. They have no metabolism, and therefore no repetitive movement, and no means for reproduction outside of a host cell.

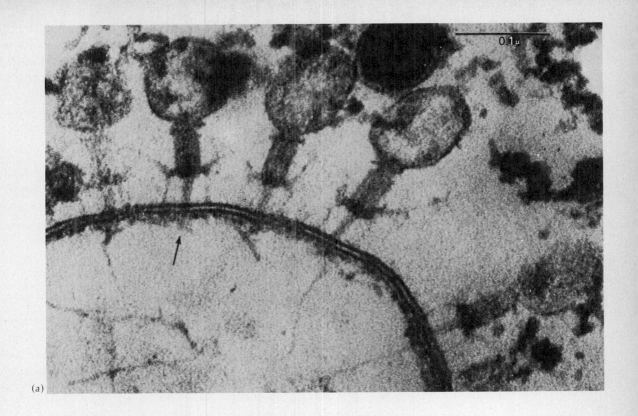

(a)

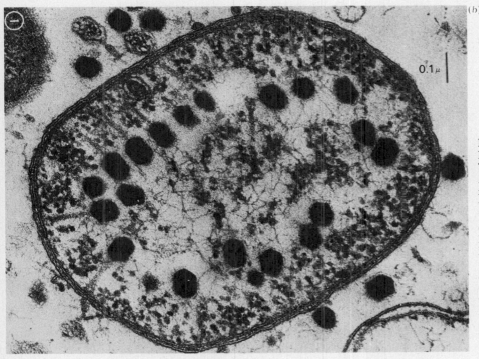

(b)

FIGURE 1–26. Injection and Replication of Phage T4. **(a)** The phages are bound to an *E. coli* cell wall by their tail fibers. The sheaths have contracted and their cores have penetrated through the wall of the host (arrow). Fibers of DNA can be seen entering the host cell. **(b)** New phages are synthesized. Note the regular arrangement of particles: tails, outside; heads, inside. [Courtesy of L. D. Simon, (a) from *New Scientist and Science Journal*, 25 March 1971, p. 671; (b) from *Virology*, **38**:285 (1969).]

41

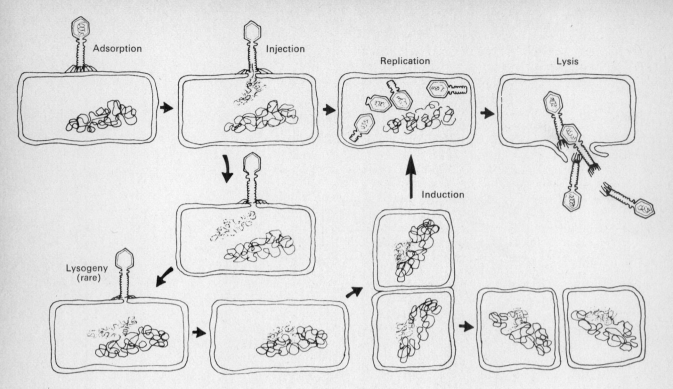

FIGURE 1–27. Life Cycle of Bacterial Viruses. Some bacteriophages (but not the T series) occasionally adopt an alternate life cycle in which injected DNA is incorporated into the host's DNA as a *provirus* and is replicated for many generations as part of the host. This condition is called *lysogeny*. Occasionally, the virus genes in one of these cells may become active and cause viral replication just as if the cell were newly infected.

Whether one wishes to say that they are alive is largely a matter of definition, for they occupy the nebulous borderline between living and nonliving. But living or not, they are both interesting and important.

The chemical nature of viruses remained a mystery until Wendell Stanley, who later received the Nobel prize, purified and crystallized tobacco mosaic virus (TMV) in 1935. (This and other milestones in virology are listed in Table 1.2.) Stanley reported that TMV is extremely simple, consisting almost entirely of protein, although it was soon found that some ribonucleic acid (RNA) is also present. Later work established that the average particle is 94.5% protein, in the form of about 2100 identical building blocks, or subunits (see Fig. 1–28). The subunits are strung together in a long helix with a pitch of 23 Å per turn of $16\frac{1}{3}$ subunits. This arrangement produces a rod-shaped structure about 180 Å in diameter and typically about 3000 Å (0.3 μ) long, with a hole down the center 35 to 40 Å in diameter. The ribonucleic acid, which contains all the genetic information of the virus, is a chain-like polymer following the helix from one end to the other at a distance of some 40 Å from the axis; it is therefore buried in, and protected by, the protein.

Most viruses are as simple chemically as the tobacco mosaic virus, though some have a morphology (structure) with more vari-

TABLE 1.2 Some milestones in virology

1798	Jenner successfully tests his smallpox vaccine.
1885	Pasteur perfects the second viral vaccine, that for rabies.
1892	Ivanovsky finds that tobacco mosaic disease can be transmitted by an agent that readily passes a ceramic filter.
1898	Löffler and Frosch demonstrate that the agent responsible for foot-and-mouth disease in cattle also passes a ceramic filter, and suggest that human diseases such as smallpox may be caused by similar agents.
1899	Beijerinck suggests that tobacco mosaic disease is not caused by a cell, but by a cellular parasite, a virus.
1911	Rous finds that a certain kind of cancer in chickens is transmitted by a filterable virus (the Rous sarcoma virus).
1915	Twort reports a bacterial disease characterized by a "glassy trans-formation" of colonies growing on agar.
1917	d'Herelle independently finds a lytic agent for a dysentery bacterium and concludes that it is a bacterial virus.
1935	Stanley crystallizes TMV, permitting a careful chemical analysis of its composition.
1950	Lwoff proves that lysogenic bacteria carry viral genes without carrying intact viruses. Though the concept had been advanced some years earlier by Sir Macfarlane Burnet, it had been highly controversial.
1952	Hershey and Chase demonstrate experimentally that phage DNA, not its protein, carries genetic information.
1955	Fraenkel-Conrat and Williams show that tobacco mosaic virus can be disassembled and reassembled in the laboratory.
1956	Gierer and Schramm demonstrate that the RNA from TMV is infective even without the protein.

ety (see Fig. 1–29 and Fig. 1–30). The tadpole-shaped T-even phages, already discussed, are among the most complicated. The T-even phages are about half protein, of a dozen or more different kinds. The remainder of the phage is a single, long molecule of DNA (see Fig. 1–25). It is this molecule that carries the genetic information and that, after injection into the host cell, controls the metabolism of the cell and directs it in the process of viral replication.

There are several degrees of complexity between tobacco mosaic virus and the T-even phages. The adenoviruses, for example, are regular icosahedrons[8] about 700 Å across (see Fig. 1–30). Members of this group cause a wide range of ailments, from sore throats in man to certain cancers in hamsters. The icosahedron is composed of spherical protein subunits, which together account for about 87% of the mass of the virus. The remaining 13% is a single

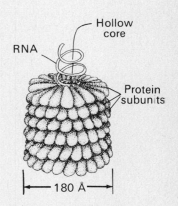

FIGURE 1–28. Partially Disassembled Tobacco Mosaic Virus. Note the strand of RNA coiling out of its protein coat.

---

8. An icosahedron is a regular polyhedron composed of 20 identical faces, each an equilateral triangle.

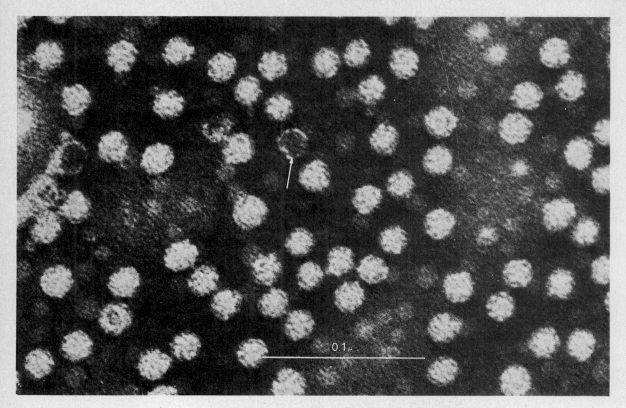

FIGURE 1–29. Two Plant Viruses. TOP: Cucumber mosaic virus, a polyhedral plant virus. Note one empty (i.e., RNA-free) particle into which the stain has penetrated (arrow). BOTTOM: Tobacco rattle virus. This RNA virus is cylindrical with a hollow core like TMV, but is unusual in that it comes in two different sizes, a long and a short rod. A complete cycle of infection and viral replication requires the presence of both versions. (Negatively stained electron micrographs courtesy of T. C. Allen, Jr., Oregon State Univ.)

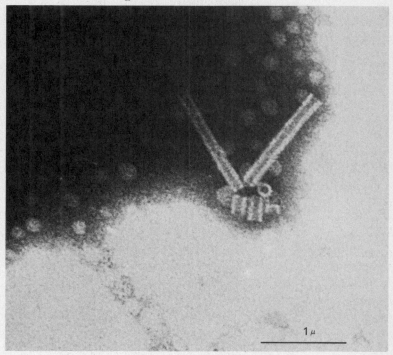

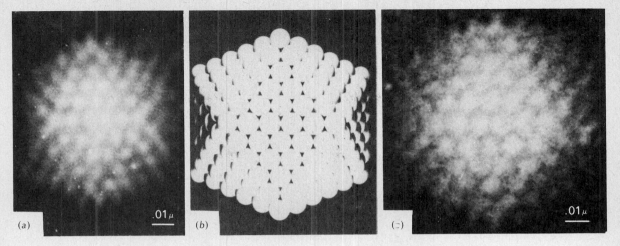

(a)   .01μ   (b)   (c)   .01μ

molecule of deoxyribonucleic acid (DNA), representing the genetic material of the virus.

Although most viruses consist almost entirely of protein and one or the other of the nucleic acids, DNA or RNA,[9] some have an outer envelope or membrane that has a structure similar to a nuclear or plasma membrane. In fact, the viral membrane is largely derived from the nuclear or plasma membrane of the host cell, being added to the virus particle as it passes through during release. The group of Herpes viruses, one of which is responsible for cold sores, have this characteristic.

LIFE CYCLE OF ANIMAL AND PLANT VIRUSES. The mode of replication of animal and plant viruses is not as well understood as that of the phages, partly because the cells themselves are less well understood. Few animal or plant viruses have tails or any other obvious organ of attachment. They gain entry to the host cell by endocytosis or by fusion of viral and cell membranes. Attachment to specific membrane receptors of the cell is needed by some animal viruses. Both viral protein and viral nucleic acid enter the host, but the protein is soon stripped away and is without further function. In spite of this mode of entry, most viruses exhibit a narrow *host range specificity*, meaning that a given virus can reproduce in only a limited number of cell types.

Once inside an appropriate host, a virus may be replicated immediately or its genes may be incorporated into the host's chromosomes for replication in a future generation. The latter situation has generated a good deal of attention, for the incorporated genes are not necessarily entirely dormant. In the bacterial counterpart

FIGURE 1–30. Two Icosahedral Animal Viruses. (a) An adenovirus particle. The spherical protein subunits, each about 70 Å in diameter, are arranged into an icosahedron. Short fibers (not seen here) extend from each apex and are necessary for infection. (b) An icosahedral model made of 252 pingpong balls. (c) Infectious canine hepatitis virus, with the same subunit structure. Both of these viruses contain DNA as their genetic material. [(a) and (b) adapted from R. W. Horne *et al.*, *J. Mol. Biol.*, **1**:84 (1959); (c) from M. C. Davies *et al.*, *Virology*, **15**:87 (1961). Courtesy of the authors and publisher.]

---

9. An infectious agent in plants consisting only of RNA with no protein coat has recently been reported. It has been termed a *viroid*. Similar agents are the suspected cause of scrapie and perhaps a small number of other animal diseases.

of this condition, lysogeny, the presence of a virus-introduced gene may be manifest as an ability to utilize some particular food molecule that the host could not use before infection. In animal cells, we are more interested in the case where virus-introduced genes cause changes in the normal control of cellular replication, for such cells may become malignant (they are said to be *transformed*), though it cannot be concluded from this that all malignancies are caused by viruses. (We shall return to the problem of malignancy in Chap. 10.)

The release of newly formed virus particles may be through the dramatic lysis characteristic of bacteriophage reproduction, or it may occur one at a time while the cell continues to live, grow, and reproduce. Leukemia viruses, for example, are released singly from infected white blood cells, while the cells reproduce uncontrollably, eventually killing the animal. In the case of avian (bird) leukemia viruses, the serum of an infected animal may have a free virus count exceeding 10^{11} virions/ml before the animal dies. The virus count is so high that the serum — normally the clear, cell-free portion of clotted blood resembles a thin, translucent milk. The leukemia viruses and certain others (e.g., the myxoviruses, often the cause of respiratory diseases such as influenza) are released from the infected cell by *budding*. When the core of a virus reaches the cytoplasmic membrane, a bulge forms and is then pinched off, freeing the mature virus particle in a process that is similar to the sequence in Fig. 1–31.

A COMPARISON OF CELLS AND VIRUSES. We have seen that viruses are much simpler than cells, and very different in both structure and life cycle. They range in size from about 0.01μ to several tenths of a micron. The largest viruses, therefore, are at least as large as the smallest bacteria (mycoplasmas, chlamydiae, and rickettsiae, certain species of which are pathogenic to man), but there is a world of difference between them. In fact, even the tiniest cells may be parasitized by viruses.

The chlamydiae are similar to viruses in that they reproduce only within living cells. However, chlamydiae are obligate cellular parasites because of stringent nutritional requirements. Viruses are obligate cellular parasites because they are not cells, and only cells can be self-reproducing. Viruses have no metabolism and no cell membrane in the usual sense. To be sure, the envelope that one finds around certain animal viruses has much the same structure as a cell membrane, but it is not a biologically active organ, only the remnant of one. In general, the molecular variety necessary to support an independent existence is simply not present in virus particles. There is no natural way in which their genes can be used to construct new virus particles without the help of a living cell.

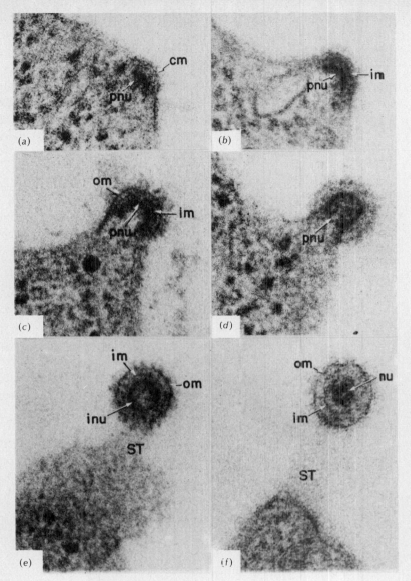

FIGURE 1-31. Reproduction of an Avian Leukemia Virus. Leukemic white blood cells of chickens continually release virus particles—called C-type particles—by budding. (a) and (b) A dense viral core (prenucleoid, pnu) appears beneath the cell membrane (cm) and begins to form a bud. (c) and (d) As the bud matures, an outer membrane (om) and inner membrane (im) become visible. The outer membrane is continuous with the cell membrane. (e) The virus is almost free, but the immature core, or nucleoid (inu), has not yet condensed. (f) The stalk (ST) that connects virus and cell breaks and frees the virus, which by then has a condensed nucleoid (nu). [Courtesy of G. de Thé, C. Becker, and J. W. Beard, *J. Natl. Cancer Inst.*, **32**:201 (1964).]

1-5 OBSERVING CELLS

Almost all of the preceding description of cells and viruses is the result of observations made with light or electron microscopes. To better appreciate the way in which cell biology depends on microscopy, we will pause briefly to consider how microscopes are constructed and used.

The first functional microscope was produced by a Dutchman, Zacharias Janssen, in the last decade of the sixteenth century. Later, Leeuwenhoek greatly improved the technique of polishing lenses. Although Leeuwenhoek was able to use his instrument to describe protozoa and bacteria, as noted earlier, the present design of the light microscope emerged only in the twentieth century. In addition, the advent of electron microscopy about midway through the century greatly extended our powers of visual observation.

RESOLUTION OF THE LIGHT MICROSCOPE. The resolution of the unaided human eye is about 0.1 mm, (100 μ). In other words, two objects within 100 μ of each other appear to be in contact. The minimum distance, d_0, at which they are resolvable as separate entities is given approximately by Abbe's relationship:

$$d_0 \approx \frac{0.6\ \lambda}{n\ \sin\ \alpha} \qquad (1\text{--}1)$$

This equation is valid for all optical instruments, including the eye. The symbol λ is the wavelength of the radiation used to form the image, n is the refractive index between the specimen and the first lens, and α is the aperture angle, or half the angle subtended by the aperture of the first lens as viewed from the specimen (see Fig. 1–32). The quantity $n \sin \alpha$ is often called the *numerical aperture.*

Abbe's relationship makes it clear that high resolution in a microscope can only be achieved by manipulating a small number of variables: the wavelength of the illuminating radiation, the refractive index, and the aperture. The aperture, of course, is limited to something less than 90°, since that would have the lens and specimen in contact with one another. In fact, 85° is about the limit in good optical microscopes. Such angles require an excellent lens. In most cases, the aperture is less because the edges of the lens, which introduce distortions, cannot be used.

Refractive index is easy to alter, but only within narrow limits. It can be increased by using oils to fill the space between the specimen and the lens. Although transparent oils have an n up to only about 1.5, that value is a big improvement over air, where $n = 1$, accounting for the popularity of the oil-immersion lenses of modern microscopes.

The wavelength of radiation is the area in which most dramatic improvement seems likely. One can, for example, use ultraviolet light instead of visible light, thus improving resolution as much as twofold. In order to do that, however, special quartz lenses must be used, since ordinary glass blocks much ultraviolet light. In addition, the eye cannot be used to view the image directly, for it does not register ultraviolet light, and may even be damaged by it. And finally, absorption by the sample itself at wavelengths below about 300 nm (0.3 μ) may become a problem.

Thus, a good light microscope, with a numerical aperture of 1.4 and using light at the short end of the visible spectrum (0.4 μ), will resolve two points at about 0.17 μ separation. This is, of course, immensely better than the unaided eye. However, while this

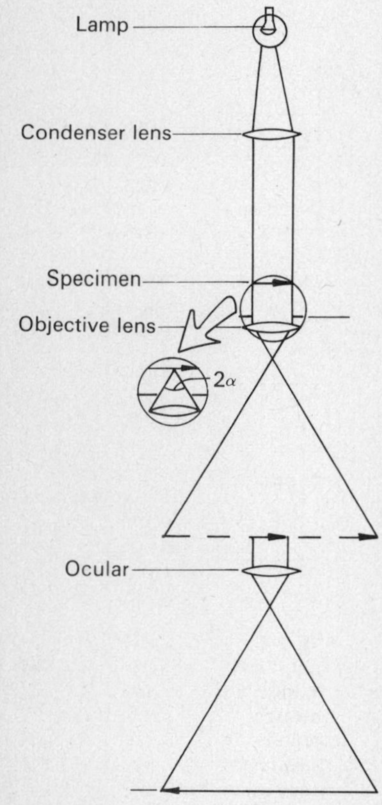

FIGURE 1–32. The Light Microscope. Note the aperture angle, α, which is related to the resolving power by equation (1–1). When using an oil immersion lens, the space between the specimen and the objective lens (circled) is filled with transparent oil.

Lamp

Condenser lens

Specimen

Objective lens

2α

Ocular

resolution permits one to see considerable detail in most cells, which may be 10 to 20 μ in diameter, there is still a great deal that cannot be seen. The ribosomes, for instance, and unaggregated threads of chromatin in the nucleus are about 0.02 μ in diameter and quite invisible to the light microscope.

In order to make full use of the resolving power that is available, special techniques have been designed to improve contrast. They include dark-field microscopy, in which the sample is seen only with oblique rays, and the use of polarized light, which accents highly ordered molecular configurations in the sample. Another powerful technique is phase-contrast microscopy (see Fig. 1–33). This technique takes advantage of the fact that different parts of a cell have regions of varying refractive indices; thus, by causing light from different regions to form patterns of destructive interference, sharp contrasts are obtained. Phase contrast microscopy is particularly useful for viewing living cells, which are otherwise relatively transparent and difficult to see. Even with these improvements in the light microscope, there is a need for greater resolving power in some investigations. That need was the impetus for development of the electron microscope.

THE ELECTRON MICROSCOPE. In principle, if one could substitute a one-angstrom X ray or γ ray for 4000 Å light, he would obtain a 4000-fold increase in resolution. In practice, there are no available

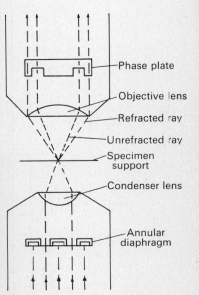

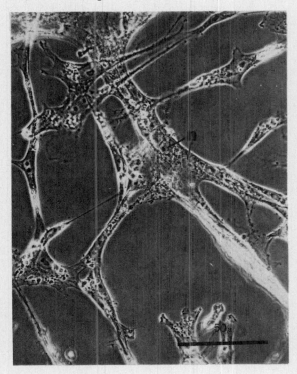

FIGURE 1–33. Phase Contrast Microscopy. ABOVE: Diagram of the phase contrast microscope. Rays that pass through a region of changing refractive index will be refracted and, after passing through the phase plate, will cause interference with unrefracted rays. This produces a halo effect, increasing contrast and making it easier to observe unstained, living cells. LEFT: Phase contrast micrograph of embryonic chick myoblasts (muscle cell precursors) after four days of culture. Nucleus (n). [Left micrograph courtesy of Y. Shimada, *J. Cell Biol.*, **48**:128 (1971).]

lenses that can be used to focus very short wavelength electromagnetic radiation. (Medical X-ray machines do not focus the beam, but use it to form shadows.) On the other hand, charged particles such as electrons respond to magnetic fields. Since particles also have associated wave properties, focusing them with magnetic lenses produces an image.

The wavelength of an electron is given, approximately, by

$$\lambda = \frac{12.3}{\sqrt{E}} \text{ angstroms} \qquad (1-2)$$

where E is the voltage through which the electron is accelerated. Electrons at 50,000 volts have a wavelength of 0.05 Å. This value does not provide quite the improvement in resolution that one would expect from equation (1-1), because the numerical aperture of electron microscopes is much smaller than that of light microscopes (see Fig. 1-34). The problem is that only the center of a magnetic field has the right properties to be useful as a lens. Hence, only a very small area in the middle of each lens is used. Although this restriction reduces the aperture angle to only a few tenths of a degree, compared to the 85 degrees or so of light microscopes, the effective resolution of an electron microscope is still about a hundred times the wavelength of the electrons. At 50,000 volts, that provides 5 Å resolution! (See Fig. 1-35.) Even higher resolution can be obtained with the electron microscope by going to greater accelerating voltages, leading to the development of megavolt electron microscopes. However, as the electrons become more energetic, more and more of them pass through the sample without being significantly deflected, leading to a loss in contrast. In addition, present methodology of preparing biological samples seldom allows effective use of even 5 Å resolution. As a result, 50,000–100,000 volts is the range commonly used.

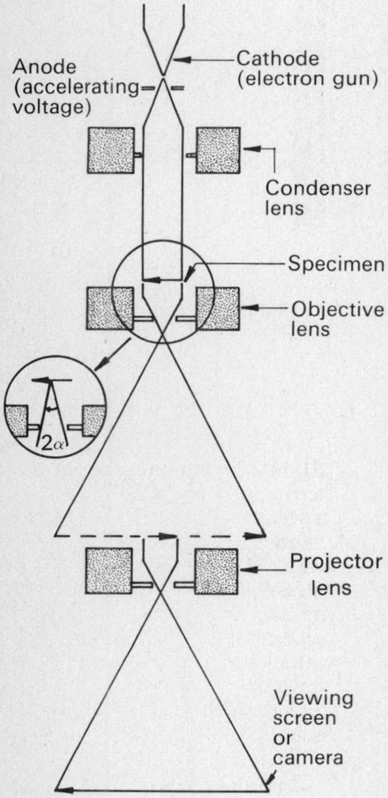

THE SCANNING ELECTRON MICROSCOPE. In recent years, considerable interest has developed in applying the scanning electron microscope to biological investigations. Although scanning instruments were available by the mid-1940s (about a decade after the appearance of commercial transmission electron microscopes), the early versions offered few advantages over other types of microscopes and much poorer resolution. By the early 1960s, however, significant improvements had been made, and industrial applications were found. Soon thereafter, biologists also found ways to employ the new instrument.

The scanning electron microscope (see Fig. 1-36) moves a thin beam of electrons back and forth across the specimen in the same way that an electron beam moves back and forth across the face of a television picture tube. In fact, a television picture tube is used to display the image, its own beam moving in synchrony with the

FIGURE 1–34. The Electron Microscope. Compare with Fig. 1–32. The entire column, from filament to screen, is maintained in a very high vacuum. Note the small aperture angle, greatly enlarged in this drawing.

Anode (accelerating voltage)
Cathode (electron gun)
Condenser lens
Specimen
Objective lens
2α
Projector lens
Viewing screen or camera

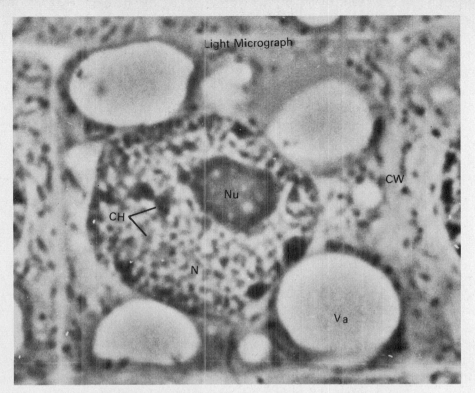

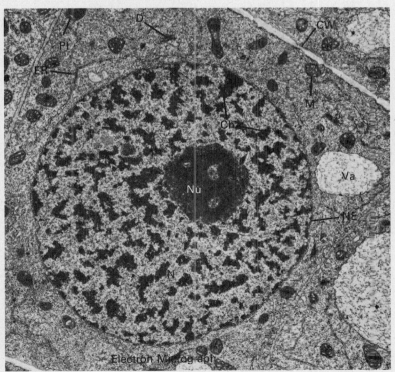

FIGURE 1–35. Resolution of Optical vs. Electron Microscopes. These images were obtained from the same piece of onion root, prepared in the same way. (The tissue was fixed with gluteraldehyde and osmium tetroxide, then embedded in an epoxy resin.) The sections were cut 1.5 μ and 0.03 μ thick for light and electron microscopy, respectively. Note the difference in resolution between the two microscopes. Nuclear diameter = 1.5 μ. Nucleus (N). Nucleolus (Nu). Chromatin (CH). Nuclear envelope (NE). Mitochondria (M). Dictyosome (D). Endoplasmic reticulum (ER). Plastid (Pl). Vacuole (Va). Cell Wall (CW). (Photographs courtesy of W. A. Jensen. From W. A. Jensen and R. B. Park, *Cell Ultrasturcture.* Belmont, Calif.: Wadsworth Publ. Co., 1967. Reprinted by permission.)

51

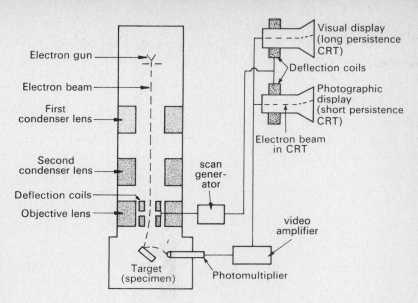

Electron gun

Electron beam

First
condenser lens

Second
condenser lens

Deflection coils

Objective lens

Target
(specimen)

Photomultiplier

scan
gener-
ator

video
amplifier

Electron beam
in CRT

Visual display
(long persistence
CRT)

Deflection coils

Photographic
display
(short persistence
CRT)

FIGURE 1–36. The Scanning Electron Microscope. The beam is swept back and forth across the specimen in synchrony with a beam moving across the face of a television picture tube. Radiation (scattered or secondary electrons, etc.) from the sample is used to modulate the observation beam, producing variations in brightness.

scanning beam. As electrons strike the specimen in the microscope, they may be scattered or secondary electrons may be knocked from the sample; in either case, electrons from the sample can be collected by a nearby photomultiplier tube. (Other types of radiations can also be induced in the specimen and detected.) Scattered electrons, for example, will vary in abundance and in accessibility to the detector according to their origin. Crevices will produce fewer detectable electrons, whereas projections will be highlighted. This variation is used to modulate the beam sweeping across the picture tube (e.g., see Fig. 1–1 or 1–2).

The resolution of the scanning electron microscope is much better than that of optical instruments, but generally poorer than that of transmission electron microscopes. One obvious limitation is the beam diameter itself. The way in which samples must be prepared (to be discussed), plus the nature of the collection and display devices, add further limitations. Nevertheless, commercial instruments operate routinely at less than 100 Å resolution (0.01 μ), and instruments with resolutions better than 5 Å — comparable to the best conventional electron microscopes — have been built.

ADVANTAGES AND DISADVANTAGES OF THE THREE TYPES OF MICROSCOPY. Although resolution of the transmission electron microscope is generally superior (at the present time at least) to that of the scanning electron microscope, and enormously greater than that of the light microscope, resolution is not the only criterion that affects the usefulness of these instruments. In fact, each of the three microscopes just described has applications in which it is clearly preferable to the others.

The traditional light microscope is easy to use and ignores the presence of water. All electron microscopes, on the other hand, are complicated by comparison, require more elaborate sample preparation, and can normally examine only dried specimens sitting in a vacuum chamber. The latter restriction results from the fact that an electron beam is scattered by air or water, although the megavolt electron microscopes have a beam that can penetrate thin layers of liquid, making it possible to observe some living specimens. In general, however, we can say that while light microscopes can easily be used to examine living objects, electron microscopes, with a very few exceptions, can only examine cells that are killed, fixed (i.e., chemically stabilized), and stained.

In addition, the ability of the light microscopist to perceive color is an important asset, for variations in color—either the natural color of the specimen and its components, or a wide variety of stains—can impart considerable information. In contrast, the transmission electron microscope is decidedly monochromatic. The ability of the scanning microscope to use several different kinds of radiation from the specimen provides some of the advantages of color, but it is a weak substitute for the countless hues to which the human eye is sensitive.

The scanning electron microscope, as it is commonly used, images surface features, for electrons do not pass through the specimen. This factor is the source of both its strengths and its weakness. Neither of the other instruments, which depend primarily on transmitted radiation, are particularly good at visualizing surfaces, although surface features can be of considerable interest. The advantage of the scanning microscope in this application is its impressive depth of field—at least 300 times greater than that of optical instruments (see Fig. 1–37).

Another advantage of the scanning electron microscope is the facility with which magnification can be changed without refocusing. For example, an operator can pick out an interesting area from a wide field at 50×, and then merely turn a knob to examine a portion of it at 50,000×. The ease with which this change can be accomplished is a result of the magnification being determined simply by the ratio of the movement of the scanning beam to that of the display beam.

When sample preparation time is critical (e.g., a pathologist's examination of a tissue while the surgeon waits by the operating table for a decision), the light microscope has no serious rival. In addition, the variety of molecular stains available for light microscopy give it a versatility unmatched by the other instruments. On the other hand, when details of intracellular structures are to be examined at the submicron level, only the transmission electron microscope can be used. But when surface features are important, particularly when small features of varying depth are to be studied, the scanning electron microscope is the instrument of choice. All

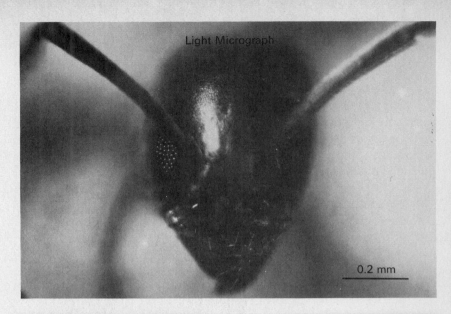

Light Micrograph

0.2 mm

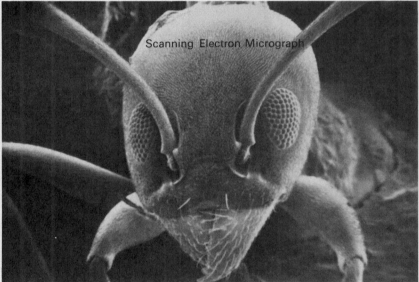

Scanning Electron Micrograph

FIGURE 1–37. Depth of Field. The light micrograph was taken, focused on an eye of the ant, then the same ant was coated with metal and examined with the scanning electron microscope. Note the impressive depth of field of the latter instrument, making it an invaluable tool for examining surface features. (Photograph courtesy of John R. Devaney, California Institute of Technology, and Kent Cambridge Scientific, Inc.)

three forms of microscopy play a prominent role in this book (see Fig. 1–38 for a comparison), although the nature of the subject causes us to rely mostly on the conventional transmission electron microscope for illustrations.

SAMPLE PREPARATION. The amount of information obtained from a microscope depends in large measure on how the specimen is prepared. Samples for the light microscope are commonly fixed (with alcohol, formalin, osmium salts, formaldehyde, and so on) to make their protein components insoluble and stable to subsequent

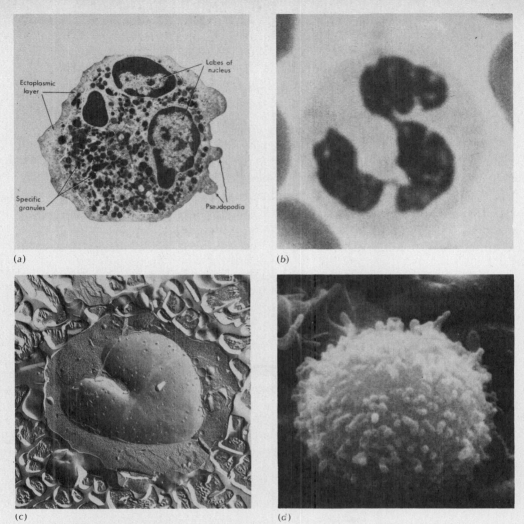

(a)

Ectoplasmic
layer

Lobes of
nucleus

Specific
granules

Pseudopodia

(b)

(c)

(d)

FIGURE 1–38. White Blood Cells: Four Views. (a) Heterophilic leucocyte from a guinea pig, as seen by transmission electron microscopy of a thin section. Note that the nucleus winds in and out of the plane of section so that it appears in three parts. (b) The corresponding cell from human blood as seen by conventional light microscopy. Although the cytoplasmic detail is lost, the nuclear shape is now readily apparent. (c) Transmission electron micrograph of human lymphocyte (another type of white blood cell) after freeze-fracture and freeze-etching (see text p. 59). Note the characteristic dent in the nucleus and the presence of nuclear pores. A thin layer of cytoplasm surrounds the nucleus. (d) A human lymphocyte by scanning electron microscopy. One can no longer see the nucleus, but villous appendages on the cell surface are now visible. These appendages identify the cell as a thymus-derived lymphocyte (T-cell). A second important class of lymphocytes, the B-cells, have a smooth surface and cannot be distinguished from T-cells by other types of microscopy. [(a) Courtesy of W. Bloom and D. W. Fawcett, *A Textbook of Histology*, 9th ed. Philadelphia: W. B. Saunders Co., 1968; (b) courtesy of Murray L. Barr, Univ. of Western Ontario; (c) courtesy of R. Scott and V. Marchesi, *Cellular Immunology*, **3**:301 (1972); (d) courtesy of Aaron Polliack, and described in *J. Exp. Med.*, **138**:607 (1973).]

procedures, then air dried and treated with a stain having an affinity for the structure to be examined. To ensure even penetration of the stain, the specimen is usually first embedded in paraffin (to provide support) and cut into thin sections. The paraffin is dissolved again with xylol before staining. Some of the dyes in common use are eosin, hematoxylin, aniline blue, crystal violet, methylene blue, methyl green, fuchsin, Congo red, rose bengal, and so on. Although most stains are specific for proteins, some are specific for nucleic acids (e.g., Feulgen's procedure for DNA, which uses a fuchsin derivative); others are specific for lipids and still others stain starches.

For rapid preparation of specimens, paraffin embedding may be replaced with rapid freezing in liquid nitrogen. Sections may be cut from a frozen specimen, dried by sublimation (freeze-drying) to preserve structural features and to avoid the flattening effect imposed by surface tension when specimens are air dried, and then stained in the usual way.

The purpose of using stains is to improve the contrast between specimen and environment. The wide variety of stains available for light microscopy has no counterpart in electron microscopy. There are, however, several different techniques by which samples are prepared for the electron microscope, each of which has its own advantages. The five most common are thin-sectioning, negative staining, metal shadowing, freeze-fracture, and whole mounts. A brief description of each follows in order to better understand the illustrations used later.

1. Thin-sectioning refers to the use of an ultramicrotome (see Fig. 1–39) to remove slices that are only a few hundred angstroms thick. It is the method used for most of the micrographs in this book. To withstand the passage of the diamond or glass knife without tearing, the specimen is first embedded in a hard plastic, such as epoxy resin, which is allowed to penetrate the sample before being polymerized. Sections are floated from the knife onto the surface of water and picked up by touching them with a copper grid (200–300 wires/inch) that has been coated with a thin plastic or carbon membrane as a support.

Thin-sectioning eliminates overlying and underlying structures that would otherwise confuse the image. However, stains must be incorporated into the tissue to improve contrast, since the structures left in these very thin sections would not otherwise scatter electrons differently enough to be seen. Typical stains are salts of tungsten, manganese, uranium, osmium, etc., all of which provide electron-dense heavy metal atoms to scatter electrons.

2. Negative staining differs from positive staining in that the specimen itself is unstained. Rather, it is embedded in a dense substance, usually phosphotungstic acid ($H_3PW_{12}O_{40}$), which is chosen for its extremely high electron density and consequent ability to scatter the beam. The portions of the specimen that

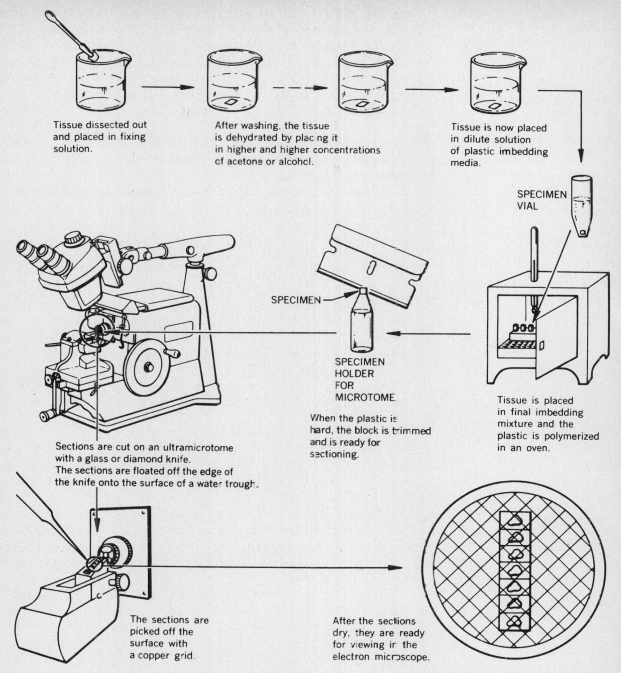

Tissue dissected out and placed in fixing solution.

After washing, the tissue is dehydrated by placing it in higher and higher concentrations of acetone or alcohol.

Tissue is now placed in dilute solution of plastic imbedding media.

SPECIMEN VIAL

SPECIMEN

SPECIMEN HOLDER FOR MICROTOME

When the plastic is hard, the block is trimmed and is ready for sectioning.

Tissue is placed in final imbedding mixture and the plastic is polymerized in an oven.

Sections are cut on an ultramicrotome with a glass or diamond knife. The sections are floated off the edge of the knife onto the surface of a water trough.

The sections are picked off the surface with a copper grid.

After the sections dry, they are ready for viewing in the electron microscope.

FIGURE 1–39. Thin-Sectioning for the Electron Microscope. (Courtesy of W. A. Jensen and R. B. Park. From *Cell Ultrastructure.* Belmont, Calif.: Wadsworth Publ. Co., 1967, by permission of the copyright owner.)

exclude phosphotungstic acid transmit electrons readily, so that their image can be seen. Because the stain penetrates various openings and crevices, some fine structure can often be observed. The technique is used mostly with viruses and other particulate material (see Figs. 1–24, 1–29, and 1–30).

3. Heavy metal shadowing is used to provide a relief of the ob-

FIGURE 1–40. Metal-Shadowing a Virus. Another way to prepare the virus for shadowing—an alternative to the procedure described here—is to place a drop of the suspension on the surface of a plastic-coated grid and then blot. A few particles adhere to the plastic. Though gentler than the diagrammed technique, blotting does not produce as even a distribution of particles. (Courtesy of W. A. Jensen and R. B. Park. From *Cell Ultrastructure.* Belmont, Calif.: Wadsworth Publ. Co., 1967, by permission.)

ject being examined. If metal deposition is carried out from one direction only, as diagrammed in Fig. 1–40, one side of the object will be coated while the other is not. Electrons pass readily through the area of lighter metal content, less readily through the plane on which the particle sits, and are scattered most severely by the side of the particle on which metal has accumulated. As a result, the particle appears as if it had been viewed in strong light (see Fig. 1–41).

In other cases, the specimen is rotated constantly while metal deposition is taking place, so that no shadows are created. Instead, metal builds up on all sides of the particle, making it stand out above the background (see Fig. 1–25). This technique is often used

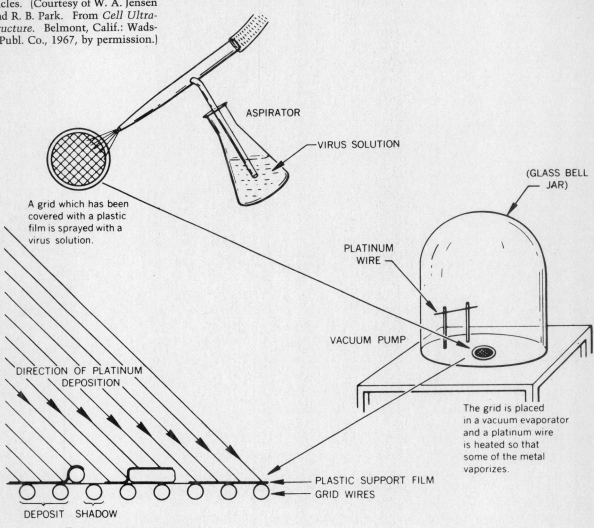

ASPIRATOR

VIRUS SOLUTION

(GLASS BELL JAR)

A grid which has been covered with a plastic film is sprayed with a virus solution.

PLATINUM WIRE

VACUUM PUMP

DIRECTION OF PLATINUM DEPOSITION

The grid is placed in a vacuum evaporator and a platinum wire is heated so that some of the metal vaporizes.

PLASTIC SUPPORT FILM
GRID WIRES

DEPOSIT   SHADOW

The metal builds up on the side toward the source of the metal vapor, and a shadow is visible on the reverse side.

FIGURE 1–41. A Metal-Shadowed Lily Virus. Shadowed with a 20:80 platinum–palladium alloy at an angle of 15°. The round object is a polystyrene latex sphere 264 nm in diameter. Note the "shadows," representing areas where electrons were least obstructed in their passage through the specimen support. Compare with negatively stained virus in Fig. 1–29. [Courtesy of T. C. Allen, Oregon State Univ. Similar to *Lily Yearbook*, **24**:29 (1971).]

with DNA, which by itself is a filament only about 20 Å thick. After "rotary shadowing," it has a diameter of a couple of hundred angstroms, and is therefore readily visible.

The layer of metal cannot be ignored when shadowed specimens are examined; it not only increases the overall size, but it may obscure some of the smaller surface features.

4. Freeze-fracture (see Figs. 1–12a and 1–38) is carried out by rapidly freezing the sample and then sectioning it in a vacuum while still at −100° C. The knife does not cut cleanly under those conditions, but tends to fracture the specimen along lines of natural weaknesses—such as the middle of a membrane running parallel to the cut. After fracture, the sample may be left in the vacuum long enough to allow some water to evaporate from and shrink softer tissues, a process called etching. The exposed face is then shadowed with metal to provide the necessary contrast; later, the organic material (i.e., the specimen itself) is removed by acids to leave a metal replica for examination in the electron microscope.

Much detail is retained in the replica, providing our best look at certain internal features of cells and our only way of seeing the face and internal features of a membrane.

5. Whole mounts (see Fig. 1–13) are often used to examine chromosomes and other relatively thick objects that can be isolated free of debris. As the name implies, the specimen is neither sectioned nor stained. Thick areas scatter electrons more strongly than thin areas, providing enough contrast to form an image. In fact, since scattering of the beam is directly proportional to the electron density of the sample, and hence to its mass, proper calibration allows estimates of specimen mass to be obtained from the image.

Samples for the scanning electron microscope are usually prepared by chemical fixing or freeze-drying to preserve shape, followed by shadowing with a layer of heavy metal. The latter step provides a conductive surface to drain off captured electrons that would otherwise cause distortions by deflecting the incoming beam. (Charge buildup can be a problem in transmission electron microscopy, too, though the thinner specimens transmit, rather than collect, most of the beam.)

1–6 THE CHOICE OF EXPERIMENTAL SYSTEMS IN BIOLOGICAL INVESTIGATIONS

> *I have often had cause to feel that my hands are cleverer than my head. That is a crude way of characterizing the dialectics of experimentation. When it is going well, it is like a quiet conversation with Nature. One asks a question and gets an answer; then one asks the next question and gets the next answer. An experiment is a device to make Nature speak intelligibly. After that one has only to listen.*
>
> GEORGE WALD, *1967 Nobel Lecture*

Unlike the physical scientist, who usually knows where to ask his questions (if not always how), the biologist has an enormous variety of living things from which to choose. His success will depend as much on choosing the right system as on asking the right question. The choice of the right system often means finding the simplest form of life that can be handled in the laboratory and yet features the process to be investigated. This consideration may lead to choices that seem, on the surface, to be of limited interest—like bacterial viruses, squid nerves, or frog muscles. However, simpler systems have fewer variables, and therefore the experimenter has a better chance of getting meaningful answers.

The information gained can then be used to formulate a model for more complicated systems. The unity of Nature being what it is, that model will, more often than not, be correct. In any case, it is much easier to prove or disprove a well-formulated model than it is to start fresh with no concept of how the complicated system might work.

For example, we are interested in how genes function in the cells of the higher plants and animals, including our own cells. But eucaryotic cells are very complex and relatively difficult to handle in the laboratory. They grow slowly, if they can be made to grow at all, and each of their genes may be found within the same cell in two alternate versions, called *alleles*. Changes in one gene may be masked by the activity of its allele. Bacteria, on the other hand, are much simpler and easier to work with, and they grow extremely fast. Although each bacterial cell may have multiple copies of the same gene, any changes in the genetic make-up of the cell will be rapidly reflected in the chemical or physical properties of the microbe as replicas of the bad gene are passed on to progeny. We might expect, then, that bacteria would serve as a better experimental system than human cells for studying the basic mechanisms of gene function. And since a virus is nothing more than a set of genes wrapped in a protective protein coat, we might also expect to learn some fundamental aspects of gene behavior by studying the manner in which virus genes control an infected cell. This is the reasoning used by some of the early molecular geneticists and, of course, their approach has been amply vindicated.

In later chapters we shall be discussing some of the specialized functions of cell types found only in higher organisms. You will find, for example, that progress in neural research was greatly facilitated by the discovery of the squid's "giant axon." This nerve cell is large enough to handle with ease, making it possible to do experiments that simply cannot be done with the tiny nerve cells found in mammals. And there is every reason to believe that the basic mechanisms found in squid nerves are the same as those found in human nerves. Although we readily admit that human systems are frequently more complex than their counterparts in simpler life forms, it is safe to say that because we have studied the lower forms of life we know a lot more about humans than we would have been likely to learn by direct observation, even with a much greater total effort.

SUMMARY

1-1 The cell theory, or cell doctrine, emerged in the nineteenth century. It holds that (1) all animals and plants are composed of cells, (2) each cell is capable of living in the absence of others, and (3) a cell can arise only from another cell. All cells have a plasma membrane to control the entry and exit of materials, a metabolic machinery for converting raw materials into needed substances, and a set of genes to guide the synthesis of new cellular components.

The nineteenth century also saw the end of vitalism, in which the activities of cells were thought to be somehow immune from the laws of physics and chemistry. We now recognize, however, that biological processes involve completely understandable chemical transformations that can be duplicated in the laboratory.

1-2 Cells are broadly divided into eucaryotes and procaryotes according to whether or not their genes are enclosed by a nuclear membrane. Procaryotes, which do not have a true nucleus, are the smaller and simpler of the two cell types. They include only bacteria and the blue-green algae. The only cellular inclusions that are universally present in procaryotes are ribosomes and a nuclear body in the form of one or more long, double-stranded molecules of DNA. External to the cytoplasmic membrane one finds, in almost all cases, a wall, and in some bacteria either flagella, pili, or both.

1-3 Eucaryotes have all the features of procaryotes and more, though only the ribosomes and plasma membrane closely resemble their procaryotic counterparts of the same name. The other major features are microfilaments, 100 Å filaments, and:

1. Units composed of microtubules or structures that look like microtubules, including flagella, cilia, basal bodies, and centrioles.
2. Bodies enclosed by a single membrane, including lysosomes, peroxisomes (microbodies), glyoxysomes, and vacuoles and vesicles of various kinds.
3. Bodies enclosed by two membranes, including nuclei, mitochondria, chloroplasts, and other plastids.
4. The endoplasmic reticulum (ER) and Golgi apparatus or dictyosome.
5. Chromatin, which is a complex of protein, DNA, and a little RNA. It is generally dispersed in the nucleus, but it condenses into chromosomes prior to cell division.

1-4 Viruses are cellular parasites, but they are not cells. They are broadly classed according to their hosts as bacterial viruses (bacteriophages), animal viruses, or plant viruses. A typical life cycle consists of the following steps:

1. Adsorption of the virus to a prospective host.
2. Separation of viral nucleic acid (which contains the genes) from viral protein. In the case of bacteriophages, the nucleic acid moves through the "tail" of the phage into the host cell, leaving the viral protein outside.
3. Replication of the virus under direction of the viral genes.
4. Release of new virions by budding or by lysis.

In some cases, viral genes can become incorporated into the host's genes following step (2), thus delaying—sometimes for many generations—step (3).

1-5 There are three important types of microscopes useful in biology: (1) the light microscope, (2) the transmission electron microscope (TEM), and (3) the scanning electron microscope (SEM). The light microscope and TEM form images from transmitted radiation (light and electrons, respectively); in contrast, most SEM's scan a beam back and forth across the specimen in synchrony with the moving beam of a television picture tube, using scattered or secondary electrons from the specimen to modulate the beam in the picture tube and hence form an image.

The light microscope (especially with phase contrast optics) can be used to observe living cells. This fact, plus the existence of a large number of available stains, gives the light microscope a versatility unmatched by the other instruments. The TEM, however, has a working resolution of 10 Å or less, which is about 200 times better than the light microscope. High resolution plus a variety of sample preparation techniques (thin-sectioning, shadowing with metal, negative staining, freeze-etching, and whole mounts) make the TEM indispensable to modern molecular biology.

The unique advantage of the SEM relative to the other two types of instruments is its enormous depth of field, with which it is possible to view the three-dimensional nature of objects.

1-6 When working with complex systems, one cannot always control all the possible variables in an experiment. The most fruitful investigations of fundamental life processes, therefore, are usually those that are carried out on the simplest forms of life in which the process is found. Once that system is understood, it can be used to formulate a model of the same process in more complex systems, often with excellent results.

STUDY GUIDE

1-1 (a) What is meant by the cell theory? (b) What is vitalism? (c) What features do all cells have in common?

1-2 (a) If a typical bacterium is a sphere, 1 μ in diameter, while a typical eucaryotic cell has the volume of a 10 μ diameter sphere, what is the ratio of cell volume to bacteria volume? [Ans: 1000] (b) What organisms produce oxygen and with what cellular process is it associated? (c) List and state the function of those structures commonly found outside the plasma membrane of procaryotes. Do the same for the internal structures.

1-3 (a) Which of the eucaryotic organelles are also found in procaryotes? Are their structures the same in the two classes of cells? (b) Describe lysosomes, microbodies, and the Golgi apparatus. (c) What features do mitochondria and chloroplasts have in common? (d) List the presumed functions of microtubules and structures composed of them. How are microfilaments different?

1-4 (a) What is a virus? Is it alive? (b) How do viruses differ from cells? (c) Describe a typical viral life cycle, including the various modes of release.

1-5 (a) What is meant by the resolution or resolving power of a microscope? (b) What is a typical resolution of a good light microscope? Transmission electron microscope? Scanning electron microscope? (c) What factors might dictate a choice among the three types of microscopes in a particular situation? (d) What ways are available to improve contrast in light microscopy? In transmission electron microscopy? (e) Five ways of preparing a specimen for the transmission electron microscope were described. What are they, and for what purposes are each employed?

1-6 An investigator who is interested in a particular biological process may be able to study that process in any number of different organisms. What kinds of considerations will influence his choice?

REFERENCES

HISTORY AND PHILOSOPHY OF CELL BIOLOGY

BROCK, THOMAS D., ed., *Milestones in Microbiology.* Englewood Cliffs, N. J.: Prentice-Hall, Inc., 1961. (Paperback.) The first section, on spontaneous generation, includes writings by Leeuwenhoek, Spallanzani, Schwann, Liebig, Pasteur, Tyndall, and Buchner, plus a commentary on their work.

DELBRÜCK, MAX, "A Physicist's Renewed Look at Biology: Twenty Years Later." *Science,* **168:**1312 (1970). 1969 Nobel Lecture.

FRUTON, J. S., and E. HIGGINS, *Molecules and Life: Historical Essays on the Interplay of Chemistry and Biology.* New York: John Wiley & Sons, 1972. Considers the period 1800–1950.

GABRIEL, M. L., and S. FOGEL, eds., *Great Experiments in Biology.* Englewood Cliffs, N. J.: Prentice-Hall, Inc., 1955. (Paperback.) Includes excerpts from papers by Hooke, Brown, Schwann, Leeuwenhoek, Koch, Pasteur, Ivanovsky, and Stanley, plus extensive chronologies.

JEFFREYS, HAROLD, *Scientific Inference* (3rd ed.). London: Cambridge Univ. Press, 1973. How scientists extrapolate from one system to another.

KOHLER, ROBERT, "The Background to Eduard Buchner's Discovery of Cell-Free Fermentation." *J. of the History of Biology,* **4:**35 (1971).

LURIA, S. E., "Molecular Biology: Past, Present, Future." *BioScience,* **24:**1289 (1970).

NEEDHAM, JOSEPH, ed., *The Chemistry of Life.* Cambridge Univ. Press, 1970. See the essay on the development of microbiology by E. Gale, and those on early biochemistry by M. Teich and R. Peters.

PORTER, J. R., "Louis Pasteur Sesquicentennial (1822–1972)." *Science,* **173:**1249 (1972). A short review of his contributions to modern biology.

ROSENBERG, E., *Cell and Molecular Biology: An Appreciation.* New York: Holt, Rinehart & Winston, 1971. (Paperback.) History of the cell theory.

CELL STRUCTURE

ECHLIN, PATRICK, "The Blue-Green Algae." *Scientific American,* June 1966. (Offprint 1044.)

FAWCETT, D. W., *The Cell: Its Organelles and Inclusions.* Philadelphia: W. B. Saunders Co., 1966. An excellent atlas.

JENSEN, W. A., and R. B. PARK, *Cell Ultrastructure.* Belmont, Calif.: Wadsworth Publishing Co., 1967. (Paperback.) A low-priced atlas of electron microscopy. Limited in scope but containing many excellent illustrations.

LEDBETTER, M. C., and K. R. PORTER, *Introduction to the Fine Structure of Plant Cells.* Berlin: Springer-Verlag, 1970. An atlas, with beautiful electron micrographs.

MOROWITZ, H. J., and M. E. TOURTELLOTTE, "The Smallest Living Cells." *Scientific American*, March 1962. (Offprint 1005.) Mycoplasmas.

PALADE, G. E., "Structure and Function at the Cellular Level." *J. Am. Med. Assoc.*, **198**:815 (1966). Lasker Award Lecture, 1966.

PORTER, K. R., and M. A. BONNEVILLE, *Fine Structure of Cells and Tissues.* Philadelphia: Lea & Febiger, 1968.

CELL ORGANELLES AND THEIR FUNCTION

BORISY, G., J. OLMSTED, J. MARCUM, and C. ALLEN, "Microtubule Assembly *in Vitro.*" *Federation Proceedings*, **33**:167 (1974).

BRYAN, J., "Biochemical Properties of Microtubules." *Federation Proceedings*, **33**:152 (1974).

BURDETT, I. D. J., "Bacterial Mesosomes." *Science Progress (Oxford)*, **60**:527 (1972).

DE DUVE, CHRISTIAN, "The Lysosome." *Scientific American*, May 1963. (Offprint 156.)

———, "The Peroxisome: A New Cytoplasmic Organelle." *Proc. Roy. Soc.* (London) *B*, **173**:71 (1969).

FREY-WYSSLING, A., *Comparative Organellography of the Cytoplasm.* New York: Springer-Verlag, 1973. Membranes, microtubules, microfilaments. Not an elementary treatment.

NEUTRA, M., and C. P. LEBLOND, "The Golgi Apparatus." *Scientific American*, February 1969. (Offprint 1134.)

NOVIKOFF, ALEX B., and E. HOLTZMAN, *Cells and Organelles.* New York: Holt, Rinehart & Winston, 1970. (Paperback.)

RICHARDSON, M., "Microbodies (Glyoxysomes and Peroxisomes) in Plants." *Science Progress (Oxford)*, **61**:41 (1974).

SATIR, PETER, "Cilia." *Scientific American*, February 1961. (Offprint 79.)

SLEIGH, MICHAEL A., ed., *Cilia and Flagella.* New York: Academic Press, Inc., 1974. Review articles.

VIGIL, E. L., "Structure and Function of Plant Microbodies." *Sub-Cellular Biochem.*, **2**:237 (1973).

WESSELLS, NORMAN K., "How Living Cells Change Shape." *Scientific American*, October 1971. (Offprint 1233.) Function of microtubules and microfilaments.

VIRUSES

CURTIS, HELENA, *The Viruses.* Garden City, New York: The Natural History Press, 1966. (Paperback.)

DALES, S., "Early Events in Cell–Animal Virus Interactions." *Bact. Revs.*, **37**:103 (1973).

DALTON, A. J., and F. HAGUENAU, eds., *Ultrastructure of Animal Viruses and Bacteriophages: An Atlas.* New York: Academic Press, Inc., 1973.

FRASER, DEAN, *Viruses and Molecular Biology.* New York: The Macmillan Co., 1967. (Paperback.)

HAHON, NICHOLAS, ed., *Selected Papers on Virology.* Englewood Cliffs, N. J.: Prentice-Hall, Inc., 1964. (Paperback.) Contains many of the most important early papers on animal, plant, and bacterial viruses.

HORNE, R. W., "The Structure of Viruses." *Scientific American*, January 1963. (Offprint 147.)

JACOB, F., and E. L. WOLLMAN, "Viruses and Genes." *Scientific American*, June 1961. (Offprint 89.) Lysogeny and viral transformation.

LWOFF, ANDRÉ, "Interaction Among Virus, Cell, and Organism." *Science*, **152**:1216 (1966). Nobel Lecture (1965) on lysogeny.

STENT, GUNTHER, *Papers on Bacterial Viruses* (2d ed.). Boston: Little, Brown & Co., 1965. See the introduction and reprinted works of d'Herelle, Twort, Bordet, Gratia, Ellis, Delbrück, and Lwoff.

TAMM, IGOR, "The Replication of Viruses." *American Scientist*, **56**:189 (1968). Animal viruses. Reprinted from the *Rockefeller Univ. Review*, November–December 1967.

ELECTRON MICROSCOPY

CARR, K. E., "Applications of Scanning Electron Microscopy in Biology." *Int. Rev. Cytology*, **30**:183 (1971).

DA SILVA, P., AND D. BRANTON, "Membrane Splitting in Freeze-etching." *J. Cell Biol.*, **45**:598 (1970). Proof that freeze fracture of red cell membranes splits the membrane without exposing either surface.

EVERHART, T. E., and T. L. HAYES, "The Scanning Electron Microscope." *Scientific American*, January, 1972, p. 54.

GRIMSTONE, A. V., and R. J. SKAER, *A Guidebook to Microscopical Methods.* Cambridge University Press, 1972. (Paperback.) Both light and electron microscopy.

HAGGIS, G. H., *The Electron Microscope in Molecular Biology.* New York: John Wiley & Sons, 1966. Considers techniques, then examples.

HORNE, ROBERT W., "Electron Microscopy of Viruses." *Science Progress (Oxford)*, **52**:525 (1964). Negative staining.

PALADE, GEORGE E., "Albert Claude and the Beginnings of Biological Electron Microscopy." *J. Cell Biol.*, **50**:5D (1971).

KOEHLER, J. K., *Advanced Techniques in Biological Electron Microscopy.* Berlin: Springer-Verlag, 1973.

ROBARDS, A. W., "Ultrastructural Methods for Looking at Frozen Cells." *Science Progress (Oxford)*, **61**:1 (1974). Freeze-etching and freeze-fracture.

CHAPTER 2

Chemical Bonds and Chemical Equilibrium

2-1   GIBBS FREE ENERGY AND RELATED PARAMETERS   66
2-2   BOND ENERGY AND CHEMICAL EQUILIBRIUM   68
      Equilibrium Constants
      Bond Energy
2-3   THE BIOLOGICALLY IMPORTANT WEAK BONDS   72
      Hydrogen Bonds
      Ionic Bonds
      Hydrophobic Bonds
      Short-Range Bonds
2-4   ENERGY CHANGES IN CHEMICAL REACTIONS   80
      The Importance of Entropy
      The Significance of $\Delta G°$
      The Relationship Between Concentration and Free Energy
      Chemical Reactions
2-5   ACIDS AND BASES   85
      The Behavior of Acids in Water
      Buffers
      The Behavior of Bases in Water
2-6   THE STATE OF INTRACELLULAR WATER   88
SUMMARY   90
STUDY GUIDE   91
REFERENCES   91

In order to gain some appreciation for why cells behave as they do, it is necessary to understand something about the nature of their molecular building blocks, how those elements are held together, and the energetics of cellular reactions. Chemical energetics will be introduced here and used to establish the concept of weak chemical bonds. In Chap. 3, the biological importance of those weak bonds will become clear as macromolecular structure is introduced. Then, in Chaps. 4 and 5, we will show how cells obtain and utilize energy, and how they channel it to achieve the necessary level of efficiency. First, however, some of the concepts and nomenclature of thermodynamics will be introduced.

2–1 GIBBS FREE ENERGY AND RELATED PARAMETERS

The kind of energy in which we are most interested here is the *free energy*, or *Gibbs free energy*, designated as G or sometimes F. A change in this parameter, indicated by ΔG, represents the amount of energy available to do work when both temperature and pressure are held constant, the conditions under which almost all biological reactions take place.

Ordinarily, the amount of energy available to do work is less than the total energy content of the system, the rest of the energy being tied up in the motion of molecules. If a change in total energy at constant temperature is given as ΔH, where H is called the *enthalpy* or heat content, then the amount of energy available to do work is

$$\Delta G = \Delta H - T\Delta S \qquad (2\text{--}1)$$

The new term, S, is called the *entropy*. It is a measure of the randomness of the system: the more disordered the system becomes, the greater its entropy. Since the delta (Δ) always means "final state minus initial state" (e.g., $\Delta G = G_2 - G_1$ in the transition from energy level one to energy level two), a drop in total energy, $\Delta H < 0$, may be partially or completely offset by a decrease in randomness ($\Delta S < 0$), leaving little or no energy available to do work.

Reactions in which heat is released (i.e., $\Delta H < 0$) are called *exothermic*. They are characterized by a warming of their environment as heat leaves the system. On the other hand, reactions that cool their surroundings do so because they are absorbing heat energy. Such reactions are called *endothermic* ($\Delta H > 0$). (See Fig. 2-1.) A similar nomenclature is used for free energy changes: reactions in which there is a release of free energy are *exergonic* ($\Delta G < 0$), while reactions that require energy to be added to the system are *endergonic* ($\Delta G > 0$).

There is a natural tendency for a system to move toward its state of lowest energy, as when a particle falls to the floor. There is also a natural tendency to lose heat to the environment and to become more disordered. We see this latter tendency in the dispersal of a dye in water, for instance. Another way of stating the above is that most naturally occurring reactions have $\Delta G < 0$, $\Delta H < 0$, and $\Delta S > 0$. A negative free energy change by itself (i.e., $\Delta G < 0$) is sufficient to ensure that the reaction will go spontaneously (though not necessarily at a measurable rate); by definition such a reaction could be used to do work on another system—drive an engine, lift a weight, etc.

The overriding consideration, then, is that all systems tend to minimize their free energy content. It follows that all spontaneous reactions will have a negative change in free energy. From equation (2-1) it is seen that such a reaction may be favorable either because of a large increase in entropy or because of a drop in heat content (enthalpy). Both may be true at the same time, of course, or one term may be dominated by the other (see Table 2.1). In other words, useful work, measured by the negative increment in G, may be contributed by an enthalpy drop or entropy increase, but a portion of one may be offset by an unfavorable change in the other.

Equation (2-1) is really a partial statement of the *first law of thermodynamics*, which says that energy can neither be created

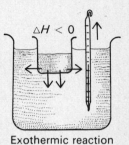

Exothermic reaction

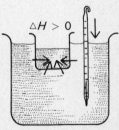

Endothermic reaction

FIGURE 2-1. Heat Transfer. An exothermic reaction is one in which heat energy is released ($\Delta H < 0$), reflected in a warming of the environment. An endothermic reaction absorbs heat energy ($\Delta H > 0$), cooling the reaction mixture, its vessel, and its environment.

TABLE 2.1 The relationship between free energy, enthalpy, and entropy

$$\Delta G = \Delta H - T \Delta S$$

ENTHALPY CHANGE	ENTROPY CHANGE	CONDITIONS FOR A SPONTANEOUS REACTION ($\Delta G < 0$)	CONCLUSION
$\Delta H > 0$	$\Delta S > 0$	when $T \Delta S > \Delta H$	The reaction can be *entropy driven*, but a portion of the entropy increase is converted to enthalpy.
$\Delta H < 0$	$\Delta S < 0$	when $\lvert \Delta H \rvert > \lvert T \Delta S \rvert$	The reaction can be *enthalpy driven*, but some of the released heat is consumed in bringing order to the system.
$\Delta H < 0$	$\Delta S > 0$	any values of ΔH and ΔS	Both entropy and enthalpy contribute to the energy available to do useful work.
$\Delta H > 0$	$\Delta S < 0$	never	Both enthalpy and entropy changes are unfavorable.

nor destroyed—only changed from one form to another. The terms G, H, and TS each represent a form of energy. Equation (2–1) tells us that while the normal tendency for heat to flow from a system of greater to a system of lesser temperature can be tapped to do useful work ($\Delta G < 0$), energy is *required* to put a stop to the normal tendency of matter to further disperse itself. As a result, the entropy of the universe is constantly increasing. Since entropy results from the normal thermal (Brownian) motion of molecules, the entropy content of a system can be zero only at 0 °K, when all molecular motion stops, which is one way of stating the *third law of thermodynamics*.

The first and third laws, just mentioned, are complemented by the *second law of thermodynamics*, which states that all systems tend toward an equilibrium. It follows naturally from the statement that systems tend to minimize their free energy content; when that minimum is reached, no further reaction takes place—i.e., $\Delta G = 0$ and the system is in equilibrium.

2–2 BOND ENERGY AND CHEMICAL EQUILIBRIUM

Ludwig Boltzmann (1844–1906) derived an equation that predicts the distribution of energy at various levels in a system at equilibrium, assuming only that the equilibrium distribution is always the most probable one. If we consider a system of just two energy states, one in which a chemical bond is made and the other in which it is broken, the Boltzmann equation will predict the fraction of samples to be found in each.

EQUILIBRIUM CONSTANTS. Let us give the name "state A" to the situation in which a chemical bond is intact and "state B" to the situation when the same bond is broken. Then Boltzmann's distribution predicts that the number of samples in state B (n_B) will be related to the number in state A (n_A) by

$$n_B = n_A e^{-\Delta G^\circ / RT}$$

where R is the universal gas constant (1.98 cal/deg-mole), T the absolute temperature (°K), e is the number 2.718 (the base for natural logarithms), and ΔG° is the energy required to break a "mole" (meaning Avogadro's number, 6.023×10^{23}) of bonds or the energy released when the same number of bonds are formed. (It always takes energy to break a bond, otherwise everything would simply fly apart into atoms.) Since ΔG° is characteristic of the kind of bond in question, and since R and T are constant, the ratio n_B/n_A is also a constant, called the *equilibrium constant* of the system:

$$K_{eq} = \frac{n_B}{n_A} = e^{-\Delta G^\circ / RT} \tag{2–2}$$

It is usually more convenient to talk about concentrations than absolute numbers of particles, but since a concentration is proportional to the number of particles in a standard volume (milliliter or liter), n_B/n_A can be replaced with the concentration ratio [B]/[A] without changing anything else. (Molar concentrations are indicated by brackets, [ ].) Thus, equation (2-2) becomes

$$K_{eq} = \frac{[B]_{eq}}{[A]_{eq}} = e^{-\Delta G^\circ/RT} \qquad (2\text{-}3)$$

or, using natural logarithms (i.e., base e),

$$\Delta G^\circ = -RT \ln K_{eq} \qquad (2\text{-}4)^{1}$$

Note that we designate the energy per mole as ΔG°, a quantity that is called the *standard state free energy change*. This is the quantity of energy released when a particle "falls" from level B to level A (as in the formation of bonds), or the amount of energy required to raise the particle from A back to B again (break the bonds). (See Fig. 2-2.) However, ΔG° describes a system that is in equilibrium, a system which, therefore, cannot be harnessed to do useful work. In accordance with the concept of equilibrium, the traffic between levels A and B is the same in both directions, with no net change in the population at either level and therefore no way of using the system to drive a machine. On the other hand, a system that is not in equilibrium can be made to do work during the transition toward its equilibrium position. The amount of work that can be done at constant temperature and pressure was defined earlier as the free energy change, written as ΔG. It follows that ΔG and ΔG° are related, in this case by the equation

$$\Delta G = \Delta G^\circ + RT \ln \frac{[B]}{[A]} \qquad (2\text{-}5)$$

Note that when $[B] = [B]_{eq}$ and $[A] = [A]_{eq}$ (i.e., when the system reaches equilibrium), combining equations (2-4) and (2-5) yields $\Delta G = 0$, as required, since a system at equilibrium can do no useful work on another system.

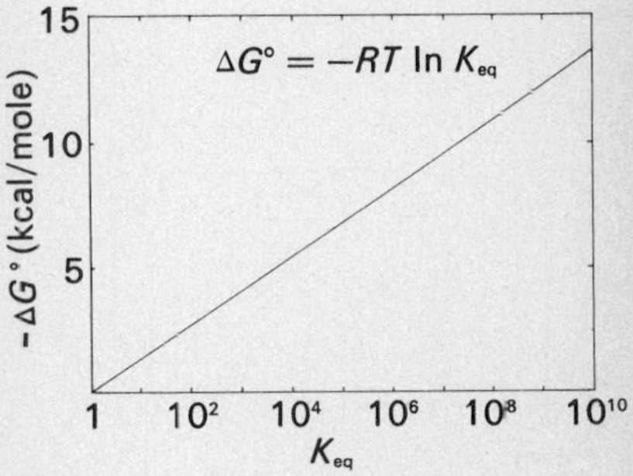

BOND ENERGY. Since, in the case of chemical bonds ΔG° is the amount of free energy per mole that is required to break them or the amount released when they are formed, ΔG° could be called the *bond energy*. Although that term is more often applied to the

FIGURE 2-2. The Relationship Between Standard State Free Energy Change (ΔG°) and the Equilibrium Constant of a Reaction at 298 °K.

---

1. In working with equation (2-4), it is often convenient to change from base e logarithms (natural logarithms, designated ln) to base 10 logarithms, designated log. Remember that $2.303 \log x = \ln x$. In addition, if we let $R = 1.98$ cal/deg-mole and $T = 298$ °K (25 °C),

$$-RT \ln K_{eq} = -1360 \log K_{eq}$$

enthalpy change than to the free energy change, if we presume that the potentially bonded substances are clamped in space, unable to drift apart when unbonded, then the difference between enthalpy and free energy can usually be ignored as the entropy change in this situation is zero [see equation (2–1)].

The Boltzmann equation, then, says that no matter how strong the bond between two atoms, a certain fraction of these bonds will always be broken. To put it another way, there is always a finite probability of finding any particular bond ruptured at any moment. The greater the bond energy, the more stable the bond, to be sure; but there is no such thing as a permanent chemical bond. However, we will find it useful to make a distinction between the more stable ("strong") bonds and the less stable ("weak") bonds.

A strong bond, for our purposes, will mean a *covalent bond*. These are, of course, the bonds that are almost universally present in substances and that are commonly designated by a single solid line or dash between the bonded atoms (e.g., H—O—H). They were first described by G. N. Lewis (1875–1946) of the University of California, who attributed their character to the sharing of one or more electron pairs by the two atoms involved. (Two shared electron pairs form a double bond, indicated by =, and so forth.)

Each atom has a particularly stable number of electrons that it can achieve by either losing or gaining a few. Hydrogen, for example, has one proton and one electron, but tends to assume the more stable helium configuration with two electrons. It can do so by accepting an electron from another substance to create the hydride anion, H^-, or by sharing an electron pair with the other substance, each of the atoms donating one electron to the shared pair. Thus, in hydrogen gas, H—H, each proton has, effectively, the more stable two-electron configuration, even though there is only a single electron pair in the molecule. This tendency to share the same electron pair bonds the two atoms together. It would take about 100 kcal to break Avogadro's number (a "mole") of H—H bonds, a bond energy that is not unusual for covalent bonds (see Table 2.2).

Other kinds of interactions, which we shall enumerate, are also important in holding atoms and molecules together. They are the weak bonds, so named because they generally have energies of only a few kilocalories per mole in biological systems. The important feature of weak bonds is that they are easily broken in the usual physiological environment. Therefore, structures that are stabilized by weak bonds can be easily rearranged—i.e., they are easy to assemble and disassemble. This simple fact, more than any other, is what makes life possible, for if there is anything that is characteristic of living organisms, it is change.

A bond with an energy of only a few kilocalories per mole has a good chance of being broken at any instant in time. The fraction

TABLE 2.2 Some covalent bond energies[a]

SINGLE BONDS	DOUBLE BONDS	TRIPLE BONDS
O—H 110 kcal/mole		
H—H 104		
P—O 100	P=O 120	
C—H 99		
N—H 93		
C—O 84	C=O 170	
C—C 83	C=C 146	C≡C 195
S—H 81		
C—N 70	C=N 147	C≡N 212
C—S 62		
N—O 53	N=O 145	
S—S 51		
N—N 38		
O—O 33		

[a] Energies are given as standard state enthalpies. From L. Pauling, *The Nature of the Chemical Bond*, 3d ed. (Ithaca, N.Y.: Cornell, 1960) and T. L. Cottrell, *The Strengths of Chemical Bonds*, 2d ed. (London: Butterworths Scientific Publ., 1958).

broken can be calculated from equation (2–3), which predicts, for example, that a bond energy of 1.4 kcal/mole results in an equilibrium constant (at 300 °K) of 0.1; an energy of 2.8 kcal/mole means K_{eq} is 0.01, and so on. The fraction of bonds in the broken state follows by simple algebra.

EXAMPLE.   Let $\Delta G° = 1.4$ kcal/mole
$$T = 300 \text{ °K}$$
$$R = 1.98 \text{ cal/deg-mole}$$

Then
$$K_{eq} = e^{-\Delta G°/RT} = e^{-1400/(300)(1.98)}$$
$$= e^{-2.36} = 0.095$$

or $K_{eq} = [B]_{eq}/[A]_{eq} = 0.095$, where $[B]_{eq}$ is the concentration of broken bonds and $[A]_{eq}$ is the concentration of formed bonds.

The fraction of bonds in the broken state is

$$[B]_{eq}/([B]_{eq} + [A]_{eq})$$
$$= [B]_{eq}/([B]_{eq} + 10.5[B]_{eq})$$
$$= 1/11.5$$

In other words, if a 1.4 kcal/mole bond exists between two molecules, then that bond will be broken about 9% of the time. This calculation assumes, as stated earlier, that the two participating groups are clamped into position, so that breaking the bond does not allow them to drift apart.

A covalent bond with a $\Delta G°$ of 50 kcal/mole, in contrast, would have an equilibrium constant of about 10^{-37}. There would be very

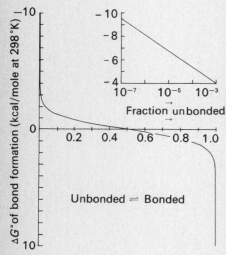

FIGURE 2-3. The Stability of Bonds. The stability of a bond is directly related to its energy. Note that bonds with energies ($\Delta G°$ in this case) greater than 5-10 kcal/mole are exceedingly stable, with little tendency to rupture spontaneously.

little spontaneous rupture in such a case. In fact, at moderate temperatures, it would take more than 10^{13} *moles* of such bonds before we could expect to find even one in the broken state. (See Fig. 2-3 for a graphical representation of this point.) If all change in molecular configuration had to depend on the spontaneous making and breaking of covalent bonds, life would be vastly more difficult.

Weak bonds are important to us because they stabilize the shape of many kinds of molecules, and because in many instances they hold together the individual molecules of larger structures as well. The protective coat of most viruses, for example, is a polymer of protein subunits held together by noncovalent (weak) interactions.

2-3 THE BIOLOGICALLY IMPORTANT WEAK BONDS

The weak chemical bonds may be classified as hydrogen bonds, ionic bonds, hydrophobic bonds, and bonds due to short-range forces. Because of their tremendous importance in biological systems, each will be considered briefly before we go on to discuss their functional role.

HYDROGEN BONDS. Hydrogen bonds are the result of proton (H^+) sharing. The energy necessary to break a mole of hydrogen bonds in an aqueous environment is about 5 kcal, give or take a couple depending on the circumstances. In biological structures, we usually find the shared proton between two nitrogens, between two oxygens, or between one of each, but always closer to the atom to which it really "belongs."

Oxygen and nitrogen atoms tend to hold electrons very tightly, and may retain a partial negative charge even while participating in a covalent bond. We say that such atoms are very *electronegative*. When either oxygen or nitrogen is covalently bonded to hydrogen, the result is a small *dipole*, or separation of unlike charges within the same molecule. In other words, the shared electron pair spends more time on the oxygen or nitrogen than on the hydrogen, leaving the latter with a partial positive charge. The oxygen or nitrogen, of course, carries a partial negative charge of equal magnitude since the overall structure must be electrically neutral.

Electrical charge originates with the nucleus and produces a field that drops off as the inverse square of distance from the nucleus. The smaller the atom, the more intense the field will be at the point of closest physical approach. Since hydrogen is the smallest atom, consisting of a single proton and a single electron, a given quantity of charge produces an electrostatic field of greater intensity than with other atoms. Hence, a hydrogen atom that is covalently bound to an oxygen or nitrogen still has a significant positive electrostatic field. This fact raises the possibility of producing an attractive force to an atom with a charge of opposite polarity, such as a second nitrogen or oxygen. Since the charge left on the hydrogen is small, the attraction to the second atom is weak, but it is by no means insignificant. That attraction is the basis for the hydrogen bond.

The positively charged proton, sitting between two negative atoms, shields them from each other. If the proton gets out of line, the negative atoms begin to repel each other again, forcing the proton back. Thus, we find that the three interacting atoms in an H-bond lie very nearly on a straight line, although deviations of up to 30° have been measured in a few cases.

Water is a good hydrogen bonder. Since it consists only of oxygen and hydrogen, there is maximum opportunity for such interactions. The degree of hydrogen bonding increases as water is cooled, until finally it can assume an open but ordered structure—i.e., the water solidifies into ice. Ice is a lattice of water molecules, with each oxygen participating in two H-bonds in addition to two covalent bonds (see Fig. 2–4).

When ice is warmed to the melting point, its three-dimensional lattice begins to break up. Considerable heat energy, called the *heat of fusion*, is necessary to rupture enough H-bonds to turn ice into a liquid. The disruption results in a more compact configuration (greater density), but since only about 15% of its bonds need be broken to melt ice, there is still a considerable amount of structure left in the resulting water. As the water is warmed to 4 °C, enough additional structure is destroyed to reduce the volume a little further. Above 4 °C water expands as it is warmed, apparently because the natural tendency for molecules to space themselves further apart as their thermal energy increases more than compensates for the additional destruction of lattice structure.

Water maintains a considerable amount of structure at all temperatures, however, right up to its boiling point. Then, moving a molecule from the liquid to the vapor state requires the breaking of still more H-bonds, resulting in a high *heat of vaporization*.

The simple hydrogen-oxygen structure of water not only is responsible for its peculiar physical properties—the expansion at very low temperatures and the high heats of fusion and vaporization—but contributes to the fact that water is a very good solvent. Water will dissolve appreciable quantities of almost any molecule that carries a net charge or that has *polar groups* (dipoles). That is, any molecule with an asymmetrically distributed electron cloud, or permanent dipole (e.g., O—H), is likely to be soluble in water because of attractive forces between it and the water molecules, which are themselves permanent dipoles. The interactions may be true hydrogen bonds, but other kinds of weak bonds are also important. It is because of such interactions that substances enter

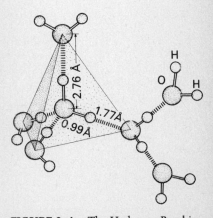

FIGURE 2–4. The Hydrogen Bond in Water. The hydrogen bond is in its most stable position when the three atoms involved (hydrogen and the electronegative atoms on either side) are linearly arranged, since the partial positive charge of hydrogen then shields the two electronegative atoms from one another. This factor, plus the normal bond angles of oxygen, gives ice (and, to a certain degree, liquid water) a tetrahedral arrangement.

aqueous solution. We said earlier that a system tends to seek its lowest energy. If more bonding energy can be released $(\Delta G < 0)$ by moving a substance into the aqueous phase of a system, then that is what will happen.

The hydrogen-bonding capacity of water is also responsible for the "weakness" of the hydrogen bond; any hydrogen bond that gets broken in an aqueous environment may be replaced with a hydrogen bond to water of nearly equivalent energy. Thus, it is not necessarily true that H-bonds are "weak"—some of them are quite strong—but it is the ease with which they may be replaced that makes hydrogen-bonded structures subject to change when water or other potential replacements are present.

IONIC BONDS. Ionic bonds result from the attractive force between completely ionized groups of opposite charge. In a vacuum, the ionic bond is very strong and is actually an extreme example of covalent bonding. The forces that hold Na^+ and Cl^- into a salt crystal are strong ionic bonds, resulting in a melting point (801 °C) that is much higher than the melting point of ice. In solution, however, the polar character of water shields charged groups from each other so that the ionic bond can no longer be considered strong (see Fig. 2–5). In fact, ionic bonds have energies of only about 5 kcal/mole in solution.

Water shields an ion because each water molecule is a small dipole. In the liquid state, the molecules are free to orient themselves about the ion, to present their pole of opposite sign to the charged group. This alignment of water has two results: (1) the electrostatic field of the ion is partly neutralized; and (2) some of the water molecules are immobilized, forming a *shell of hydration* about the ion. Typically, an ion of either charge will immobilize from four to six water molecules. For many purposes, the ion is now much larger than it was, since the entire unit—the ion plus its shell of hydration—will move as one (see Table 2.3). This *bound water* has properties somewhere between those of free water and of ice.

Consider the effect of the alkali ions, Li^+, Na^+, and K^+ on water. Water will be warmed appreciably as LiCl dissolves in it, implying that $\Delta H < 0$ for the process. Since it takes energy (enthalpy) to

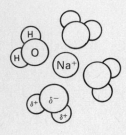

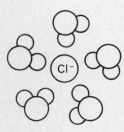

FIGURE 2–5. The Hydration of Ions. Because of their tendency to attract and hold water molecules, each of which behaves as a small dipole, ions in aqueous solution are partially shielded from one another. As a consequence, ionic interactions *in vivo* tend to be weak—in contrast, for example, to the extremely strong ionic bond in a crystal.

TABLE 2.3 The effect of hydration on ionic size

	Li^+	Na^+	K^+
Anhydrous diameter in a crystal lattice	1.6 Å	2.0 Å	2.3 Å
Hydrated diameter in solution	7.3	5.6	3.8
$\dfrac{\text{(hydrated volume)}}{\text{(anhydrous volume)}}$	95.0	22.0	4.5

break the strong bonds in the LiCl crystal, new bonds must be formed that more than compensate for the energy used: those would be the ion–water bonds as the water molecules orient themselves in the strong electrostatic field of the tiny Li^+ ion and in the weaker field of the larger Cl^- ion. On the other hand, water will be cooled slightly ($\Delta H > 0$) as NaCl dissolves, and the effect is even more pronounced when KCl is dissolved. Apparently, the new ion–water bonds do not have a total energy as great as the lost ionic bonds in their respective crystals. This is understandable if one considers that the physical size of the ions is $K^+ > Na^+ > Li^+$, whereas their charge is the same. The electrostatic fields of Na^+ and K^+ must be weaker than that of Li^+, and so they must be less capable of orienting water. Even though the bonds that were made to Na^+ and K^+ in the crystalline state would also be less strong than those to Li^+ (again because of size), the lost bond energies are not fully compensated by new ion–water bonds. In fact, in order for NaCl and KCl to be soluble ($\Delta G < 0$), there must be a positive entropy change (dissolution of the crystal) such that $\Delta H - T\Delta S$ is negative in spite of the positive ΔH.

One concludes that Li^+ has a very large shell of hydration, Na^+ less, and K^+ still less. This difference is amply demonstrated in a number of ways. For example, their relative velocities in water toward a negative electrode (at 25 °C) are 4.01, 5.19, and 7.62 μ/sec per unit field strength (volt/cm), respectively, with K^+ being the fastest in spite of having the largest atomic size. And in passing through pores in membranes, one also finds that K^+ behaves as if it were the smallest ion and Li^+ as if it were the largest, even though their actual physical sizes (without the shell of hydration) are the reverse of that.

The strength of an ionic bond in solution, then, is a measure of the difference in energy between the actual ion–ion bond and the total energy of the ion–water bonds that replace it—a situation similar to that of the hydrogen bond. One way of measuring the relative ability of a substance to respond in this way to the electrostatic field of an ion is to find its *dielectric constant*. The dielectric constant is a measure of the effectiveness with which a substance diminishes the electrostatic force between charged groups. The force is given by Coulomb's law,

$$F = \frac{q_1 q_2}{r^2 d} \tag{2-6}$$

where q_1 and q_2 are the magnitude of the charge on the two groups, and r the distance between them. The dielectric constant is given as d. Because it orients itself about charged groups, water has a high dielectric constant, equal to 80 at 20 °C relative to air or a vacuum with $d = 1$. This means that two oppositely charged ions attract each other only 1/80 as strongly when they are in water as when they are in air or in a vacuum. Thus, although ionic bonds in solution are weak, in a low-dielectric environment such as a salt crystal they are very strong.

The dielectric constant is thus a measure of the polar character of a molecule and of the ease with which it can orient itself in an electric field. While water has a dielectric constant of 80 at 20 °C, the value is very low (e.g., 1 to 5) for wood, glass, oils, fats, and other nonpolar substances. In contrast, glycerol (CH_2OH—$CHOH$—CH_2OH) and ethanol (CH_3—CH_2OH) have dielectric constants of 42 and 24 at 25 °C, respectively, because the —OH groups of these molecules are all small dipoles. We also find that glycerol and ethanol are soluble in water in all proportions (i.e., they are miscible with water) because of the large number of electrostatic interactions that they may make with water. Molecules with low dielectric constants do not interact with water and thus are not soluble in it to any appreciable extent. These are the nonpolar molecules involved in the next class of weak interactions, the so-called hydrophobic bonds.

HYDROPHOBIC BONDS. "Hydrophobic bond" is actually a misnomer, because what we really are referring to is the tendency of nonpolar groups to aggregate (associate with each other) when in the presence of water. This hydrophobic, or "water-fearing," behavior minimizes the nonpolar surface exposed to water and hence minimizes the unfavorable interaction that these groups have with it. The "unfavorable interaction" leads to a structuring of the water through an increase in hydrogen bonding. Water thus forms a lattice about the intruding group—as if the water molecules were clinging together to avoid contact with the nonpolar substance.

If you have been a careful reader, you may be a little puzzled at this point. On the one hand, you were told that the making of new bonds is a desirable situation, leading to a state of lower energy. On the other hand, you are now asked to believe that the increased number of water–water bonds caused by the introduction of hydrophobic groups is an undesirable event. To understand this apparent paradox, you will have to recall the fact that entropy may either contribute to or detract from the free energy change of a reaction (see equation (2–1) and Table 2.1).

To say that nonpolar molecules are insoluble in water is just another way of saying that the free energy change, ΔG, is positive (unfavorable) for the process of dissolving such molecules. And yet we find that the small amount that does dissolve causes an evolution of heat ($\Delta H < 0$). According to equation (2–1), a drop in heat content can only accompany an unfavorable reaction when $\Delta S < 0$, and then just if the decrease in entropy is more than enough to offset the favorable enthalpy change.

The heat evolved when a nonpolar molecule enters solution must come from the formation of new bonds. But since nonpolar molecules cannot bond to water, we conclude that there must be an increase in the number of water–water bonds. This increased water structure means a decrease in entropy, and the positive ΔG for moving the nonpolar substance into solution tells us that the loss of entropy due to the structuring of water must more than compensate for the energy released by the formation of new hydrogen bonds. This unfavorable entropy change is the only feature keeping ΔG positive, and as such, it is the reason why nonpolar groups are insoluble.

That nonpolar groups are insoluble because of an entropy barrier has a very important consequence. The negative sign of ΔS means that raising the temperature makes ΔG even more positive. In other words, the nonpolar groups become even less soluble with increasing temperature and have an even greater tendency to get out of solution if they are already in it. Accordingly, the strength of hydrophobic bonds *increases* with temperature. In contrast, most other bonds are weakened by increasing temperature, since the more violent agitation of the molecules strains the forces holding them together. (Ionic bonds sometimes get stronger with increasing temperature, too, depending on the amount of entropy increase when the shell of hydration of the ions is disturbed during bond formation.)

There are many examples of hydrophobic bonding in biological systems, but just to bring the concept closer to home, consider an everyday situation. Ordinary detergents, including soap, are long, nonpolar hydrocarbon ($-CH_2-CH_2-CH_2-$, etc.) chains, terminated at one end by a charged group, usually an acidic one such as the sulfate ($-OSO_3^-$) or carboxylate ($-COO^-$) ion. The

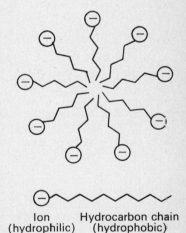

$$CH_3-(CH_2)_n-\overset{\displaystyle O}{\overset{\displaystyle \|}{C}}-O^-$$

Anion of a fatty acid (soap)

$$CH_3-(CH_2)_n-O-\overset{\displaystyle O}{\underset{\displaystyle O}{\overset{\displaystyle \|}{\underset{\displaystyle \|}{S}}}}-O^-$$

An unbranched synthetic detergent

charged end is highly water soluble, whereas the hydrocarbon "tail" does not enter readily into an aqueous environment. Although the charged end is soluble enough to drag its tail into the water, a critical concentration may be reached that allows the tails to solve their solubility problem by aggregating. They then form little spheres, or *micelles*, that have a hydrocarbon interior and a surface of charged groups (see Fig. 2–6). Several hundred molecules may participate in a single micelle, effectively excluding water and creating their own nonpolar environment. The micelles, of course, are stabilized by hydrophobic bonds.

If a droplet of fat encounters a micelle, it will enter the interior where it is soluble. A detergent solution thus presents two sol-

Ion   Hydrocarbon chain
(hydrophilic)   (hydrophobic)

FIGURE 2–6. Hydrophobic Bonds. The tendency for nonpolar groups to exclude water from their midst leads to the hydrophobic bond. Detergent molecules, which have a charge at one end of a long nonpolar hydrocarbon chain, exhibit hydrophobic bonding when their nonpolar "tails" aggregate to form spherical micelles.

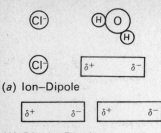

(a) Ion–Dipole

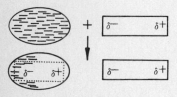

(b) Dipole–Dipole

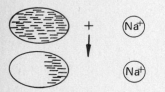

(c) Dipole–Induced Dipole

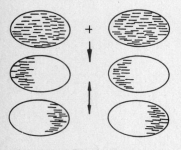

(d) Ion–Induced Dipole

(e) Induced Dipole–Induced Dipole
(charge fluctuation forces)

FIGURE 2–7. Short-Range Bonds. **(a)** The ion–dipole interaction is the kind of behavior exhibited by water (a dipole) in the vicinity of an ion (see also Fig. 2–5). **(b)** Two permanent dipoles tend to orient and attract one another, much as two permanent bar magnets would do. **(c)** Since a dipole creates an electrostatic field, it can induce a dipole in a neutral atom. The two will then attract one another. **(d)** An ion can also attract neutral molecules by inducing dipoles in them. **(e)** The weakest interactions are dispersion forces between two neutral atoms or molecules having generally symmetrical charge distributions.

vents, water and a hydrocarbon. The latter substance, confined to the interior of micelles, allows detergent solutions to dissolve fatty material like greases and oils.

Hydrophobic bonds, though important, are not strong. Depending on the temperature and area of the nonpolar groups, they may have energies ranging from less than 1 kcal/mole to 2 or 3 kcal/mole.

SHORT-RANGE BONDS. The fourth class of weak interactions includes bonds due to short-range forces (see Fig. 2–7). Such forces are negligible until the potentially bonded groups get very near each other, and they decrease rapidly as the interacting groups are moved apart: The effective bond energy will decrease as $1/r^4$ to $1/r^6$, where r is the distance between the interacting groups. Like the hydrophobic bonds, these short-range bonds are individually weak; however, if many occur between two molecules a very stable structure may result. A typical bond energy involving such forces is only about 1 kcal/mole.

The short range forces can be divided into two groups: those in which an ion interacts with a neutral molecule, called an *ion-dipole interaction;* and those in which two neutral atoms or molecules interact, called a *van der Waals interaction* (after the Dutch scientist J. C. van der Waals).

You have already been introduced to the ion–dipole interaction in the discussion of ionic bonds, for it is the ability of ions to form ion–dipole interactions with water (i.e., to structure water, the molecules of which are permanent dipoles) that reduces forces between the ions themselves to the point where the attractions can be classified as "weak." However, one need not have a permanent dipole to get an interaction with ions, for a positive ion will attract the electrons of neutral atoms while a negative ion repels them. In either case, the previously symmetrical charge distribution on the neutral atom is distorted, turning it into a dipole called an *induced dipole.* The alignment of the induced dipole is such that attractive forces between it and the ion are created. These forces fall off rapidly with increasing distance, dissipating as $1/r^5$. The effective bond energy between two such groups, then, decreases as $1/r^4$.

The second group of short-range forces, the van der Waals forces, are attractions between neutral atoms or molecules. They include: (1) dipole-dipole interactions, (2) dipole-induced dipole interactions, and (3) London dispersion forces, or induced dipole-induced dipole interactions. In each case, the force between the groups decreases with $1/r^7$ and the bond energy with $1/r^6$.

Two molecules with asymmetric charge distributions (dipoles) will attract each other if they are close and oriented with ends of opposite polarity facing each other. But since the charges at the two ends of a dipole are equal and opposite, the attraction between two dipoles falls off rapidly as the intermolecular distance becomes larger. When separated, the dipoles lose alignment, tumbling around in a random way and causing the rapid dissipation of interaction energy.

Peter Debye suggested in 1920 that since a dipole does establish an electrostatic field, it can induce an oppositely oriented dipole in a molecule with a symmetrical charge distribution, just as an ion can do, but not as successfully. If the negative pole of the dipole is near the uncharged molecule, the electrons of the uncharged molecule will experience a net repulsion and will spend more time than usual on the side away from the dipole. Similarly, if the positive pole of the dipole is nearer a neutral atom, the electrons of the atom will be attracted to the side facing the dipole. In either case, the neutral atom or molecule becomes an induced dipole. These interactions are usually weaker than the interactions between permanent dipoles, but they are stronger than the third type of van der Waals forces, the dispersion forces.

In 1930, F. London extended the concept of induced dipoles to neutral atoms. Although on the average the charge distribution about such atoms is spherically symmetrical, at any one moment in time the distribution will be distorted. You can convince yourself of this by thinking of the hydrogen atom, which has one proton and one electron. Although hydrogen is neutral, with the electron spending equal time on all sides, if you could stop the motion of the electron for an instant you would have a complete dipole (a "snapshot dipole"). In other words, even a neutral atom represents a fluctuating dipole. If two neutral atoms are brought close together, the transitory dipole nature of one will, for brief instants, induce a dipole of opposite polarity in the other atom, and vice versa. The result is that the two atoms experience a slight mutual attraction. London calculated the magnitude of this attraction from quantum mechanical considerations; in his honor such forces are sometimes called *London dispersion forces*.

Short-range forces, especially the dispersion forces, also contribute to the hydrophobic bond, which was presented earlier as a purely entropy-driven interaction. The nonpolar groups that aggregate to avoid water must experience dispersion forces, stabilizing their association and hence increasing the energy of the hydrophobic bond.

The concept of dispersion forces is useful in still another context, for although neutral atoms attract each other with a force that increases rapidly as they approach, a point will be reached when that attraction is offset by an overlap in the electron clouds of two atoms. This overlap produces a repulsive force that becomes extremely strong as the atoms move still closer. At some given separation the overlap repulsion and the van der Waals attraction will exactly cancel, and in the absence of any disturbing factor, that is the separation we can expect the two atoms to maintain. When two identical atoms are in that position, half the nucleus-to-nucleus distance defines the *van der Waals radius* of the atoms, a parameter that is a useful way of describing their physical size.

2–4 ENERGY CHANGES IN CHEMICAL REACTIONS

From the foregoing discussion, one might get the impression that structures stabilized by weak chemical bonds are more resistant to change than they really are. This impression is a result of our neglect of entropy, except in the case of hydrophobic bonds.

THE IMPORTANCE OF ENTROPY. In earlier examples, it was required that potentially bonded groups be fixed in space. Such is rarely the case, of course, for in reality molecules containing the participating groups are usually free to drift apart once the bond is broken. This freedom obviously introduces a positive entropy term, making ΔG smaller. In other words, the existence of entropy means that it takes less free energy to break a bond, and that less can be recovered as ΔG when the bond forms. Reestablishing a ruptured bond will, in most cases, require a collision between two molecules free in solution; the equilibrium constant for the reaction reflects this requirement by including the concentration of each species, as we shall see.

As an example of the importance of entropy, consider the hydrogen bond in the following situation (from Laskowski and Scheraga, 1954: see References).

Here we have a part of a large protein structure holding the two organic side chains in place. (They are portions of the amino acids glutamate and tyrosine.) The groups cannot drift apart when the hydrogen bond (O · · · H) is broken, but breaking the H-bond does permit rotational freedom about each single bond. When the hydrogen bond is formed, rotation is almost completely stopped. Formation of the hydrogen bond results in a release of about 6 kcal/mole of bond energy ($\Delta H° = -6$ kcal/mole); but this is offset by a decrease in entropy because of the loss of rotational freedom, a decrease that requires virtually all of the 6 kcal. Thus, the change in free energy $(\Delta G°)$ is very close to zero, making the equilibrium constant about unity for the formation of this bond.[2]

The preceding example is not unusual. In general, part of the energy released in the formation of a bond goes into the entropy term instead of being made available to do work (ΔG). In some cases, such as hydrophobic bond formation, the entropy term is actually the dominant one. We are not justified, therefore, in neglecting molecular motion, except in very special cases.

THE SIGNIFICANCE OF $\Delta G°$. The parameter of greatest interest to us in any reaction is the Gibbs free energy; changes in it represent the amount of energy available to do work at constant temperature and pressure, which are the conditions encountered in a living cell. In particular, we are interested in the standard state free energy change, $\Delta G°$, for reasons that will now be explained.

First, we have already seen how the standard state free energy change is related to the equilibrium constant, equations (2-3) and (2-4), and hence to the long term stability of a bond or molecule. Thus, with a value for $\Delta G°$, one can predict the ratio of products to reactants when the system is allowed to reach equilibrium. In a system at equilibrium, $\Delta G°$ represents the energy required to move a single molecule from the product state to the reactant state, although energy is usually expressed on a per mole basis. A "single molecule" is specified because the movement must be accomplished without upsetting the equilibrium.

There are two unrealistic limitations to the concept of equilibrium and free energy change as it has been presented so far: (1) Most reactions in the cell never reach a true equilibrium position, because the cell is in a constant state of flux. And (2) only the simplest possible reaction has been considered: a one-dimensional movement of a molecule or particle between two states, A and B. In other words, we have assumed that higher levels of energy could be reached from the reference state in only one way — by breaking a bond, by rising vertically upwards, etc. These restrictions will now be eliminated.

THE RELATIONSHIP BETWEEN CONCENTRATION AND FREE ENERGY. According to the accepted convention in defining energy, a concentrated solution should have a higher content of energy than a more dilute solution, since the natural tendency is for solute to move from the former to the latter when the two solutions are brought together. In addition, to be consistent with the Boltzmann concept of energy distributions, the concentration (which is a measure

---

2. In other words, since $\Delta H° \approx T\Delta S°$, both $\Delta G°$ and hence $-RT \ln K$ must be approximately zero. But $\ln K = 0$ when $K = 1$. This relationship can be understood qualitatively if you remember that all motion is thermally inspired, and hence that stopping such motion at $T > 0$ °K must require energy.

of the number of molecules in a unit volume) should be proportional to the exponential term $e^{+G/RT}$. The plus sign on the exponent is required to allow the energy to increase with an increase in concentration. Then, by agreeing to consider only ratios of concentrations and the difference in free energy between two concentrations, we obtain the relationship

$$\frac{[A]}{[A]^{\circ}} = e^{(G_A - G^{\circ}_A)/RT} \tag{2-7}$$

Here the free energy G_A has been assigned to the concentration [A], and the free energy G°_A to an arbitrarily chosen reference, or standard state, concentration $[A]^{\circ}$. When dealing with substances on the molar scale, we will find that the most convenient reference concentration is a one molar solution.

The value of G_A or G°_A could be the amount of energy necessary to assemble the substance A from its elements in their native form and then bring it to the appropriate concentration in solution (the "free energy of formation"). Whether it is actually assigned in this or some other way is not important here, however, because only changes in free energy interest us. Since equation (2-7) is independently true for each substance in the solution, one should be able to use it to calculate the free energy change for any chemical reaction, such as a simple isomerization or interconversion

$$A \rightleftharpoons B \tag{2-8}$$

We shall now proceed to do that.

CHEMICAL REACTIONS. Taking the natural logarithm of both sides of equation (2-7) and rearranging the results yields

$$G_A = G^{\circ}_A + RT \ln \frac{[A]}{[A]^{\circ}} \tag{2-9}$$

A similar expression must hold for the product molecule, B:

$$G_B = G^{\circ}_B + RT \ln \frac{[B]}{[B]^{\circ}} \tag{2-10}$$

The free energy change for the reaction $A \rightleftharpoons B$, according to standard nomenclature, is given as the energy of the product state minus that of the reactant state. Since the product, by definition, is on the right, we have

$$\Delta G = G_B - G_A$$

or from equation (2-10) minus equation (2-9)[3]:

$$\Delta G = (G^{\circ}_B - G^{\circ}_A) + RT \ln \frac{[B]/[B]^{\circ}}{[A]/[A]^{\circ}} \tag{2-11}$$

---

3. Recall that logarithms may be combined according to the rule $\ln x - \ln y = \ln x/y$.

This may be further simplified by substituting ΔG° for $G_B^\circ - G_A^\circ$, and by noting that $[A]^\circ = [B]^\circ = 1$ mole/liter by definition. Thus, equation (2–11) becomes

$$\Delta G = \Delta G^\circ + RT \ln \frac{[B]}{[A]} \qquad (2\text{–}12)$$

(This simplification is why one mole/liter is a useful choice for the reference concentration.) Note that when A and B are in the standard state,

$$\ln \frac{[B]}{[A]} \longrightarrow \ln \frac{[B]^\circ}{[A]^\circ} = \ln \frac{1}{1} = 0$$

so that $\Delta G = \Delta G^\circ$, as required. And note further that when the system is allowed to reach equilibrium,

$$\Delta G = \Delta G^\circ + RT \ln \frac{[B]_{eq}}{[A]_{eq}} = 0 \qquad (2\text{–}13)$$

Thus, from equation (2–12) the free energy change varies with the initial concentrations of reactants and products, being equal to the standard state free energy change only when the reaction is started with all components in the standard state. If, for example, the isomerization reaction were started with a very high concentration of B, then the logarithmic term would be positive, leading to a positive (unfavorable) free energy change. This unfavorable change is entirely consistent with our intuition, for it tells us that a reaction that starts with too much B should not proceed further toward B, but should instead convert some of the B back to A. In other words, if the further reaction A → B is unfavorable, then B → A must be favorable. By convention, the free energy change for the reverse reaction is given by equation (2–12) with the positions of A and B reversed, which makes positive terms negative and negative terms positive.

On the other hand, if the isomerization reaction were started with all, or almost all, A, then only the log in equation (2–12) would be negative. (The logarithm of any number less than one is negative.) The reaction A ⇌ B must be energetically favorable ($\Delta G < 0$) under those conditions. Again, that is exactly what one would expect.

Starting the reaction with equal concentrations of A and B produces a free energy change equal to ΔG° if the solution is ideal, just as though both substances were initially in their one molar standard state. And, as noted earlier, if A and B are at their equilibrium concentrations, the logarithmic term just cancels ΔG°, making the free energy change zero. In other words, nothing will happen because the reaction is already at equilibrium; in that condition there will be no further change in the concentration of either reactant or product, and therefore no change in the free energy.

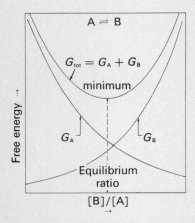

FIGURE 2–8. Equilibrium and Free Energy. All systems tend toward an equilibrium position characterized by a minimum in the total free energy.

It is important to notice that the absolute value of the free energy change increases with deviations from the equilibrium position in either direction. The sign of the free energy change will depend only on whether we are talking about the reaction A → B or the reaction B → A. In either case, there will be a spontaneous movement toward equilibrium, which is the position where the free energy change is zero (see Fig. 2–8). This tendency for a reaction always to move toward its equilibrium position, or position of lowest free energy, is what the second law of thermodynamics says will happen. Note that although the free energy *change* may be either positive or negative, the actual value of G becomes smaller as equilibrium is approached. Whether ΔG is negative or positive in a particular case depends only on what is called reactants and what is called products. The reaction can be harnessed to do useful work in either case, right up to the time at which it reaches equilibrium.[4]

From this discussion and Section 2–2, the standard state free energy change, $\Delta G°$, of a reaction: (1) provides a knowledge of the equilibrium constant, and hence of the long term stability of the system, though it is important to note that $\Delta G°$ says nothing at all about how long it will take to reach an equilibrium position; and (2) is the constant necessary to calculate free energy changes in nonequilibrium situations. Furthermore, note that $\Delta G°$ is numerically equal to the free energy change that would accompany the reaction with all reactants and products initially in the standard state.

The above considerations apply equally well to more general reactions, such as the one in which a moles of A plus b moles of B react to give c moles of C and d moles of D:

$$a\text{A} + b\text{B} \rightleftharpoons c\text{C} + d\text{D} \tag{2–14}$$

The free energy change for the reaction, following the earlier nomenclature and derivations, would be given as

$$\Delta G = \Delta G° + RT \ln \frac{[\text{C}]^c[\text{D}]^d}{[\text{A}]^a[\text{B}]^b} \tag{2–15}$$

Note that when each component is in the standard state, equation (2–15) reduces to $\Delta G = \Delta G°$, as required. When the system is allowed to reach equilibrium, the free energy change, ΔG, must be zero. Since

$$\Delta G° = -RT \ln K_{eq} \tag{2–16}$$

the equilibrium constant for this more general reaction must be given as

---

4. From the rules of calculus, when $\Delta G = 0$, G must be either a maximum or a minimum. In general, if $dx/dy = \Delta x/\Delta y = 0$, then Δx must be zero and x must be at a local maximum or minimum.

$$K_{eq} = \frac{[C]_{eq}{}^c [D]_{eq}{}^d}{[A]_{eq}{}^a [B]_{eq}{}^b} \qquad (2\text{--}17)$$

We shall have an opportunity to make use of this equation later, especially in Chap. 4.

2–5 ACIDS AND BASES

Certain chemical groups have a tendency to lose a hydrogen ion (H^+) in aqueous solution, whereas other groups have a tendency to attract hydrogen ions. The former are called acids and the latter are bases. This is the Brønsted criterion of acids and bases, although it is not the only way of defining them. Because we are accustomed to thinking in terms of hydrogen ion concentration as a measure of acidity, we will generally use Brønsted's nomenclature.

THE BEHAVIOR OF ACIDS IN WATER. An acid has a tendency to transfer a proton (hydrogen ion) to water, creating a hydronium ion, H_3O^+. The reaction is given as

$$HA + H_2O \rightleftharpoons H_3O^+ + A^- \qquad K_{eq} = \frac{[H_3O^+][A^-]}{[HA][H_2O]} \qquad (2\text{--}18)$$

Here HA is the acid (e.g., a carboxylic acid, $-\overset{\overset{\displaystyle O}{\|}}{C}-OH$) and A^- its anionic form ($-\overset{\overset{\displaystyle O}{\|}}{C}-O^-$), also called its conjugate base because of the ability of A^- to accept a proton. For dilute solutions, water concentration is almost always taken to be 55.55 molar, which is the concentration of pure water, and so it is convenient to combine this value with K_{eq} to give a new constant, K_a:

$$K_a = [H_2O]K_{eq} = \frac{[H_3O^+][A^-]}{[HA]} \qquad (2\text{--}19)$$

Taking the logarithm (base 10) of both sides of this equation yields

$$\log K_a = \log [H_3O^+] + \log \frac{[A^-]}{[HA]} \qquad (2\text{--}20)$$

But $-\log [H_3O^+]$ is simply the definiton of pH.[5] By analogy we define

$$pK_a = -\log K_a$$

---

5. One commonly sees pH defined as $-\log[H^+]$ and the acid reaction written as $HA \rightarrow H^+ + A^-$. This is just a shorthand notation; its use should cause no confusion as long as it is recognized as such.

so that equation (2–20) becomes

$$pH = pK_a + \log \frac{[A^-]}{[HA]} \qquad (2\text{–}21)$$

This equation, sometimes called the *Henderson-Hasselbalch* equation, states that the pH of a solution can be predicted if both the pK_a of an acid and its degree of ionization are known. Conversely, the degree of ionization of an acid can be predicted from the pH and a knowledge of its pK_a. The pK_a, since it describes an equilibrium position, is a constant for a given acid in a given environment.

BUFFERS. The logarithmic relationship between pH and degree of ionization of an acid leads to a *buffering capacity* in the region of the pK_a. In other words, there is a tendency to compensate for added acid or base by shifting the equilibrium, equation (2–18), in the opposite direction via simple mass action. Thus, when the pH is equal to the pK_a, the acid is half ionized. At one pH unit below the pK_a, the ratio $[A^-]/[HA]$ is 0.1 (i.e., the acid is only about 9% ionized); at one unit above the pK_a, the ratio is 10. Outside that range, buffering capacity falls off very rapidly since the acid is almost entirely in one form or the other (see Fig. 2–9).

Acids that have a pK_a near physiological pH (around pH 7) are of particular interest to us, because they may stabilize the pH of the cell. Phosphoric acid is an example. This acid has three ionizable groups (see Fig. 2–10), with equilibria as follows:

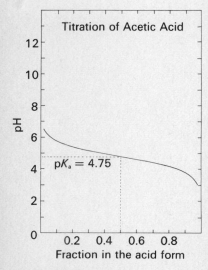

$$\begin{array}{ll}
H_2O + H_3PO_4 \rightleftharpoons H_2PO_4^- + H_3O^+ & pK_a \approx 2 \\
H_2O + H_2PO_4^- \rightleftharpoons HPO_4^{2-} + H_3O^+ & pK_a \approx 7 \\
H_2O + HPO_4^{2-} \rightleftharpoons PO_4^{3-} + H_3O^+ & pK_a \approx 12
\end{array}$$

$$(2\text{–}22)$$

Since the salts of phosphoric acid and phosphate derivatives are present in the cell in considerable quantities, we can expect them to be important in maintaining physiological pH. The second of the above reactions has a pK_a of 7.2 in very dilute solutions, but about 6.8 under conditions approximating those of the cytoplasm. Thus, the most common intracellular phosphoric acid ion (also called inorganic phosphate or orthophosphate, abbreviated P_i) is HPO_4^{2-}.

THE BEHAVIOR OF BASES IN WATER. The base reaction

$$B + H_2O \rightleftharpoons BH^+ + OH^- \qquad (2\text{–}23)$$

FIGURE 2–9. The Titration Curve of Acetic Acid. The equilibrium constant for the dissociation of acetic acid gives it a single pK_a of 4.75. Since pH is measured on a logarithmic scale, the molecule goes from 91% acid to 9% acid in the range pH 3.75 to 5.75.

is also well represented in cells, for example by amino groups:

$$R-NH_2 + H_2O \rightleftharpoons R-NH_3^+ + OH^-$$

The K_a for bases is written as

$$K_a = \frac{[H_3O^+][B]}{[BH^+]} \qquad (2\text{-}24)$$

Hence,

$$pH = pK_a + \log \frac{[B]}{[BH^+]} \qquad (2\text{-}25)$$

Thus, bases can be discussed in the same terms used for acids. That is, a base is half ionized when the pH is equal to its pK_a, and so on. Clearly, a base must be a good buffer at pH's near its pK_a for the same reason that an acid behaves in that way. As the pH is lowered, an acid becomes less highly ionized while a base becomes more ionized. In both cases, the lower pH's correspond to a more highly protonated form.

Water itself may act either as an acid, by donating a proton, or as a base, by accepting a proton. Both conditions are described by the reaction

$$2H_2O \rightleftharpoons H_3O^+ + OH^-$$

from which we define

$$K_{eq} = \frac{[H_3O^+][OH^-]}{[H_2O]^2}$$

and

$$K_w = [H_2O]^2 K_{eq} = [H_3O^+][OH^-] \qquad (2\text{-}26)$$

From K_w one obtains pK_w, defined as $-\log K_w$. Of course, electrical neutrality demands that $[H_3O^+]$ always be equal to $[OH^-]$ in pure water since there is no outside source of either ion. With

$$pOH = -\log[OH^-] \qquad (2\text{-}27)$$

as the counterpart of pH, taking the log of equation (2-26) gives

$$pOH + pH = pK_w \qquad (2\text{-}28)$$

Because pK_w changes markedly with temperature (from 15 at 0° to 13 at 60° — see Table 2.4), neutrality is pH 7 only at 25 °C. At physiological temperature, 37 °C, pure (neutral) water has a pH of 6.82.[6]

---

6. Actually, it is quite difficult to prepare pure water, so these predicted pH's are seldom achieved. The problem is that CO_2 from the air combines with water to form carbonic acid $(CO_2 + H_2O \rightarrow H_2CO_3 \rightarrow H^+ + HCO_3^-)$. Hence, most distilled water has a pH of 6 or less.

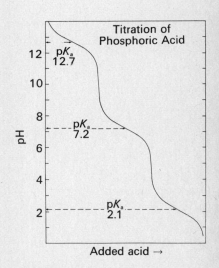

FIGURE 2-10. The Titration Curve of Phosphoric Acid. The biologically important molecule orthophosphoric acid (H_3PO_4 or just P_i) has three titratable hydrogens, and hence three pK_a's. In very dilute solution, the pK_a's are 2.12, 7.21, and 12.67, respectively.

TABLE 2.4   Dissociation constant of water at various temperatures

T, °C	pK_w
0	14.943
10	14.535
20	14.167
25	13.997
30	13.833
37	13.651
40	13.535
50	13.262
60	13.017

The dissociation constants given above for water presume that it is uncontaminated.   Water in the vicinity of other substances may behave quite differently.   For example, the complex of water and mercuric ion (Hg^{2+}) behaves as a fairly strong acid.   The reaction

$$[Hg \cdot 2H_2O]^{2+} \rightleftharpoons [Hg(H_2O)(OH)]^+ + H^+$$

has a pK_a of 3.7, which is stronger than acetic acid (p$K_a = 4.7$).

Before leaving acids and bases, it must be emphasized that there is nothing peculiar about the covalent bond made to protons in acids or bases.   The O—H bond in water, for example, has a bond energy of 100 kcal/mole.   Obviously, then, a water molecule in an isolated environment will not ionize.   That is, the probability of a spontaneous rupture in such a bond is about 10^{-81}.   Yet about 2×10^{-9} of the molecules in pure water will, in fact, be ionized at 25 °C.   Clearly, there must be a compensating factor in the form of a new bond to the released proton, leaving the system in a state of energy not so very different from that of undissociated water.   The new bond is the H_2O—H^+ bond in the hydronium ion, H_3O^+.   As a result, the dissociation of a water molecule has a standard state free energy change of about 24 kcal/mole, rather than the hundred or more one might otherwise expect.   Thus, a potential acid can only act as a proton donor when a suitable proton acceptor is present.   That acceptor, in biological systems, is water.

2–6   THE STATE OF INTRACELLULAR WATER
Ordinary water is by far the most prevalent compound in a cell, frequently representing about three-quarters of the total mass.   Its presence determines the character of the cytoplasm in numerous ways, some of which have been suggested here.   We will see many more examples in later chapters, but it might be prudent to summarize what has already been said about the substance.

The electronegative character of oxygen makes water a dipole. This in turn makes it a good dielectric, capable of shielding charged groups by forming structures about them. This shielding causes ionic interactions to be very labile, and hence we classify them as "weak" compared to covalent bonds.

The electronegativity of oxygen also makes water a good hydrogen bonder. It can both donate and accept a shared proton, interacting with other water molecules as well as with other kinds of proton donors and proton acceptors. This capacity has several consequences: (1) it makes the acid–base reaction possible, since the released proton can be bonded to water to form a hydronium ion (H_3O^+); (2) it causes water to act as a hydrogen bond competitor of other groups; and (3) it gives water a considerable amount of structure, even in the liquid state. The amount of structure is noticeable in many situations, among which is the measurement of ionic mobilities: If one measures the rate at which an ion moves between two oppositely charged electrodes in an aqueous solution, it is found that the hydrogen ion migrates five times as fast as any other positive ion (36.3 μ/sec per volt/cm compared with 7.6 for K^+ and 4.0 for Li^+). It appears to do so by combining with a water molecule and "bumping" another H^+ off the other end of the structure.

Proton Transfer

Hence, hydrogen ions migrate even faster in ice than in liquid water, a mobility that is fostered by the increased structure in spite of the lower temperature.

The structured character of water may be increased by the presence of other groups. Molecules of water become structured about ionic groups or dipoles due to ion–dipole or dipole–dipole interactions. They may also be structured by forming hydrogen bonds with both proton donors and proton acceptors. In addition, it has been mentioned that the presence of nonpolar groups increases the structure of water in their vicinity. Since the cytoplasm is replete with dipolar, ionic, and nonpolar groups, one wonders whether there is any "free" water in the cell at all. This question has prompted a number of investigations.

In the late 1960s several groups of researchers began drawing conclusions about the nature of cell water by using a technique called "nuclear magnetic resonance," or NMR. With this tool one can determine the degree of freedom experienced by hydrogens. In 1969 the results of two independent projects were announced in which it was concluded that the water in skeletal muscle cells exists in two fractions. Neither fraction has quite the freedom of ordinary liquid water, and one, 10 to 25% of the total, is very nearly icelike in character. These studies strongly support the suggestion, made on the basis of other kinds of evidence, that cytoplasmic water has considerably more structure than liquid water. Therefore, any conclusions concerning cellular reactions based on studies in free solution should take this factor into account.

SUMMARY

2–1 When a reaction releases energy in the form of heat, it is said to have released enthalpy ($\Delta H < 0$). A portion of the released heat energy may be available to do work. This available portion is called Gibbs free energy, G. The remainder of the released energy is often consumed by a decrease in randomness, or entropy, S. Any reaction in which there is a drop in free energy ($\Delta G = \Delta H - T\Delta S < 0$) can be made to do work; such reactions always occur spontaneously, though not necessarily at a measurable rate.

2–2 The equilibrium constant of a system (K_{eq}) is related to the standard state free energy change ($\Delta G°$) of the reaction involved. In the case where the system is defined as pairs of bonded atoms clamped in position, the equilibrium constant becomes the ratio of unbonded to bonded pairs, while $G°$ becomes a measure of the bond energy. Strong (covalent) bonds, with energies from about 50 to 200 kcal/mole, have very little chance of spontaneous rupture. On the other hand, weak bonds, with energies of only a few kcal/mole, may spend a significant fraction of the time in the unbonded condition.

2–3 Weak bonds can be divided into four groups:

1. *Hydrogen bonds*, in which a proton (carrying a partial positive charge) is positioned between two electronegative atoms, usually nitrogen or oxygen.
2. *Ionic bonds*, which result from the mutual attraction of two oppositely charged ions (considered weak in biological systems because of the shielding effect of water).
3. *Hydrophobic bonds*, where nonpolar groups aggregate to avoid increasing the structure of water. (These are entropy-driven interactions, characterized by a bond energy that increases with temperature.)
4. *Short-range bonds* due to ion–dipole interactions (where the dipole is either permanent or induced by the field of the ion) or due to van der Waals interac-

tions between neutral atoms or molecules. The van der Waals group consists of interactions between two dipoles, where one or both may be induced. In the case where both dipoles are induced, the interaction is also called a London dispersion force.

2–4 The free energy change for the generalized reaction

$$aA + bB \rightleftharpoons cC + dD \qquad (2\text{–}14)$$

is given by

$$\Delta G = \Delta G° + RT \ln \frac{[C]^c[D]^d}{[A]^a[B]^b} \qquad (2\text{–}15)$$

Note that when the reaction has reached equilibrium, $\Delta G = 0$; note further that ΔG will be favorable (negative) for progress toward that position from either excess reactant or excess product. This expression also tells us why $\Delta G°$ is such a useful quantity to have: (1) it provides the equilibrium constant ($\Delta G° = -RT \ln K_{eq}$), and (2) it provides the constant necessary to calculate ΔG in a nonequilibrium situation.

2–5 An acid is a molecule that tends to lose a proton (Brønsted's definition); a base tends to gain one. Acid–base reactions involve exchanges in covalent bonds, often with water as a participant. Water can both donate a proton, leaving OH^-, or accept one to create the hydronium ion, H_3O^+. Many substances tend to transfer a proton to water or accept one from it, thereby acting as acids or bases, respectively. An acid or base reaction when it is about half complete tends to buffer the solution against changes in pH that would otherwise result from the addition of more acid or base.

2–6 The properties of biological systems depend heavily on the characteristics of water, including: (1) that it is a dipole and can thus be oriented in electrostatic fields such as those of an ion; (2) that it can act as both a donor and an acceptor molecule in hydrogen bonds; and (3) its ability to donate or accept a proton in acid–base reactions. This multiplicity of possible interactions makes water an excellent solvent, but because of them a substantial fraction of protoplasmic water has a structure somewhere between that of ice and that of ordinary liquid water at the same temperature.

STUDY GUIDE

2-1 What do we mean by the phrases "entropy-driven reaction" and "enthalpy-driven reaction?"

2-2 If a bond has an energy ($\Delta G°$) of 5 kcal/mole, what percent can we expect to find broken at any one time? (Neglect entropy changes and assume $T = 300$ °K.) [Ans: 0.02%]

2-3 (a) Suppose that two molecules experience a van der Waals interaction between them equivalent to a 1 kcal/mole weak bond. At $T = 300$ °K and no entropy changes, what is the probability, P, of finding this bond in the broken state? (P is simply the ratio of broken bonds to total potential bonds.) [Ans: 0.16] (b) Suppose the two molecules are held together by three interactions like that described in part a. What is the probability that the molecules will be freed from each other at any instant? (NOTE: If the probability of an event is P, then the probability of n such events occurring simultaneously is P^n.) [Ans: 0.004] (c) Compare the probability of freeing two molecules by breaking three 1 kcal/mole bonds as in Problem 2-3b with the probability of freeing the molecules by breaking one 3 kcal/mole bond. [Ans: for the 3 kcal/mole bond, $P = 0.006$; note that for higher bond energies the difference in P between n bonds of energy $\Delta G°$ and one bond of energy $n \Delta G°$ becomes even smaller.] (d) How do the special properties of water contribute to each of the four classes of weak bonds? (e) If an ionic bond has a measured energy of 5 kcal/mole in an aqueous environment, what would be the strength of that bond in the interior of a membrane largely composed of hydrocarbon chains with a dielectric constant of about 4? [Ans: 100 kcal/mole]

2-4 (a) Why must we consider entropy when we are having a discussion about the energetics of chemical reactions? (b) What is the significance of $\Delta G°$?

2-5 (a) Given a pK_a of 6.8 for the second dissociable group of phosphoric acid, what fraction will exist as HPO_4^{2-} at pH 7.4 in a cell? (Neglect any contribution from other equilibria.) [Ans: The ratio $[HPO_4^{2-}]/[H_2PO_4^-]$ is 4; 80% of the phosphate is HPO_4^{2-}] (b) If you add 0.01 mole of HCl to one liter of distilled water (pH 7), what will be the final pH? If, instead of distilled water, you have a liter of 0.05 M phosphate (a mixture of the monobasic and dibasic salts) at pH 7, what will be the final pH after adding 0.01 mole of HCl? Assume HCl dissociates completely. Also assume that the pK_a of phosphate at this dilution is 7.0, so that $[HPO_4^{2-}] = [H_2PO_4^-] = 0.025$ M. (Hint: The answer should be obtained by working it as an equilibrium problem, but a good estimate can be obtained by recognizing that 0.01 mole of HCl will convert almost that much HPO_4^{2-} to $H_2PO_4^-$, and then applying the Henderson-Hasselbalch equation.) [Ans: pH 2; pH 6.63] (c) How could a divalent metal ion affect the pH of a solution?

2-6 (a) In what classes of reactions (discussed here) does water participate? (b) What are the special properties of water that allow it to take part in so many different kinds of interactions? (c) We find the following relative velocity of movement toward a negative electrode in a water environment:

$$H^+ >> K^+ > Na^+ > Li^+$$

The relative size of the ions is

$$K^+ > Na^+ > Li^+ > H^+$$

One would normally expect size to be inversely proportional to velocity. Explain the discrepancy.

REFERENCES

CHEMICAL BONDS AND THERMODYNAMIC PRINCIPLES

KLOTZ, IRVING M., *Energy Changes in Biochemical Reactions.* New York: Academic Press, 1967. (Paperback.)

LASKOWSKI, MICHAEL JR., and HAROLD A. SCHERAGA, "Thermodynamic Considerations of Protein Reactions: I. Modified Reactivity of Polar Groups." *J. Am. Chem. Soc.*, **76**:6305 (1954).

Morowitz, H. J., *Entropy for Biologists: An Introduction to Thermodynamics.* New York: Academic Press, 1970. (Paperback.)

THE STRUCTURE OF WATER

Erlander, S. R., "The Structure of Water." *Science J.,* **5:**60 (1969).

Fletcher, N. H., "The Freezing of Water." *Science Progress* (Oxford), **54:**227 (1966).

Scholander, P. F., "Tensile Water." *American Scientist,* **60:**584 (1972). Biologically important physical properties of water.

Solomon, A. K., The State of Water in Red Cells. *Scientific American,* February 1971. (Offprint 1213.)

CHAPTER 3

Molecular Architecture and Biological Function

3-1 THE IMPORTANCE OF CARBON TO BIOLOGICAL STRUCTURE 94
The Chemistry of Carbon
Classification of Organic Compounds

3-2 LIPIDS 101
Fatty Acids
Glycerides
Phospholipids
Steroids

3-3 CARBOHYDRATES 106
Monosaccharides
Oligosaccharides
Polysaccharides

3-4 NUCLEIC ACIDS 112
Nucleotides
Nucleotide Polymers
The Structure of DNA

3-5 PROTEINS 120
Amino Acids
The Peptide Bond
Primary Structure
Secondary Structure
Tertiary Structure
Quaternary Structure
Prosthetic Groups
Formation of Structure in vivo

SUMMARY 137

STUDY GUIDE 138

REFERENCES 139

The cell is composed almost entirely of water, assorted inorganic ions, many small organic molecules, and four types of more complex organic compounds: lipids, carbohydrates, nucleic acids, and proteins (Tables 3.1 and 3.2). Each of these four classes has characteristic physical and chemical properties and a set of well-defined biological roles to play in life processes. Although the basic unit of each is small, almost never over about 350 amu,[1] the units are usually found associated together to form very large polymers. Structures composed of carbohydrates, nucleic acids, or proteins are held together through a combination of covalent and non-covalent bonds. Lipids associate mostly through hydrophobic bonding, introduced in the last chapter.

After a quick overview of some of the relevant properties of carbon and its compounds, we will embark on a brief survey of the major biological macromolecules and the building blocks of which they are comprised. The reader who wishes more information on the subject than is contained here should consult a good biochemistry textbook, such as one of those listed at the end of this chapter.

3–1 THE IMPORTANCE OF CARBON TO BIOLOGICAL STRUCTURE

The biologically important organic molecules are composed of only a few different kinds of atoms. The carbohydrates, for example,

TABLE 3.1 Molecular composition of bacteria and some mammalian tissues

	RAT LIVER	RAT SKELETAL MUSCLE	*E. coli*
Water	69%	75%	70%
Protein	16	7	15
Glycogen	3	4	—
Phospholipids	3	2	2
Neutral lipids	2	9	—
RNA	1	1	6
DNA	0.2	0.3	1

---

1. amu, or atomic mass unit, is the weight of a hydrogen atom (actually $\frac{1}{16}$ the mass of an oxygen atom). Another name for this unit is a "dalton."

TABLE 3.2 The elementary composition of bacteria and humans

| | THE HUMAN BODY[a] | | E. coli | |
	GROSS COMPOSITION	DRY WEIGHT	GROSS COMPOSITION	DRY WEIGHT
Oxygen	65%	18%	69%	20%
Carbon	18	54	15	50
Hydrogen	10	8	11	10
Nitrogen	3	9	3	10
Phosphorus	1.0	3.0	1.2	4
Sulfur	0.25	0.75	0.3	–

[a] The body also contains appreciable quantities of calcium (1.5% of total mass), potasium (0.35%), sodium (0.15%), chloride (0.15%), and magnesium (0.05%). Iron and zinc each account for 0.004% or less.

consist almost exclusively of carbon, hydrogen, and oxygen. The other three classes of bio-organic substances contain these elements plus nitrogen. In addition, phosphorus is found in all nucleic acids and in some of the lipids, while sulfur is found in most proteins. Although there is an occasional divalent metal ion held by dative bonds (a type of covalent bond), mostly to proteins, these six elements, C, H, O, N, P, and S, out of the more than 100 elements known, account for almost the entire mass of a living organism (Tables 3.2 and 3.3).

THE CHEMISTRY OF CARBON. The structure and characteristics of organic substances depend so much on the properties of carbon that the study of this one element and its compounds is a recognized sub-discipline within the field of chemistry. The name of that sub-discipline, organic chemistry, reflects an obsolete view that the carbonaceous compounds derived from plants and animals (i.e., organically derived compounds) are endowed with a "vital force" associated only with living things. Although this idea died in the mid-nineteenth century, the fact that far more than half of all known chemical substances contain carbon, and that these million or more organic compounds are easily classed into a few

TABLE 3.3 Elementary composition of some macromolecules

	PROTEINS	CARBOHYDRATES	NUCLEIC ACIDS
Carbon	50–55%	40%	38%
Oxygen	19–24	53	31
Nitrogen	13–19	–	17
Hydrogen	6–7.3	7	3
Sulfur	0–4	–	–
Phosphorus	–	–	10

categories the members of which have similar properties, is sufficient justification for maintaining organic chemistry as a separate subject.

Carbon does not participate in any of the weak interactions described in the last chapter, except for van der Waals attractions, but its electronic configuration is such that an additional four electrons can be accommodated by each carbon nucleus. A covalent bond is formed by sharing an electron pair between two atoms, one electron of the pair having been donated by each atom. Thus, by participating in four covalent bonds, carbon gains, in effect, four additional electrons—giving it the electronic configuration of the nearest inert gas, neon. Carbon, in other words, is tetravalent. In the same way, hydrogen is normally monovalent, oxygen and sulfur divalent, nitrogen trivalent, and phosphorus pentavalent.

The fundamental importance of the tetravalent nature of carbon was first recognized by Kekule in the mid-nineteenth century. He suggested—correctly, as it turns out—that the four valency bonds of carbon are directed to the four corners of a tetrahedron having the carbon atom at its center (see Fig. 3–1). A tetrahedron has four identical sides, each an equilateral triangle. One consequence of the tetrahedral nature of carbon is that a carbon atom with four different groups bonded to it may exist in two isomeric forms, called *stereoisomers* because they are mirror images of each other, like your left and right hands (see Fig. 3–2). These *asymmetric carbons* rotate the plane of plane-polarized light. Hence, any molecule containing an asymmetric carbon is said to be *optically active*; it will have 2^n isomers, where n is the number of asymmetric carbons, for each of the carbons may have two nonidentical spatial arrangements of the four groups bonded to it. If the net effect of these asymmetric carbons is to rotate plane-polarized light clockwise (to the right) as one faces the source, the molecule is said to be *dextrorotatory*. If the plane is rotated in the opposite direction, the molecule is said to be *levorotatory*.

Two multivalent atoms may share more than one electron pair between them. If there are two shared pairs, the atoms are said to be held together by a double bond (C=C). Similarly, three shared pairs between two atoms constitute a triple bond (C≡C). Thus, while carbon may participate in four single bonds, it may also participate in some combination of single, double, and triple bonds that add up to four single bond equivalents. But, of course, the tetrahedral bond angles do not apply when a carbon atom is participating in double and triple bonds, nor can such a carbon atom be optically active. In fact, if it participates in two single and one double bond, the carbon atom will be found at the junction of a planar Y, with the slightly shorter double bond as the leg of the Y. And when carbon makes two double bonds, as in carbon dioxide (O=C=O), the three bonded atoms lie in a straight line.

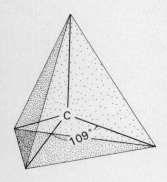

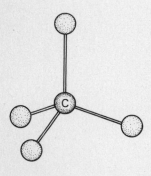

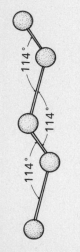

FIGURE 3–1. The Bond Angles of Carbon.

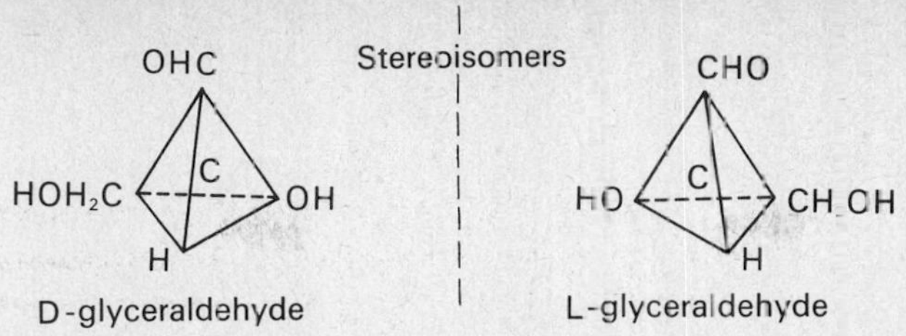

OHC   Stereoisomers   CHO

HOH₂C ←---C---→ OH   HO ←---C---→ CH₂OH

H   H

D-glyceraldehyde   L-glyceraldehyde

FIGURE 3-2. Stereoisomers. Note that these stereoisomers of glyceraldehyde are mirror images of each other. The central carbon is asymmetric because four different chemical groups are attached to it.

A double bond is always much stronger than a single bond between the same two atoms. Furthermore, double-bonded atoms are closer together, with a rigid link between them that prevents the rotation of one atom with respect to the other. (Usually there is free rotation about the axis of a single bond.) A triple bond is still shorter, stronger, and more rigid.

Some bonds, however, are neither single nor double, but something in between. Benzene, for example, can be written in two equivalent configurations, each having only tetravalent carbons.

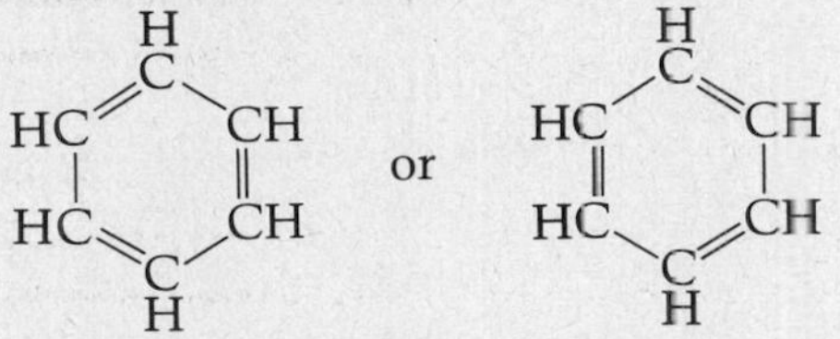

This situation is sometimes called *resonance,* a name that implies a shifting back and forth between two configurations. However, according to the concept of resonance developed by the American chemist Linus Pauling in the 1930s, all the carbon–carbon bonds in benzene are equivalent, with properties (bond energy, internuclear spacing, and so on) somewhere between those of a single and of a double bond.

Benzene has a total of six electrons (three pairs) that contribute equally to the partial double bond between its carbons. These *pi electrons* are said to be *delocalized,* for they travel freely about the molecule. The existence of resonance is indicated by drawing benzene in any of the following ways where, for the sake of simplicity, both carbons and hydrogens are omitted.

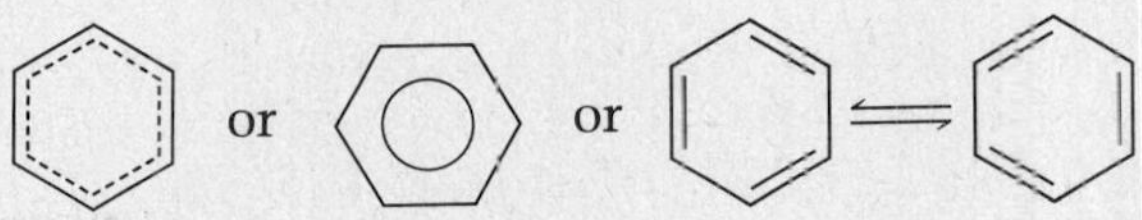

It is presumed that a carbon lies at the end of each straight line in these abbreviated structures, and that each carbon is tetravalent. If less than four bonds are shown to any carbon, the other bonds are assumed to be made to hydrogens. All of these representations indicate resonance or *conjugation*—i.e., the existence of delocalized electrons. It is also common, however, to draw only one of the fully bonded diagrams, for a structure that can be written in more than one way merely by interchanging adjacent single and double bonds should be recognized by the reader as a conjugated structure. It is this latter convention that will be adopted here.

CLASSIFICATION OF ORGANIC COMPOUNDS. It was stated earlier that, in spite of their great number, organic molecules fall easily into only a few classifications. The three main divisions are: (1) aliphatic compounds, (2) aromatic compounds, and (3) heterocyclic compounds.

Aliphatic compounds are derivatives of simple hydrocarbons (molecules comprised only of hydrogen and carbon). The simplest are the alkanes, having the general formula C_nH_{2n+2}: methane, ethane, propane, butane, pentane, and so on for $n = 1, 2, 3, 4$, etc.

$$
\begin{array}{ccc}
\text{H} & \text{H} \quad \text{H} & \text{H} \quad\text{H}\quad \text{H} \\
| & |\quad\; | & |\quad\;|\quad\;| \\
\text{H}-\text{C}-\text{H} & \text{H}-\text{C}-\text{C}-\text{H} & \text{H}-\text{C}-\text{C}-\text{C}-\text{H} \\
| & |\quad\;| & |\quad\;|\quad\;| \\
\text{H} & \text{H} \quad \text{H} & \text{H} \quad\text{H}\quad \text{H} \\
\text{Methane} & \text{Ethane} & \text{Propane}
\end{array}
$$

$$ CH_3-(CH_2)_2-CH_3 $$
n-butane

Note that butane is expressed here in a more compact way than the others. The *n* in front of butane stands for "normal," and means that the carbons are in a straight line, rather than branched as they would be in *iso*-butane.

The hydrocarbon chains of aliphatic compounds may form closed "alicyclic" structures, such as cyclohexane (C_6H_{12})

$$
\begin{array}{c}
\text{H}_2 \\
\text{C} \\
\text{H}_2\text{C} \quad\quad \text{CH}_2 \\
\text{H}_2\text{C} \quad\quad \text{CH}_2 \\
\text{C} \\
\text{H}_2
\end{array}
\quad \text{or} \quad \hexagon
$$

Note the difference between this substance and benzene, C_6H_6.

Dehydrogenation of alkanes—i.e., the removal of hydrogens consisting of one electron and one proton each—produces the unsaturated *alkenes* or olefins.

$$ CH_3-CH_2-CH_2-CH_3 \xrightarrow{-2H} CH_3-CH_2-CH=CH_2 \text{ or } CH_3-CH=CH-CH_3 $$
n-butane $\qquad\qquad$ 1-butene $\qquad\qquad$ 2-butene

Further dehydrogenation produces *alkynes*

$$CH_3-CH=CH-CH_3 \xrightarrow{-2H} CH_3-C\equiv C-CH_3$$

2-butene   2-butyne

The second major class of organic compounds, the *aromatics*, are derivatives of benzene or benzenelike molecules. The benzene ring, like the hydrocarbons, is decidedly nonpolar, and hence water insoluble. Because it is also flat (planar), molecules containing one or more benzene rings have a strong tendency to stack when permitted to come together in the presence of water. Using terms from Chap. 2, we would say that they are held together by hydrophobic bonds and van der Waals attractions. Since the pi electron clouds of stacked molecules are in contact, one also sees reference to "pi–pi interactions" among such molecules. However, the use of that term does not imply the existence of any interactions not already discussed under other lables.

The third class of organic compounds, termed *heterocyclic*, are of more interest to us here. Heterocyclic molecules are rings in which one or more atoms in the ring is something other than carbon—usually oxygen, nitrogen, or sulfur. Examples include pyridine, pyrimidine, and purine.

Pyridine   Pyrimidine   Purine

These three molecules form the basis for a host of biologically important substances. They and their derivatives are planar and subject to the same stacking forces as aromatic substances.

Organic molecules of any of the three basic classes just listed may be substituted in a variety of ways. That is, one or more hydrogens may be replaced by another atom or group (see Table 3.4). The substitution may replace a hydrogen with a hydroxyl (—OH) to create an alcohol, as in ethanol and phenol (note the "-ol" ending), derived from ethane and benzene, respectively.

$$CH_3-CH_2-OH$$
Ethanol

Phenol

A hydrogen may also be replaced by $-\overset{\displaystyle O}{\underset{\displaystyle \|}{C}}-OH$ to create a car-

TABLE 3.4 The major functional groups of organic compounds

R MAY BE ALIPHATIC, AROMATIC, OR HETEROCYCLIC.

Alcohols	R—OH	Amines	
Ethers	R—O—R	Primary	R—NH$_2$
Aldehydes	R—$\overset{\overset{\text{O}}{\|}}{\text{CH}}$	Secondary	R—NH—R
Ketones	R—$\overset{\overset{\text{O}}{\|}}{\text{C}}$—R	Tertiary	R—$\overset{\overset{\text{R}}{\|}}{\text{N}}$—R
Carboxylic acids	R—$\overset{\overset{\text{O}}{\|}}{\text{C}}$—OH	Esters	R—$\overset{\overset{\text{O}}{\|}}{\text{C}}$—O—R
		Nitro compounds	R—NO$_2$
Amides	R—$\overset{\overset{\text{O}}{\|}}{\text{C}}$—NH$_2$	Sulfonic acids	R—$\overset{\overset{\text{O}}{\|}}{\underset{\underset{\text{O}}{\|}}{\text{S}}}$—OH

boxylic acid, as in acetic, benzoic, and nicotinic acids. The latter compound, which is also called niacin, a B vitamin, is a pyridine derivative. (It is not the same as nicotine, found in tobacco.)

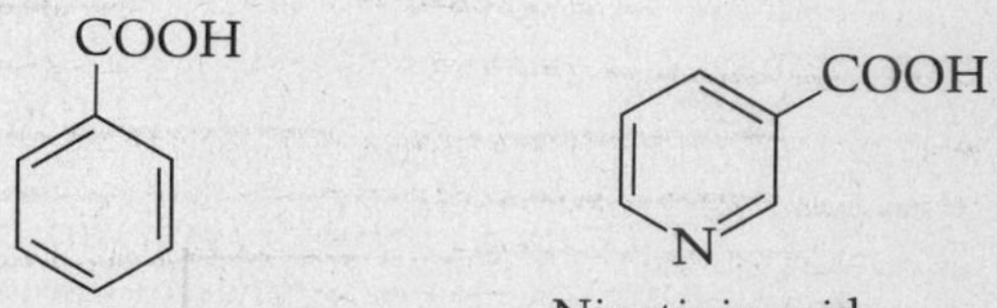

$$CH_3—\overset{\overset{\text{O}}{\|}}{\text{C}}—OH$$

Acetic acid

Benzoic acid

Nicotinic acid
(Niacin)

Another substitution is an —NH$_2$ to create an amine, and so on. Note that carboxylic acids may condense with alcohols to form esters

$$R—\overset{\overset{\text{O}}{\|}}{\text{C}}—OH + HO—R' \longrightarrow R—\overset{\overset{\text{O}}{\|}}{\text{C}}—O—R' + H_2O$$

and with amines in an amide-type linkage:

$$R—\overset{\overset{\text{O}}{\|}}{\text{C}}—OH + H_2N—R \longrightarrow R—\overset{\overset{\text{O}}{\|}}{\text{C}}—NH—R + H_2O$$

Because the reactivity of the functional groups is predictable, as in the above two examples, the behavior of organic molecules in various situations is also largely predictable.

The classification of naturally occurring organic compounds is complicated by the fact that they may have more than one functional group, or be part aliphatic, part aromatic, and so on. Hence, another classification scheme is superimposed on those already mentioned, giving rise to the names lipids, carbohydrates, nucleic acids, and proteins. Each of these will now be described. A common feature of these natural compounds is their tendency to form large aggregates, in some cases through covalent bonds and in some cases through weak chemical interactions.

3-2 LIPIDS

Lipids are small organic molecules that tend to be insoluble in water but soluble in organic solvents. In other words, they tend to be hydrophobic or to contain significant hydrophobic regions, a result of their nonpolar nature and low dielectric constant (generally 2 to 4—see Chap. 2). Lipids form a broad and rather ill-defined class of molecules that includes fats and oils, waxes, and certain of the hormones and vitamins, among other things. Any organic molecule that is poorly soluble in water and not easily classed as a carbohydrate, nucleic acid, protein, or a derivative or breakdown product thereof, is apt to be called a lipid.

The biological role of lipids is as diverse as their structure. For example, fats and oils form the most important source of energy in the ordinary diet, and they serve to store energy in the body. The lipid hormones allow a gland in one part of the body to control the metabolic activity of a tissue in another part. Many of the vitamins, which are essential dietary supplements with a wide range of functions, are also lipid. So are the basic structural units of biological membranes. Most lipids can be divided into six groups: fatty acids, glycerol esters (glycerides and phospholipids), sphingolipids, waxes, terpenes, and steroids (see Table 3.5). Only the fatty acids, glycerides, phospholipids, and steroids will be discussed here.

FATTY ACIDS. Fatty acids are carboxylic acids containing from four to twenty or more carbons in a hydrocarbon chain. The chain may be either saturated (entirely —CH_2—) or unsaturated (containing double bonds). Fatty acids are lipids because the nonpolar character of their hydrocarbon "tails" dominates their properties. Note, however, that their short-chain counterparts, such as acetic acid, CH_3—COOH, are very soluble in water and so are not classed as lipids.

The most common fatty acids in animals are the 16- and 18-carbon saturated and unsaturated fatty acids given in Table 3.6. (Note that unsaturated fatty acids are not conjugated structures.) Though there are some exceptions, nearly all the naturally occurring fatty acids in animals (95-99%) contain an even number of carbons, a result of being synthesized from acetate, which has two

TABLE 3.5   Summary of the lipids

NAME	BASIC STRUCTURE	BIOLOGICAL ROLE
Fatty acids	$CH_3-(CH_2)_n-COOH$	important source of food energy; constituent of other lipids, including the prostaglandins
Glycerol esters	$H_2C-CH-CH_2$ $\quad\lvert\quad\lvert\quad\lvert$ $\quad OH\ OH\ OH$ Glycerol	
Glycerides	one, two, or three fatty acids esterified to glycerol	source of food energy; fats (triglycerides) serve also as thermal and mechanical insulators
Phospholipids	glycerol with one hydroxyl esterified to phosphate (which may make a second ester bond to another constituent) and fatty acids at its other hydroxyls	major constituent of all membranes
Sphingolipids	derivatives of sphingosine $\quad\quad\quad\quad\quad OH\ NH_2$ $\quad\quad\quad\quad\quad\lvert\quad\lvert$ $CH_3(CH_2)_{12}-CH=CH-CH-CH-CH_2OH$ Sphingosine	constituent of some membranes
Waxes	fatty acid ester of any alcohol except glycerol	source of energy; the harder waxes (beeswax) may also be structural components
Terpenes	polymers of isoprene $\quad\quad\quad CH_3$ $\quad\quad\quad\lvert$ $CH_2=C-CH=CH_2$ Isoprene	several vitamins (A, E, K) and some oils are terpenes
Steroids		constituent of membranes; some of the hormones (e.g., sex hormones) are steroids, as is vitamin D

carbons. In general, the fatty acids derived from plant tissues have more double bonds and are more likely to have an odd number of carbons. "Vegetable oil," for example, is largely a mixture of unsaturated fatty acids.

Ordinary soap is the Na^+ or K^+ salt of fatty acids. These salts are relatively soluble, but the Mg^{2+}, Fe^{2+}, and Ca^{2+} salts of fatty acids are insoluble, which is why "hard" water (water that con-

TABLE 3.6 Some common fatty acids

FORMULA	COMMON NAME	FORMAL NAME	STRUCTURE
$C_{16}H_{32}O_2$	palmitic	n-hexadecanoic	$CH_3(CH_2)_{14}COOH$
$C_{18}H_{36}O_2$	stearic	n-octadecanoic	$CH_3(CH_2)_{16}COOH$
$C_{16}H_{30}O_2$	palmitoleic	9-hexadecenoic	$CH_3(CH_2)_5CH\!\!=\!\!CH(CH_2)_7COOH$
$C_{18}H_{34}O_2$	oleic	cis-9-octadecenoic	$CH_3(CH_2)_7CH\!\!=\!\!CH(CH_2)_7COOH$
$C_{18}H_{32}O_2$	linoleic	cis,cis-9-12-octadecadienoic	$CH_3(CH_2)_4(CH\!\!=\!\!CHCH_2)_2(CH_2)_6COOH$
$C_{18}H_{30}O_2$	linolenic	9,12,15-octadecatrienoic	$CH_3CH_2(CH\!\!=\!\!CHCH_2)_3(CH_2)_6COOH$

tains divalent cations) precipitates soap. The detergent action of
fatty acids, as explained in the previous chapter, is the result of
aggregation to micelles having a hydrocarbon interior formed from
the side chains and a charged surface formed from the carboxylate
ions. The charged surface keeps the micelles in solution, but their
interior provides a solvent for fats and oils, which can thus be
"emulsified." Synthetic detergents are generally sulfates
$(R\!\!-\!\!OSO_3^-$, $pK_a \approx 2)$ instead of carboxylates. Sulfates are much
stronger acids (the pK_a of —COOH is 4.75), and hence are more
soluble for a given chain length.

Fatty acids with double bonds in the terminal seven carbons (the
ones distal to the carboxyl group) cannot be made by mammals,
but are required by them. Linoleic and linolenic acids (Table 3.6)
are examples of *essential fatty acids*. The essential fatty acids are
precursors of *prostaglandins*, a group of twenty-carbon fatty acids
containing a five-membered ring and various substitutions. Al-

Prostaglandin E_1

though they were discovered in the 1830s (in prostate secretions,
hence their name), the structure and real importance of prostaglandins only became known in the 1960s. They have now been
found in small quantities in a variety of different tissues, where
they affect a perplexing array of biological processes, such as the
regulation of blood pressure, the contraction of smooth muscles,
and so on. It takes only 10^{-9} g/ml to cause contraction of smooth
muscle, for example, making prostaglandins among the most po-
tent biological effectors found to date.

TABLE 3.7 Glycerol esters

H₂ H H₂	NAME	

$$\begin{array}{ccc} \text{H}_2 & \text{H} & \text{H}_2 \\ \text{C}-\text{C}-\text{C} \\ | & | & | \\ \text{O} & \text{O} & \text{O} \\ | & | & | \end{array}$$

			NAME	
H	H	H	glycerol	
FA[a]	H	H	monoglyceride	
FA	FA	H	diglyceride	Glycerides
FA	FA	FA	triglyceride	
FA	FA	P	phosphatidic acid (PA)	
FA	FA	P-choline	phosphatidyl choline (lecithin)	
FA	FA	P-ethanolamine	phosphatidyl ethanolamine (cephalin)	Phosphatides
FA	FA	P-serine	phosphatidyl serine (cephalin)	
FA	FA	P-inositol	phosphoinositide	
PA	H	PA	cardiolipin	

[a] FA = fatty acid; P = phosphate; PA = phosphatidic acid, attached through a phosphate ester.

GLYCERIDES. When glycerol is esterified[2] to one fatty acid, leaving two of the hydroxyls unreacted, the compound is called a *monoglyceride*. If two of the hydroxyls are reacted in this way, the result is a *diglyceride*; the corresponding triester is a *triglyceride* (see Table 3.7). Mixtures of triglycerides aggregate by hydrophobic bonding to form globules of macromolecular dimensions. Such is the nature of the ordinary neutral fat found in the adipose cells one might trim from steak before broiling. (It is this association with fat that gave the fatty acids their name.) Neutral fat, or "depot fat," provides the body with insulation and serves as a reservoir of energy, since both the glycerol and fatty acid components may be oxidized by cells in a series of energy-yielding (exergonic) reactions.

Fat always contains a mixture of fatty acids. In fact, even the three fatty acids of a single triglyceride are not usually the same. Hence, the physical properties of the glycerides, which are dominated by the properties of their fatty acids, vary considerably.

---

2. As noted earlier, an ester is the condensation product (meaning water is removed) of an acid and an alcohol, though the participating acid is not necessarily a carboxylic acid—e.g.,

$$R_1-OH + HO-\overset{\overset{\displaystyle O}{\|}}{\underset{\underset{\displaystyle OH}{|}}{P}}-O-R_2 \longrightarrow R_1-O-\overset{\overset{\displaystyle O}{\|}}{\underset{\underset{\displaystyle OH}{|}}{P}}-O-R_2 + H_2O$$

The reverse reaction, where a compound is split by the addition of water, is called *hydrolysis* ("lysis by water").

However, the glycerides are all quite insoluble (a few mg/liter) because of a total lack of charged groups and a paucity of polar ones. Hence, few glycerides (or fatty acids) are found free in the blood, although they can be transported in the form of hydrophobic complexes with serum proteins, such as albumin. Glycerides may also be rendered more soluble by replacing one of the fatty acids with a phosphate or phosphate derivative, thus forming a phospholipid. (See, for example, Fig. 3–3.)

PHOSPHOLIPIDS. *Phosphatidic acid* is a diglyceride with the third hydroxyl esterified to phosphoric acid (see Table 3.7). Phosphoric acid is a strong acid, so the substitution improves water solubility. The rest of the molecule remains hydrophobic, however, and so phosphatidic acid forms micelles in the same way as fatty acids. This property facilitates the incorporation of phosphatidic acid and other phospholipids into membranes (see Chap. 6).

The phosphate of phosphatidic acid is capable of forming a second ester bond. It reacts with choline, ethanolamine, or serine to form a phosphatide; it can react with inositol (at the hydroxyl underlined in the following figure) to create a phosphoinositide; or two phosphatidic acids may be joined through phosphate ester

FIGURE 3–3. Phospholipids. The structural diagram and space-filling model represents phosphatidyl choline, also called lecithin. (Molecular models, except where otherwise noted, are from R. C. Bohinski, *Modern Concepts in Biochemistry*. Boston: Allyn and Bacon, Inc., 1973.)

$$H_2C-O-\overset{\overset{O}{\|}}{C}-CH_2-CH_2-CH_2-CH_2-CH_2-CH_2-CH_2-CH_2-CH_2-CH_2-CH_3$$

$$HC-O-\overset{\overset{O}{\|}}{C}-CH_2-CH_2-CH_2-CH_2-CH_2-CH_2-CH_2-CH_2-CH_2-CH_2-CH_3$$

$$CH_3-\overset{\overset{CH_3}{|}}{\underset{\underset{CH_3}{|}}{N^+}}-CH_2-CH_2-O-\overset{\overset{O}{\|}}{\underset{\underset{O^-}{|}}{P}}-O-CH_2$$

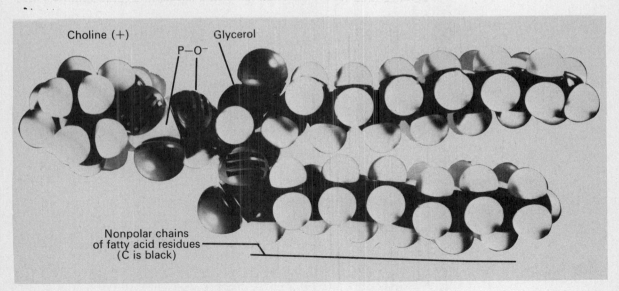

Choline (+)   Glycerol
P–O⁻
Nonpolar chains
of fatty acid residues
(C is black)

bonds to the first and third hydroxyls of a common glycerol to form *cardiolipin* (see Table 3.7). All of these compounds are constituents of various membranes, especially in nervous tissue.

$$HO—CH_2—CH_2—NH_2$$
Ethanolamine

$$HO—CH_2—CH_2—N^+(CH_3)_3$$
Choline

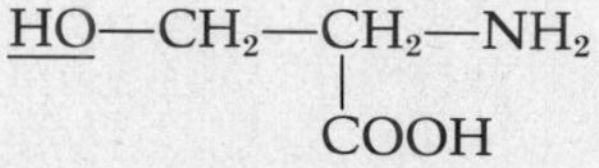

$$HO—CH_2—CH_2—NH_2$$
$$|$$
$$COOH$$
Serine

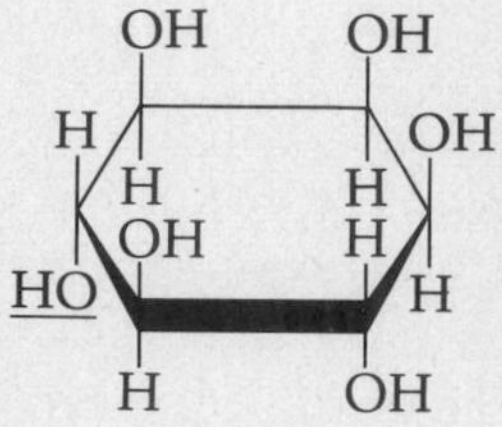

myo-Inositol

STEROIDS. Steroids are an important class of lipids characterized by the basic four-ring structure shown in Table 3.5 and, even more than most lipids, extreme water insolubility and consequent solubility in other lipids. Sterols have the same basic structure as steroids, augmented by a free alcoholic group (—OH) that renders them slightly soluble in water—perhaps 0.25% by weight in a typical case. Steroids and sterols are important constituents of many cell membranes. In addition, a number of hormones, including the male and female sex hormones (androgens and estrogens, respectively) are steroids; so is the pharmacologically important substance, cortisone, used to suppress allergic and auto-allergic reactions. The infamous cholesterol, a component of the fatty plaques that can occlude arteries, is the biological precursor of the steroids just mentioned. As its name implies, cholesterol itself is a sterol.

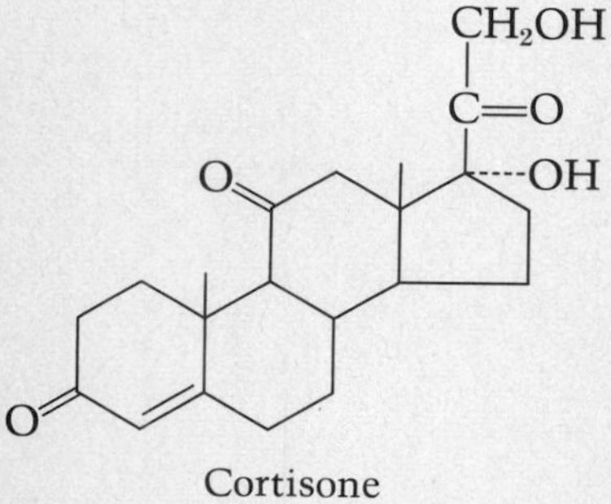

Cortisone

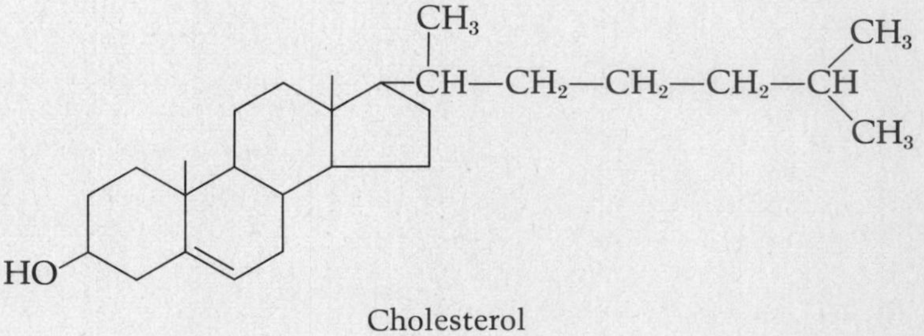

Cholesterol

3–3 CARBOHYDRATES

Carbohydrates are molecules that generally conform to the empirical formula $(CH_2O)_n$—hence the name "hydrate of carbon," or carbohydrate. The most common carbohydrates are the sugars, or *saccharides*, which are found as simple sugars (monosaccharides), as short chains (oligosaccharides), and as large polymers, the polysaccharides.

MONOSACCHARIDES. The monosaccharides are polyhydroxyl aldehydes (HC=O) or ketones (C=O). Typically, each carbon other than the carbonyl (C=O) will have an attached hydroxyl. If the carbonyl is at the end of the carbon chain, the resulting aldehyde is designated an *aldose;* if it is at an interior position, making the molecule a ketone, the sugar is referred to as a *ketose.* As aldehydes are much more reactive than ketones, aldehyde sugars are sometimes called "reducing sugars." Since hydroxyl groups are very polar, all the monosaccharides are very soluble in water.

Monosaccharides are named by specifying the position of the carbonyl, the number of carbons, and finally the ending "-ose." An "aldohexose," for example, is a six-carbon sugar with the carbonyl at the end of the carbon chain, while a "2-ketopentose" has its carbonyl at the second of five carbons.

The aldoses of greatest biological interest are the trioses, pentoses, and hexoses. These have, respectively, three, five, and six carbons. The most common ketoses are the ketotrioses and the pentuloses, hexuloses, and heptuloses—with three, five, six, and seven carbons, respectively. (Note the inserted "ul," indicating the ketose family.)

The monosaccharides usually exist as stereoisomers. The two isomers are designated as "D" or "L" by analogy to D- and L-glyceraldehyde, which are aldotrioses:

```
  H—C=O              H—C=O
     |                  |
  H—C—OH           HO—C—H
     |                  |
 H₂—C—OH           H₂—C—OH
```
D-glyceraldehyde   L-glyceraldehyde

Note that the middle carbon of glyceraldehyde has four different groups attached to it, and hence is asymmetric. (It is the substance chosen for Fig. 3–2.) If the asymmetric carbon most distant from the carbonyl of a sugar (the fifth carbon in a hexose) is like that of D-glyceraldehyde, the sugar is prefixed with a D. There is a similar rule for the L sugars, but D monosaccharides are more common by far.

Most of the monosaccharides are optically active. The dextrorotatory molecules are designated *d* or (+), while levorotatory molecules are designated *l* or (−). It is important to remember that the capital D and L refer to structure, whereas the small *d* and *l* refer to optical activity, established before the structure was determined. Thus, one sees reference to D(+)-glucose, also called dextrose, and D(−)-fructose, also called levulose.

For the sake of simplicity, one often writes the sugars in their linear form; in fact, however, the more common configuration is the heterocyclic isomer, which has an oxygen bridge between two of the carbons as shown in the following examples:

D-Glucose   α-D-Glucose   α-D-Glucopyranose

D-Fructose   β-D-Fructose   β-D-Fructofuranose

D-Ribose   α-D-Ribose   α-D-Ribofuranose

Ring formation introduces a new asymmetric carbon at position one, so the nomenclature must reflect the additional pair of isomers. If the hydroxyl at position one is below the plane of the molecule, it is the alpha *anomer;* otherwise we have the beta anomer (see Fig. 3–4). Thus, one can have α-D-glucose or β-D-glucose as well as the two anomers of L-glucose.

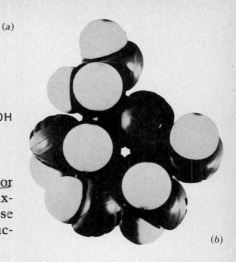

α-D-glucose   β-D-glucose

FIGURE 3–4. The Spatial Conformation of D-Glucose. Axial bonds, which are those bonds perpendicular to the plane formed by the center four carbons, are drawn vertically in the diagram.

The ring forms of the saccharides are referred to as *furanoses* or *pyranoses,* by analogy to furan and pyran, which are five- and six-membered rings, respectively. Thus, one may refer to α-D-glucose and β-D-fructose as, respectively, α-D-glucopyranose and β-D-fructofuranose.

Furan   Pyran

In addition to the monosaccharides themselves, a number of their derivatives are also biologically important. Among the most common is L-ascorbic acid, or vitamin C, an essential carbohydrate that almost all animals except primates can make from glucose. The hydroxyl at the number three position of ascorbic acid ionizes with a $pK_a = 4.2$ to form the ascorbate ion, which is also used as vitamin C—for example, sodium ascorbate.

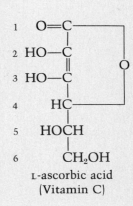

L-ascorbic acid
(Vitamin C)

OLIGOSACCHARIDES. Monosaccharides may be joined by condensation, removing a hydroxyl from one saccharide and a hydrogen from another to form water.

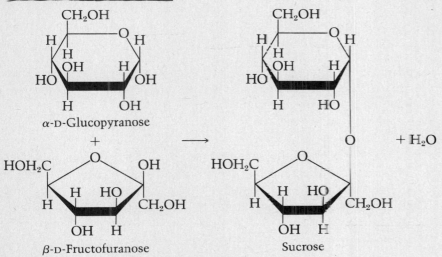

α-D-Glucopyranose

+   ⟶   O   + H₂O

β-D-Fructofuranose   Sucrose

When two monosaccharides are joined, the result is a disaccharide; three form a trisaccharide, and so on. Small chains are called *oligosaccharides* to distinguish them from long chains, or *polysaccharides*. The dividing line between oligosaccharide and a polysaccharide is rather arbitrary, as "oligo-" simply means "few."

The most familiar disaccharide is ordinary table sugar, or *sucrose,* widely distributed in plants and obtained commercially from sugar cane or sugar beets. It consists of α-D-glucose and β-D-fructose, linked $1 \rightarrow 2$. That is, the number one carbon of glucose is linked by an oxygen bridge to the number two carbon of fructose. (In each case, this is the first carbon clockwise from the ring oxygen, as shown.)

Another common disaccharide is lactose, or milk sugar. Lactose, which accounts for about 5% of the solid portion of the milk of most mammals, is composed of β-D-galactose linked $1 \rightarrow 4$ to β-D-glucose. Since both monosaccharides of lactose are in the pyranose form, the disaccharide could also be called 4-O-(β-D-galactopyranosyl)-β-D-glucopyranose. Most people would prefer to say "lactose."

Lactose

POLYSACCHARIDES. The most important polysaccharides are polymers of glucose or glucose derivatives. These include cellulose, chitin, glycogen, and the starches.

More than half of all the carbon in higher plants is in *cellulose*, an insoluble, rigid, structural polymer of several hundred to several thousand glucose units. More than 90% of cotton, for example, is cellulose. It consists of β-D-glucose units with the oxygen bridge between the number one and number four positions. The corresponding β-1,4 disaccharide is called cellobiose.

Repeating cellobiose unit of cellulose

(The kinks in the bonds between the monosaccharides are there only to permit each monosaccharide to be drawn in the same orientation.)

Chitin is similarly constructed, but it consists of N-acetyl-β-D-glucosamine units instead of glucose units:

Repeating N-acetyl-glucosamine unit of chitin

Chitin is a structural polymer, abundant in lower plants and invertebrates. As it is comprised of amino sugars, it is also classed as a *mucopolysaccharide*.

While cellulose is a linear polymer of glucose, *glycogen* is a branched polymer of glucose. It consists of α-D-glucose units, mostly linked $1 \to 4$, but highly branched via frequent $1 \to 6$ linkages, each of which starts other α-1,4 chains of some eight to twelve glucose units. A complete molecule of glycogen will consist of several thousand glucose monomers, which, because of the branching, form a much more compact structure than cellulose.

Glycogen is found mostly in the muscles and livers of animals, and is sometimes called "animal starch." The true starches, however, are plant products that come in two forms, amylose and amylopectin, which are found together in granules. *Amylose* is an unbranched α-1,4 polymer of glucose. (The disaccharide of the α-1,4 polymers is maltose.) The remainder of the starch granule is *amylopectin*, a molecule with the same structure as glycogen, though not so highly branched—each branch being some 24–30 units in length. (Amylopectin should not be confused with the pectin of fruits and berries, which is a polymer of a galactose derivative.)

Repeating maltose unit of the starches

Glycogen and the starches serve as depots for glucose. They give the cell a way of storing this important food molecule in a form that is compact, yet accessible to specialized enzymes. Cellulose, on the other hand, is a structural component that is relatively rigid and quite insoluble. The only important difference, however, between cellulose and the starches, plant or animal, is that cellulose is a β-1,4 polymer of glucose; the starches are basically α-1,4 polymers. On the surface, this may not seem like an important distinction, but in fact the bond angles of carbon and oxygen are such that α-1,4 polymers are helical, with about six glucose units per turn, while β-1,4 polymers are linear. Hence, cellulose is found in fibers of 100–200 parallel chains, with few of its hydroxyls exposed to the surrounding solvent. This configuration accounts for both its rigidity and its insolubility. In contrast, the helical α-1,4 polymers (glycogen and starch) provide open spaces for solvent penetration. As a result, glycogen and the starches are more flexible and more soluble (see Fig. 3–5).

The difference between alpha and beta linkages has another biological implication, for while man has enzymes that can hydrolyze the α-1 $\rightarrow$ 4 glucoside bond, he does not have any that will hydrolyze the corresponding beta linkage. Therefore, we derive no food value from cellulose, in spite of the fact that it is a polymer of nothing more than ordinary α-D-glucose. Certain microorganisms do contain such enzymes. Grazing animals utilize these organisms to break down the cellulose in their diet, and then live by digesting the microorganisms and their metabolic by-products. As carnivores, we take advantage of the food value of cellulose by eating the grazing animals, but since we are only getting it third-hand, the advantages to be gained from finding a more direct route from plant to man are obvious. This is one of the possibilities for extending the capacity of our agriculture to feed an ever-increasing population.

The polysaccharides, like the lipid polymers, tend to be mixtures of various size polymers. However, the next two classes of macromolecules have discrete molecular sizes. In addition, their biological functions require very definite spatial configurations. These are the proteins and the nucleic acids.

3–4 NUCLEIC ACIDS

Nucleic acids are the informational molecules of the cell. As will be shown in Chap. 8, they carry essentially all the information needed to construct a cellular duplicate and, in addition, are important components of the machinery necessary for the construction of proteins. Information is coded into deoxyribonucleic acid, DNA, and is retrieved and interpreted with the help of ribonucleic acid, RNA. The main products of this cooperation are *enzymes*,

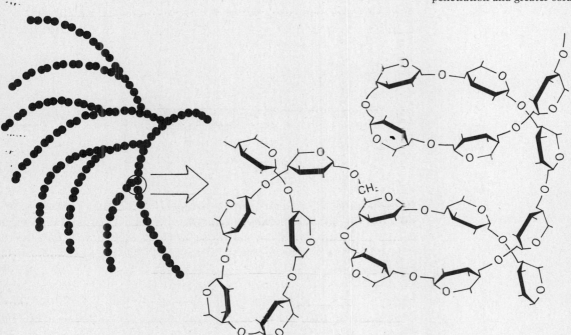

Cellulose (β,1 → 4 linkages)

FIGURE 3–5. Cellulose and the Starches. The β, 1 → 4 linkage in cellulose leads to the formation of insoluble fibers. The α, 1 → 4 linkages found in starches (amylose, amylopectin, and glycogen) allows solvent penetration and greater solubility.

which are specific biological catalysts that determine the kind and amount of the various chemical reactions necessary for the support of life and reproduction.

Nucleic acids are polymers of simpler compounds called *nucleotides*. Each nucleotide is the phosphoric acid ester of a five-carbon sugar that is also linked to an organic base. Because phosphoric acid is a strong acid, and because nucleotide polymers were first associated with the nucleus of the cell, the nucleotide polymers are called "nucleic acids."

NUCLEOTIDES. The sugar component of the nucleotides is either ribose or 2-deoxyribose (abbreviated simply deoxyribose). The choice of the sugar alone determines whether the completed polymer is RNA or DNA, for that is the only major difference between the two classes of nucleic acid.

113

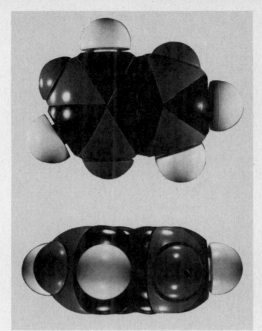

FIGURE 3–6. Purine (left) and Pyrimidine.

The hydroxyl on the 1′ carbon (read "one prime carbon"—the "prime" identifies the position as being on the sugar) is replaced by a weak organic base, a derivative of either purine or pyrimidine. (See Fig. 3–6 and the accompanying diagrams.) This combination of sugar and base is called a *nucleoside,* or more specifically a ribonucleoside if the sugar is ribose, or deoxyribonucleoside (deoxynucleoside for short) if the sugar is 2-deoxyribose. When phosphate is esterified to the sugar, usually at the 3′ or 5′ position, the nucleoside becomes a *nucleotide.*

The organic bases derived from pyrimidine are uracil, thymine, and cytosine, collectively called "the pyrimidines." Likewise there are purines—adenine and guanine—that are simple derivatives of their parent molecule. Normally one finds thymine linked only to deoxyribose and uracil only to ribose, while the other three bases are found in both kinds of nucleosides and nucleotides.

Purine

Adenine

Guanine

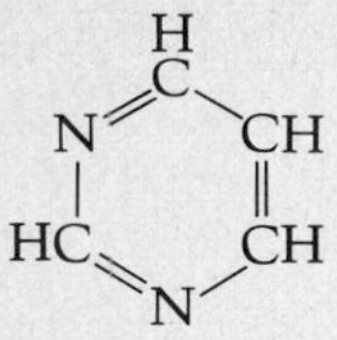

Pyrimidine

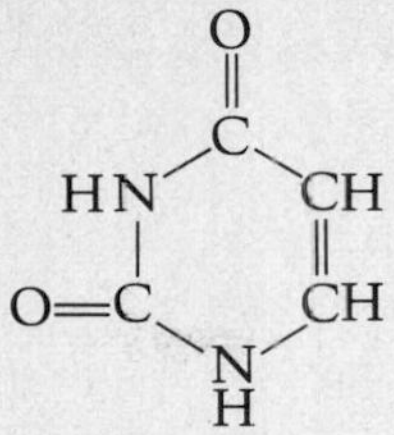

Uracil

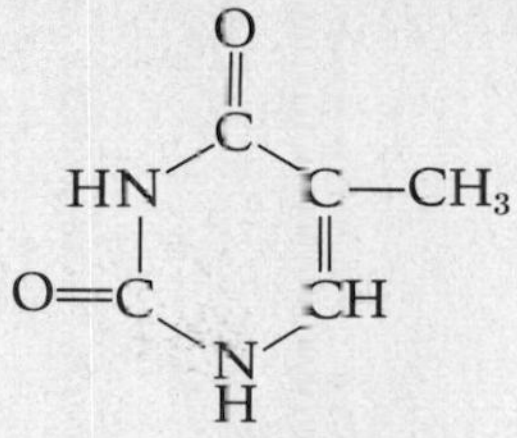

Thymine

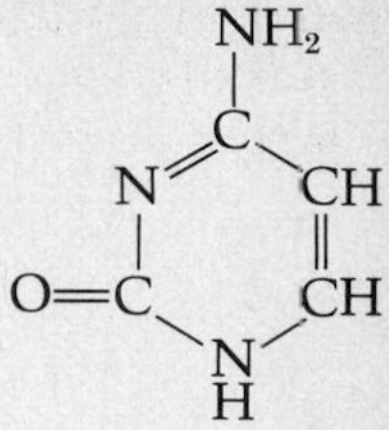

Cytosine

The nomenclature of the nucleosides and nucleotides is summarized in Table 3.8. Hereafter, the abbreviations given there will be used. For example, the condensation of adenine with one of the sugars, ribose or deoxyribose, forms adenosine or deoxyadenosine, respectively. When phosphate is esterified to the 5' carbon of one of these nucleosides, it becomes an adenine nucleotide, either adenosine monophosphate (AMP) or deoxyadenosine monophosphate (dAMP). With phosphate at the 3' position, the nucleotide would be 3'-adenosine monophosphate, or 3'AMP, and so on. (Although it may be found at either or at both positions, the phosphate is assumed to be at the 5' position unless otherwise

TABLE 3.8   Nomenclature of the nucleosides and nucleotides

BASE		NUCLEOSIDES[a] (BASE + SUGAR)		NUCLEOTIDES[b] (BASE + SUGAR + PHOSPHATE)
Pyrimidines	uracil	$\xrightarrow[\text{only}]{+\text{ ribose}}$	uridine	$\xrightarrow{+P_i}$ UMP uridine monophosphate uridylic acid
	thymine	$\xrightarrow[\text{only}]{+\text{ deoxyribose}}$	deoxythymidine	$\xrightarrow{+P_i}$ dTMP deoxythymidine monophosphate thymidylic acid
	cytosine	$\xrightarrow[\text{deoxyribose}]{+\text{ ribose or}}$	(deoxy)cytidine	$\xrightarrow{+P_i}$ (d)CMP (deoxy)cytidine monophosphate (deoxy)cytidylic acid
Purines	guanine	$\xrightarrow[\text{deoxyribose}]{+\text{ ribose or}}$	(deoxy)guanosine	$\xrightarrow{+P_i}$ (d)GMP (deoxy)guanosine monophosphate (deoxy)guanylic acid
	adenine	$\xrightarrow[\text{deoxyribose}]{+\text{ ribose or}}$	(deoxy)adenosine	$\xrightarrow{+P_i}$ (d)AMP (deoxy)adenosine monophosphate (deoxy)adenylic acid

[a] Note that the names for the pyrimidine nucleosides end with "-dine," whereas the purine nucleosides end with "-sine."
[b] Unless otherwise specified, phosphate is esterified to the 5' hydroxyl of the ribose or deoxyribose. For example, AMP means 5'-AMP; therefore, 3'-AMP (where the phosphate is at the 3' instead of 5' hydroxyl) would never be written just AMP.

stated.) If the acid portion is pyrophosphate $\left(\begin{array}{c} \\ -O-\overset{\overset{\displaystyle O}{\|}}{\underset{\underset{\displaystyle O^-}{|}}{P}}-O-\overset{\overset{\displaystyle O}{\|}}{\underset{\underset{\displaystyle O^-}{|}}{P}}-O^- \\ \end{array}\right)$

instead of phosphate, the corresponding nucleotide is adenosine diphosphate or ADP. If there is a string of three phosphates at the 5′ position, the nucleotide is called adenosine triphosphate, or ATP. The corresponding deoxynucleotides are abbreviated dAMP, dADP, and dATP. As we shall see, beginning in the next chapter, the nucleotides by themselves—particularly the adenine nucleotides—play very important biological roles. Our interest in them for the moment, however, is in their polymerization to form macromolecules, the nucleic acids DNA and RNA.

NUCLEOTIDE POLYMERS. Nucleic acids are formed from the corresponding nucleoside triphosphates. As each nucleotide is added to the chain its terminal pyrophosphate (PP_i for "inorganic pyrophosphate") is lost, leaving the first phosphate in a sugar–phosphate-sugar *diester*, so called because the phosphate is esterified with two alcohol-like hydroxyl groups at the same time—the 5′ hydroxyl of one sugar and the 3′ hydroxyl of the next.

Nucleotide addition to DNA

Long chains of nucleotides can be constructed in this way, with a "backbone" of alternating phosphate and sugar and an organic base at each 1' position. If the sugar is ribose, the polymer is ribonucleic acid, or RNA; if the sugar in the backbone is deoxyribose, the polymer is deoxyribonucleic acid, or DNA. The genetic information of all cells is stored in the sequence of bases in DNA and expressed with the aid of RNA, a concept that will get more attention in Chap. 8.

THE STRUCTURE OF DNA. Deoxyribonucleic acid consists of two antiparallel strands of deoxyribose nucleotides, wound about each other to form a helix (see Figs. 3–7 and 3–8). This model was first proposed in 1953 by J. D. Watson and Francis Crick, working at Cambridge University in England. It was based on various chem-

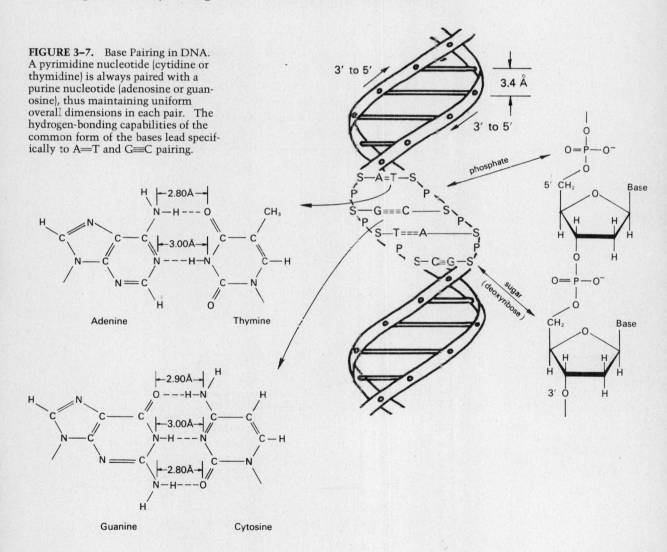

FIGURE 3–7. Base Pairing in DNA. A pyrimidine nucleotide (cytidine or thymidine) is always paired with a purine nucleotide (adenosine or guanosine), thus maintaining uniform overall dimensions in each pair. The hydrogen-bonding capabilities of the common form of the bases lead specifically to A==T and G≡C pairing.

Adenine   Thymine

Guanine   Cytosine

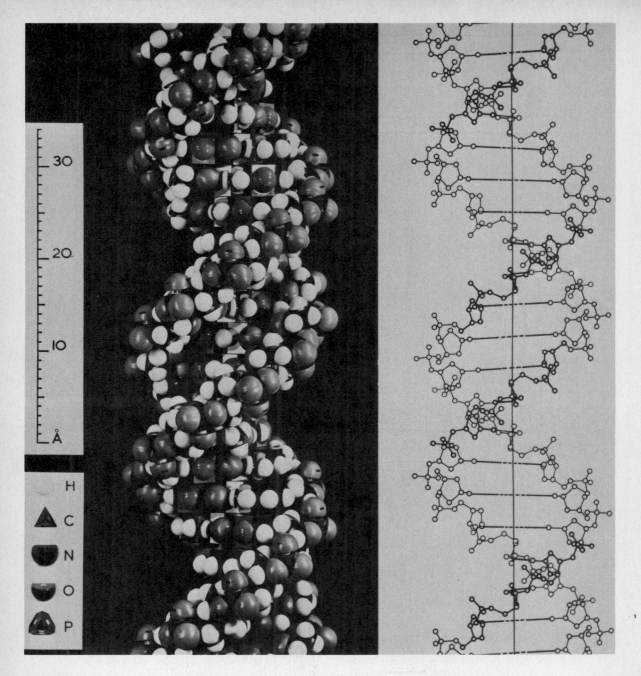

Scale markings: 30, 20, 10, Å

Legend: H, C, N, O, P

FIGURE 3–8. Molecular Models of the DNA Double Helix. (Courtesy of Prof. M. H. F. Wilkins.)

ical and physical data, but mostly on the results of studies carried out at the King's College, London, laboratory of Maurice Wilkins. Wilkins' data consisted of measurements on how fibers of DNA scatter X rays (Fig. 3–9). Watson, Crick, and Wilkins shared a 1961 Nobel Prize for this work.

Other characteristics of the DNA model include a specific

FIGURE 3–9. X-Ray Diffraction Pattern of a DNA Fiber. Note the large "X," a characteristic scattering pattern for helices. The fibers are not highly ordered, hence the "fuzziness." (Courtesy of Prof. M. H. F. Wilkins, Medical Research Council Biophysics Unit, Univ. of London, King's College.)

pairing of bases via hydrogen bonds, with two hydrogen bonds between adenine (A) and thymine (T), and three hydrogen bonds between guanine (G) and cytosine (C). These base pairs lie at the center of the helix with their planes perpendicular to the long axis. Because each pair consists of one pyrimidine and one purine, all pairs are about the same size, giving the helix a smooth contour. The helix makes one turn for each ten base pairs, or one turn for each 34 Å along the axis (see Fig. 3–8). (Since this original model was published, other forms of DNA have been found with tilted bases and somewhat different dimensions.) The helical arrangement is stabilized by hydrophobic bonding and by van der Waals forces between the bases which, because of their platelike stacking and 3.4 Å spacing, have a considerable surface area exposed to each other. The importance of these "stacking interactions" to the structure of DNA is clearly demonstrated by the fact that even certain single-stranded polymers of deoxyribonucleotides form helices, relying entirely on base stacking for stability.

The DNA double helix is quite stable in solution because, although it is held together with weak bonds only, there are a great many of them. This helical configuration gives the molecule considerable rigidity, reflected in the fact that even dilute solutions (0.01% or less) become very viscous as the long rigid cylinders get in each other's way. This rigidity also makes the molecules vulnerable to mechanical breakage; just a vigorous stir of a DNA solution will result in shearing phosphodiester bonds of long molecules, particularly at the center of the helix.

Double-helical RNA's are also known, but RNA's are found in a much wider variety of configurations, including intrastrand loops and hairpins. RNA will be considered more in Chapter 8.

3–5 PROTEINS

The name *protein* is derived from a Greek word meaning "first." This name is appropriate, for it could be argued that proteins are of first importance to the continuing functioning of the cell. They provide many of the structural elements of a cell, and they help to bind cells together into tissues. Proteins also catalyze most of the chemical reactions that occur in the cell. Some proteins act as contractile elements, to make movement possible; others control the activity of genes, transport needed material across membranes, and carry certain substances from one part of an animal to another. Proteins, in the form of antibodies, protect animals from disease and, in the form of interferon, mount an intracellular attack against viruses that have escaped destruction by antibodies and other defenses. And finally, some hormones are proteins.

It is not surprising that the performance of so many different tasks requires many kinds of proteins, varying markedly in size and shape as well as in function. This diversity is possible because proteins are polymers of usually twenty different kinds of building blocks called *amino acids*—the name is short for "α-amino carboxylic acid." (The alpha carbon is in the number two position—i.e., adjacent to the carboxyl group.) The amino acids differ from one another in the nature of their side chain, R; the general

structure is $H_2N-\underset{\underset{R}{|}}{\overset{\overset{H}{|}}{C}}-COOH.$ Since a protein may contain any-

where from about 50 to more than 1,000 amino acids, an enormous variety of proteins is possible. For example, one of the simplest proteins, the hormone insulin, has 51 amino acids. With 20 to choose from at each of these 51 positions, a total of 20^{51}, or about 10^{66} different proteins could be made. As we shall see, it is the sequence of amino acids that determines the shape and biological function of a protein, as well as its physical and chemical properties.

AMINO ACIDS.   The amino acids can be broadly classified into two groups, hydrophobic and hydrophilic, according to the nature of their side chain (see Table 3.9).   The hydrophilic amino acids—those whose side chains interact favorably with water—include several that have either a second free amine or a second free carboxyl group.   Since the amino and carboxyl groups on the alpha carbon are not normally available for interaction with a solvent, the solubility characteristics of a protein will depend on the kinds of amino acid side chains on its surface and on the arrangement of those side chains—i.e., a protein may have areas that are clearly hydrophilic in nature and other areas that are clearly hydrophobic.

Note that one of the common amino acids, proline, has its side chain folded back and joined to the alpha amino group.   (Hence, it is an *imino*, rather than an amino acid.)   Later, it will be pointed out how this deviation has important implications for protein structure.   Note also that with the exceptions of cysteine and methionine, which contain one sulfur atom each, the twenty most common amino acids are composed entirely of carbon, hydrogen, oxygen, and nitrogen.

The alpha carbon of each of the amino acids—with the exception of glycine, which has a single hydrogen for a side chain—has four different groups attached to it.   Therefore, all the amino acids but glycine exist in two optically active, asymmetric forms that are the mirror images of each other.   One form is designated D and the other L by analogy to the convention used in naming sugars (see Fig. 3–10).   As with the sugars, however, there is no correlation between the D and L designation and whether the molecule is dextrorotatory (rotates the plane of light to the right) or levorotatory.   For example, L-alanine is dextrorotatory in water, while L-tyrosine is levorotatory, leading to the designations L(+)-alanine and L(−)-tyrosine, respectively.   The situation is further complicated in that threonine and isoleucine each have a second asymmetric carbon, so that four "diastereo" isomers exist for each, designated L, D, L-allo and D-allo.

It is important to note that only the L-amino acids occur in proteins.   Obviously, the mechanisms whereby amino acids are synthesized and utilized can distinguish between the two isomers. The D-amino acids are found largely in microorganisms, particularly in the cell walls of bacteria and in several of the antibiotics.

Another important feature of the amino acids is the existence of both a basic and an acidic group at the alpha carbon.   The amino group typically has a pK_a (see Chap. 2) between 9 and 10, while the alpha carboxyl group has a pK_a that is usually close to 2 (very low

TABLE 3.9 The common amino acids[a]

Amino Acids with Polar Side Chains

H₂N—CH—COOH | CH₂ | COOH
Aspartic acid
asp
$pK_a = 3.65$

H₂N—CH—COOH | CH₂ | CH₂ | COOH
Glutamic acid
glu
$pK_a = 4.25$

H₂N—CH—COOH | CH₂ | SH
Cysteine
cys
$pK_a = 8.18$

H₂N—CH—COOH | (CH₂)₄ | NH₂
Lysine
lys
$pK_a = 10.53$

H₂N—CH—COOH | (CH₂)₃ | NH | C=NH | NH₂
Arginine
arg
$pK_a = 12.48$

H₂N—CH—COOH | CH₂ | (imidazole ring) N   NH
Histidine
his
$pK_a = 6.00$

H₂N—CH—COOH | CH₂ | C=O | NH₂
Asparagine
asn

H₂N—CH—COOH | (CH₂)₂ | C=O | NH₂
Glutamine
gln

H₂N—CH—COOH | HC—OH | CH₃
Threonine
thr

H₂N—CH—COOH | CH₂ | (benzene ring) | OH
Tyrosine
tyr
$pK_a = 10.07$

H₂N—CH—COOH | CH₂ | OH
Serine
ser

Amino Acids with Nonpolar Side Chains

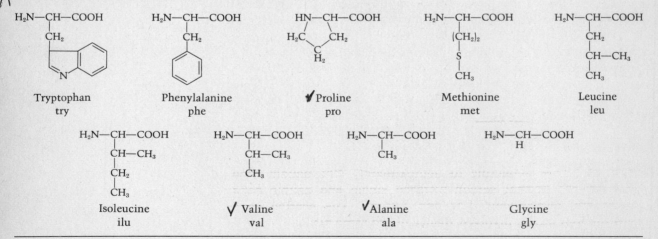

H₂N—CH—COOH | CH₂ | (indole ring) N
Tryptophan
try

H₂N—CH—COOH | CH₂ | (benzene ring)
Phenylalanine
phe

HN—CH—COOH | H₂C   CH₂ | C H₂
✔ **Proline**
pro

H₂N—CH—COOH | (CH₂)₂ | S | CH₃
Methionine
met

H₂N—CH—COOH | CH₂ | CH—CH₃ | CH₃
Leucine
leu

H₂N—CH—COOH | CH—CH₃ | CH₂ | CH₃
Isoleucine
ilu

H₂N—CH—COOH | CH—CH₃ | CH₃
✔ **Valine**
val

H₂N—CH—COOH | CH₃
✔ **Alanine**
ala

H₂N—CH—COOH | H
Glycine
gly

[a] The usual three-letter abbreviation is given, along with the pK_a of side chain groups that may carry significant charge at physiological pH.

for carboxyls). Thus, at physiological pH's, the free amino acids exist largely as *zwitterions*, or double ions, having both a quater-

nary amine ($-\overset{+}{N}H_3$) and a carboxylate $\left(\begin{matrix} O \\ \parallel \\ -C-O^- \end{matrix}\right)$ group.

$$H_3\overset{+}{N}-\overset{\overset{\displaystyle H}{|}}{\underset{\underset{\displaystyle H}{|}}{C}}-\overset{\overset{\displaystyle O}{\parallel}}{C}-O^-$$

Glycine as a zwitterion

THE PEPTIDE BOND. Amino acids can be linked by a condensation in which an —OH is lost from the carboxyl group of one amino acid along with a hydrogen from the amino group of a second, forming a molecule of water and leaving the two amino acids linked via an amidelike *peptide bond* (see Fig. 3-11)

$$H_2N-\overset{\overset{\displaystyle H}{|}}{\underset{\underset{\displaystyle R_1}{|}}{C}}-\overset{\overset{\displaystyle O}{\parallel}}{C}-OH + HN-\overset{\overset{\displaystyle H}{|}}{\underset{\underset{\displaystyle R_2}{|}}{C}}-\overset{\overset{\displaystyle O}{\parallel}}{C}-OH \longrightarrow$$

$$H_2N-\overset{\overset{\displaystyle H}{|}}{\underset{\underset{\displaystyle R_1}{|}}{C}}-\overset{\overset{\displaystyle O}{\parallel}}{C}-\overset{\overset{\displaystyle H}{|}}{N}-\overset{\overset{\displaystyle H}{|}}{\underset{\underset{\displaystyle R_2}{|}}{C}}-\overset{\overset{\displaystyle O}{\parallel}}{C}-OH + H_2O$$

The peptide group is planar and conjugated, as if it were about

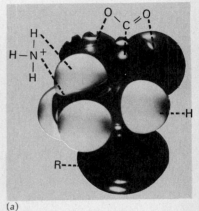

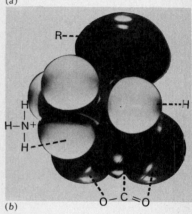

(a)

(b)

FIGURE 3-10.   D and L Amino Acids. Note that they are mirror images. The large knobs on the outside represent side chains.

60% in the structure shown $\left(\begin{matrix} O \\ \parallel \\ -C-N- \\ | \\ H \end{matrix}\right)$, and about 40% in

the alternate resonant form $\left(\begin{matrix} O^- \\ | \\ -C=\overset{+}{N}- \\ | \\ H \end{matrix}\right)$. The C—N bond has a

length of 1.32 Å, reflecting its approximately 40% double bond character: the normal carbon–nitrogen single and double bond lengths are, respectively, 1.43 and 1.26 Å. The peptide bond holds the four atoms involved rigidly in the same plane, but there is generally free rotation about the bonds to the alpha carbons on either side of the peptide group. (Proline is an exception, since the

C—N bond to its alpha carbon is part of a ring structure.) This flexibility at the alpha carbon permits chains of amino acids to form a variety of configurations.

Two amino acids joined by a peptide bond form a dipeptide, three form a tripeptide, and so on. Small polymers are termed *oligopeptides*, while larger ones are referred to as *polypeptides*. A protein can be defined as one or more polypeptide chains (see Fig. 3–12) folded into a definite spatial configuration. Its spatial configuration, or *conformation*, which is the key to a protein's biological activity, is a consequence of the sequence of amino acids in its polypeptide chain(s). It is convenient to describe the conformation of a protein in terms of four levels of organization: primary, secondary, tertiary, and quaternary.

PRIMARY STRUCTURE. The primary structure (abbreviated 1° structure) of a protein is its amino acid sequence. The first protein to have its primary structure determined was insulin, a hormone that regulates glucose metabolism in mammals. Insulin consists of two polypeptide chains of 21 and 30 amino acid residues, called the A and B chains, respectively (see Fig. 3–13). (An amino acid residue is that which is left when the elements of water are removed during polymerization.) The sequence of residues within each chain was determined by F. Sanger and his colleagues at Cambridge in a long and difficult series of experiments completed in 1955. Sanger was awarded the Nobel prize in 1958 for this accomplishment.

The determination of the primary structure of insulin was possible because all insulin molecules isolated from a given animal have exactly the same amino acid sequence. In fact, except for occasional exceptional individuals, all the insulin molecules isolated from animals of a given species will be exactly the same. This constancy of primary structure is true for other proteins as well, and is something that was suspected, but not really proved, until Sanger began his studies. (In Chap. 8 we shall learn that this constancy is because the primary structure of a protein is determined by the sequence of deoxynucleotides in a gene.)

The sequence of amino acids is determined by reproducibly

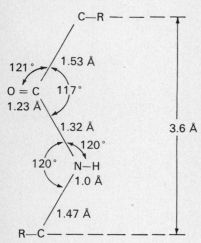

FIGURE 3–11. The Peptide Bond.

FIGURE 3–12. A Polypeptide Chain of L-Amino Acids, Fully Extended. The side chains and hydrogens of the alpha carbons are omitted.

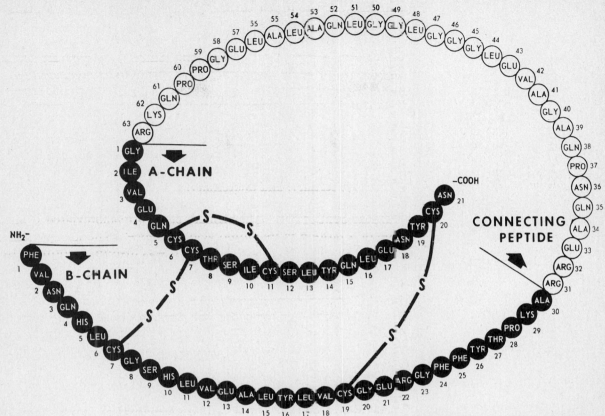

FIGURE 3-13. The Conversion of Proinsulin to Insulin. The molecule is synthesized as a single polypeptide chain of 74 amino acids. It is activated by removing a 33-peptide fragment connecting the A and B chains. [Courtesy of R. E. Chance, *Recent Prog. in Hormone Res.*, **25**:272 (1969).]

breaking a protein into oligopeptides, and then finding the sequence within each oligopeptide by stepwise degradation from one end or the other, identifying each amino acid as it comes off. By comparing overlapping oligopeptides, the primary structure of the protein can be reconstructed.

SECONDARY STRUCTURE. The next level of protein structure, which is called secondary (2°), is the result of proton sharing between the oxygen of one peptide group and the nitrogen of another—that is, hydrogen bonding. Only a few configurations are possible because of the rigidity of the peptide bond, and because the first three atoms of each hydrogen bonded group (NH · · · O=C) generally lie in a straight line with an oxygen–nitrogen distance of 2.79 Å.

In the late 1940s Linus Pauling and R. B. Corey, at the California Institute of Technology, found four basic polypeptide configurations that satisfied the known chemical and physical data. Two of these four configurations are helices, stabilized by hydrogen bonds from each peptide group to the peptide group third from it in each direction (see Fig. 3-14). Both a left-handed and a right-handed

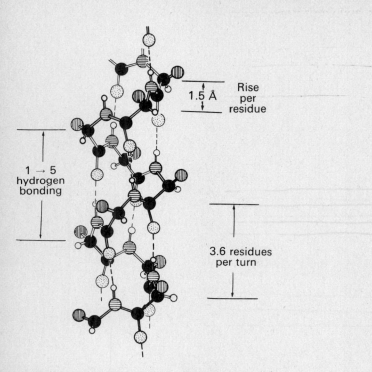

Rise
per
residue

1.5 Å

1 → 5
hydrogen
bonding

3.6 residues
per turn

FIGURE 3–14a. Right-Handed Alpha
Helix.

FIGURE 3–14b. Space-Filling Model
of an Alpha Helix, Without Side
Chains.

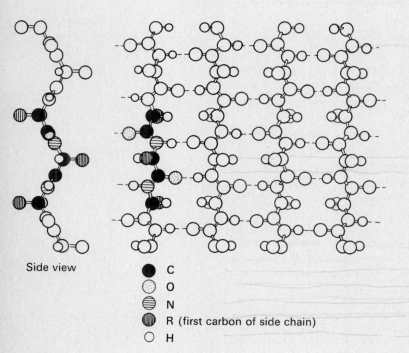

Side view

- ● C
- ○ O
- ⊜ N
- ▥ R (first carbon of side chain)
- ○ H

FIGURE 3–14c. An Antiparallel Beta
Structure.

126

helix are allowed, rotating counterclockwise and clockwise, respectively. The other two configurations that satisfied the known data are called *pleated sheets*, a structure that was first suggested by W. T. Astbury in the 1930s (Fig. 3–14c). The polypeptide chains of pleated sheets are stretched out and lie side by side, either parallel or antiparallel to one another. The bonded groups may be portions

FIGURE 3–15a. Scanning Electron Micrograph of Human Forearm Skin. The surface is composed of dead cells comprised mostly of keratin. [Courtesy of E. O. Bernstein and C. B. Jones, *Science*, **166**:252 (1969). Copyright by the A.A.A.S.]

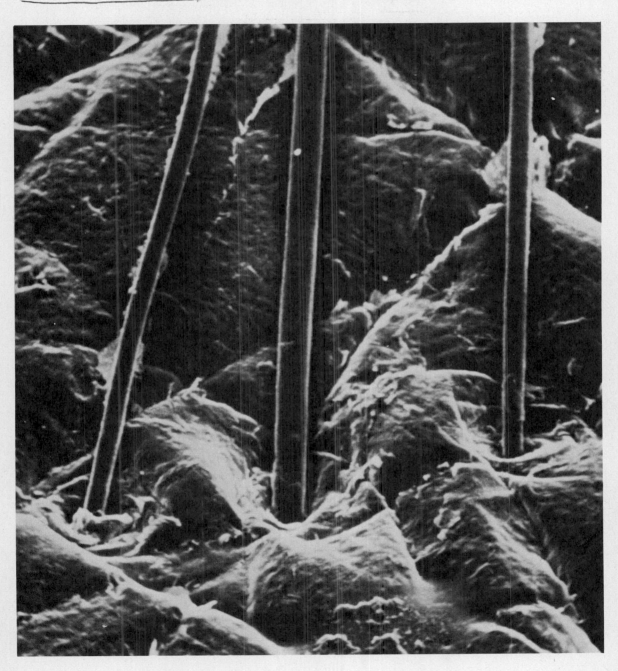

of the same chain folded back on itself, or they may be separate chains.

The helical and pleated sheet arrangements are known as "alpha" and "beta" structures, respectively, because they were first identified in alpha and beta keratins. Keratin is the fibrous protein of human skin and of hair, wool, fur, nails, beaks, feathers, hooves, and so on (see Fig. 3–15). Hair and wool, for example, are composed of dead cells containing alpha keratin strands. These strands are highly elastic as a result of their ability to assume the more extended beta configuration when stretched and a tendency to return to a helix when tension is removed. The return is fostered by interchain —S—S— bonds, the rearrangement of which forms the basis for the permanent-wave process used on human hair.

Nature seems to have relied mostly on the right-handed alpha helix and the antiparallel beta sheet for secondary structure, although a number of other configurations have also been identified. The alpha helix, as shown in Fig. 3–14, has 3.6 residues per turn of 5.4 Å, or 1.5 Å along the axis for each amino acid residue. This spacing allows the carbonyl (C=O) groups to lie parallel to the long axis and to point almost straight at the amino groups to which they are hydrogen bonded. A typical beta sheet, on the other hand,

FIGURE 3–15b. Detail of the Middle Hair from Fig. 3–15a. The shaft is also composed of keratinized cells. [Courtesy of E. O. Bernstein and C. B. Jones, *Science*, **166**:252 (1969). Copyright by the A.A.A.S.]

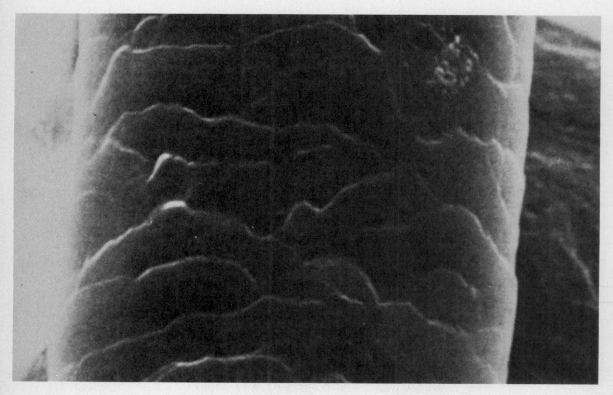

PROTEINS
Section 3-5

will have a spacing of about 4.7 Å between polypeptide chains and one amino acid residue per chain for each 3.5 Å along the axis—a distance per amino acid that is more than double the figure for alpha helices. Because of this difference, wool can be stretched to about twice its original length as it changes from the alpha to the beta form under tension. In general, the alpha helix lends elasticity to a structure, while the interchain bonding of the beta sheet provides rigidity.

One important example of a protein that does not conform to these guidelines is *collagen*. Collagen is the widespread connective tissue fiber that, among other things, forms scars and tendons. A collagen fibril is the side-by-side aggregate of long, slender molecules of *tropocollagen*, each about 2800 Å × 15 Å with a mass of about 300,000 amu. Tropocollagen, which is secreted by connective tissue cells called *fibroblasts* or *fibrocytes* (Fig. 1–17), consists of three polypeptide chains hydrogen bonded to each other and wound into a triple helix (see Fig. 3–16).

Collagen has an unusual secondary structure. Almost every third amino acid in the tropocollagen chains is glycine, and a large fraction of the remainder is either proline or a derivative of proline called hydroxyproline. (Hydroxylation is one of several minor

FIGURE 3–15c. Cross-Section of Merino Wool. LEFT: Part of a peripheral cell and its external covering, consisting of a cuticle of scale cells. RIGHT: Detail of a cell interior, revealing numerous microfibrils, each of which is composed of several α-helical strands. [(Left) from R. B. D. Fraser, T. P. MacRae, and G. E. Rogers, *Keratins: Their Composition, Structure and Biosynthesis*, 1972. Courtesy of R. B. D. Fraser, and Charles C. Thomas, Publisher, Springfield, Illinois. (Right) courtesy of G. E. Rogers and B. K. Filshie, *J. Mol. Biol.*, **3**:784 (1961).]

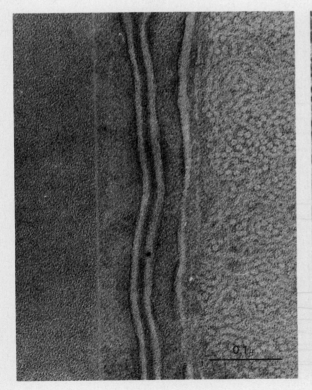
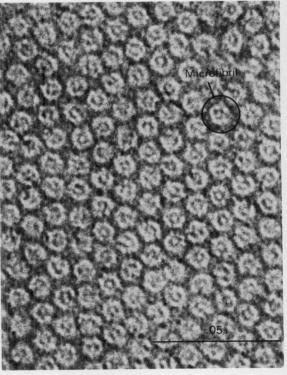

Three-stranded
tropocollagen

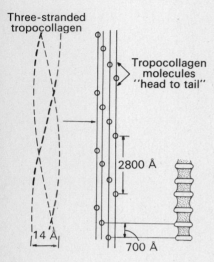

FIGURE 3–16. Collagen. The polypeptide chains are synthesized, hydroxylated at their prolines, and released to form a three-stranded helix. Further chemical modification occurs by the addition of mono- and disaccharides (galactose and glucose), after which the subunits are shortened by 15–20% through proteolysis, rendering them less soluble and prone to aggregate to form collagen fibrils. The subunits overlap one another by three-quarters of their length, providing a 700 Å repeat distance to the enlargements seen in the electron micrograph. (Micrograph of chromium shadowed human skin collagen courtesy of Dr. Jerome Gross.)

Hydroxyproline

modifications to which amino acids are subject.) The geometry of proline and hydroxyproline precludes their participation in an alpha helix—the restricted rotation at the alpha carbon "points" their nitrogen in the wrong direction. The three polypeptide strands of tropocollagen are interlinked by hydrogen bonds at every opportunity, however, forming a tight unit with few polar groups left to interact with water. Thus, native collagen is very insoluble, although it can be solubilized by boiling: heat separates the strands to produce gelatin. Gelatin strands do not form compact structures because of the limitations of the proline and hydroxyproline bond angles, but there is a marked tendency for them to interact nonspecifically through interchain hydrogen bonds, so that a gel may be formed at concentrations of about 1% or more.

TERTIARY STRUCTURE. A typical protein will consist of sections of secondary structure interspersed with apparently unordered segments called "random coils." The way that regions of secondary structure are oriented with respect to each other is referred to as the tertiary structure (3°) of a protein. For example, a protein with a high alpha helix content will typically have several well-defined helices that could, through rotation about single bonds, be placed in any of several configurations. "Tertiary structure" is a description of their actual arrangement.

One does not expect to find a protein in one continuous helix or pleated sheet for two reasons: (1) Either of the two secondary structures may occasionally result in unfavorable interactions between amino acid side chains—for example, the juxtaposition of two side chains of like charge or two very bulky side chains in a space not adequate for them. (2) Long continuous helices are also prevented by the presence of proline, which cannot conform to the required geometry. Its appearance in a helix introduces a kink so that the only position proline can occupy in a helix is the amino end.[3] Proline, then, is a "helix breaker," so that a high proline content means there will be little opportunity for alpha helix.

The various segments of a helix and a pleated sheet (the secondary structures) may be held in position by favorable side chain interactions, including all of the weak bonds mentioned in Chap 2 and, in addition, an occasional covalent crosslink in the form of a disulfide (—S—S—) bridge between cysteines, or even an occasional isopeptide bond, defined as a peptide bond between a side chain amine of lysine or arginine paired with a side chain carboxyl of glutamate or aspartate. (See Fig. 3–17.) However, although covalent bonds may help *stabilize* a tertiary structure, there is considerable evidence to indicate that they do not necessarily *determine* it. It is often possible to completely *denature* a protein—i.e., remove all secondary and higher structure—and still find that the polypeptide chains can spontaneously refold *in vitro* to their original configuration and biological activity. Primary structure, it seems, is enough to determine the higher levels of organization (see Fig. 3–18).

The first protein to have its secondary and tertiary structure determined was myoglobin, a 153-amino acid, oxygen–binding protein found in red muscle and largely responsible for the color of that tissue (see Fig. 3–19). The work was done at Cambridge under the direction of J. C. Kendrew. The X ray analysis, to a resolution of 2 Å, was announced in 1961, and was based on about 40,000 X ray reflections made by the molecule and four chemical derivatives

---

3. One refers to an "N-terminal" or "C-terminal" amino acid of polypeptide chains. These are the first and thirtieth amino acids in the B chain of insulin, for example (see Fig. 3–13). We might also say that phenylalanine occupies the "amino end" and alanine the "carboxyl end" of the chain, a nomenclature that is unambiguous since a chain will never have more than one amino acid with an unbonded alpha amino group, nor more than one with an unbonded alpha carboxylate group.

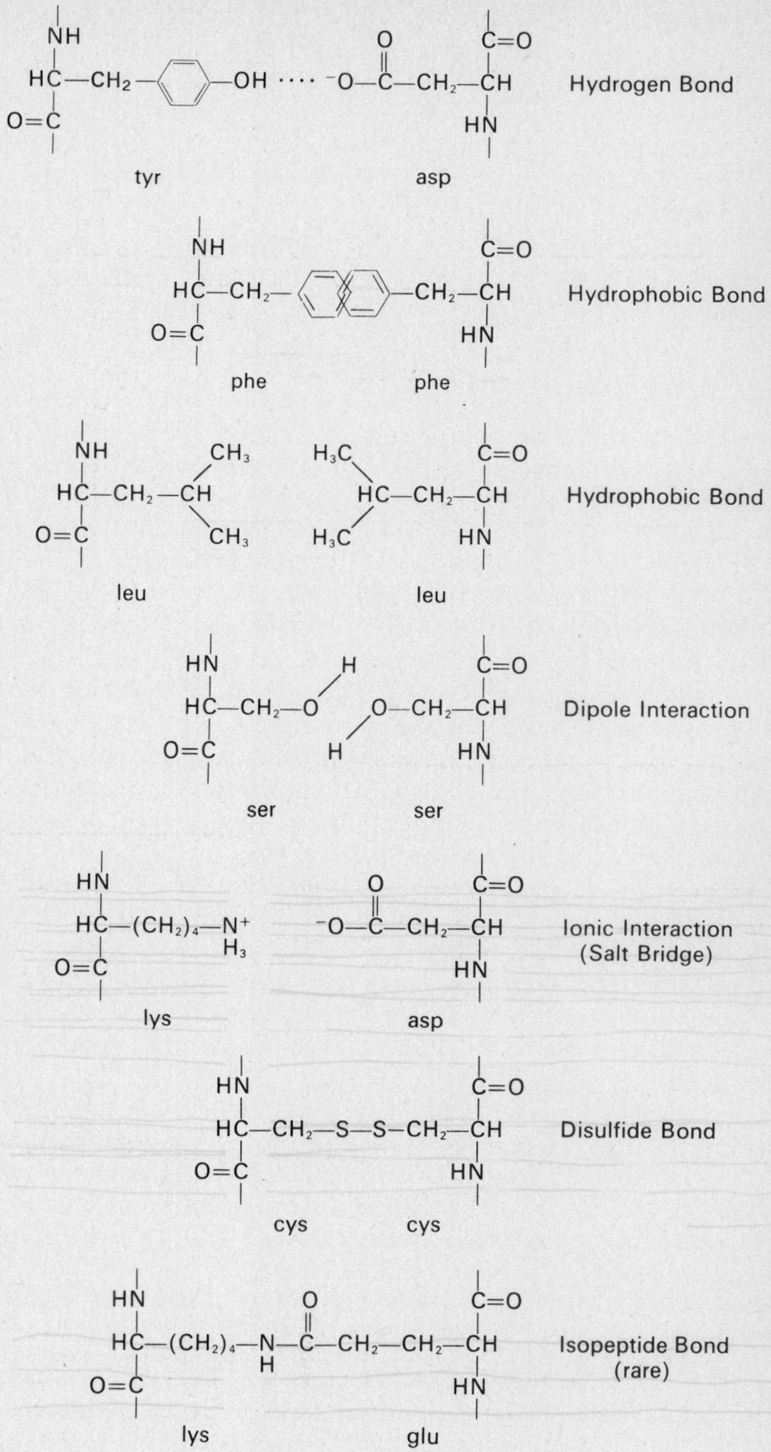

FIGURE 3–17. Bonds That Stabilize
Tertiary Structure.

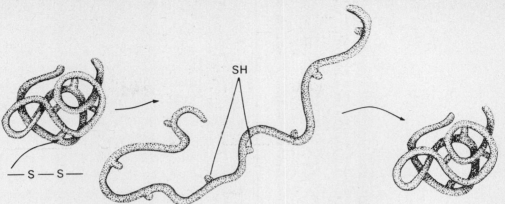

of it (see Fig. 3–20). To produce just the final calculation alone took a high-speed (for those days) digital computer about twelve hours. The procedures have been improved a great deal since then, including the advent of automatic, computer-controlled machines. However, determination of the tertiary structure of proteins still remains a major task.

QUATERNARY STRUCTURE. The highest level of protein structure, called quaternary (4°), refers to the *subunit structure* of proteins. Proteins that have more than one polypeptide chain may also have more than one independently folded unit, each of at least one chain. The independently folded units are called *subunits;* the

FIGURE 3–18. The Spontaneous Formation of Tertiary Structure. Many proteins can be completely denatured and renatured *in vitro*, implying that the native configuration is the most stable. The disulfide bridges in these cases may help to stabilize a structure, but they obviously do not determine it. [Described by C. B. Anfinsen and E. Haber, *J. Biol. Chem.*, **236**:1361 (1961), and by F. H. White, Jr., *ibid.*, p. 1353.]

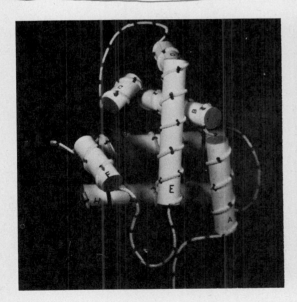

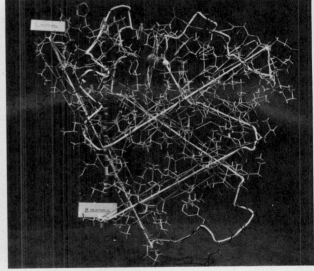

FIGURE 3–19. The Tertiary Structure of Myoglobin. The pipestem model is designed to show the relationship of the helical segments to each other.

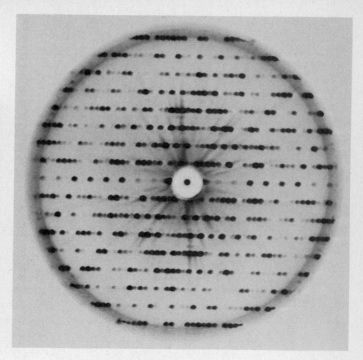

FIGURE 3–20. The X-ray Crystallography of Proteins. LEFT: Crystals of a bacterial protein (the enzyme, isocitrate dehydrogenase) were chemically fixed with glutaraldehyde, coated with carbon and gold, then photographed in a scanning electron microscope to reveal their shape. RIGHT: In X-ray diffraction, a single large, unfixed crystal is mounted in an X-ray beam and the scattering pattern examined to deduce molecular structure. This pattern was obtained from a crystal of another enzyme, pork adenylate kinase. [Scanning electron micrograph courtesy of W. Burke, J. Swafford, and H. Reeves, *Science*, **181**:59 (1973) copyright by the A.A.A.S.; diffraction pattern courtesy of I. Schirmer, H. Schirmer, and G. E. Schulz, Max-Planck-Inst. für Medizinische Forschung, Heidelberg, Germany.]

spatial arrangement of these subunits is the *quaternary structure of the protein*.

Whether a protein with two polypeptide chains has one or two subunits depends on the relationship of the chains to each other. If each chain has its own tertiary structure, and if it could maintain that structure in the absence of the other chain, then there are two subunits. On the other hand, if it is impossible to separate the chains without changing their tertiary structure, then they are part of the same subunit. Accordingly, insulin has no quaternary structure; it is a single unit composed of two polypeptide chains. Because there is some ambiguity involved in these definitions, it has been suggested that the smallest unit of independent biological function be called a *protomer*. In most cases, the number of polypeptide chains is also the number of subunits and the number of protomers; however, in other cases it may take two or more subunits to make a protomer.

Subunits may be held together by weak interactions of the kind described in Chap. 2, by covalent (disulfide) linkages, or by both.[4] Frequently, the subunits of a protein are identical to each other and are constructed from information contained in the same gene. This permits large structures to be assembled without an undue

---

4. Multisubunit proteins designed for intracellular use rely on weak interactions to hold their subunits together. Intersubunit disulfide bonds seem to be found only in proteins that are secreted from the cell in which they are made—e.g., in blood serum proteins.

expenditure of genetic material. The protein portion (i.e., the coat) of tobacco mosaic virus, for example, consists of more than 2,000 identical protein subunits, each with a mass of about 17,000 amu. The use of small subunits in TMV construction offers the same advantage as building a house of bricks instead of casting it in one piece. (See Fig. 1–28.)

The subunits of a protein are not always identical to one another, however. Hemoglobin is an example of a protein with more than one kind of subunit. It is a red protein of 65,000 amu, found in the red blood cells (erythrocytes), where it is responsible for carrying oxygen from the lungs to the tissues. Human hemoglobin consists of four chains, each folded into a separate subunit. The configuration in adults is two "α chains" of 141 amino acid residues each and two "β chains" of 146 amino acid residues each. The molecule is thus designated an $\alpha_2\beta_2$ tetramer.

Hemoglobin was the first protein to have the three-dimensional features of its quaternary structure determined by X ray diffraction. The work was accomplished in the laboratory of M. F. Perutz at Cambridge, a project that began some 15 years earlier than the myoglobin project in Kendrew's laboratory. However, since hemoglobin is four times as large as myoglobin, progress came slowly. In fact, the myoglobin work was of very great value to the hemoglobin project (and vice versa) because, as it turns out, each of the four chains of hemoglobin closely resembles myoglobin in tertiary structure. The two groups of X ray crystallographers cooperated closely and, as a result, the preliminary three-dimensional model for hemoglobin was presented at about the same time as that of myoglobin. The placement of the subunits in hemoglobin can be seen in Fig. 3–21, which presents both the oxygenated and deoxygenated forms.

It should be noted that neither myoglobin nor hemoglobin contains any disulfide bonds, again emphasizing the importance of weak bonds in stabilizing protein structure.

PROSTHETIC GROUPS. So far we have considered only the structure of the polypeptide chains of proteins. However, many proteins contain organic or inorganic units that are an integral part of the molecule but which are not composed of amino acids. These additions are called *prosthetic groups;* their presence defines a *conjugated protein*. Very often prosthetic groups are simple metal ions, particularly Fe^{2+}, Zn^{2+}, Mn^{2+}, Mg^{2+}, or Cu^{2+}, which are bonded to oxygens and nitrogens. The prosthetic group may also be lipid, in which case the conjugate is called a *lipoprotein;* or it may be a carbohydrate, forming a *glycoprotein*. It is not always clear what role prosthetic groups play. The presence of carbohydrates is particularly puzzling, for a great many proteins (especially those that are secreted from the cell in which they are made) seem to contain at least a little.

oxy

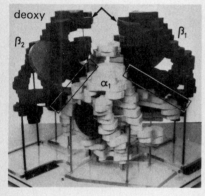

deoxy

β_2 β_1

α_1

FIGURE 3–21. The Quaternary Structure of Hemoglobin. LEFT: The change in quaternary structure upon oxygenation. (Note spacing at arrows.) The boxes mark areas of contact between unlike chains. RIGHT: A demonstration of how the change is achieved: on deoxygenation, the α_1 subunit turns about its axis by 9.4°; the β_1 subunit turns about its axis by 7.4°. Both rotations are clockwise relative to the center. A corresponding change takes place in the α_2 and β_2 subunits. [Photos courtesy of M. F. Perutz: Left from *J. Mol. Biol.*, **28**:117 (1967); Right from *Proc. Roy. Soc.* (London) *B*, **173**:113 (1969). By permission of Academic Press, London, and The Royal Society.]

Hemoglobin and myoglobin are examples of conjugated proteins. Each subunit of hemoglobin, and each myoglobin molecule, contains a heme group: a ring structure with a central iron ion (Fe^{2+}). The heme group is stuck in a hydrophobic crevice of the protein in both hemoglobin and myoglobin, but the hydrophobic oxygen mol-

Heme

ecule is able to penetrate readily into this area and displace a water molecule bound to the iron of the heme.

This same prosthetic group, the heme, is found in several other proteins as well. For example, the enzyme catalase, which catalyzes the decomposition of hydrogen peroxide (H_2O_2) into water and oxygen, also contains a heme, the iron of which plays a central role in the catalytic process. The cytochromes are also heme-containing proteins, responsible for transferring electrons by a cyclical change in the valence state of the iron from Fe^{3+} to Fe^{2+} and back again. With the substitution of Mg^{2+} for the iron, plus some changes in the side chains, heme becomes chlorophyll, which is the green pigment of plants and the key to the process of photosynthesis. The biological function of these and other conjugated proteins will be discussed in more detail later.

FORMATION OF STRUCTURE IN VIVO. Even a medium-sized protein of 30,000 amu will contain about 300 amino acid residues. With free rotation about all single bonds, it should be obvious that a great many different spatial configurations might be possible. Yet very little deviation is observed in the finished product, for the biological role of a protein is a direct consequence of its three-dimensional shape. And that shape must be determined by the primary structure of the protein, at least in those cases where the native structure is spontaneously regained after denaturation *in vitro*. This situation implies that the native configuration is the configuration of lowest energy—i.e., the configuration that would allow the maximum number of bonds if all bonds had the same energy.

It is common, however, to have two or more configurations (*conformations*) that are separated from each other by relatively small differences in energy, and to find that a protein can exist as an equilibrium mixture of these different conformations. This aspect of protein structure will be considered in more detail in the next chapter, for it is fundamental to an understanding of a very important class of proteins, the enzymes.

SUMMARY

3–1 A cell is typically about three-quarters water. The remainder is largely in the form of macromolecules, either lipid, carbohydrate, nucleic acid, or protein, each of which has a relatively small basic unit that may be joined to others by covalent and/or non-covalent bonds. The structure and properties of the macromolecules are dictated largely by the behavior of carbon: its tendency to form four single covalent bonds or their equivalent in double and triple bonds, its tetrahedral bond angles, and its tendency to form conjugated structures when single and double bonds are made to the same carbon. The three main divisions of organic (carbonaceous) compounds are the aliphatic (e.g., $—CH_2—CH_2—$), aromatic (e.g., benzene), and heterocyclic substances, the latter being rings in which one or more of the atoms in the ring is something other than carbon.

3-2 Lipids are a diverse group of nonpolar molecules characterized by water insolubility. Three of the major classes of lipid are: (1) Glycerides, which are esters of glycerol plus one to three molecules of fatty acid; glycerides serve as a major source of dietary energy. (2) Phospholipids, which are glycerides with one glycerol hydroxyl esterified to phosphate; they are a major constituent of biological membranes. (3) Steroids, which include a number of hormones. There are still other groups of lipids, for almost any organic molecule that is not classed as a protein, carbohydrate, or nucleic acid, and that is relatively insoluble in water, is apt to be called a lipid.

3-3 Carbohydrates generally conform to the empirical formula $(CH_2O)_n$. Glucose and its isomers, for example, are all $C_6H_{12}O_6$. Glucose is a simple sugar, or monosaccharide. It can be linked through an oxygen (—O—) bridge to a second monosaccharide, forming a disaccharide such as lactose or sucrose. It can also form higher polymers, the most important of these being the structural polymer cellulose, which is an insoluble, linear β-1,4 polymer, and the starches, which are open, soluble, helical α-1,4 polymers. The starches include the highly branched (through $1 \rightarrow 6$ linkages) animal starch called glycogen, the less highly branched plant starch, amylopectin, and an unbranched plant starch called amylose. All serve as temporary storage granules for glucose.

3-4 Nucleotides and their polymers, the nucleic acids RNA and DNA, are built around one of two five-carbon sugars: ribose for the RNA nucleotides, 2-deoxyribose for the DNA nucleotides. Each nucleotide carries a phosphate (or diphosphate or triphosphate), generally at the 5' carbon of the sugar, and an organic base at the 1' carbon. (Without the phosphates, a nucleotide is called a nucleoside.) The bases are either purine derivatives (commonly adenine or guanine) or pyrimidine derivatives—generally uracil (only in RNA), thymine (only in DNA), or cytosine.

The common nucleosides and nucleotides have names derived from their bases, with the ending "-dine" for the pyrimidines and "-sine" for the purines (e.g., adenosine).

The deoxynucleotides may be polymerized through 3'-5' phosphodiester linkages to form DNA, which is normally a double-stranded structure with the bases stacked in pairs, hydrogen bonded to each other and with their common plane at right angles to the long axis. Each pair of bases contains a purine from one strand and a pyrimidine from the other strand: specifically, A with T, G with C, and vice versa.

RNA's, on the other hand, are generally single stranded, but may form intrastrand loops and hairpins, also by complementary (A═U, G≡C) base pairing.

3-5 Proteins are polymers of some twenty different amino acids linked by the planar, rigid, peptide bond. The individual amino acids may have either hydrophilic (including ionic) or hydrophobic side chains. The pattern of side chains left on the surface of the completed protein, and hence exposed to the solvent, will determine the solution properties of the molecule. Thus, some proteins may be distinctly hydrophobic, a feature needed in those that function in a lipid environment such as a membrane. Most proteins, however, are hydrophilic and quite soluble.

The structure of a protein is discussed in terms of four levels of organization. *Primary structure* refers to the sequence of amino acids in a polypeptide chain. *Secondary structure* is the chain configuration imposed by hydrogen bonds between peptide groups. The most common secondary structures are the alpha structure, which is a helix, and the beta structure, which is a planar pleated sheet. *Tertiary structure* describes the arrangement in space of segments of secondary structures—e.g., the way that several helices within a given molecule are arranged with respect to one another. The fourth level of structure, *quaternary structure*, is the arrangement of individually folded chains or groups of chains called subunits. All of the higher orders of a protein's structure are determined largely by its primary structure and by the weak interactions that it allows.

STUDY GUIDE

3-1 (a) The radioactive isotopes most commonly used in biological research are ^{14}C, ^{3}H (tritium), ^{32}P, and ^{35}S, all of which emit relatively weak beta rays (electrons). A cell grown in the presence of any of these isotopes (e.g., with $H^{32}PO_4{}^{2-}$ or $^{35}SO_4{}^{2-}$) will incorporate it into cellular constituents just as it would the corresponding nonradioactive element. If you wished to "label" (make radioactive) the proteins of a cell but avoid labelling the nucleic acids, which isotope would you choose? If you wished to label the nucleic acids without labelling the proteins in any significant way, which would you choose? **(b)** What are the most important features of carbon relative to its role in the formation of biological molecules? **(c)** Define the major classes of organic compounds listed here.

3–2 (a) Why are the fatty acids relatively insoluble when their two-carbon counterpart, acetic acid, is highly soluble? **(b)** Which would be more soluble, a monoglyceride or a triglyceride? A diglyceride or phosphatidic acid? Why? **(c)** Some of the hormones are hydrophilic proteins. The cellular receptor site that potentiates their action in a target tissue is generally on the surface of a cell. The receptor site for the steroid hormones, however, appears to be within the cytoplasm. Can you offer a reasonable explanation for why this difference should have evolved? (*Hint:* Consider the major molecular constituent of cell membranes, and the solubility properties of proteins and steroids.)

3–3 (a) Define: aldose, ketose, aldopentose, ketopentose, pentulose, pyranose, and furanose. **(b)** What is the difference between D and L glucose? Between α and β glucose? **(c)** Why is cellulose insoluble, whereas the starches (glycogen, amylose, and amylopectin) are soluble? Since cellulose and the starches are all polymers of glucose, why can we utilize the starches for food but not cellulose?

3–4 (a) What is the difference between a nucleoside and a nucleotide? **(b)** What is a phosphodiester linkage, and why is it called that? **(c)** What are the synonyms (including abbreviations and full name) for ATP? **(d)** What are the forces responsible for the helical structure of DNA? For its double-stranded structure? **(e)** A typical human cell contains about 4×10^{12} amu (6×10^{-12} g) of DNA. If that amount of DNA were in a single double-stranded helix, how long would it be? (The average mass of a nucleotide in DNA is about 300 amu.) [Ans: about 7 ft]

3–5 (a) Define: amino acid; peptide bond; primary, secondary, tertiary, and quaternary structure; and prosthetic group. **(b)** Describe the most important features of the peptide bond. **(c)** Some proteins can be completely denatured *in vitro* (robbed of secondary, tertiary, and quaternary structure), but will spontaneously regain their native configuration when their environment is returned to normal. To what do you attribute this? What factors might prevent a protein from behaving in this way?

REFERENCES

GENERAL
BOHINSKI, ROBERT C., *Modern Concepts in Biochemistry.* Boston: Allyn and Bacon, 1973.

CONN, E. E., and P. K. STUMPF, *Outlines of Biochemistry* (3d ed.). New York: John Wiley & Sons, 1972.

LEHNINGER, A. L., *Biochemistry: The Molecular Basis of Cell Structure and Function.* New York: Worth Publishing Co., 1970.

WHITE, ABRAHAM, P. HANDLER, and E. L. SMITH, *Principles of Biochemistry* (5th ed.). New York: McGraw-Hill Book Co., 1973.

PROSTAGLANDINS
HORTON, E. W., "The Prostaglandins." *Proc. Royal Society* (London) *B*, **182**:411 (1973).

PIKE, JOHN E., "Prostaglandins." *Scientific American*, November 1971. (Offprint 1235.)

NUCLEIC ACIDS
CRICK, F. H. C., "The Structure of the Hereditary Material." *Scientific American*, October 1954. [Offprint 5.]

———, "Nucleic Acids." *Scientific American*, September 1957. (Offprint 54.)

HERSHEY, A. D., "Idiosyncrasies of DNA Structure." *Science*, **168**:1425 (1970). Nobel Lecture, 1969.

PROTEINS
ANFINSEN, C. B. "The Formation and Stabilization of Protein Structure." *Biochem. J.*, **128**:24 (1972). A very readable review by this 1972 Nobel Prize winner.

DICKERSON, R. E., and IRVING GEIS, *The Structure and Action of Proteins.* Menlo Park, Calif: W. A. Benjamin, Inc., 1973. (Paperback.) A beautifully illustrated introduction to the subject.

FRASER, R. D. B., "Keratins." *Scientific American*, August 1969. (Offprint 1155.)

GRAY, W. R., L. B. SANDBERG, and J. A. FOSTER, "Molecular Model for Elastin Structure and Function." *Nature*, **246**:461 (1973). An important extracellular protein.

GROSS, J., "Collagen." *Scientific American*, May 1961. [Offprint 88.]

KENDREW, JOHN C., "The Three-Dimensional Structure of a Protein Molecule." *Scientific American*, December 1961. (Offprint 121.) Myoglobin.

MILLER, E. J., and V. J. MATUKAS, "Biosynthesis of Collagen." *Federation Proceedings*, **33**:1197 (1974). Synthesis and assembly.

MOORE, S., and W. H. STEIN, "Chemical Structures of Pancreatic Ribonuclease and Deoxyribonuclease." *Science*, **180**:458 (1973). Nobel lecture, 1972.

PERUTZ, M. F., "The Hemoglobin Molecule." *Scientific American*, November 1964. (Offprint 196.)

WEINSTOCK, M., and C. P. LeBLOND, "Formation of Collagen." *Federation Proceedings*, **33**:1205 (1974). Electron micrographic study.

CHAPTER 4

The Energetics and Control of Cellular Reactions

4–1 ENZYMES   142
Energy of Activation
ᴠEnzymes as Catalysts
Enzyme-Substrate Interaction
The Mechanisms of Enzymatic Catalysis skip

4–2 ENZYME KINETICS   149 skip
The Michaelis-Menten Equation
Experimental Determination of v_{max} and K_M
The Hill Equation
The Advantages of Cooperativity
The Basis for Cooperativity
A Choice of Models

4–3 FEEDBACK REGULATION   155 pick up here
The Feedback Principle
Competitive Inhibition
Noncompetitive Inhibition
Allosteric Regulation

4–4 ATP AND COUPLED REACTIONS   162
The Role of ATP
Coupled Reactions

4–5 GLYCOLYSIS   166
The Glycolytic Pathway
The Regulation of Glycolysis

4–6 BIOLOGICAL OXIDATIONS   172

SUMMARY   175

STUDY GUIDE   176

REFERENCES   177

Thousands of different chemical reactions take place in a living cell. Although many are thermodynamically favorable, some are not. The unfavorable reactions must be "driven," often by coupling them to favorable reactions in such a way that one cannot occur without the other. However, very few reactions of any kind would occur spontaneously at rates fast enough to maintain the cell. This problem results from the existence of energy barriers and is solved by the presence of cellular catalysts, the enzymes.

This chapter is concerned with the nature of enzymes, how they function, and some examples of enzymatically catalyzed reactions.

4–1 ENZYMES

Every high school chemistry student knows that hydrogen and oxygen gases can react with explosive force to form water:

$$H_2 + \tfrac{1}{2} O_2 \longrightarrow H_2O$$

The explosion is the result of a very large negative free energy and enthalpy change for the reaction, which is another way of saying that the equilibrium position of the reaction strongly favors the formation of water. Yet if the two gases are mixed at ordinary temperatures, nothing seems to happen, although water is actually forming at an extremely slow rate. But when a match is dropped into the container, water forms immediately, in a violent reaction.

ENERGY OF ACTIVATION. One can explain the drastic change in reaction rate caused by dropping a match into a container of hydrogen and oxygen by saying that the bonds between the hydrogen atoms, and those between the oxygen atoms, must be broken before new, hydrogen-oxygen bonds can form. Although oversimplified, this explanation does introduce the concept of an energy barrier (see Fig. 4–1), for it calls for a certain amount of energy to be put into the system to break up the hydrogen and oxygen molecules before a reaction can take place. This energy is called the *energy of activation*, designated $\Delta E^{\ddagger}$ or $\Delta G^{\ddagger}$.

An energy of activation is supplied through normal thermal motion. Therefore, the chance of achieving it in a collision increases with temperature. Since the standard state enthalpy change (ΔH°)

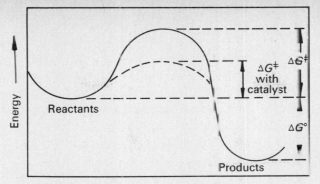

FIGURE 4–1. Energy of Activation. Even a highly favorable reaction may proceed very slowly if there is a significant energy barrier. The input of energy needed to initiate a reaction will be recovered along with the free energy difference between reactants and products. Note that the reaction shown here is essentially irreversible: To proceed in the forward direction, the barrier $\Delta G^{\ddagger}$ must be jumped; to proceed in the reverse direction, the much higher barrier $(\Delta G^{\circ} + \Delta G^{\ddagger})$ must be overcome.

for the formation of water from hydrogen and oxygen is -68.3 kcal/mole at 298 °K, the reaction itself could be a source of considerable heat. At ordinary temperatures, water forms too slowly to have any real effect on the average temperature; however, when a match is introduced, the area around it may be raised to the critical temperature, causing a reaction that releases heat rapidly. The reaction can thus be propagated throughout the container, leading to an explosion.

Virtually all reactions that take place in a cell have significant energies of activation, and hence they, too, would require elevated temperatures to support any significant rate of product formation. That option, of course, is not open to cells. Instead, the reaction rates are enhanced and controlled by using enzymes to lower their energies of activation, a process known as *catalysis*.

ENZYMES AS CATALYSTS. All enzymes are proteins, though many of them also contain prosthetic groups of other types of material.[1] Enzymes are true catalysts, in that: (1) they are effective in small amounts; (2) they are unchanged by the reaction; and (3) they do not affect the ultimate equilibrium concentrations, but only reduce the required activation energy and hence increase the rate of reaction. In addition to these three properties common to all catalysts, enzymes are unique in that each exhibits a very great specificity for a particular reaction. This specificity provides a way of balancing the metabolism of a cell, since the relative rates of each reaction can be controlled by adjusting the amount and/or catalytic efficiency of the enzyme that is specific for it.

The specificity and catalytic properties of enzymes are a result of their structure. This fact is easy to demonstrate, since, like other proteins, enzymes may be "denatured" by a variety of chemical and physical agents. When their native spatial configuration is lost (denaturation), so is their biological activity.

---

1. The complex of an enzyme and its prosthetic group is called a *holoenzyme*. The protein portion alone is an inactive *apoenzyme*.

ENZYME-SUBSTRATE INTERACTION. There is an area on each enzyme, called its *active site*, that is particularly sensitive to chemical or conformational change, for it is there that reactants are bound. The interaction between a reactant (also called a *substrate*) and an enzyme is fostered by a complementary arrangement of mutually attractive atoms or groups of atoms, e.g., a negatively charged amino acid side chain on the enzyme with a positively charged group on the substrate. Typically, several weak bonds will be formed between enzyme and substrate, creating a relatively stable complex.

An enzyme-substrate interaction might be possible because of a complementary shape between the substrate and a rather rigid enzyme molecule. This is the *lock and key theory*, introduced by Emil Fischer in 1894, long before it was even realized that enzymes are proteins. More recently, J. Wyman and D. W. Allen suggested in 1952 that most proteins have a variety of very similar conformations of nearly equivalent stability and that therefore we should think of proteins as being relatively flexible, with the ability to move from one to another conformation. Extending this idea, D. E. Koshland, Jr., in 1958 proposed his *induced fit theory* of enzyme-substrate interaction, which suggests that the most energetically favorable configuration of the enzyme when substrate is present may not be the same as when it is absent. (See Fig. 4-2.) The active site, instead of being rigidly complementary to a substrate, might yield to substrate in much the same way as a catcher's mitt yields to a baseball. The net effect could be to hold the substrate more tightly once a contact has been made by random collision.

FIGURE 4-2. Enzyme-Substrate Interaction. **(a)** According to the lock and key theory, enzyme and substrate have complementary configurations. In this diagram, matching spatial conformations permit ionic and hydrophobic interactions to take place between the enzyme and its specific substrate. **(b)** Alternatively, the enzyme can be viewed as a more flexible entity, so that much of the contact is due to a change in conformation of the enzyme fostered by the presence of substrate.

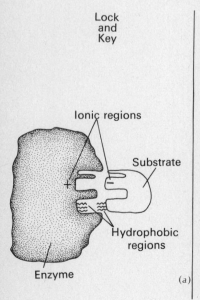

Lock
and
Key

Ionic regions

Substrate

Hydrophobic
regions

Enzyme

(a)

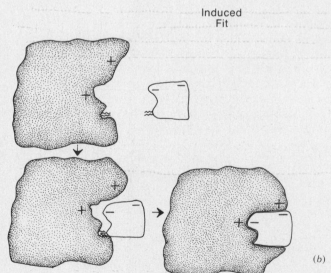

Induced
Fit

(b)

A still more recent concept, called *allostery,* has the enzyme existing as an equilibrium mixture of two or more discrete (rather than continuously variable) conformations. The predominant conformation will depend upon whether or not substrate is bound. These ideas will be discussed in more detail later.

In support of the induced fit and allosteric models, we note that conformational changes have been observed in a number of enzymes when their substrates are introduced (as when oxygen is bound to hemoglobin—see Chap. 3). Very often, too, the enzyme-substrate complex has physical and chemical properties (such as resistance to denaturation) that are measurably different from those of the enzyme alone. These changes are not observed with every enzyme, however, so it is probable that the lock and key theory adequately describes some enzyme-substrate interactions, while remaining inadequate for others.

THE MECHANISMS OF ENZYMATIC CATALYSIS. Once a substrate is bound, catalysis takes place. Again, there does not seem to be a single simple explanation for the phenomenon, but rather a number of possible mechanisms, some of which are more important than others in any given case. We may arbitrarily classify the more probable mechanisms into four groups: (1) catalysis due to an enhancement of collision frequency or orientation between bound substrates; (2) mechanisms that result from the solvent properties of the active site; (3) catalysis through the formation of one or more covalent intermediates between substrate and the enzyme itself; and (4) catalysis resulting from strains induced in the substrate by conformational changes occurring in the enzyme after the substrate is bound. (See Fig. 4–3.)

1. Catalysis due to an enhanced collision frequency or orientation is perhaps the easiest to understand. Suppose the reaction to be catalyzed is A + B → AB. The rate at which this reaction occurs will be proportional to the number of collisions between A and B, a number that must increase with the concentration of reactants. (More precisely, the reaction rate is proportional to the product of the reactant concentrations.) If A and B are bound to adjacent sites on the surface of the enzyme but still retain some freedom of movement, then one should find a greatly increased collision rate simply because the effective local concentration of A and B is very high. In addition, the orientation of the two molecules at the moment of collision is not random, but depends on the architecture of the active site and how substrates are bound to it. Since a reaction involves only certain atoms, orientation of reactants with respect to each other at the moment of collision might very well be important.

   Attributing reaction rate increase to a favorable collision frequency and orientation of substrates at the active site is attractively simple. However, calculations by Koshland and others show that, although the localized concentration of reactants may indeed be effectively increased

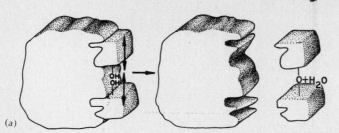

(a)

Collision Frequency and Orientation

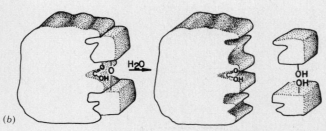

(b)

Solvent Properties of the Active Site

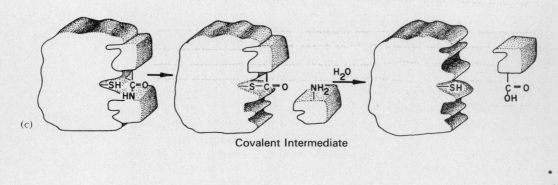

(c)

Covalent Intermediate

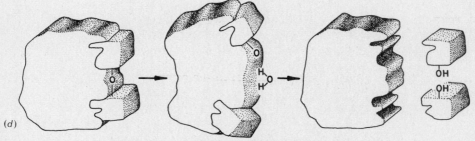

(d)

Conformational Strain

by the enzyme, the low concentration of the enzyme itself may more than offset this particular advantage. In other words, even though collision between substrates is more probable after they are bound, the enzyme-substrate collision necessary for binding is infrequent enough to prevent any dramatic change in rate due to this effect alone. However, Koshland has also demonstrated, by synthesizing organic compounds that bind substrates and catalyze reactions, that orientation during collision may have dramatic effects on reaction rates, a phenomenon that he calls *orbital steering*. Therefore, the ability of enzymes to bind substrates in a certain position relative to one another, and to hold them there while they continue to collide, must be considered as a plausible explanation for catalytic effectiveness in certain cases.

2. A second explanation for enzymatic catalysis relies on the fact that a substrate bound to an active site may find itself in a solvent with very different properties from those of water. For example, if the active site has a number of nonpolar amino acid side chains protruding into it, the bound substrate will be effectively trapped in a hydrocarbon environment of very low dielectric constant. Since the attraction between two charges of opposite sign, or the repulsion between two charges of like sign, is inversely proportional to the dielectric constant of the medium between them, charged groups on the substrate may begin to attract or repel one another strongly enough to set up strains in the substrate, making it more reactive. Conversely, there are indications that some active sites bind the substrate into a region even more polar—that is, with a higher dielectric constant—than water, thus shielding charged groups from each other. This factor, too, could foster certain kinds of reactions.

Other types of solvent effects are possible. Most hydrolyses, for example, are fostered by strong acids or bases because the reaction involves the addition of a proton and hydroxyl ion to the substrate. This action could be duplicated at an active site if the latter included acidic or basic groups in the form of proton donors or acceptors. The effect would be to alter the pH in the vicinity of the substrate.

The active site can also include electron donors or acceptors—for example iron, which can be reversibly cycled between Fe^{2+} and Fe^{3+} by removing or adding an electron. An electron donor would favor the reduction of a substrate whereas an electron acceptor would favor its oxidation, as we will later discuss in more detail. Obviously, in such cases the electron excess or deficit must be corrected before the enzyme can function again, a task that may be aided by certain *coenzymes*, which are relatively small organic molecules that are altered during the course of an enzymatic reaction and then recycled back to their original condition.

3. In some cases, the enzyme may participate chemically in the catalysis by forming a short-lived covalent intermediate. The formation of a covalent enzyme–substrate complex may occur more readily than the

FIGURE 4-3. Catalytic Mechanisms. (a) The positioning of bound substrates may both increase their rate of collision and control their orientation during collision. (b) The reaction may be fostered by the solvent properties of the active site—e.g., by acid catalysis from a proton donor. (c) The enzyme may form an unstable covalent intermediate with the substrate. (d) Conformational changes in the enzyme after substrate binding may induce strains in the substrate and increase its reactivity.

conversion of substrate to product. The catalysis would then depend on this complex being highly unstable and decomposing to yield a product.

For example, the enzyme papain[2] (from papaya fruit) hydrolyzes peptide bonds through formation of an intermediate with the sulfhydryl (—SH) of one of its own amino acids, cysteine:

$$
\begin{array}{ccc}
NH_2 & NH_2 & NH_2 \\
| & | & | \\
HC\text{—}R_1 & HCR_1 & HCR_1 \\
| & | & | \\
Enz\text{—}S + C{=}O \longleftrightarrow Enz\text{—}S\text{—}C{=}O \xrightarrow{H_2O} Enz\text{—}S + C{=}O \\
| \quad | & | & | \quad | \\
H \quad HN & HNH & H \quad OH \\
| & | & \\
HC\text{—}R_2 & HCR_2 & \\
| & | & \\
COOH & COOH &
\end{array}
$$

The steps must each have a low activation energy if the reaction is to proceed rapidly, but the overall energy change can be no different from what it would have been had the enzyme not participated, for the reactants and products are the same in either case. Many hydrolyses take place by analogous reactions, although not all of them use cysteine. Chymotrypsin and trypsin, for example, catalyze the hydrolysis of peptide bonds through the intercession of a serine hydroxyl, —CH$_2$—OH, instead of the sulfhydryl group.

4. The fourth theory of enzyme activity, sometimes called the "rack and strain theory," is a little harder to confirm experimentally. It suggests that a conformational change in the enzyme might literally pull a bound substrate apart. The theory presumes that the most stable conformation of an enzyme is different when a substrate is bound to it. If intraenzyme or enzyme–solvent bonds are replaced with enzyme–substrate bonds of different energies, several kcal/mole might be made available to induce a conformational change. If this change is impeded by the particular shape of a bound substrate, the substrate might become more reactive as strains are induced in it. Once a reaction takes place, however, the products diffuse away, leaving the enzyme to regain its original conformation in preparation for a new molecule of substrate.

Some enzymatically catalyzed reactions can be reasonably explained with one or another of the above mechanisms. Others require a combination of them, and still others remain unexplained. In addition, there are mechanisms that we have not considered, as the intent here is not to be exhaustive, but merely to emphasize that there is nothing mysterious about the way in which enzymes function. Their mechanisms are straightforward, and can be understood using only simple chemical principles, even if we have not yet determined the precise mechanism in each and every case.

---

2. All enzymes have a formal name that ends in the suffix "-ase." There are, however, a few enzymes whose historical name is still widely accepted even though it deviates from this rule. Papain, trypsin, pepsin, and lysozyme are a few examples.

4–2 ENZYME KINETICS

The simplest type of reaction is an isomerization, or interconversion, represented by the expression

$$S \rightleftharpoons P$$

where S stands for substrate (reactant) and P for product. If left undisturbed, this reaction, like all reactions, will eventually come to an equilibrium described by an equilibrium constant, K_{eq}, and associated standard state free energy change, $\Delta G°$:

$$\Delta G° = -RT \ln K_{eq} = -RT \ln \frac{[P]_{eq}}{[S]_{eq}}$$

As before, R represents the gas constant, T the absolute temperature, and the brackets indicate concentration.

The equilibrium concentrations of substrate and product will not be affected by the presence of an enzyme, but the rate at which equilibrium is reached can be greatly accelerated by the appropriate enzyme. Since few cellular reactions are at equilibrium, it is the transition period that is most interesting. A mathematical description of an enzymatically catalyzed reaction during this period was presented in 1903 by V. Henri, and modified in 1913 by L. Michaelis and M. L. Menten.

THE MICHAELIS-MENTEN EQUATION. An enzymatically catalyzed reaction consists of three steps: (1) the formation of an enzyme-substrate complex, (2) catalysis, converting substrate to product, and (3) release of product.

$$E + S \xrightarrow{\text{binding}} ES \xrightarrow{\text{catalysis}} EP \xrightarrow{\text{release}} E + P \qquad (4\text{--}1)$$

Catalysis is often very rapid compared to the other two events; hence, either binding or release becomes the "rate-limiting" step in the reaction. Since each step is also reversible, the reaction can be written

$$E + S \rightleftharpoons ES \rightleftharpoons E + P \qquad (4\text{--}2)$$

A simple equation can be derived describing the rate at which substrate is utilized (i.e., the velocity, v, of the reaction) in terms of readily measured variables.[3] That equation, the *Michaelis-Menten equation*, is

$$v = \frac{v_{max}}{1 + K_M/[S]} \qquad (4\text{--}3)$$

Here v_{max} (often designated by V) represents the *maximum velocity* of the reaction, or the maximum number of substrate molecules

---

3. Note that these small v's have units of concentration/time. In the terminology of calculus, each is formally equivalent to a time derivative (e.g., $v \equiv -d[S]/dt$).

that a single enzyme can convert to product in a unit of time. That maximum is attained only at an infinite concentration of substrate because only then will the enzyme always be saturated with substrate. K_M represents the *Michaelis constant*. It is equivalent to the concentration of substrate needed to achieve half of the maximum velocity. (Set $v = \frac{1}{2}v_{max}$ in equation (4–3) and rearrange the equation to demonstrate that relationship.)

The Michaelis-Menten equation is derived from the following assumptions and is therefore limited in its applicability to conditions where these assumptions are valid: (1) catalysis is rapid relative to the rates of binding and release, so that equation (4–1) can be replaced by equation (4–2); (2) for every molecule of substrate bound to the enzyme a molecule of product is immediately released—the so-called *steady state hypothesis*; and (3) the rate at which product is bound back to the enzyme in the reverse reaction is negligible. The first assumption, as explained earlier, is generally valid for enzymes. The second assumption, steady state, is usually reached within milliseconds of initiating a reaction. And the third assumption, the absence of a back reaction, can be assured experimentally by starting with zero concentration of product and terminating the period of observation before any significant concentration of product appears. Since a collision between enzyme and product is a necessary prerequisite to the reformation of an enzyme–product complex, keeping the product concentration low minimizes the number of collisions and hence the rate of reverse reaction.

The two constants, v_{max} and K_M, do not completely describe an enzymatically catalyzed reaction; by themselves they predict neither the equilibrium nor the standard state free energy change for a reaction. Nevertheless, they are extremely valuable. Since the conditions in which the Michaelis-Menten equation is valid are easily attained *in vitro*, and often approximated *in vivo*, v_{max} and K_M permit one to predict the velocity of a reaction for a given availability of substrate.

EXPERIMENTAL DETERMINATION OF v_{max} AND K_M. Taking the reciprocal of both sides of equation (4–3), we get

$$\frac{1}{v} = \frac{K_M}{v_{max}} \frac{1}{[S]} + \frac{1}{v_{max}} \qquad (4\text{–}4)$$

This equation suggests that a plot—sometimes called a Lineweaver-Burk plot—of $1/v$ vs. $1/[S]$ should produce a straight line with a slope of K_M/v_{max} and an intercept on the $1/v$ axis of $1/v_{max}$. (See Fig. 4–4a.) In addition, the extrapolated intercept on the independent (substrate) axis will have a value equal to $-1/K_M$. (Set $1/v$ equal to zero in equation (4–4) to convince yourself of that.) In this way, both the Michaelis constant and the maximum velocity

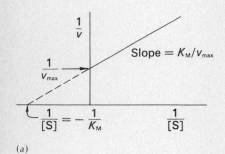

(a)

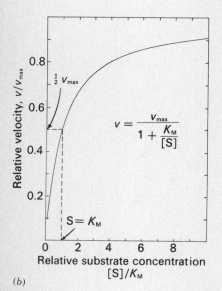

(b)

FIGURE 4–4. Michaelis-Menten Kinetics. **(a)** Double reciprocal plot, also called a Lineweaver-Burk plot—see equation **(4–4)**. It represents a convenient way of obtaining the kinetic constants, K_M and v_{max}. **(b)** Hyperbolic velocity plot—see equation **(4–3)**.

may be obtained from a relatively small number of observations made under conditions where the maximum velocity is not closely approached.

As an alternative to a Lineweaver-Burk plot, the velocity itself may be plotted against [S]. Such a curve is hyperbolic, approaching v_{max} asymptotically as [S] approaches infinity (see Fig. 4–4b). If measurements are made over a wide range of substrate concentrations, with some of the concentrations high enough to make the position of the asymptote clear, both v_{max} and K_M may be obtained. The position of the asymptote on the v axis is, of course, v_{max}. The value of K_M is read from the graph as [S] when $v = \frac{1}{2}v_{max}$.

Maximum velocities determined in these ways are often in the neighborhood of a few thousand molecules converted per minute per molecule of enzyme (its *turnover number*), but the turnover number of a few enzymes may exceed a million or more. Michaelis constants are often on the order of $10^{-4} M$.

The description up to now has assumed that the enzyme has only one active site or, if there are more than one, that the sites behave independently of each other. But most enzymes have more than one active site; and in many cases, what goes on at one site profoundly influences the catalytic activity of the other sites, even if the other sites are too far away from the first to cause any direct physical interference. Such an interaction, which is attributed to *cooperative effects* among the sites, was first noticed in the oxygen-carrying molecule, hemoglobin.

THE HILL EQUATION. Although hemoglobin is not an enzyme, it is considered an "honorary enzyme" because the binding of ligands (small molecules like oxygen, carbon monoxide, and so forth) by hemoglobin is analogous to the binding of substrate by a true enzyme. The function of hemoglobin is mainly to pick up oxygen in the lungs and transport it to some other point where it is discharged.

Since one oxygen molecule can be bound by each of the four heme groups in a reaction that mimics enzyme–substrate binding, one might expect the oxygen-binding curve of hemoglobin (the percent saturation of ligand sites versus the partial pressure of oxygen) to have the same characteristics as the velocity-versus-substrate curve of an enzyme. This expectation follows from the observation that the velocity of an enzymatic reaction, in the absence of back reaction, is proportional to the concentration of enzyme–substrate complex. However, rather than the hyperbolic curve thus predicted, the actual oxygen-binding curve of hemoglobin is sigmoidal (S-shaped) and is adequately described, over a limited concentration range, by the empirical relationship known as the *Hill equation* (see Fig. 4–5):

$$\frac{y}{1-y} = K(pO_2)^n \qquad\qquad (4\text{--}5)$$

The fraction of binding sites occupied, equivalent to the fractional velocity v/v_{max} of an enzyme, is given the symbol y. The substrate concentration is represented as the partial pressure of oxygen, pO_2, and K is merely a constant of proportionality. The superscript, n, is called the *Hill coefficient*.

By rearranging equation (4–5), we can make it look very much like the Michaelis-Menten equation, except for n. In fact, when n is one, equation (4–5) *is* the Michaelis-Menten equation:

$$\boxed{\frac{1}{v} = \frac{1}{Kv_{max}}\frac{1}{[S]^n} + \frac{1}{v_{max}}} \qquad\qquad (4\text{--}6)$$

Here y has been replaced by v/v_{max} and pO_2 by S to complete the analogy. In the special case where $n = 1$, the constant K is equivalent to $1/K_M$, but otherwise K is a somewhat more complicated term than the Michaelis constant.

The combination of n and K will be such that y (or v) is less at any given substrate concentration than would be expected in the absence of cooperativity. When n is greater than one, the protein is said to exhibit *positive cooperativity;* an n less than one indicates *negative cooperativity*. The latter is unusual, and will cause the v or y versus [S] curve to be a flattened hyperbola rather than a sigmoid. Positive cooperativity means that the addition of each molecule of substrate to an enzyme makes the addition of the next molecule of substrate to the same enzyme easier; negative cooperativity implies that the addition of the next molecule is more difficult.

Clearly, the value of n, the Hill coefficient, is a measure of cooperativity, for the larger its value the sharper the sigmoidicity.[4] The Hill coefficient is measured from plots of $\log [y/(1 - y)]$ versus $\log [S]$, which results from recasting equation (4–5) into the following form:

$$\log\left(\frac{y}{1-y}\right) = n \log [S] + \log K \qquad\qquad (4\text{--}7)$$

The slope of such a plot is equal to n.

THE ADVANTAGES OF COOPERATIVITY. An examination of Fig. 4–5 shows the value of cooperativity to the biological function of hemoglobin. In the absence of cooperativity, each of the four subunits of hemoglobin would bind oxygen the way myoglobin

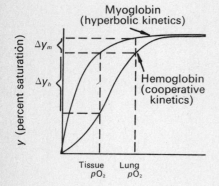

FIGURE 4–5. The Advantages of Cooperativity. The advantages conferred by cooperativity can be seen by comparing hemoglobin with myoglobin. The drop in oxygen tension (change in pO_2) between the lungs and tissues would cause a change, Δy_m, in the binding of oxygen by myoglobin. Hemoglobin, on the other hand, loses Δy_h of its oxygen, a much larger fraction.

4. Theoretically, the value of n can never be greater than the number of binding sites per molecule. A value of n exactly equal to the number of binding sites would be perfect cooperativity—i.e., the protein could only have n substrate molecules bound to it at any given time, or none at all. Obviously this limiting case cannot be attained, for the ligands will always be added one at a time.

binds oxygen—upper curve in the graph. But note the steepness of the binding curve of hemoglobin in the region of physiological oxygen tension, a steepness achieved by changing from the hyperbolic to a sigmoidal shape. The degree of oxygen saturation of hemoglobin is very sensitive to the availability of oxygen: it drops a larger fraction of its load in the tissues than it could without cooperativity, while still reaching almost full saturation in the lungs. In addition, it is clear from the graph that, at oxygen levels characteristic of the tissues, myoglobin has a much higher affinity for oxygen than does hemoglobin. The difference fosters an orderly transfer of oxygen from hemoglobin to myoglobin, the function of which is to accept oxygen from the blood and to store it in the tissues until needed by mitochondria.

Sensitivity to small changes in substrate concentration is an important feature of enzymes that exhibit cooperative effects. In addition, however, cooperativity can be exploited by molecules other than substrate to achieve a particularly effective kind of control over enzymatic rates, to be discussed later.

THE BASIS FOR COOPERATIVITY. When cooperative effects were first noticed in hemoglobin in the early part of this century, they were attributed to direct electrostatic interactions among the binding sites. It is now known that direct interaction is unlikely, since the binding sites (the heme groups) are approximately at the corners of a tetrahedron composed of the four roughly spherical subunits that make up the molecule (see Fig. 3–21). Because of the physical separation of the binding sites, then, other ideas had to be proposed to explain the apparent interaction among the sites.

The first real clue to the nature of cooperativity came in the early 1930s, when F. Haurowitz noticed that crystals of deoxyhemoglobin broke when exposed to oxygen. Further studies revealed that oxyhemoglobin and deoxyhemoglobin have different crystal structures, implying that the molecules have different conformations, or spatial configurations of the polypeptide chains.

X ray work by M. F. Perutz and his colleagues confirmed that the binding of oxygen is accompanied by a change in the structure of hemoglobin. There are two current explanations for the phenomenon. One, D. E. Koshland's *sequential* or "induced fit" theory, introduced earlier in this chapter, envisions a rather flexible protein in which the most thermodynamically stable configuration of protein plus ligand is different from that of protein alone. Since changes occurring in one subunit of the molecule might induce some change in the other subunits, the binding of one ligand may increase the affinity of other binding sites for ligand. Thus, when one site is occupied, the others will bind ligand faster than did the first, a condition that provides the necessary cooperativity and leads to sigmoidal binding curves.

The other widely known model for cooperative effects is the *concerted* model published in 1965 by J. Monod, J. Wyman, and J.-P. Changeux of the Institute Pasteur in Paris. They proposed a natural equilibrium between (at least) two states (see Fig. 4–6): (1) a state in which all the binding sites have maximum affinity for ligand, and (2) a state where all the binding sites have a minimum affinity. This equilibrium is an isomerization between states of differing energies, the exact difference of which will determine the amount of protein existing in either, according to the rules discussed in Chap. 2. The binding of one ligand is sufficient to stabilize a molecule in the active state, effectively removing it from the equilibrium. By mass action, then, equilibrium is shifted toward the active form as more molecules bind ligand. Those molecules of hemoglobin or enzyme having only one occupied site will have a high affinity for more ligand, so again one has a cooperative effect, manifest by a sigmoidal binding curve.

A CHOICE OF MODELS. In either type of model for cooperativity, the binding curve is such that relatively little substrate is bound at low ligand concentration. However, as the ligand concentration is increased, some is accepted by the subunits. Eventually a critical point is reached at which nearly all enzyme molecules are shifted entirely to the active state, leading to a rapid increase in percent saturation with only a small additional increase in ligand availability. The curve resembles a standard amplifier curve, with a region in which small changes in ligand concentration cause very large changes in percent saturation.

The two theories of cooperativity have many points in common, of course. The main distinction is that Monod, Wyman, and Changeux assumed that all of the binding sites must be in either one or the other configuration, whereas Koshland assumed a more flexible molecule and a less rigid interaction among the subunits. Some enzymes exhibiting cooperative effects seem better described by one model, while others seem more closely to fit the alternate model. Since cooperative effects are widespread, there is no obvi-

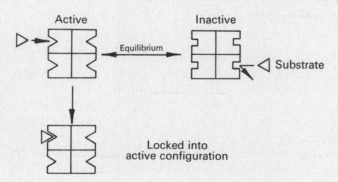

FIGURE 4–6. The Two-State Model for Cooperativity. The enzyme is normally an equilibrium mixture of two configurations. In one configuration, all sites have a high affinity for substrate; in the other configuration all sites have a low affinity. The presence of a single molecule of bound substrate (in this model) locks the enzyme in the active form, causing the remaining sites on that molecule to become fully active.

ous reason why nature should have adopted a single mechanism in all cases, so it may be that each model exactly describes some systems while other systems are not adequately described by either. But regardless of whether one adopts the two-state theory or the induced fit theory for cooperativity, the fundamental conclusion is the same: proteins exhibiting cooperative binding effects will have one conformation on the average in the presence of ligand and another in its absence. The change from inactive to active occurs before the protein is saturated with ligand, so that unoccupied sites have a maximum affinity for more ligand once the binding process begins.

We have assumed that in all cases of cooperativity, the enzyme is composed of several subunits. That is observed to be the case, although multiple binding sites on a single subunit should be able theoretically to produce similar effects. Enzymes with only a single subunit and only one binding site should always have hyperbolic binding of substrate with no hint of sigmoid character, as in Fig. 4-4. Thus, when the hemoglobin tetramer is dissociated, it is found that each isolated subunit by itself not only resembels myoglobin physically, but behaves like myoglobin (see Fig. 4-5). When hemoglobin is reduced only to the alpha–beta dimers, however, it still retains its cooperativity.

With this brief introduction to the kinetics of catalysis, we are in a position to understand the mechanisms by which the catalytic rate of an enzyme is controlled. It is through such control that the cell is able to respond promptly to fluctuations in the availability of various nutrients and to adjust its metabolism accordingly. Enzymes exhibiting cooperative effects are especially important in this control.

4-3 FEEDBACK REGULATION

All enzymes function at a rate that varies with the availability of substrates and with the concentration of product. Many, however, are subject to control by less obvious mechanisms as well. In particular, certain key enzymes are regulated by molecules that are neither substrate nor product, a situation that is most often seen in feedback regulation.

THE FEEDBACK PRINCIPLE. Feedback control, as applied to metabolism, means that the intracellular level of a substance regulates the rate of its own synthesis by controlling the amount and/or the activity of one or more enzymes within the pathway that produces it. For example, it was reported in the mid-1950s that the biosynthesis of several amino acids in microorganisms can be suppressed by their addition to the growth medium. Further in-

155

vestigations showed that the enzymes normally needed to produce the added amino acids were no longer being made by the microorganisms, and that those enzyme molecules already present in the cells were being inhibited. There are now many similar examples known, where the catalytic rate of enzymes can be inhibited by small molecules that are neither a substrate nor a product of the affected enzyme.

Consider the following hypothetical biosynthetic pathway:

$$A \xrightarrow{E_1} B \xrightarrow{E_2} C \xrightarrow{E_3} D \qquad (4\text{--}8)$$

The intracellular level of the product, D, may control its own rate of synthesis by acting on E_1, the enzyme that catalyzes the first step unique to its production. When the level of D rises, its presence causes E_1 to work more slowly; when the level of D falls, E_1 is allowed to work at a faster rate. The situation is similar to the function of a float valve: a rising liquid level causes the float valve to inhibit the incoming flow, maintaining a preset level of liquid.

The regulation of E_1 by D, the end product of the pathway, is possible only if D can bind to the enzyme. It may do so at the active site, thus interfering with the ability of the site to catalyze the transformation of A to B. Or D may bind elsewhere on the protein and affect the rate of catalysis in more subtle ways. In either case, D is called an *inhibitor* of the enzyme.

COMPETITIVE INHIBITION. Enzymes gain their specificity from the particular conformation of an active site, a conformation that is designed to recognize and bind to substrate molecules. However, some molecules that merely resemble the substrate also bind to the active site. This results in a *competitive inhibition* of the enzyme.

In competitive inhibition, the inhibitor (I) is competing with the substrate (S) for the active site (see Fig. 4–7). Thus, in addition to the usual enzyme-substrate reaction

$$E + S \rightleftharpoons ES \rightleftharpoons E + P$$

there is the competing reaction

$$E + I \rightleftharpoons EI$$

described by an inhibitor constant

$$K_I = \frac{[E][I]}{[EI]} \qquad (4\text{--}9)$$

Instead of the Michaelis-Menten expression, equation (4–3) or (4–4), the reaction is described by

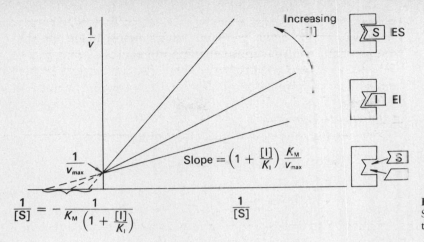

$$\frac{1}{[S]} = -\frac{1}{K_M\left(1 + \frac{[I]}{K_I}\right)}$$

Slope $= \left(1 + \frac{[I]}{K_I}\right)\frac{K_M}{v_{max}}$

FIGURE 4-7. Competitive Inhibition. Substrate and inhibitor compete for the same site.

$$\frac{1}{v} = \left(1 + \frac{[I]}{K_I}\right)\left(\frac{K_M}{v_{max}}\right)\left(\frac{1}{[S]}\right) + \frac{1}{v_{max}} \qquad (4\text{-}10)$$

This equation implies that the only effect of competitive inhibition is to change the slope of the double reciprocal plot. The higher the inhibitor concentration (or the lower the value of K_I), the steeper will be the slope. The maximum velocity, however, will be unchanged; this is not surprising, since v_{max} is defined in terms of an infinite substrate concentration, and under those conditions a finite amount of inhibitor should be unable to compete for binding sites. The double reciprocal plots are still linear, so that K_M, v_{max}, and K_I may be obtained from the slope and intercepts of two such plots carried out at different inhibitor concentrations.

Competitive inhibition is very common *in vivo*, since the product of a reaction is often a competitive inhibitor of the substrate, which it may still resemble. Thus, an enzyme-catalyzed reaction may be slowed by the presence of an increasing concentration of product long before equilibrium is approached. This is undoubtedly an important way of maintaining the proper balance between reactant and product for various reactions in which attainment of actual equilibrium concentrations would be undesirable.

NONCOMPETITIVE INHIBITION. In addition to inhibition by binding at the active site, it is possible for small molecules to inhibit enzymes by binding at other sites. Such an event may result in inhibition if the molecules physically interfere with (sterically hinder) the catalytic site, a situation that is represented schematically in Fig. 4-8. Small molecules that behave in this way are called *noncompetitive inhibitors*.

In noncompetitive inhibition, it is assumed that substrate and

157

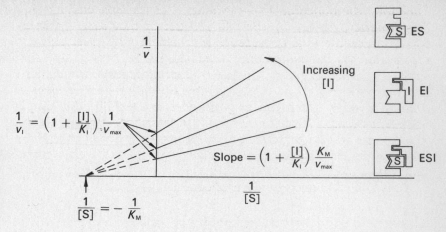

FIGURE 4–8. Noncompetitive Inhibition. Substrate and inhibitor bind to separate sites. In this particular model, when inhibitor is bound, the substrate cannot be released as product. Other modes of inhibitor action can also be imagined.

inhibitor occupy separate sites, so that both may be bound at the same time. Thus, in addition to the reaction

$$E + S \rightleftharpoons ES \rightleftharpoons E + P$$

there are three competing reactions that must be considered:

$$E + I \rightleftharpoons EI$$
$$EI + S \rightleftharpoons ESI$$
and $\quad\quad ES + I \rightleftharpoons ESI$

These three reactions are described by the same inhibitor constant,

$$K_I = \frac{[E][I]}{[EI]} = \frac{[EI][S]}{[ESI]} = \frac{[ES][I]}{[ESI]} \quad\quad (4\text{–}11)$$

because they really represent the same reaction, namely the binding of inhibitor to its specific site. The double reciprocal form of the Michaelis-Menten equation is modified by these reactions to become

$$\frac{1}{v} = \left(1 + \frac{[I]}{K_I}\right)\left[\left(\frac{K_M}{v_{max}}\right)\left(\frac{1}{[S]}\right) + \frac{1}{v_{max}}\right] \quad\quad (4\text{–}12)$$

Thus, both the slope and the intercept of the double reciprocal plot are a function of inhibitor concentration if the inhibition is noncompetitive. (Only the slope was changed in competitive inhibition, providing a way to distinguish experimentally between competitive and noncompetitive inhibition.) Note that a noncompetitive inhibitor does not change the intercept on the substrate axis: If the lines are extrapolated back to infinite velocity $(1/v = 0)$, they all meet at the point where the value of $1/[S]$ is numerically equal to $-1/K_M$.

Noncompetitive inhibition is characterized by a smoothly decreasing value of the reaction velocity as the inhibitor concentration is increased. In fact, a plot of $1/v$ against [I] will be linear.

Therefore, the noncompetitive inhibition of an early enzyme in a biosynthetic pathway by the end product of that same pathway could furnish the kind of feedback mechanism needed for efficient regulation. As the end product builds up, the activity of an inhibited enzyme decreases; as the end product disappears, the enzyme is again free to function. Although this is exactly what happens in many pathways, often the enzyme does not exhibit the simple Michaelis-Menten kinetics we have just described. Rather, Nature has found a way to sensitize certain enzymes to small changes in inhibitor concentration through cooperative effects. This condition, called *allosteric inhibition*, is a phenomenon of great importance to the efficient regulation of cellular reactions.

ALLOSTERIC REGULATION. As explained earlier, cooperativity is due to the existence of at least two conformational states: one in which the binding sites have a high affinity for substrate and another in which the binding sites have a low affinity for substrate. Suppose now that a second kind of binding site exists, separate from the active site, at which the presence of a ligand stabilizes the enzyme in one or the other of its possible conformations. If the enzyme is stabilized in the inactive state, its activity will be inhibited by the second ligand, which need be neither substrate nor product. This kind of noncompetitive inhibition is called *allosteric inhibition*, which means "inhibition without steric hindrance" of the active site. On the other hand, a ligand might stabilize the enzyme in its active state, in which case the enzyme will function at a more rapid rate. Ligands that have either of these properties are said to be *allosteric effectors* (see Fig. 4–9). A *positive effector* is one that stabilizes the enzyme in the active

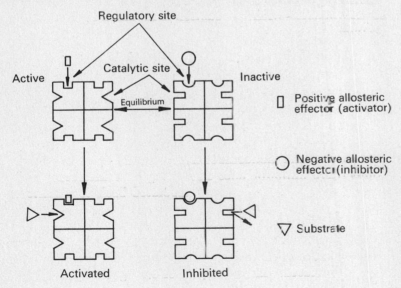

FIGURE 4–9. Allosteric Regulation. When the catalytic sites are active, the regulatory sites will accept a positive allosteric effector (activator). Binding of the activator locks the enzyme into this state and prevents a transition back to the less active state. Conversely, a negative effector (inhibitor) can bind when the enzyme is in its inactive conformation, and in so doing stabilizes that condition.

159

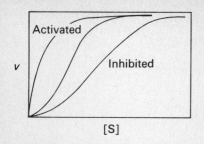

FIGURE 4-10. The Effects of Allosteric Regulators on Reaction Velocity. In the presence of a positive effector (activator), the Hill coefficient in equation (4–6) may be reduced to unity, whereupon the enzyme becomes indistinguishable from a one-site, Michaelis-Menten type enzyme. On the other hand, the presence of a negative effector (inhibitor) flattens the curve, through changes in both n and K.

state. It is, therefore, an "activator." Conversely, allosteric inhibitors are *negative effectors*. The sites at which allosteric effectors bind are called *regulatory* or *allosteric* sites, to distinguish them from the catalytic, or active, sites.

Some effectors act by changing K in the Hill equation, (4–6) or (4–7); others change the Hill coefficient, n; and some do both. A positive effector often has the property of reducing the Hill coefficient to 1, in which case the velocity curve becomes hyperbolic, as we have seen. The enzyme, in this condition, is indistinguishable from a normal one-subunit enzyme, since it obeys Michaelis-Menten kinetics (see Fig. 4–10).

Aspartate transcarbamylase (ATCase) is an example of an allosteric enzyme. It catalyzes the condensation of the amino acid aspartate with the simple molecule carbamyl phosphate to form carbamyl aspartate (see Fig. 4–11a). This condensation is the first step unique to the biosynthesis of the pyrimidine nucleotides, CTP and UTP, in *E. coli*. ATCase is therefore analogous to the hypothetical control enzyme, E_1, in the model feedback system introduced at the beginning of this section, in that its activity is regulated by an end product of the pathway, CTP.

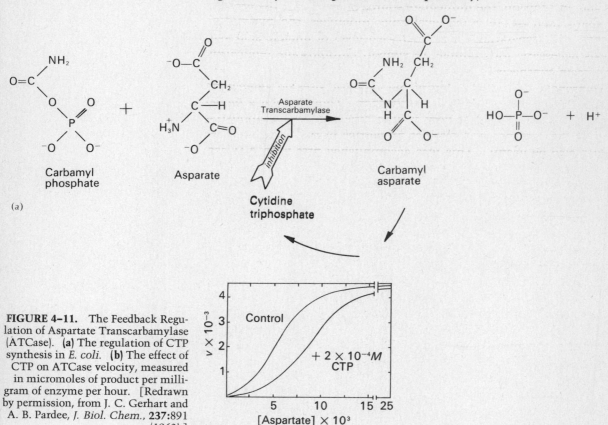

FIGURE 4–11. The Feedback Regulation of Aspartate Transcarbamylase (ATCase). **(a)** The regulation of CTP synthesis in *E. coli*. **(b)** The effect of CTP on ATCase velocity, measured in micromoles of product per milligram of enzyme per hour. [Redrawn by permission, from J. C. Gerhart and A. B. Pardee, *J. Biol. Chem.*, **237**:891 (1962).]

160

If one of the substrates of ATCase, carbamyl phosphate, is held at a constant high level, the velocity plot formed when the other substrate, aspartate, is varied is distinctly sigmoidal (see Fig. 4-11b). The presence of CTP flattens the curve, reducing the enzyme's velocity for any given substrate concentration. However, when ATP is present instead of CTP, the velocity curve becomes hyperbolic. We conclude that CTP is a negative allosteric effector (inhibitor) and that ATP is a positive effector (activator) of the enzyme. The shape of the velocity curve of ATCase depends on which effector is present or, if both are present, the shape of the curve will depend on their relative concentrations.

The control of ATCase activity by CTP and ATP makes a good deal of sense. Presumably there is some optimum amount of CTP that should be present in the cytoplasm to allow unhindered nucleic acid synthesis. When the level of CTP gets too high, the activity of ATCase is reduced until the CTP concentration drops to its stable value again. Similarly, if the level of CTP falls too low, the degree of inhibition of ATCase will be lessened until the optimum level is restored. ATP fits into this scheme very nicely, because ATP is itself one of the precursors of nucleic acid synthesis and, in addition, the immediate source of energy for the process. A sharp increase in ATP level thus mobilizes the pyrimidine pathway in preparation for an increased level of nucleic acid synthesis necessary, for example, to support chromosome replication or protein synthesis.

To propose separate regulatory and catalytic sites on an enzyme is one thing; to prove it is quite another. Though such a model fits the experimental data very well, a direct physical demonstration of the presence and distinct properties of two kinds of sites would also be useful. This was accomplished first with ATCase. In the early 1960s, Arthur Pardee and John Gerhart carried out a series of experiments with the enzyme in which it was demonstrated that inhibition by CTP could be completely prevented by first treating the protein with a mercurial that reacts with —SH groups, see Fig. 4-12. In the presence of the mercurial, the enzyme exhibits normal Michaelis-Menten kinetics no matter how much CTP is added.

Studies with an ultracentrifuge revealed that the mercurial-treated ATCase is smaller than the native enzyme, and heterogeneous. Following this lead, Gerhart, and H. K. Schachman at the University of California in Berkeley found that mercurials cause the enzyme to dissociate into two kinds of subunits. By purifying the subunits, they found that one kind has all the catalytic activity of the native enzyme—i.e., it exhibits Michaelis-Menten kinetics with substrates, but is not controlled by allosteric effectors. The second kind of subunit binds the allosteric effectors, but has no catalytic activity.

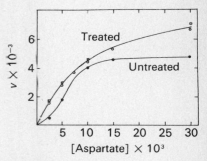

FIGURE 4-12. The Desensitization of Aspartate Transcarbamylase. Heating for four minutes at 60 °C before assaying, or treatment with mercurials, desensitizes the enzyme to CTP inhibition by dissociating the three regulatory from the two catalytic subunits. [Redrawn by permission, from J. C. Gerhart and A. B. Pardee, J Biol. Chem., **237**:891 (1962).]

Gerhart and Schachman concluded that the two kinds of sub-units of ATCase have different functions, namely regulation and catalysis. The presence of CTP bound to the regulatory subunits induces (or stabilizes) a conformation in which the catalytic sub-units are relatively inactive. In other words, it takes a much higher substrate concentration to achieve the same catalytic rate when CTP is present. On the other hand, when ATP is bound to the regulatory subunits, the activity of the catalytic subunits is increased. This, of course, is in full agreement with our model allosteric enzyme.

We cannot conclude that all allosteric enzymes have both regula-tory and catalytic subunits. In fact, they do not. The existence of both kinds of sites on the same subunit is quite sufficient to provide allosteric control. However, the existence of two distinct classes of subunits in the case of ATCase emphasizes the necessity for explaining allosteric effects in terms of conformational changes rather than by some kind of direct interaction among the sites. It is very difficult to imagine how the binding of ATP to one subunit of ATCase could change the catalytic activity of another quite sep-arate chain (one with a physically distant binding site), unless a conformational shift is propagated through the protein. Such con-formational shifts can be detected by several different physical techniques. Allosteric control is distinguished from competitive and noncompetitive inhibition by its reliance on such changes.

4-4 ATP AND COUPLED REACTIONS

All enzymes catalyze reactions that may approach equilibrium from either side, though the two rates may be vastly different. Hence, we speak of "substrates" and "products" interchangeably, according to a convention that designates the former as those parti-cipants appearing on the left of the reaction equation, while prod-ucts are found on the right. The dominant direction *in vivo* is most often, but not always, the direction of favorable (negative) free energy change, so that the equilibrium position results in a higher concentration of products than reactants. However, even a reaction that would otherwise be extremely unfavorable can be driven by an obligate coupling to a very favorable one. The cou-pling is ensured by the way in which the responsible enzyme is constructed. The favorable reaction is most often the hydrolysis of ATP.

THE ROLE OF ATP. The capture of light or the breakdown of food-stuffs is carried out in a way that makes energy available to the cell. The synthesis of needed organic molecules or the support of movement requires energy. The link between the two is ATP (see Fig. 4–13) and a small number of other compounds with similar

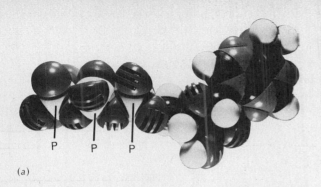

(a)

(b)

FIGURE 4–13. ATP. (a) A space-filling model. (b) Diagrammatic representation of adenosine-5'-triphosphate.

properties. Energy-yielding reactions synthesize ATP from ADP and inorganic phosphate (P_i); energy-requiring reactions hydrolyze ATP back to ADP and P_i again. If the supply of available ATP begins to decline, either the rate of energy-yielding reactions is increased, or the rate of energy-requiring reactions is decreased, or both. The opposite adjustments are made when the supply of ATP rises above its optimum value. The functional parts of the ATP molecule are the two pyrophosphate (or phosphoric acid anhydride P—O—P) linkages, for the hydrolysis of either is energetically favorable.

It is customary to talk about the two pyrophosphate bonds in ATP as "high energy bonds," a concept introduced by Fritz Lipmann in 1941. But before you jump to erroneous conclusions concerning the nature of these bonds, we should hasten to point out that the term "high energy" is somewhat misleading. The bonds themselves are not unusual, except that the *hydrolysis* of them is relatively favorable. However, the availability of this energy makes it immensely useful to cells.

The amount of energy ($\Delta G°$) released by the hydrolysis of either of the two phosphate anhydride bonds of ATP is in the neighborhood of -8 kcal/mole, depending on the exact conditions of temperature, pH, salt concentration, and so on. A value that is often used to represent the reaction at pH 7 is -7.3 kcal/mole. The corresponding values for the hydrolyses of other "high energy bonds" are given in Table 4.1 for comparison.

COUPLED REACTIONS. Suppose that a substance A can be transformed to a second substance B with a large drop in free energy. If A is stable—in other words, if there is a large free energy barrier to the reaction—B will not appear at any appreciable rate unless a catalyst is present. That catalyst could be an enzyme constructed in such a way that it is not active unless an ADP and P_i are also bound to its surface. It may also be constructed in such a way that the transformation from A to B simultaneously condenses P_i to ADP, forming ATP. In this way the enzyme, which we shall call

163

TABLE 4.1   Some standard state free energies of hydrolysis

	$\Delta G°$
phosphoenolpyruvate $\rightarrow$ pyruvate + P_i	-13.0 kcal/mole
phosphocreatine $\rightarrow$ creatine + P_i	-10.3
acetylcoenzyme A $\rightarrow$ acetate + CoA	-7.5
sucrose $\rightarrow$ glucose + fructose	-7.0
polynucleotide$_n$ $\rightarrow$ polynucleotide$_{n-1}$ + nucleotide	-6.0
glucose-1-phosphate $\rightarrow$ glucose + P_i	-5.6
glycogen$_n$ $\rightarrow$ glycogen$_{n-1}$ + glucose	-5.0
glucose-6-phosphate $\rightarrow$ glucose + P_i	-3.9
α-glycerol phosphate $\rightarrow$ glycerol + P_i	-2.4
polypeptide$_n$ $\rightarrow$ 2 polypeptides$_{n/2}$	-0.5

E_1, sees to it that in order for A to be changed to B, ATP must also be created. The two reactions are said to be "coupled" by the enzyme.

Let us assign a standard state free energy change to each reaction:

$$\begin{array}{ll} A \longrightarrow B & \Delta G_1^\circ = -12 \text{ kcal/mole} \\ \underline{ADP + P_i \longrightarrow ATP} & \underline{\Delta G_2^\circ = +7.3 \text{ kcal/mole}} \\ A + ADP + P_i \longrightarrow B + ATP & \Delta G^\circ = \Delta G_1^\circ + \Delta G_2^\circ \\ & \quad\quad = -4.7 \text{ kcal/mole} \end{array} \qquad (4\text{--}13)$$

One could say that 7.3 kcal/mole was "captured" by the phosphorylation of ADP. The overall equilibrium for the reactions reflects the total energy change of $-12 + 7.3 = -4.7$ kcal/mole. Obviously the combined reactions are still favorable, but the requirement for an obligate coupling of the two steps has conserved about two-thirds of the energy that would have been lost to heat and entropy if the first reaction had been allowed to take place alone.

Now suppose there is an essential, but energetically unfavorable, reaction C $\rightarrow$ D. Let $\Delta G°$ for this reaction be $+6$ kcal/mole. No significant amount of C will be converted to D unless the reaction is coupled to a second reaction whose free energy change is negative. Since the hydrolysis of ATP is a favorable reaction, a second enzyme, E_2, might be designed to catalyze the combined reaction:

$$\begin{array}{ll} C \longrightarrow D & \Delta G_3^\circ = +6 \text{ kcal/mole} \\ \underline{ATP \longrightarrow ADP + P_i} & \underline{\Delta G_4^\circ = -7.3 \text{ kcal/mole}} \\ C + ATP \longrightarrow D + ADP + P_i & \Delta G^\circ = \Delta G_3^\circ + \Delta G_4^\circ \\ & \quad\quad = -1.3 \text{ kcal/mole} \end{array} \qquad (4\text{--}14)$$

Like equation (4–13), this reaction is also favorable, meaning that the equilibrium will produce more products than reactants.

By comparing equations (4–13) and (4–14), one can see that part of the energy released by the conversion of A to B was captured in the form of ATP and used to drive the second reaction. The overall reaction, then, is

$$A + C \longrightarrow B + D$$

$$\Delta G° = -6 \text{ kcal/mole}$$

The intermediate role played by ADP/ATP is indicated by writing the two reactions as follows:

In this way it is clear that the energetically unfavorable, or "uphill," reaction C → D is driven indirectly by the favorable, or "downhill," reaction A → B.

Actually there is usually more than one step involved in each of the two reactions, producing a phosphorylated ("activated") intermediate. These intermediates are most common in the "uphill" reactions, so that

may be carried out by the two steps

$$ATP + C \longrightarrow ADP\text{—}C + P_i$$
$$ADP\text{—}C \longrightarrow ADP - D$$

or by the set

$$C + ATP \longrightarrow C\text{—}P + ADP$$
$$C\text{—}P \longrightarrow D + P_i$$

In still another variation, there may be a phosphorylated enzyme as an intermediate:

$$ATP + E_2 \longrightarrow ADP + E_2\text{—}P$$
$$E_2\text{—}P + C \longrightarrow D + P_i + E_2$$

The result is the same in each case, of course, as is the role of ATP. In the first two variations, however, we have the possibility of using more than one enzyme—i.e., one for each of the two steps.

To couple the two reactions A → B and C → D directly with one enzyme would be far too rigid. Using ATP as an intermediate source of energy permits a reaction occurring in one part of the cell to be used to drive a number of different reactions taking place at scattered points elsewhere in the cell. Controlling the flow of energy through the cell then becomes a matter of balancing the production of ATP on the one hand with its utilization on the other. Though maintaining this balance is an intricate task, it is in no way as complicated as direct coupling would have to be; we need only require that the ATP yielding reactions work at a variable rate. When the available supply of ATP shrinks, the reactions that create it by breaking down foodstuffs or capturing light are urged on to greater activity. When the ATP supply is plentiful, these ATP yielding reactions will be slowed.

Even granting the necessity for an intermediate such as ATP, there is still the need to explain why ATP itself was chosen to fill the role in all cells, from whatever source. We can not say for sure why ATP was chosen, of course, except to point out that it does have the required characteristics: (1) it has a large negative free energy of hydrolysis; (2) it is stable enough to prevent accidentally wasting that energy in a nonenzymatic reaction; and (3) it is complicated enough to be unambiguously recognized by the enzymes that use it. The other nucleoside triphosphates share all three of these virtues with ATP. Although most of them do participate in a few coupled reactions, the burden lies mostly on ATP. The choice of ATP may originally have been accidental, or it may have been because ATP is more readily formed from primordial components. However, once it was made, the lack of a competitive advantage and the great difficulty of switching to another nucleotide are probably enough to have assured evolutionary continuity.

4–5 GLYCOLYSIS

Reactions that require energy and that result in the biosynthesis of needed compounds are called *anabolic*. Reactions that provide this energy at the expense of the degradation of food molecules are called *catabolic*. *Metabolism* could refer to either or both. Readers who desire detailed information about the many anabolic and catabolic pathways should consult a biochemistry text, for it is not our mission to provide that information here. However, we will discuss one catabolic pathway to gain some insight into the flow of energy through a cell and the way in which that flow is regulated.

All our discussions will center about the metabolism of glucose. First, we will outline its anaerobic degradation in animal cells to demonstrate the way in which coupled reactions are employed. Then, in Chap. 5, the role of mitochondria in extending the glycolytic pathway through the oxidation of glucose to carbon dioxide and water will be discussed. In that chapter also, we will see how glucose is synthesized from carbon dioxide and water by chloroplasts (photosynthesis).

This preoccupation with glucose metabolism is appropriate for both historical and practical reasons: (1) Glycolysis was among the first metabolic pathways known, as it is nearly identical to alcoholic fermentation, to which most early biochemists' efforts were directed. (2) Humans obtain a substantial fraction of their daily caloric intake from glucose, mostly polymerized as starches. (3) Certain animal cells rely very heavily, or even exclusively, on glucose (blood sugar) for their energy. These cells include white muscle cells, red blood cells, and nerve cells of all types. (4) And finally, it is easy to tie glucose into the rest of the metabolic scheme.

THE GLYCOLYTIC PATHWAY. When glucose is degraded to lactate, the process is called *glycolysis* (i.e., "glycogen lysis"). The slight variation that provides alcohol as the product is called *alcoholic fermentation.* The individual steps in both pathways have been known since about 1940, with some of the most important contributions coming from Gustav Embden and Otto Meyerhof. In their honor, the scheme is sometimes called the *Embden-Meyerhof pathway*. For details of the pathway, see Fig. 4-14 and Table 4.2.

The production of two molecules of lactate from one molecule of glucose is highly exergonic (energetically favorable), with a standard state free energy change of about -47.4 kcal/mole of glucose at pH 7.

$$\text{glucose} \longrightarrow 2 \text{ lactate}^- + 2H^+ \qquad \textbf{(4-15)} \qquad (\Delta G^\circ = -47.4 \text{ kcal/mole})$$

A portion of this free energy is conserved in the phosphorylation of two molecules of ADP to form two ATPs. Hence, glycolysis represents a way of deriving ATP from glucose—the only way animal cells can do so without oxygen. Glycolysis is virtually the sole source of energy for mammalian red blood cells (they make no metabolic use of the oxygen they carry to other tissues), it is an important source of energy to muscles, and it may be utilized by assorted other tissues when oxygen is scarce. The pathway is also used heavily by a few microorganisms (e.g., lactic acid bacteria), but microorganisms generally produce end products other than lactate.

The transformation from glucose to lactate does not take place in a single step, but in a series of eleven reactions, each catalyzed by its own specific enzyme. This stepwise degradation is characteristic of all metabolic pathways, for it allows the cell to make better use of the relatively modest 7.3 kcal/mole energy change accompanying ATP synthesis or hydrolysis. Hence, the synthesis of a single ATP may conserve a high proportion of the free energy released in a reaction, or the hydrolysis of a single ATP may be used to drive an unfavorable step. In addition, some of the intermediates of one pathway will often serve as convenient starting materials for a second pathway. And, as we shall see, the stepwise pathway, with its several specific enzymes, can be more easily and flexibly controlled.

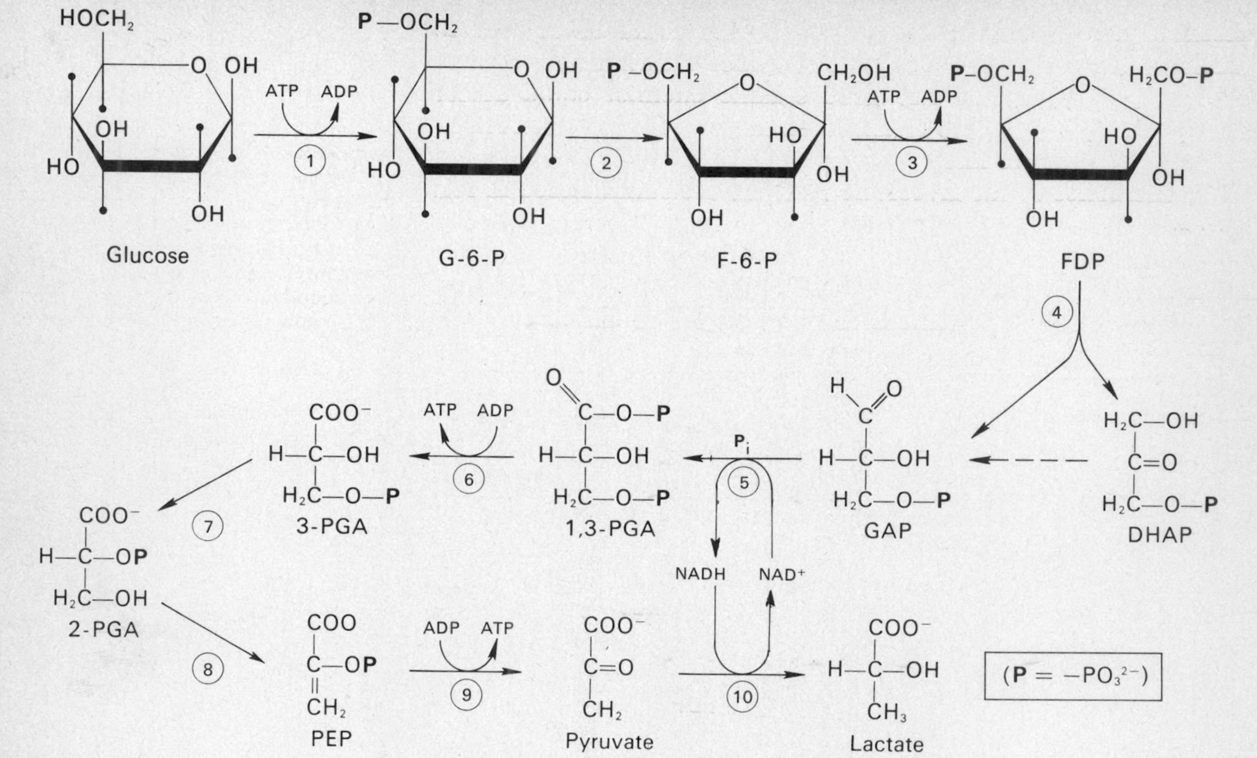

FIGURE 4–14. Glycolysis. See Table 4.2 for a list of the abbreviations and more details on the individual reactions.

But aside from these advantages of stepwise degradation, we should admit that probably some multistep mechanism is essential in most cases just because there is no simple chemical reaction that cells could use to effect the transformation in one step under physiological conditions. On the other hand, the pathway as it is

TABLE 4.2 Glycolysis

STEP	REACTION[a]	$\Delta G°$ KCAL/MOLE	ENZYME NAME
1	glucose + ATP → G-6-P + ADP	−3.42	glucokinase or hexokinase
2	G-6-P → F-6-P	+0.50	phosphoglucose isomerase
3	F-6-P + ATP → FDP + ADP + H⁺	−3.4	phosphofructokinase
4	FDP → DHAP + GAP	+5.73	aldolase or FDP aldolase
	DHAP → GAP	+1.83	triosephosphate isomerase
5	GAP + P_i + NAD⁺ → 1,3-PGA + NADH + H⁺	+1.50	glyceraldehyde phosphate dehydrogenase
6	1,3-PGA + H⁺ + ADP → 3-PGA + ATP	−6.78	phosphoglycerate kinase
7	3-PGA → 2-PGA	+1.06	phosphoglyceromutase
8	2-PGA → PEP	−1.08	enolase
9	PEP + H⁺ + ADP → pyruvate + ATP	−5.72	pyruvate kinase
10	pyruvate + H⁺ + NADH → NAD⁺ + lactate	−6.00	lactate dehydrogenase

[a] G-6-P glucose-6-phosphate   FDP fructose-1,6-diphosphate   GAP glyceraldehyde-3-phosphate   PEP phosphoenolpyruvate
F-6-P fructose-6-phosphate   DHAP dihydroxyacetone phosphate   1,3-PGA 1,3-diphosphoglycerate

presented here is not the only scheme that could be devised to transform glucose to lactate. In fact, minor variations do exist in different cell types, and any good organic chemist could devise other perfectly reasonable alternatives that nature has not seen fit to adopt, as far as we know.

The energetics of the glycolytic pathway are summarized in Fig. 4–15 and Table 4.2. The vertical axis in Fig. 4–15 is calibrated in kcal/mole, starting from glucose. The rise or fall between any two intermediates represents the standard state free energy change ($\Delta G°$) when the first intermediate is converted to the next. The

FIGURE 4–15. The Energetics of Glycolysis. The solid line shows the free energy changes as they exist in the coupled reactions. The dotted line shows free energy changes among the main intermediates only, without ATP coupling and with direct (air) oxidation and reduction.

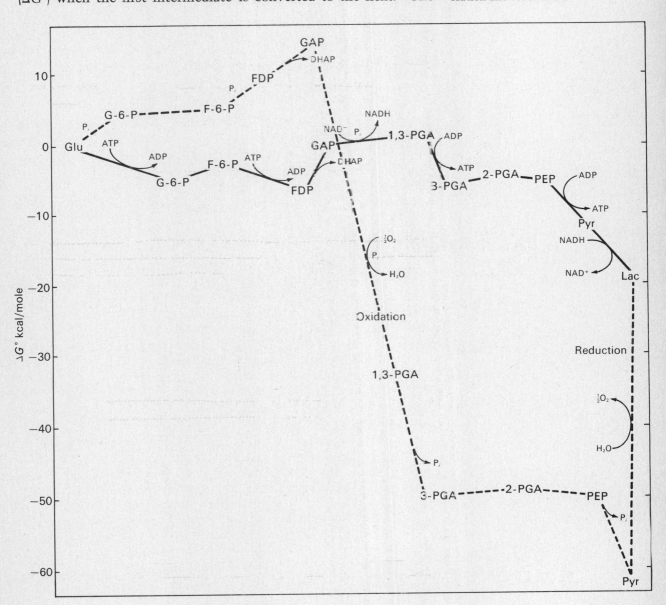

solid line shows the relationship as it actually exists in the cell, while the dotted line shows what it would be without an obligate coupling to ATP or, in the case of steps 5 and 10, without the coenzyme NAD^+ (to be discussed later). Note that both curves reach the same final position, for both represent the net reaction

$\Delta G° = -17.6$ kcal/mole

$$\text{glucose} + P_i \longrightarrow \text{DHAP} + \text{lactate}^- + H^+ \qquad (4\text{-}16)$$

In other words, there is no net gain or loss of ATP (two gained and two lost) and no net oxidation or reduction (one of each, as we shall see).

It is the utilization of dihydroxyacetone phosphate (DHAP) that accounts for the usefulness of glycolysis, since the conversion of this triose phosphate to its isomer, glyceraldehyde phosphate, permits DHAP to proceed on to lactate:

$\Delta G° = -15.2$ kcal/mole

$$\text{DHAP} + 2\text{ADP} + P_i + H^+ \longrightarrow \text{lactate}^- + 2\text{ATP} \qquad (4\text{-}17)$$

The two reactions together, equations (4-16) and (4-17), are what we call glycolysis:

$\Delta G° = -17.6 - 15.2$
$\quad = -32.8$ kcal/mole

$$\text{glucose} + 2\text{ADP} + 2P_i \longrightarrow 2 \text{ lactate}^- + 2\text{ATP} \qquad (4\text{-}18)$$

Note that $\Delta G°$ for equation (4-18) is 14.6 kcal/mole less negative than $\Delta G°$ for (4-15). The difference represents the net synthesis of two moles of ATP from two moles of ADP at $\Delta G° = +7.3$ kcal/mole each. Hence, glycolysis conserves 14.6/47.4 or about one-third of the available standard state free energy.

One of the more interesting features of glycolysis is the requirement that energy in the form of 2ATPs be put into the system before any recovery of energy is possible. Without this "pump priming," glycolysis would have about a 15 kcal/mole energy barrier—from glucose to GAP—even though the overall reaction is favorable. With the ATP input, however, GAP lies some 0.6 kcal/mole "below" glucose on the energy scale.

The energy yielding reactions of glycolysis are at steps 6 and 9 in Table 4.2 and Fig. 4-14. Step 6 is the hydrolysis of the phosphoric–carboxylic acid anhydride, 1,3-phosphoglyceric acid. Like the hydrolysis of the phosphate anhydride bonds of ATP, this reaction is very favorable. Step 9 is the reaction that converts phosphoenolpyruvate (PEP) to pyruvate. Here, a combination of factors are at work, the most important of which are charge repulsion between the phosphate and nearby carboxylate, and the fact that

phosphate locks PEP into its more unfavorable, enol $\begin{bmatrix} | \\ C-O \\ \| \\ C \end{bmatrix}$,

configuration. Once hydrolyzed, the immediate product, enolpyruvate, is free to isomerize to the much more stable keto

170

$$\left[\begin{array}{c} | \\ C=O \\ | \\ C \end{array} \right]$$ form, a transition that accounts for some 8 kcal/mole of the

observed total free energy change.

$$\begin{array}{cccc}
COO^- & COO^- & COO^- & \\
| & | & | & \\
C-OPO_3^{2-} & \xrightarrow[-P_i]{H_2O} & C-OH & \longrightarrow & C=O \\
\| & \| & | & \\
CH_2 & CH_2 & CH_3 & \\
\text{PEP} & \text{Enolpyruvate} & \text{Pyruvate} &
\end{array}$$ (4–19) $\Delta G° = -13$ kcal/mole

Both parts of this step are catalyzed by the same enzyme, pyruvate kinase. ("Kinase" in the name of an enzyme means that the reaction *in vivo* is coupled to the synthesis or hydrolysis of ATP.)

THE REGULATION OF GLYCOLYSIS. Like all other metabolic pathways, glycolysis is subject to regulation according to the principles developed earlier in this chapter. There are two sites at which control is particularly important: steps 3 and 9 in Table 4.2 and Fig. 4–14.

The first of these two steps is catalyzed by the enzyme phosphofructokinase, an allosteric control enzyme whose inhibitors are ATP and citrate and whose most important activator is AMP. ATP and AMP effect conjugate controls; when the ATP level in a cell is high, its AMP level is low, and vice versa. This relationship between ATP and AMP is due to the ubiquitous presence of an enzyme called adenylate kinase that catalyzes phosphate transfer from one ADP to another,

$$2ADP \rightleftharpoons AMP + ATP \qquad (4\text{–}20)$$

(The equilibrium constant is near one for this reaction.) Thus, as ATP levels fall, mass action will tend to create more ATP at the expense of ADP, causing AMP levels to rise. And as ATP levels rise, mass action works the other way to decrease AMP levels. Since the function of glycolysis is to create ATP from the breakdown of glucose, it is important that glycolysis be inhibited when ATP is already plentiful and that it be stimulated when ATP is scarce, which is precisely what the controls on phosphofructokinase tend to foster. (The inhibition of the enzyme by citrate is equally important, for reasons that will become clear when the citric acid cycle is discussed in the next chapter.)

The second control point in glycolysis is the reaction catalyzed by pyruvate kinase. Here, however, the situation is complicated by the existence of tissue-specific *isozymes*, which are different proteins having the same enzymatic activity. Mammalian tissues have three important isozymes of pyruvate kinase, only two of which will be mentioned here. They are the muscle isozyme and the liver isozyme, named for the tissues where they are found in greatest abundance.

Muscle pyruvate kinase is not an allosteric control enzyme (at least in the sense used here), for it always exhibits standard Michaelis-Menten kinetics with its substrates. It is, however, subject to inhibition by ATP, apparently by competition for the active site. (Since ATP is also the product of the reaction, this is an example of *product inhibition*.) Thus, the activity of pyruvate kinase is reduced when ATP levels are high which, as pointed out above, is when the rate of glycolysis should be reduced.

Liver pyruvate kinase is much more strongly inhibited by ATP, but, in addition, it is an allosteric control enzyme whose positive effector (activator) is fructose diphosphate (FDP). This sensitivity tends to keep the lower part of the glycolytic pathway operating at a rate that is consistent with the activity of the upper part of the pathway, allowing for the removal or addition of intermediates between the two control points by interconnecting pathways. If pyruvate kinase activity lags, there will be an accumulation of FDP, which lies at an "energy minimum" in the glycolytic sequence (Fig. 4–15). The increase in FDP activates pyruvate kinase until a more satisfactory level of its activity is reached. A full discussion of why this control is not so important to muscle and why muscle should require a different isozyme of pyruvate kinase is beyond the scope of this work; it is related to the fact that liver cells have an extremely complicated metabolism, with many active and interconnected pathways, while the enzymatic systems of muscle are dedicated mostly to the production of ATP, the energy of which is needed to power contractions. Instead of going into more detail on these points now, consider instead, two more reactions from the glycolytic sequence of Table 4.2 and Fig. 4–14, steps 5 and 10. The first of these is an oxidation of a glycolytic intermediate, while the second is a reduction.

4–6 BIOLOGICAL OXIDATIONS

An *oxidation* is the removal of one or more electrons from an atom or molecule; a *reduction* is the addition of electrons. Obviously, the former requires an electron acceptor and the latter an electron donor, for electrons do not just float loosely about in aqueous systems. In other words, a compound can only be oxidized if, at the same time, some other substance is reduced, and vice versa. The combined process is called an oxidation–reduction reaction, or *redox* reaction. The two substances involved are referred to as a "redox pair." In biological systems, many oxidations are accomplished by removing the same number of protons as electrons, thus leaving the charge of the oxidized molecule unaffected. This kind of oxidation is also called a *dehydrogenation.* Conversely, reductions are often hydrogenations, or the addition of neutral hydrogen.

Step 5 in glycolysis, the formation of 1,3-diphosphoglycerate from glyceraldehyde phosphate, represents the oxidation of an aldehyde to an acid. The last step in the pathway involves the reduction of a keto group in pyruvate to an alcoholic group in lactate.

The glycolytic pathway, then, contains one oxidation and one reduction, the net effect of which is to transfer two electrons from glyceraldehyde phosphate to pyruvate. The vehicle for this transfer is the *coenzyme* nicotinamide adenine dinucleotide, or NAD^+, the nicotinamide portion of which is made from the B vitamin nicotinic acid or niacin (not, unfortunately, the same as nor interconvertible with nicotine).

Nicotine   Niacin

NAD^+ is one of a small group of biological electron carriers that also includes its close relative $NADP^+$ and derivatives of another B vitamin, riboflavin (see Fig. 4–16). Because they can be oxidized in one place and reduced in another, NAD^+ and the other electron carrying coenzymes are capable of shuffling electrons around the cytoplasm in much the same way that ATP carries phosphate.

Since NAD^+ is present only in tiny amounts in the cell, the function of step 10 in glycolysis, the reduction of pyruvate to lactate, is to regenerate the NAD^+ needed by step 5 (GAP to 1,3-PGA). Any other mechanism for regenerating NAD^- would also serve this purpose, and in various organisms other methods are used. Some microorganisms, for example, accomplish the regeneration by reducing (and decarboxylating) pyruvate to ethanol instead of to lactate.

Pyruvate   Carbon dioxide   Ethanol
            and Acetaldehyde

The most favorable way of regenerating NAD^+, however, is to use molecular oxygen as an electron acceptor. Oxygen is readily

173

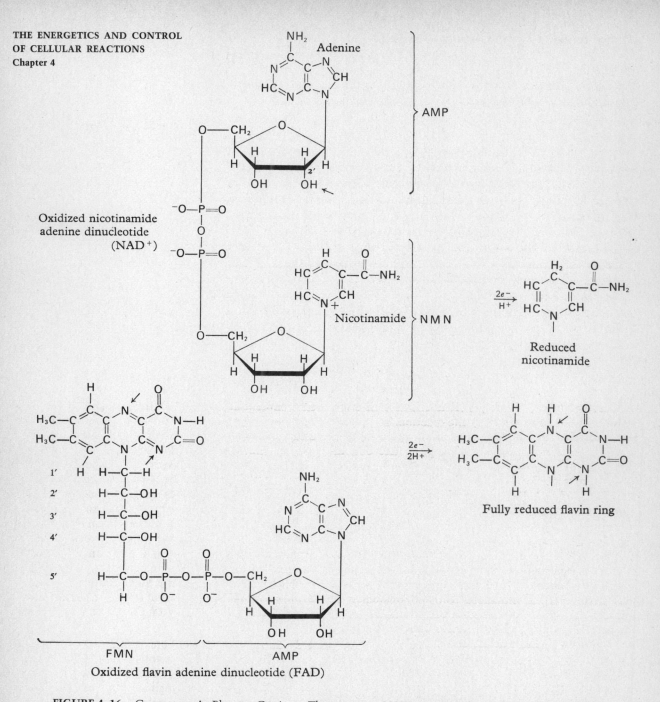

FIGURE 4–16. Coenzymes As Electron Carriers. The coenzyme NAD^+ consists of two nucleotide monophosphates: nicotinamide mononucleotide (NMN^+) and AMP. If the adenosine carries an additional phosphate at its 2′ position, the coenzyme is NAD-phosphate, or $NADP^+$. Two electrons and a proton reduce NAD^+ to NADH, or $NADP^+$ to NADPH. (NAD^+ and $NADP^+$ also go by older names, diphosphopyridine nucleotide or DPN^+, and triphosphopyridine nucleotide or TPN^+, respectively.) Riboflavin is found as flavin mononucleotide, FMN, and as the dinucleotide, FAD. Two electrons and two protons reduce these species to $FMNH_2$ and $FADH_2$, respectively. The flavin nucleotides are often found as covalently linked prosthetic groups of *flavoproteins*.

reduced to water in the following reaction:

$$\tfrac{1}{2}O_2 + 2e^- + 2H^+ \longrightarrow H_2O \qquad (4\text{-}21)$$

Because of the ease with which oxygen is reduced, the transfer of electrons from NADH to oxygen is a very exergonic process:

$$NADH + H^+ + \tfrac{1}{2}O_2 \longrightarrow NAD^+ + H_2O \qquad (4\text{-}22)$$

$\Delta G° = -52.6$ kcal/mole.

The size of this free energy change is reflected in Fig. 4-15 by the huge difference in $\Delta G°$ between step 5, as it occurs, and its counterpart in which the oxidation is carried out directly by oxygen instead of by NAD^+. We see the same difference when we compare step 10 with its counterpart. Most cells take advantage of the large negative free energy change of equation (4-22) by coupling the reaction to the synthesis of several molecules of ATP in a sequence of reactions called the *electron transport chain*. The sequence starts with the oxidation of NADH and ends with the reduction of water. In between is a series of reactions having manageably small changes in free energy. The electron transport chain, found in the mitochondria of eucaryotes, will occupy a portion of our attention in the next chapter.

SUMMARY

4-1 By definition, a catalyst (1) is effective in trace amounts, (2) is unchanged by the reaction, and (3) affects only the rate, not the equilibrium position, of a reaction. Enzymes have all these properties but, in addition, exhibit a very great selectivity in their catalysis, a result of the fact that their active site is designed to bind and affect only molecules with a particular atomic arrangement.

Enzymes increase the rate of reactions by effectively lowering the needed amount of activation energy. Their function depends on a particular spatial configuration, especially at their active sites where substrates are bound. The actual catalytic mechanism may involve (1) an enhanced collision frequency or improved collision orientation of bound substrates, (2) the solvent properties of the active site (more or less polar, effectively more acidic or basic), (3) an unstable covalent enzyme–substrate intermediate, or (4) strains induced in the substrate by conformational shifts in the protein after the substrate is bound.

4-2 The simplest kind of enzyme is one that exhibits Michaelis-Menten kinetics [see equation (4-3) or (4-4)]. Two useful parameters of such an enzyme are K_M, its Michaelis constant, and v_{max}, its maximum velocity. The maximum velocity represents the greatest rate at which an enzyme could work in a given

environment at infinite substrate concentration. The Michaelis constant is the substrate concentration necessary to achieve exactly half of that maximum rate.

Some enzymes with more than one active site have a velocity profile that is sigmoidal rather than hyperbolic. Although the Michaelis-Menten scheme assumes that binding sites are independent of each other, the existence of sigmoidal kinetics is taken as evidence for a cooperative interaction among the sites. Within a certain range of substrate concentration, such enzymes exhibit a very large change in reaction velocity for a given change in substrate. The basis for cooperativity is a conformational change in the enzyme such that the binding of substrate (ligand) to one site increases the affinity of the other sites for substrate.

4-3 In feedback regulation, the catalytic rate of an enzyme is controlled by a substance (amino acid, nucleotide, etc.) that it helps to make. Control is generally accomplished through an inhibition of the enzyme, so that increasing levels of inhibitor depress the activity of the enzyme while falling levels release the inhibition. Thus, the inhibitor concentration tends to be maintained at a stable value.

In the simplest kind of enzyme inhibition, competitive inhibition, substrate and inhibitor compete for

175

the same site. In noncompetitive inhibition, the inhibitor is bound to a site separate from the active site. A special kind of noncompetitive inhibition, allosteric inhibition, results from the cooperative kinetics of some enzymes: an inhibitor—bound at a regulatory rather than a catalytic site—may foster a conformational change that reduces the binding capacity for substrate. Allosteric inhibition has its counterpart in allosteric activation, where binding of a regulatory substance causes an enzyme to prefer the conformation in which its active sites are most attractive to substrate.

4–4 Energy yielding reactions, such as the capture of light or the breakdown of foodstuffs, usually result in the phosphorylation of ADP to form ATP. The object is to store chemical energy in a retrievable form so that unfavorable reactions can be "driven" by coupling them to ATP hydrolysis. ATP can serve in this capacity because the hydrolysis of its pyrophosphate-type (acid anhydride) linkages is energetically very favorable.

4–5 Metabolic pathways always involve numerous steps, each catalyzed by its own specific enzyme and characterized by a free energy change that is rarely more than a few (e.g., 10) kcal/mole, negative or positive. This stepwise progression has the following advantages: (1) It permits an efficient coupling to ATP, to conserve energy from favorable reactions or to drive unfavorable ones; (2) it provides intermediates that may be used by other pathways; (3) it provides for flex-

ible control, achieved by adjusting the concentration or activity of individual enzymes.

Glycolysis is an example of one metabolic (in this case catabolic) pathway. It is the anaerobic transformation of glucose to lactic acid (lactate ion at physiological pH). Glycolysis is the only way cells from higher organisms can obtain useful energy (ATP) from glucose when oxygen is not available or cannot be used. Two of the steps in glycolysis are coupled to the hydrolysis of ATP, and two to its synthesis. The latter two operate on three-carbon fragments from the original glucose molecule, however, so that a net synthesis of two ATP is realized per glucose—a conservation of about one-third of the available free energy.

4–6 An oxidation is the removal of one or more electrons from a substance; a reduction is the addition of electrons. The glycolytic pathway has two oxidation–reduction steps: the first produces NADH from NAD^+; the second restores the NAD^+ again. Other mechanisms can also be used to regenerate NAD^+, however, including the reduction of pyruvate to ethanol (alcoholic fermentation) instead of to lactate. But the most efficient way of regenerating NAD^+ is to transfer electrons from NADH to oxygen via the electron transport chain, creating water and producing ATP from energy released by the transfer. NAD^+, then, serves as an electron carrier, using the electrons produced by an oxidation in one place to reduce another substance in another place. It is one of several classes of electron carriers available to the cell.

STUDY GUIDE

4–1 (a) What is an "energy of activation?" How does it affect the rate of a reaction, and why? (b) What is a catalyst? Does it affect $\Delta G°$ or $\Delta G°^{\ddagger}$ of a reaction? (c) To what does an enzyme owe its catalytic specificity? (d) Describe the four possible catalytic mechanisms mentioned. Which are consistent with the lock and key theory, and which with induced fit?

4–2 (a) Define K_M and v_{max} in terms of the Michaelis–Menten equation and in words that give a physical meaning to the values. (b) Enolase catalyzes the dehydration of 2-phosphoglycerate (2-PGA), forming phosphoenolpyruvate (PEP):

$$\begin{array}{c} COO^- \\ | \\ HC-OPO_3^{2-} \\ | \\ HOCH_2 \end{array} \longrightarrow \begin{array}{c} COO^- \\ | \\ C-OPO_3^{2-} + H_2O \\ || \\ CH_2 \end{array}$$

The following data might be collected on this enzyme at 25 °C (a mM is a millimole/liter or 10^{-3} M):

v	[2-PGA]
0.010 micromoles/min	0.01 mM
0.018	0.02
0.030	0.04
0.042	0.07
0.050	0.10
0.060	0.16
0.075	0.40
0.083	1.00

The reaction vessel contained one microgram of enzyme in a volume of one milliliter. Make a Lineweaver-Burk plot, and then calculate as follows: (1)

Find the "specific activity" (v_{max}) in international units per mg. (1 I.U. = 1 micromole of substrate consumed per minute.) (2) Find the K_M for 2-PGA. [Ans: 90 units/mg; 0.08 mM] (c) What is meant by "cooperativity" in an enzyme, and to what do we attribute it? (d) What advantages are conferred by cooperative kinetics?

4-3 (a) What is the purpose of feedback regulation? How does it work? (b) What distinguishes competitive from noncompetitive inhibition? How is allosteric inhibition different from other noncompetitive inhibitions? (c) Define "positive allosteric effector" and "negative allosteric effector." How do they work? (d) The following table contains three sets of data such as might be obtained from an enzyme-catalyzed reaction with 10 μg enzyme (mol. wt. = 10^5 g/mole) in 1 ml of assay mixture. The first set of velocities is for the enzyme–substrate reaction alone, while the other two represent the reaction in the presence of two different inhibitors. Graph v vs. [S] and $1/v$ vs. $1/$[S] with all the data presented on each graph. (That is, each graph will have three curves.) (1) What are the turnover number and K_M of the enzyme? (Specify units.) (2) Does either inhibitor compete with the substrate for the catalytic site? If so,

which, and how do you know? (3) If the inhibitor concentration is 1 mM in each case, what are the values for their respective K_I's? [Ans: (1) 10^5 moles of substrate per mole of enzyme per min, 3 mM; (3) 0.66 mM for A, 0.5 mM for B]

4-4 (a) What is a coupled reaction, and why is it useful? (b) What properties of ATP make it a useful intermediate in the exchange of energy? (c) Assume that a person consumes 2400 kcal/day in food energy, a third of which is captured in the form of ATP. At 8 kcal/mole for ATP hydrolysis and 550 g/mole for its molecular weight, what will be the total amount, in grams, of ATP synthesized per day? [Ans: 55,000 g] (d) There is roughly 0.005 mole of ATP per kg of body weight. Using this figure, how much ATP is found in a 70 kg man at any one time, and how many times per day, on the average, is a given ATP molecule cycled to ADP and back (see preceding problem)? [Ans: 192 g; about 286 times]

4-5 (a) What advantages are derived by the cell from breaking a metabolic pathway into numerous steps? (b) What would be the energy yield from glycolysis, in terms of net ATP synthesis, in an organism with a mutant (inactive) triose phosphate isomerase? (c) What reasons can be given for the extremely favorable free energy change accompanying hydrolysis of phosphoenolpyruvate to pyruvate? (d) How is glycolysis regulated?

4-6 (a) Define oxidization; reduction; dehydrogenation; electron carrier. (b) Name two common electron carriers.

[S] (mM):	1	2	5	10	20
(v μmoles/min no inhibitor)	2.5	4.0	6.3	7.6	9.0
(v μmoles/min with inhibitor A)	1.17	2.10	4.00	5.70	7.20
(v μmoles/min with inhibitor B)	0.77	1.25	2.00	2.50	2.86

REFERENCES

ENZYMES AND THE MECHANISM OF CATALYSIS

FOTTRELL, PATRICK F., "Functions and Applications of Isoenzymes." *Science Progress* (Oxford), 55:543 (1967).

GUTFREUND, H., and J. R. KNOWLES, "The Foundations of Enzyme Action." *Essays in Biochem.*, 3:25 (1967).

HOARE, D. G., "Significance of Molecular Alignment and Orbital Steering in Mechanisms for Enzymatic Catalysis," *Nature*, 236:437 (1972).

KOSHLAND, D. E., JR., K. W. CARRAWAY, G. A. DAFFORN, J. D. GASS, and D. R. STORM, "The Importance of Orientation Factors in Enzymatic Reactions." *Cold Spring Harbor Symp. Quant. Biol.*, 36:13 (1971).

MCELROY, W. D., M. DELUCA, and JAMES TRAVIS, "Molecular Uniformity in Biological Catalysis." *Science*, 157:151 (1967). Conformational changes that occur when substrates are bound to enzymes.

NEURATH, H., K. A. WALSH, and W. P. WINTER, "Evolution of Structure and Function of Proteases." *Science*, 158:1639 (1967)

PHILLIPS, DAVID C., "The Three-Dimensional Structure of an Enzyme Molecule." *Scientific American*, November 1966. (Offprint 1055.) Lysozyme and its catalytic activity.

REUBEN, JACQUES, "Substrate Anchoring and the Catalytic Power of Enzymes.' *Proc. Nat. Acad. Sci.* (U. S.), 68:563 (1971).

Sumner, James B., "The Isolation and Crystallization of the Enzyme Urease," *J. Biol. Chem.*, **69**:435 (1926). Reprinted in *Great Experiments in Biology*, edited by M. L. Gabriel and S. Fogel. Englewood Cliffs, N. J.: Prentice Hall, 1955, p. 50, The first isolation of an enzyme.

Wang, Jui H., "Facilitated Proton Transfer in Enzyme Catalysis." *Science*, **161**:328 (1968).

REGULATORY MECHANISMS

Atkinson, D. E., "Biological Feedback Control at the Molecular Level." *Science*, **150**:851 (1965).

Gerhart, J. C., "A Discussion of the Regulatory Properties of Aspartate Transcarbamylase from *Escherichia coli*." *Current Topics in Cellular Regulation*, **2**:276 (1970).

Gerhart, J. C., and A. B. Pardee, "The Enzymology of Control by Feedback Inhibition." *J. Biol. Chem.*, **237**:891 (1962).

Gerhart, John C., and H. K. Schachman, "Distinct Subunits for the Regulation and Catalytic Activity of Aspartate Transcarbamylase." *Biochemistry*, **4**:1054 (1965).

Koshland, D. E., Jr., "A Molecular Model for the Regulatory Behavior of Enzymes." In *The Harvey Lectures, 1969–1970 (Series 65)*. New York: Academic Press, 1971, p. 33.

———, "Protein Shape and Biological Control." *Scientific American*, October 1973. (Offprint 1280.) Conformational changes in proteins.

Monod, Jacques, "From Enzymatic Adaptation to Allosteric Transitions." *Science*, **154**:475 (1966). Nobel Lecture, 1965.

Monod, J., J.-P. Changeux, and F. Jacob, "Allosteric Proteins and Cellular Control Systems." *J. Mol. Biol.*, **6**:306 (1963).

Newsholme, E. A., and C. Short, *Regulation in Metabolism*. New York: John Wiley and Sons, Inc., 1973. Emphasis on enzymatic regulation.

Pardee, A. B., "Control of Metabolic Reactions by Feedback Inhibition." In *The Harvey Lectures, 1969–1970 (Ser. 65)*. New York: Academic Press, 1971, p. 59.

ATP

Alberty, Robert A., "Effect of pH and Metal Ion Concentration on the Equilibrium Hydrolysis of Adenosine Triphosphate to Adenosine Diphosphate." *J. Biol. Chem.*, **243**:1337 (1968).

———, "Thermodynamics of the Hydrolysis of Adenosine Triphosphate." *J. Chemical Education*, **46**:713 (1969).

Kalckar, Herman M., "High Energy Phosphate Bonds: Optional or Obligatory." In *Phage and the Origins of Molecular Biology*, edited by J. Cairns, G. Stent, and J. Watson. Cold Spring Harbor (New York) Lab. of Quant. Biol., 1966, p. 43.

———, "Lipmann and the 'Squiggle.'" In *Current Aspects of Biochemical Energetics*, edited by N. Kaplan and E. Kennedy. New York: Academic Press, 1966, p. 1. On Lipmann's advocacy of a "high energy" phosphate.

Lipmann, Fritz, "Metabolic Regulation and Utilization of Phosphate Bond Energy." *Adv. in Enzymology*, **1**:99 (1941). The original concept of a "high energy phosphate."

van Niel, C. B., "Lipmann's Concept of the Metabolic Generation and Utilization of Phosphate Bond Energy: A Historical Appreciation." In *Current Aspects of Biochemical Energetics*, edited by N. O. Kaplan and E. P. Kennedy. New York: Academic Press, 1966, p. 9.

CHAPTER 5

Mitochondria and Chloroplasts

5-1   THE MITOCHONDRION   180
5-2   ELECTRON TRANSPORT AND OXIDATIVE
       PHOSPHORYLATION   185
       The Electron Transport Chain
       Oxidative Phosphorylation
       Chemical Coupling
       Electrochemical Coupling
       Conformational Coupling
       The Stoichiometry of Oxidative Phosphorylation
5-3   THE CITRIC ACID CYCLE   194
       Entry to the Cycle
       The Cyclic Oxidation of Acetate
       The Advantages of Aerobic Catabolism
5-4   THE RELATIONSHIP BETWEEN MITOCHONDRIAL
       STRUCTURE AND FUNCTION   199
5-5   THE CHLOROPLAST   200   skip
5-6   THE PHOTOSYNTHETIC LIGHT REACTION   207   skip
       The Photoelectric Effect
       Light Capture in Chloroplasts
       The Dual Pigment System
       Photophosphorylation
       The Stoichiometry and Efficiency of Photosynthesis
5-7   THE PHOTOSYNTHETIC DARK REACTION   213   skip
       Carbon Dioxide Fixation
       The Calvin Cycle
       Alternate Pathways of Carbon Fixation
5-8   PHOTOSYNTHESIS IN PROCARYOTIC CELLS   219   skip
5-9   ORIGIN OF MITOCHONDRIA AND CHLOROPLASTS   221
SUMMARY   226
STUDY GUIDE   227
REFERENCES   228

The typical eucaryotic cell depends on mitochondria or chloroplasts to supply most of its energy. Chloroplasts, found only in plant cells, capture light energy and transform it to chemical energy in the form of ATP and reduced coenzymes, a process called photosynthesis. Mitochondria, formerly called *chondriosomes* or, in muscle, *sarcosomes,* are found in both animal and plant cells. They oxidize reduced coenzymes and certain other organic molecules, using oxygen from the air as an electron acceptor. Oxidation releases energy, a portion of which is captured in the form of ATP.

Many of the reactions associated with mitochondria and chloroplasts are also found in some procaryotes, where they are catalyzed by elements associated with the plasma membrane. Most procaryotes can oxidize organic material, and the blue-green algae and a few species of bacteria are capable of photosynthesis. On the other hand, there are a few specialized eucaryotic cells that can do neither. Mature mammalian red blood cells, for example, have no mitochondria and so can obtain energy only from glycolysis.

Thus, in spite of some exceptions, the mitochondria and chloroplasts are in general the "powerhouses" of most eucaryotic cells. As you will see, these two organelles have a great deal in common.

5-1 THE MITOCHONDRION

Mitochondria were first isolated in 1888 by A. Kölliker, who demonstrated that they are enclosed by their own membrane and are not directly connected to any portion of the cell itself. A couple of years later, Altmann developed a staining procedure that permits mitochondria to be observed in intact cells. He concluded that they are autonomous organisms living within the cytoplasm of their host. Though many doubted this theory, there is now the suspicion that mitochondria are, in fact, the highly evolved descendants of bacteria that long ago lived symbiotically within other cells.

In 1912, B. F. Kingsbury suggested that mitochondria might be the sites of cellular respiration—that is, the sites of oxygen utilization. Although he was quite right, it was not until about 1950 that mitochondria received full recognition for what they are: not only the site of respiration, but also the source of nearly all the ATP produced in aerobic animal cells (see Table 5.1).

Since mitochondria are a major source of cellular ATP, we should not be surprised to learn that the number of them per cell, as well as their intracellular location, varies with the type of cell and with its metabolic state. Cells with a large ATP requirement tend to have more mitochondria (e.g., 1,000 in a liver cell) and, what is more, the mitochondria are likely to be clustered where the ATP is needed most. For example, heavily utilized muscles have cells containing large numbers of mitochondria, most of which are found aligned adjacent to the contractile elements (see Fig. 5–1).

Mitochondria are semiautonomous and self-replicating. That is

TABLE 5.1. Some Early Milestones in Respiratory Research

ca 1500	Leonardo da Vinci likens animal nutrition to the burning of a candle.
1770–74	Priestly shows that oxygen is consumed by animals.
1780	Lavoisier and Laplace conclude that the respiration of animals is an oxidation ("burning").
1857	Kölliker discovers mitochondria in muscle.
1872	Pluger finds that oxygen absorbed by the lungs is actually consumed in part by all tissues.
1886	MacMunn discovers the cytochromes (originally "histo-hematins").
1888	Mitochondria are isolated by Kölliker.
1890	Altmann finds a specific stain for mitochondria and suggests that they are autonomous organelles.
1912	Warburg demonstrates that iron is essential to respiration.
1923	Keilin shows that cytochromes have an altered oxidation state during respiratory activity.
1928–33	Warburg determines the basic structure of heme.
1931	Engelhardt shows that phosphorylation is coupled to oxygen consumption.
1933	Keilin succeeds in the partial reconstitution of an electron transport chain.
1937	Krebs formulates the citric acid cycle.
1937–41	Kalckar and Belitser each devise ways of quantitatively studying oxidative phosphorylation.
1939–41	Lipmann proposes a central metabolic role for ATP.
1947–50	Lipmann and Kaplan determine the structure of coenzyme A.
1948–50	Kennedy and Lehninger show that the citric acid cycle, oxidative phosphorylation, and fatty acid oxidation take place in the mitochondria.
1951	Lehninger shows that oxidative phosphorylation requires electron transport.
1954	Palade describes the *cristae mitochondriales*, or crista membranes.

their replication is not necessarily synchronized with that of the cell itself, leading to fluctuations in their number. In addition, they contain DNA that apparently directs the synthesis of some,

FIGURE 5–1. Mitochondrial Placement. One is apt to find mitochondria (M) at locations where ATP requirements are greatest. In this frog muscle, for example, they are aligned adjacent to the contractile elements. [Courtesy of C. Franzini-Armstrong, *J. Cell Biol.*, **47**:488 (1970).]

but not all, of the mitochondrial proteins. However, though they may replicate independently of the cell and contain some genes, mitochondria definitely are not complete life forms. Rather, they are a true part of the cell in which they are found, and their characteristics are largely determined by the cell's nuclear genes.

Mitochondria are about the size of many common bacteria. They usually appear as cylinders 0.5 to 1 μ in diameter and 3 to 10 μ long, but both their shape and dimensions are highly variable. Some appear round, others Y-shaped, and in rare cases they may be as long as 40 μ (see Fig. 5–2).

Mitochondria are bounded by a 60 Å outer membrane, separated from an inner membrane of the same thickness by a gap, the *outer space*, of about 80 Å. This second membrane is folded inward at

FIGURE 5–2. Mitochondrial Size. Even within the same cell, there can be a very wide range of mitochondrial sizes. This intestinal absorptive cell from a rat shows a tenfold range, from 0.5 to 5 μ in the long axis. [Courtesy of H. I. Friedman and R. R. Cardell, Jr., *J. Cell Biol.*, **52**:15 (1972).]

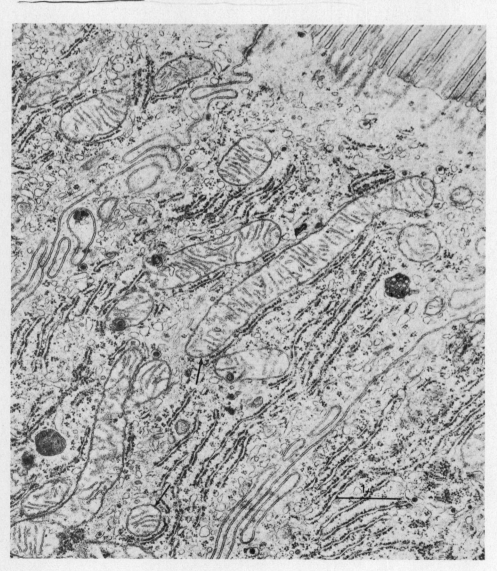

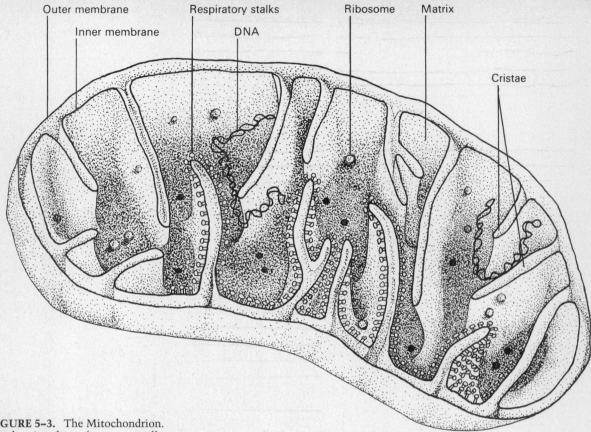

FIGURE 5–3. The Mitochondrion. The drawing shows features usually associated with mitochondria. Note that cristae are formed from folds in the inner membrane. The subunits are respiratory assemblies.

numerous points to form *cristae* that project toward the interior (*matrix*) of the organelle (see Figs. 5–3 and 5–4). These folds divide the matrix into compartments that are only partially separated from one another. The gap between the two layers of a crista is an open core called the *intracristal space*, a region that is continuous with the outer space already defined.

The cristae have closely packed mushroomlike projections that point into the matrix. It has been proposed that each projection is part of a 200 Å *elementary particle*, or *respiratory assembly*, con-

FIGURE 5–4. Mitochondrial Structure. **(a)** From a strain of tumor cells (Ehrlich ascites) grown in the abdominal cavity of a mouse. **(b)** From the adrenal cortex of a rat. The cristae of adrenal cortical mitochondria (arrows, shown in cross-section) are tubular ("tubulovesicular") rather than plate-like. Ribosome-sized granules can be seen in the matrix and in the surrounding cytoplasm. The abundance of smooth endoplasmic reticulum (SER) and lipid (L) reflects the synthesis and secretion of steroids. **(c)** Freeze-etched view of two rabbit heart mitochondria. Note cristae in the specimen to the left. The outer membrane (double arrow) of the mitochondrion on the right is torn away at one point to reveal its particle-studded inner membrane (single arrow). The circled arrow indicates the direction from which metal was deposited. [(a) courtesy of E. E. Gordon and J. Bernstein, *Biochim. Biophys. Acta,* **205**:464 (1970); (b) courtesy of D. S. Friend and G. E. Brassil, *J. Cell Biol.,* **46**:252 (1970); (c) courtesy of J. M. Wrigglesworth, L. Packer, and D. Branton, *Biochim. Biophys. Acta,* **205**:125 (1970).]

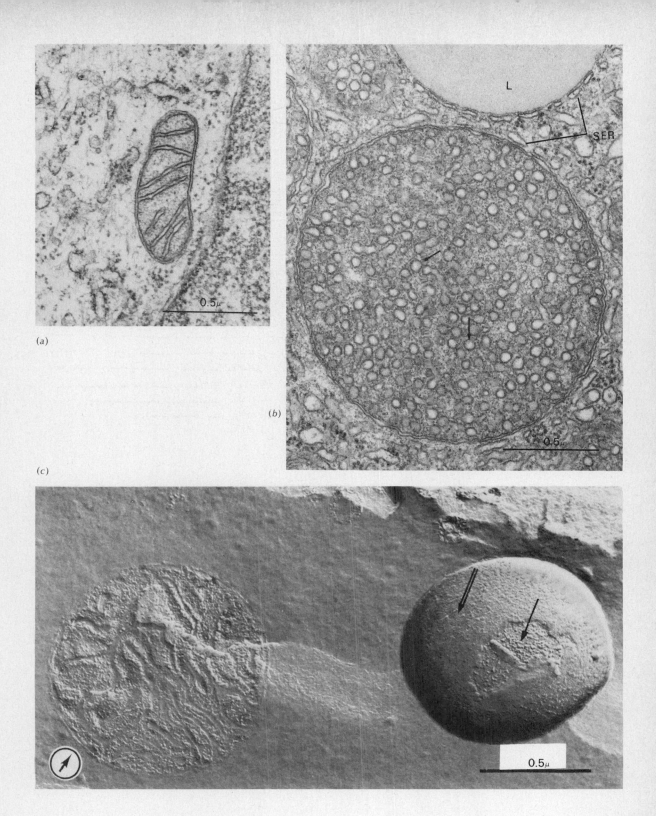

(a)

(b)

L

SER

(c)

185

sisting of an 80 to 100 Å head piece (or *inner membrane sphere*), a stalk some 30 to 40 Å in diameter, and a base piece that is part of the crista membrane itself (see Fig. 5–5). These units contain many of the elements involved in respiration and ATP synthesis, and are assisted in this task by soluble enzymes found mostly in the matrix. The matrix also contains DNA and ribosomes, both of which are distinctly different from their counterparts found elsewhere in the eucaryotic cell.

The job of the mitochondrion is to oxidize coenzymes, pyruvate, fatty acids, and certain of the amino acids, using oxygen as an electron acceptor. Carbon dioxide, water, and ATP are produced. We will begin with a discussion of the mechanism whereby "reducing power," in the form of NADH and $FADH_2$, is utilized by mitochondria to generate ATP.

5–2 ELECTRON TRANSPORT AND OXIDATIVE PHOSPHORYLATION

It was pointed out in the last chapter that oxidation could be a source of considerable energy. It was noted that if the glycolytic pathway were stopped at pyruvate, there would be a net yield of

FIGURE 5–5. Structure of the Cristae. LEFT: Negative-stained side view of isolated crista from a beef heart mitochondrion. Note respiratory assemblies (EP). They are not usually seen in sectioned whole mitochondria (Fig. 5–4), leading to the suggestion that the stalks are normally retracted within the main part of the membrane. RIGHT: Rat liver mitochondrial crista. Face view in a freeze-etched (unfixed, unstained) preparation. Note that the particles frequently appear in rows (arrow). Circled arrow shows direction of metal deposition. [(Left photo), courtesy of H. Fernández-Morán, T. Oda, P. V. Blair, and D. E. Green, *J. Cell Biol.*, **22**:63 (1964); (right photo), courtesy of J. M. Wrigglesworth, L. Packer and D. Branton, *Biochim. Biophys. Acta*, **205**:125 (1970).]

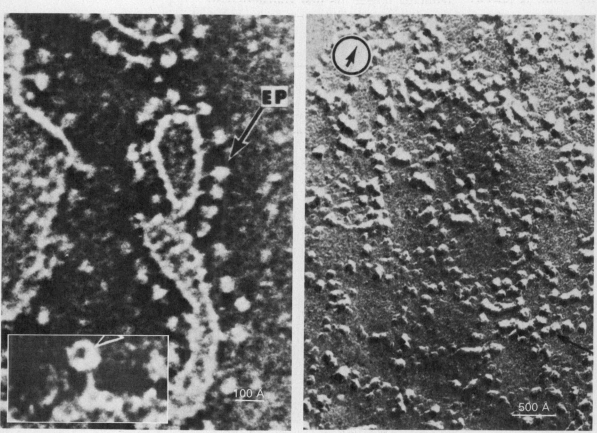

two ATP and two NADH for every glucose molecule consumed. This is an important observation, because the oxidation of NADH, using molecular oxygen as an electron acceptor, has a $\Delta G°$ of -52 kcal/mole. This oxidation (the regeneration of NAD^+), is accomplished in the mitochondria of eucaryotic cells and is used to drive the production of ATP.

THE ELECTRON TRANSPORT CHAIN. The transfer of electrons from NADH to oxygen takes place in a multistep reaction catalyzed by a group of molecules known collectively as the *electron transport chain* or *respiratory chain*. The components of the chain are associated with the cristae of mitochondria and the plasma membrane of aerobic procaryotes. In addition to $NAD^+/NADH$, the activity of the respiratory chain depends on the reversible oxidation of flavins (riboflavin derivatives), quinones, cytochromes, and the nonheme metal ions of other proteins.

The reaction catalyzed by the electron transport chain is either

$$NADH + H^+ + \tfrac{1}{2}O_2 \longrightarrow NAD^+ + H_2O \qquad (5-1) \qquad \Delta G° = -52.6 \text{ kcal/mole}$$

or

$$FADH_2 + \tfrac{1}{2}O_2 \longrightarrow FAD + H_2O \qquad (5-2) \qquad \Delta G° = -43.4 \text{ kcal/mole}$$

If the input is NADH, "reducing power" is first transferred to a flavoprotein (a flavin covalently bound to a protein) by the redox reaction

$$NADH + H^+ + FAD \longrightarrow NAD^+ + FADH_2 \qquad (5-3) \qquad \Delta G° = -9.2 \text{ kcal/mole}$$

The reduced flavin, whether it arises from the oxidation of NADH or by some other reaction, is reoxidized by the next component in the chain; the reduced form of that component gets oxidized by the next element, and so on. The effect is to transfer electrons from one molecule to another through small steps of $\Delta G°$ until the electrons are passed to molecular oxygen, completing equation (5-1) or (5-2) above. If the free energy change for all the individual steps is added, the total will be the value shown.

The exact order of components in the electron transport chain may not be rigidly fixed. A widely accepted, if somewhat simplified, scheme is as follows:

$$(5-4)$$

Co Q (coenzyme Q) is a quinone, called ubiquinone, that is similar to the plastoquinones of chloroplasts. It is followed by heme-containing proteins called cytochromes, the last of which, the a/a_3 complex, is called *cytochrome oxidase* because of its ability, alone among the cytochromes, to reduce molecular oxygen—i.e., to be oxidized by it. (Some of the subunits in this complex seem to be specified by nuclear genes and some by mitochondrial genes.)

	R_1	R_2	R_3	n	Oxidized Quinone	Reduced Quinone
Ubiquinones	CH_3O-	CH_3O-	CH_3-	6-10		
Plastoquinones	CH_3-	CH_3-	$H-$	6-10		

The electron transport chain, then, is the name given to the collection of molecules which, through reversible oxidations and reductions, pass electrons from reduced coenzymes to oxygen. Several of the individual redox steps are highly exergonic, as is the combined reaction. A portion of this released energy is used to drive the phosphorylation of ADP in a process that is called *oxidative phosphorylation* in order to emphasize its dependence on oxygen and to distinguish it from *substrate level phosphorylations*, such as those found in the glycolytic pathway.

OXIDATIVE PHOSPHORYLATION. The function of the electron transport chain is to use oxygen as an electron acceptor in the regeneration of FAD or NAD^+ and to make energy available for ATP synthesis. In intact mitochondria, oxygen is not consumed unless ADP and inorganic phosphate are present. This dependence represents an obvious control mechanism, for it means that the rate of respiration is limited by the amount of ATP required. In addition, it indicates that electron transport and mitochondrial ATP synthesis are tightly coupled processes. They are, however, separate processes, as one can demonstrate with certain poisons—like dinitrophenol—that allow electron transport to continue without ATP production.[1]

In spite of several decades of intense investigation, we do not know exactly how the activity of the electron transport chain causes ADP to be phosphorylated. One of the difficulties is that

---

1. When the existence of uncouplers was first discovered, certain enterprising companies felt that the less toxic ones might be the perfect reducing aid—food could be literally burned up, releasing the energy as heat instead of making ATP available to synthesize fat. The principle is sound enough, for it is the way in which adipose tissue regularly generates body heat. Unfortunately, however, all the really effective synthetic uncouplers are either terribly toxic or—including some that were marketed for a while—tend to promote the growth of tumors.

all the individual steps cannot yet be isolated and observed *in vitro*. These technical difficulties are due in part to many of the components being insoluble and intimately connected with the inner membranes.

There are three basic theories that attempt to explain how electron transport is coupled to ADP phosphorylation. In chronological order, these theories attribute phosphorylation to the following mechanisms: (1) a direct chemical coupling to certain redox reactions in the chain; (2) the collapse of a voltage and/or pH gradient established through electron transport; and (3) a conformational change in macromolecular components of the mitochondrial membranes, the unstable (high energy) state of which is achieved as a result of electron transport. These basic themes have several variations. They are not accepted with equal enthusiasm by specialists in the area, but each appears to be the best explanation for certain experimental observations while remaining apparently inconsistent with others.

CHEMICAL COUPLING. The oldest theory of oxidative phosphorylation dates back to the work of E. C. Slater and others in the early 1950s. It proposes only direct enzymatic coupling between particular redox steps and phosphorylations. There is reason to believe that the respiratory redox reactions would have to be coupled to the formation of a phosphorylated intermediate other than to ATP itself. The phosphate would then be passed to ADP, regenerating the intermediate. In spite of an intense search, this postulated intermediate has yet to be isolated, a fact that need not discourage proponents of the chemical coupling theory because only trace amounts would be required. Until the intermediate is found, however, there can be no experimental verification of chemical coupling. (This elusive intermediate is thought to be required by the other mechanisms also, but it is less central to those theories.)

The theory of direct enzymatic coupling has several advantages over its competitors. Not the least of these is that it does not propose any new mechanisms but only new applications of reactions for which there are several well-characterized examples. In effect, it proposes nothing more than an enzymatically coupled oxidation and phosphorylation not so very different from the combined phosphorylation and oxidation of glyceraldehyde phosphate to form 1,3-diphosphoglycerate, a reaction from the glycolytic pathway. There are at least three steps in the electron transport chain that are energetically favorable enough to make direct coupling practical: NADH to flavin, cytochrome b to c_1, and cytochrome a/a_3 to oxygen.

ELECTROCHEMICAL COUPLING. The second theory of oxidative phosphorylation originated in 1961 with an English scientist,

P. Mitchell. Referred to as the *chemiosmotic* or *electrochemical theory*, it capitalizes on the observation that whereas electrons are passed smoothly from one component to the next in the respiratory chain, protons are not. Some of the steps require protons and some release them. One can see this in the chain as it was drawn earlier, but additional proton transport might also be achieved by slightly rearranging the order of reactions in the chain or introducing new steps at strategic points.

Mitchell suggested that proton-requiring and proton-releasing reactions are associated with membranes in such a way that proton fixation occurs always on one side while release occurs always on the other. If these membranes (the cristae) are otherwise impermeable to hydrogen ions, then the effect of respiratory chain activity will be to "pump" protons from one side to the other, thereby creating a proton (pH) gradient across the membrane along with a charge imbalance or voltage gradient.

Voltage and pH gradients are each a potential source of energy. Since it takes energy to create them, one should be able to "capture" almost the same amount from their collapse. Mitchell suggested that voltage and pH gradients might drive the phosphorylation of ADP through a membrane-bound ATPase. (An ATPase is an enzyme capable of hydrolyzing or synthesizing ATP.) The enzyme must be arranged through a membrane in such a way that an electrical and/or proton gradient can tilt the ATP/ADP equilibrium in the direction of ATP synthesis.

Since ATP hydrolysis releases protons at physiological pH

$$ATP^{4-} + H_2O \rightleftharpoons ADP^{3-} + P_i^{2-} + H^+ \qquad (5\text{--}5)$$

if a membrane were erected between reactants and products, then an increase in hydrogen ions on the right would shift the equilibrium toward the left. Alternatively, one could imagine that the enzyme is placed within a membrane so that the active site for the protons is by itself on one side while the phosphates are all on the other. Then the synthesis of ATP would reduce the proton gradient, and hence be driven by it just as the hydrolysis of ATP could create the gradient. In addition, the reaction causes a change in the voltage across the membrane, a factor that could also be used as a source of energy to force ATP synthesis, as in the following schematic:

Side with higher initial pH and/or negative potential $P_i^{2-} + ADP^{3-}$ → | Membrane ATPase | ⟶ H^+ / H_2O *Side with low initial pH and/or positive potential*
ATP^{4-} ⟵

The reaction, as it is shown here, will make the left side less negative and the right side less positive, equivalent to transferring one electron from left to right. This reduction of charge difference could be the force that drives the reaction. Note, however, that a low pH on the right would have the same effect.

The electrochemical theory, then, suggests that the activity of the respiratory chain causes the formation of pH and electrical gradients across membranes, and that those gradients provide the energy needed to generate ATP.

CONFORMATIONAL COUPLING. The third theory of oxidative phosphorylation is based largely on the work of David E. Green of the University of Wisconsin. It assumes that macromolecular components within the mitochondrion have at least two conformational configurations with a significant difference in energy between them. The high energy configuration results from electron transport. The free energy change in dropping from a configuration of higher energy to a configuration of lower energy is postulated to be sufficient to force the phosphorylation of ADP. The synthesis of ATP would be carried out by an ATPase that is an integral part of the conformational unit. Since we discussed earlier (Chap. 4) how conformational changes might play important roles in enzymatic catalysis, this postulated mechanism should not seem completely foreign to us. (See Fig. 5–6.)

To summarize the situation concerning oxidative phosphorylation, we are faced with choosing among three theories (and numerous variations thereof), each of which attempts to explain how the free energy released by oxidizing NADH or FADH$_2$ could be put to use in the generation of ATP. As diagrammed in Fig. 5–7, the direct coupling theory proposes one or more enzymatically linked reactions; the electrochemical hypothesis suggests that electrical and concentration gradients are the coupling factors; and the conformational theory supposes that conformational isomerization is the link between respiratory chain activity and ATP synthesis. Each of these proposals has its advocates and protagonists, though the chemiosmotic (electrochemical) theory has probably received the most attention in recent years. Because both conformational changes and ion gradients seem in many experiments to be intimately related in some way to the process of coupling, perhaps both chemiosmotic gradients and conformational changes participate in transferring energy from electron transport to ADP phosphorylation. David Green, who originally proposed the conformational model, has suggested as much, referring to the combination as the *electromechanical model*.

THE STOICHIOMETRY OF OXIDATIVE PHOSPHORYLATION. Oxidative phosphorylation is a "tightly coupled" process in intact mitochondria. As mentioned earlier, oxygen consumption by mitochondria requires the presence of ADP and P$_i$. When an ADP/ATP equilibrium is reached, added ATP will drive the reac-

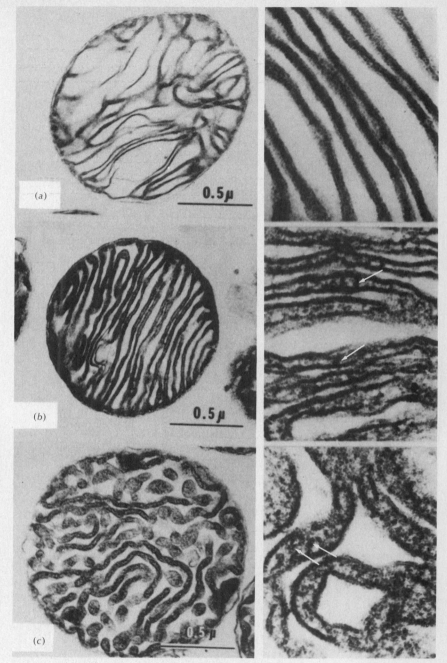

FIGURE 5–6. Conformational Changes in Mitochondrial Cristae. **(a)** The isolated beef-heart mitochondrion is inactive, with swollen cristae that press tightly against each other. (The light area is the intracristal and outer space.) **(b)** Electron transport seems to cause the stalks of the respiratory assemblies (arrows) to push adjacent cristae apart. **(c)** The addition of phosphate to active mitochondria causes an "energized twisted" configuration in which the stalks and head pieces are clearly evident (arrow). [Courtesy of D. E. Green and J. H. Young, *American Scientist*, **59**:92 (1971). High resolution micrographs were taken by T. Wakahayashi.]

tion back the other way—that is, some of the added ATP will be hydrolyzed and the energy used to reduce NAD^+ and create oxygen from water. This observation implies that there is some definite ratio of phosphorylation to oxygen uptake—in other words, a fixed value for "n" in the following reaction, which is the sum of equations (5–1) and (5–5), the latter taken n times:

$$\tfrac{1}{2}O_2 + NADH + (n + 1)H^+ + nADP^{3-} + nP_i^{2-} \longrightarrow$$
$$NAD^+ + (n + 1)H_2O + nATP^{4-} \qquad (5-6)$$

The value for the stoichiometry coefficient, n, which is also the
P/O ratio — the amount of P_i used per oxygen atom reduced — has
not been precisely determined. In isolated mitochondria, n has a
value near 3, but there is some question as to whether one can expect the same value *in vivo*. If the enzymatic coupling hypothesis
is correct, with a direct link between respiration and phosphorylation, n may in fact have the same value *in vitro* as it does *in vivo*.
If, however, the link is indirect, as in the other two theories, then
the environment of the mitochondria might very well alter n.
Since mitochondria can not be isolated without some damage, and
since the *in vitro* environment never really approximates cytoplasm, the value of 3 becomes suspect. Because of these uncertainties, we must admit that n could conceivably be more than 3
and that the process of oxidative phosphorylation could be much
more efficient than is generally assumed. However, whenever we
need a definite number for the amount of ATP produced per
NADH input, this classical value of 3 will be adopted.

The immediate electron acceptor for NADH is a flavoprotein,
with the reaction strongly favoring reduction of the flavin (see
equation (5-3)). The standard state free energy change of

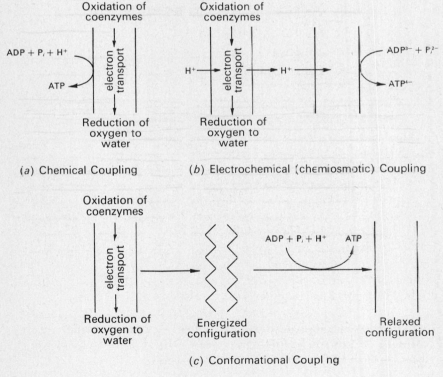

(a) Chemical Coupling (b) Electrochemical (chemiosmotic) Coupling

(c) Conformational Coupling

FIGURE 5-7. The Coupling of Electron Transport to Oxidative Phosphorylation in Mitochondrial Membranes. The coupling is variously attributed to one (or sometimes a combination) of the following: **(a)** a direct
chemical coupling similar to other enzymatically coupled reactions; **(b)**
electrochemical (or chemiosmotic)
coupling, in which voltage and/or proton gradients are produced by electron
transport and utilized to drive ATP
synthesis; and **(c)** conformational coupling, in which an energized configuration of mitochondrial constituents
is fostered by electron transport and
used to synthesize ATP.

−9.2 kcal/mole, plus some experimental measurements, has led to the suggestion that only two ADP are phosphorylated when the input to the respiratory chain is a flavin coenzyme instead of NADH. This ratio, 2 ATP per $FADH_2$, will be used along with 3 ATP per NADH in subsequent calculations.

5–3 THE CITRIC ACID CYCLE

Most of the reduced coenzymes used in oxidative phosphorylation originate within the mitochondrion itself from a cyclic reaction sequence known as the *citric acid cycle* or, since citrate is a tricarboxylic acid, the *tricarboxylic acid (TCA) cycle*. It is also known as *Krebs cycle* in honor of H. A. Krebs, who first proposed it in 1937.

ENTRY TO THE CYCLE. The sequence starts by condensing acetate (obtained from carbohydrates, amino acids, or fatty acids) with a molecule called oxalacetate or oxaloacetate, thus forming citrate.

Acetate   Citrate   Oxalacetate

A total of four oxidations and two decarboxylations follow, regenerating oxalacetate ready to start a new round. The effect is to oxidize acetate completely to carbon dioxide and water, thereby producing NADH and $FADH_2$ for entry into the electron transport chain.

β-Mercaptoethylamine   Pantothenic acid (a vitamin)

Coenzyme A

The complete oxidation (to CO_2) of fatty acids, carbohydrates, and amino acids requires that they first be converted to acetylcoenzyme A (acetyl–SCoA) for entry into the citric acid cycle. Only the carbohydrates will be considered as a source of acetyl–SCoA, as we have already followed the degradation of the most common carbohydrate, glucose, as far as pyruvate. That degradation was via the Embden-Meyerhof pathway, an anaerobic sequence of reactions that takes place in the cytoplasm outside the mitochondria. When oxygen is available, most cells use the pathway not for glycolysis but to make possible the complete oxidation of glucose.

The change from glycolysis to an aerobic pathway is accomplished by routing NADH and pyruvate to mitochondria instead of using those molecules to produce lactate. In mitochondria, pyruvate becomes the substrate for *pyruvate dehydrogenase*, a complex of about fifty (in *E. coli*) enzyme molecules of three different varieties (Fig. 5–8). The overall reaction results in the *oxidative decarboxylation* of pyruvate, and in its coupling to coenzyme A preparatory to entry into the citric acid cycle.

$$\begin{array}{c} COO^- \\ | \\ C{=}O + HSCoA + NAD^+ \\ | \\ CH_3 \end{array} \longrightarrow \begin{array}{c} SCoA \\ | \\ C{=}O + CO_2 + NADH \\ | \\ CH_3 \end{array} \quad (5\text{-}7) \qquad \Delta G^\circ = -9.4 \text{ kcal/mole}$$

The relatively large drop in free energy makes this reaction virtually irreversible.

THE CYCLIC OXIDATION OF ACETATE. The citric acid cycle itself results in the complete oxidation of acetate, accepted in the form of acetyl-SCoA. The reactions are shown in Fig. 5–9 and Table 5.2. Their sum, for each turn of the cycle, is

$$\text{acetyl–SCoA} \longrightarrow 2CO_2 + HSCoA \qquad (5\text{-}8)$$

$$\text{coupled to} \begin{cases} 3NAD^+ \longrightarrow 3NADH \\ FAD \longrightarrow FADH_2 \text{ (as a reduced flavoprotein)} \\ GDP + P_i \longrightarrow GTP \end{cases}$$

The overall standard state free energy change is -13.0 kcal/mole at pH 7 in the presence of magnesium ions.

Note that after condensing acetate with oxalacetate to form citrate (step 1), and its isomerization to isocitrate (step 2), two *oxidative decarboxylations* reduce the six-carbon tricarboxylic acid, isocitrate, to the four-carbon dicarboxylic acid, succinate, producing two NADH.

The newly produced succinate is left coupled to coenzyme A. The free energy of hydrolysis of the succinate-SCoA thiol ester is used to phosphorylate GDP in step 5 of the cycle. GTP can be

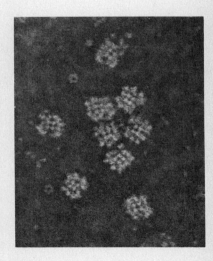

FIGURE 5–8. Pyruvate Dehydrogenase Complex from a Bacterium, *E. Coli*. Negatively stained. The complex diameter is about 300 Å. [Courtesy of L. J. Reed, from *The Enzymes*, Vol. 1, 3d ed., P. D. Boyer, ed. New York: Academic Press, (1970), p. 213.]

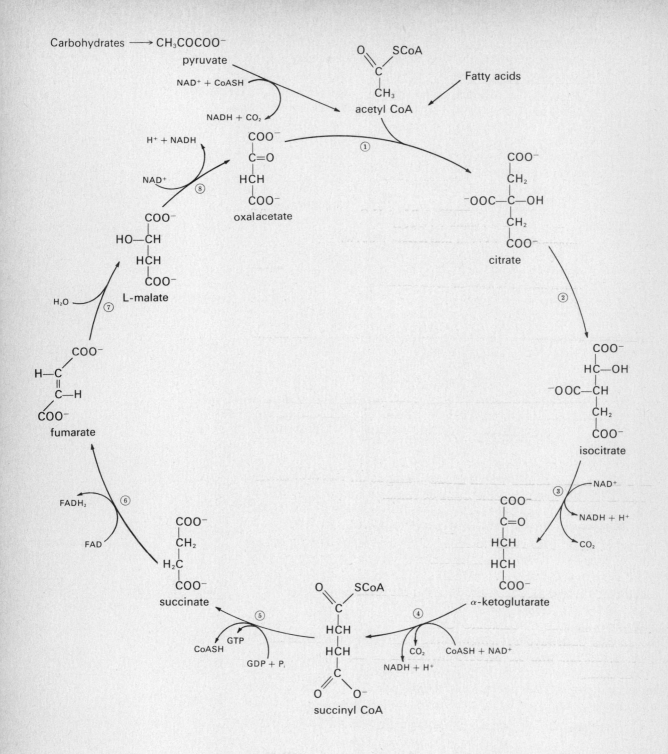

FIGURE 5–9. The Citric Acid Cycle. Also known as the tricarboxylic acid cycle or Krebs cycle. The reactions are described in Table 5.2.

TABLE 5.2.   The Citric Acid Cycle

STEP	REACTION	$\Delta G°$ (pH 7)	ENZYME
1	acetyl-SCoA + oxalacetate^{2-} + $H_2O \rightarrow$ citrate^{3-} + CoASH + H^+	-9.08 kcal/mole	citrate condensing enzyme
2	citrate$^{3-} \rightarrow$ isocitrate^{3-}	$+1.59$	aconitase[a]
3	isocitrate^{3-} + NAD$^+ \rightarrow$ α-ketoglutarate^{2-} + NADH + CO_2	-1.70	isocitrate dehydrogenase
4	α-ketoglutarate^{2-} − CoASH + NAD$^+ \rightarrow$ succinyl-SCoA$^-$ + CO_2 + NADH	-8.82	α-ketoglutarate dehydrogenase[b]
5	succinyl-SCoA$^-$ + GDP^{3-} + $P_i{}^{2-}$ + $H_2O \rightarrow$ succinate^{2-} + GTP^{4-} + CoASH	-0.8	succinyl thiokinase
6	succinate^{3-} + FAD $\rightarrow$ fumarate^{2-} + FADH$_2$	0	succinate dehydrogenase
7	fumarate^{2-} + $H_2O \rightarrow$ L-malate^{3-}	-0.88	fumarase
8	L-malate^{2-} + NAD$^+ \rightarrow$ oxalacetate^{2-} + NADH + H^+	$+6.69$ -13.0 kcal/mole	malate dehydrogenase

[a] It was originally thought that Step 2 involves aconitate (hydrated citrate) as an intermediate. Hence the name aconitase for the enzyme.
[b] α-keto glutarate dehydrogenase is a complex containing the same three enzymes as the pyruvate dehydrogenase complex.

used in a number of different energy-requiring reactions (e.g., protein synthesis) or to produce ATP via a mixed-nucleotide kinase:

$$GTP + ADP \rightleftharpoons ATP + GDP \qquad (5-9)$$

Succinate is oxidized to fumarate in step 6 by an enzyme that has a covalently bound flavin group acting as an electron carrier. Water is added to the double bond of fumarate to produce malate (step 7), and the newly created hydroxyl of malate is oxidized to a carbonyl (step 8). This latter step, the fourth and last oxidation of the cycle, again uses NAD$^+$/NADH as the electron carrier.

Thus, the cycle comes full circle, from oxalacetate to oxalacetate, with two carbon atoms added in the form of acetate and two carbon atoms lost (not the same two) in the form of carbon dioxide.

THE ADVANTAGES OF AEROBIC CATABOLISM.   We have now followed two pathways by which energy can be obtained from glucose: glycolysis and its aerobic extension in the mitochondrion. Let us compare the efficiency of these processes. Glycolysis, the anaerobic degradation of glucose, was summarized in the last chapter as

$$glucose + 2ADP + 2P_i \longrightarrow 2 \text{ lactate} - 2ATP \qquad (5-10)$$

The aerobic oxidation of glucose, on the other hand, consists of the following reactions: (1) cytoplasmic oxidation and degradation to pyruvate;

$$\text{glucose} + 2ADP + 2P_i + 2NAD \longrightarrow$$
$$2 \text{ pyruvate} + 2ATP + 2NADH \qquad (5\text{--}11)$$

followed by (2) the pyruvate dehydrogenase complex in the mitochondrion;

$$2 \text{ pyruvate} + 2NAD + 2CoASH \longrightarrow$$
$$2 \text{ acetyl-SCoA} + 2NADH + 2CO_2 \qquad (5\text{--}12)$$

and finally (3) by the degradation of acetyl-SCoA via the citric acid cycle, also in the mitochondrion:

$$2 \text{ acetyl-SCoA} + 2GDP + 2P_i + 2FAD + 6NAD \longrightarrow$$
$$4CO_2 + 2CoASH + 2GTP + 2FADH_2 + 6NADH \qquad (5\text{--}13)$$

In summary, the aerobic oxidation of glucose, equations (5–11) through (5–13), combined with the respiratory chain oxidation of the coenzymes, gives (with glucose written as $C_6H_{12}O_6$)

$$C_6H_{12}O_6 + 6O_2 \longrightarrow 6CO_2 + 6H_2O \qquad (5\text{--}14)$$

and is accompanied by the production of 2 ATP, 2 GTP, 2 FADH$_2$, and 10 NADH. In terms of net ATP equivalents, using the stoichiometry factors introduced previously, glucose oxidation has an energy yield of 38 ATP or, at 8 kcal/mole of ATP, 304 kcal.

When we compare the 38 ATP available from the aerobic oxidation of glucose with the 2 ATP available from glycolysis, the advantages of aerobic growth become obvious. (Pasteur first noted the much greater gain in final mass, per gram of glucose consumed, when yeast cultures grow aerobically instead of anaerobically.) Since the combustion of glucose has a $\Delta G°$ of -686 kcal/mole, the overall process of cellular oxidation of glucose is about 45% efficient.

It should be pointed out, however, that the difference between aerobic and anaerobic oxidation of glucose has a somewhat different meaning in higher animals than it does in microorganisms. Although it is true that lactate, as the end product of glycolysis, is normally excreted from the cell as waste, in multicellular organisms this does not automatically mean that it is passed out of the animal. Rather, lactate produced in red blood cells or vigorously exercising skeletal muscle is carried by the bloodstream to other tissues, such as the liver, where it can be oxidized to pyruvate. From there it can enter the citric acid cycle or be converted back to glucose. The difference in energy between glycolysis and the aerobic oxidation of glucose is therefore not wasted as far as the animal is concerned: it is only the individual cell that fails to derive as much energy from the carbohydrate as it might. This kind of cooperation among various cell types is, of course, one of the advantages of being multicellular.

5–4 THE RELATIONSHIP BETWEEN MITOCHONDRIAL STRUCTURE AND FUNCTION

The outer space (which is continuous with the intracristal space), the matrix, and the two mitochondrial membranes each contribute in a specific way to mitochondrial function (see Fig. 5–10.)

The outer membrane can be removed from the mitochondrion by certain detergents or enzymes, leaving the rest of the organelle intact. Under these circumstances, the cristae disappear as the matrix swells to stretch the inner membrane. In addition, soluble enzymes from the outer space are released, allowing their identification. Among them are some nucleoside phosphorylases, including adenylate kinase which catalyzes an interconversion of the adenine nucleotides.

$$2ADP \rightleftharpoons ATP + AMP \qquad (5\text{–}15)$$

Whereas the outer membrane strongly resembles the plasma membrane and endoplasmic reticulum, the inner mitochondrial membrane is strikingly different in both appearance and composition. It contains, for example, much less lipid, much more protein, and a considerable number of enzymes. A substantial fraction of the inner membrane is a positively charged protein of about 22,000 amu. It rapidly polymerizes when isolated and, because of its charge, readily binds negatively charged constituents such as phospholipids or most cytochromes. (While cytochromes a, b, and c_1 are negatively charged, cytochrome c is positively charged at physiological pH. It does, however, appear to be associated with the inner membrane *in vivo* through hydrophobic interactions.) About a quarter of the total protein of the inner membrane is known to be enzymatically active, consisting in part of the cy-

FIGURE 5–10. Structure–Function Relationship in Mitochondria.

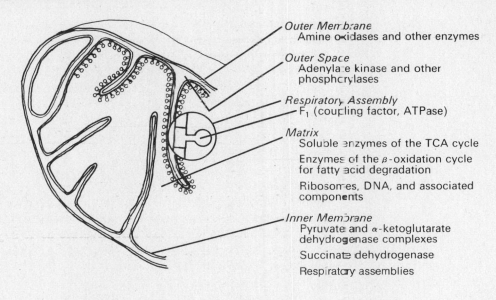

Outer Membrane
Amine oxidases and other enzymes

Outer Space
Adenylate kinase and other phosphorylases

Respiratory Assembly
F_1 (coupling factor, ATPase)

Matrix
Soluble enzymes of the TCA cycle

Enzymes of the β-oxidation cycle for fatty acid degradation

Ribosomes, DNA, and associated components

Inner Membrane
Pyruvate and α-ketoglutarate dehydrogenase complexes

Succinate dehydrogenase

Respiratory assemblies

tochromes and pyruvate dehydrogenase. The inner membrane also includes the proteins responsible for ATP synthesis.

There is reason to believe that proteins of the electron transport chain and oxidative phosphorylation are arranged in discrete clusters or subunits within the inner membrane. These clusters are the *respiratory assemblies* mentioned earlier. Each mushroomlike projection from a respiratory assembly contains *coupling factors* responsible for linking ATP synthesis to electron transport. One of these factors, isolated by E. Racker, is called factor one (F_1) and appears to be the actual ATPase itself. Mitochondria or mitochondrial fragments that lack this protein cannot couple ATP synthesis to electron transport and do not have inner membrane spheres. Adding purified F_1 restores ATP synthesis and the characteristic appearance, implying that F_1 is the sphere itself. Isolated F_1 readily catalyzes ATP hydrolysis, a reaction that presumably occurs in the opposite direction *in vivo*.

The matrix of the mitochondrion is almost a gel, containing roughly 50% protein, including most of the enzymes of the citric acid cycle. The matrix also contains the enzymes responsible for fatty acid oxidation and for a limited amount of protein synthesis, using DNA and ribosomes also found therein.

Thus, the reactions of the mitochondrion are highly compartmentalized. The separation is fostered by the selective permeability of the inner membrane. An isolated inner membrane appears to pass only water, small neutral molecules, and the smaller fatty acids. It does not readily pass nucleotides, coenzymes, or amino acids, and hence can retain these compounds within the matrix at relatively high concentrations. They do get in and out, but only at controlled rates, thanks to the presence of specific transport mechanisms in the inner membrane (see Chap. 6). The high matrix concentration of a given substance should result in an increase in the rate of those reactions that consume it, a factor that is probably important to mitochondrial function. The outer space, on the other hand, more nearly reflects the composition of the cytoplasm as a whole, since the outer membrane is relatively permeable to a wide variety of substances.

This pattern of compartmentalization within an organelle and division of labor between membrane-bound and soluble enzymes is also found in chloroplasts.

5–5 THE CHLOROPLAST

Chloroplasts are a subdivision of the plant organelles called plastids. Like the other plastids, they are formed by differentiation from proplastids, which are colorless, membrane-enclosed bodies having little internal structure. (See Figs. 5–11 and 5–12.) The chloroplasts, in turn, can themselves grow, divide, and in some

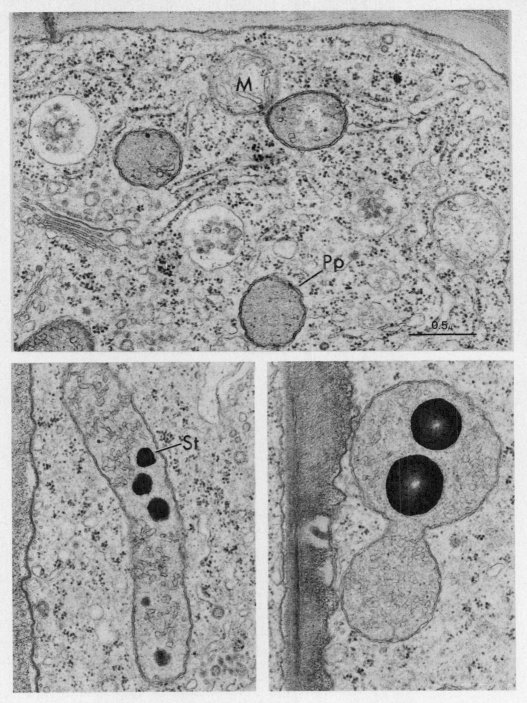

FIGURE 5–11. The Plastids. TOP: Proplastids (Pp) in a sieve element from a bean seedling. Note mitochondrion (M). Note also the numerous ribosomes and (at left) the dictyosome. BOTTOM: Sieve element plastids containing starch grains (St) and numerous cisternae. The plastid on the right appears to be in the process of replication. [Courtesy of B. A. Palevitz and E. H. Newcomb, *J. Cell Biol.*, **45**:383 (1970).]

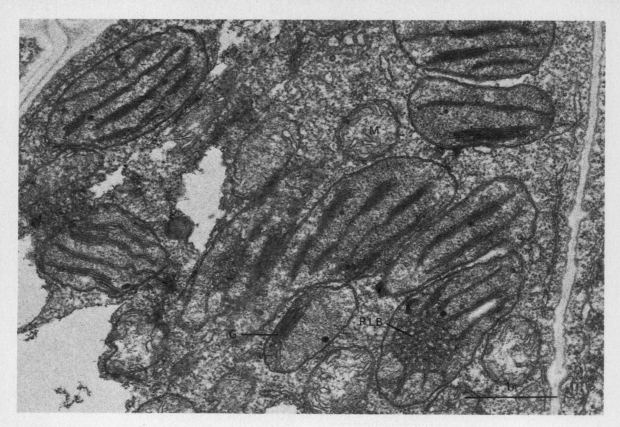

FIGURE 5–12. Immature Chloroplasts. In addition to their capacity for self-replication, chloroplasts may arise from proplastids. An early stage in that process is illustrated here (sugar cane mesophyll cell), showing some elementary grana (G) and a prolamellar body (PLB). Note mitochondrion (M). [Courtesy of W. M. Laetsch, *Sci. Prog.* (Oxford), **57**:323 (1969).]

cases even differentiate to other plastids. This differentiation is observed in the ripening of a tomato as many of the chloroplasts lose their chlorophyll, and hence their green color, to become bright red chromoplasts. In addition to chloroplasts and chromoplasts, the plastid group also includes leucoplasts, which are storage depots for oils, starch, and protein.

Chloroplasts bear a superficial resemblance to mitochondria, although they are usually somewhat larger, typically 2–3 microns thick by 5–10 microns long. They are also biconvex like a lens (see Fig. 5–13). But like mitochondria, chloroplasts have numerous internal membranes and are semiautonomous structures within the cytoplasm of the host cell. That is, their reproduction is not necessarily closely synchronized with the cell itself, since they contain genetic material capable of directing the synthesis of some of their own proteins. They can even be transferred in the laboratory from one cell to another and still continue to function.

As noted in Table 5.3, the ability to capture light energy has been associated by scientists with the green parts of plants since the eighteenth century. The green color comes from chlorophyll, which is localized within the chloroplasts and is the key to pho-

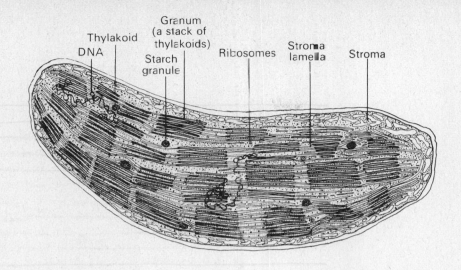

Thylakoid DNA Granum (a stack of thylakoids) Starch granule Ribosomes Stroma lamella Stroma

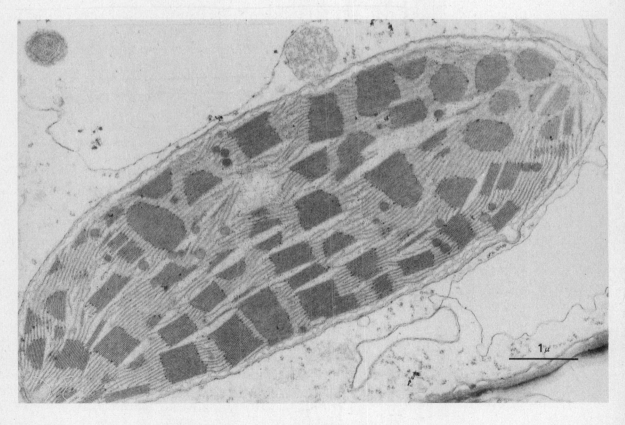

1 μ

FIGURE 5–13. The Mature Chloroplast. TOP: Diagrammatic representation. BOTTOM: A maize leaf chloroplast. [Micrograph from *Brookhaven Symp. Biol.*, **19** 353 (1966), courtesy of L. K. Shumway, Program in Genetics and Dept. of Botany, Washington State Univ.]

TABLE 5-3.   The Early History of Photosynthesis

1727	Hales concludes that plants are nourished, in part, from the atmosphere.
1772	Priestly discovers the evolution of oxygen by plants. He also shows that oxygen is consumed by animals.
1779–96	Ingen–Housz shows that carbon dioxide is consumed by plants as oxygen is evolved, and that light is essential to the process. He associated these reactions with the green parts of plants.
1804	de Saussure notes and measures the fixed stoichiometry between CO_2 consumption and oxygen evolution in plants.
1837	Dutrochet recognizes that chlorophyll is essential to oxygen evolution by plants.
1862	Sachs proves that starch is synthesized by plants in a light-dependent reaction (photosynthesis).
1882	Englemann shows that red light is the most efficient stimulator of photosynthesis.
1883	Meyer first describes details of chloroplast structure.
1913	Wilstätter and Stoll isolate chlorophyll and later determine its structure.
1919	Warburg finds that the efficiency of photosynthesis is higher with intermittent light.
1922	Warburg and Negelein first measure the efficiency of photosynthesis.
1923	Thunberg recognizes that carbon dioxide is reduced and water oxidized during photosynthesis.
1936	Wood and Werkman distinguish between the "light reaction" and the "dark reaction."
1938	Hill finds that isolated chloroplasts evolve oxygen when illuminated, provided that an appropriate electron acceptor is also made available.
1941	Ruben, Randall, Kamen, and Hyde show that the oxygen liberated during photosynthesis comes from water.
1948	Calvin and Benson show that phosphoglycerate is an early product of CO_2 fixation.

tosynthesis. Not all chloroplasts are green, however, as they also contain pigments of other colors—in some cases present in sufficient quantity to mask the green of the chlorophyll.

In 1883, A. Meyer reported that the internal structure of most chloroplasts of higher plants consists of dense cylinders, which he named *grana*, embedded in a lighter material, the *stroma*. More recent work, using the electron microscope, reveals that there are 40 to 80 grana in a typical chloroplast, each composed of layers of membranes that are paired off to form discs or sacks called *thylakoids* (see Fig. 5–14). Each granum usually consists of five to thirty discs, 0.25–0.80 μ in diameter with a dense core about 0.01 μ (100 Å) thick. The membranes or *lamellae* of the discs seem at times to be interconnected, sometimes extending through

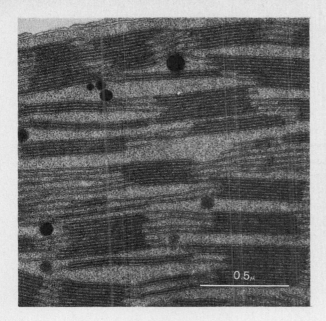

the stroma to adjacent grana. It should be emphasized, however, that this description is not applicable to all chloroplasts. One exception, the dimorphic chloroplast, will be considered later. Another exception is found in algae, as can be seen in Fig. 5–15.

The thylakoid membranes are embedded with a paracrystalline array of small particles called *quantasomes*, and sometimes with additional, even smaller particles. Quantasomes appear to be 180 Å long by 160 Å wide and 100 Å thick, and to consist of four smaller subunits (see Fig. 5–14). Each quantasome has a mass of about 2×10^6 amu, about equally divided between lipid and protein. Chlorophyll itself accounts for a large part of the lipid. A certain fraction of the chlorophyll is bound to protein, although the protein fraction also includes several cytochromes, plastocyanin (a copper-containing protein), and ferredoxin (a protein that contains nonheme iron ions).

The question is how these and other substances are organized to carry on photosynthesis. That process, which in eucaryotes can be summarized by writing the reverse of equation (5–14)

$$6CO_2 + 6H_2O + \text{light energy} \longrightarrow C_6H_{12}O_6 + 6O_2 \quad \textbf{(5–16)}$$

consists of two parts, a *light reaction* and a *dark reaction*. In the former, light energy is used to oxidize water to oxygen, passing the electrons to $NADP^+$ to create NADPH and, at the same time, generating some ATP. The light reaction seems to be centered in the quantasomes. The dark reaction, which is catalyzed by soluble enzymes found in the stroma, consists of those steps by which the "reducing power" of NADPH is used to turn CO_2 into carbohydrate.

FIGURE 5–14. Thylakoids. The chloroplast granum is a stack of discs. LEFT: Side view of several grana. RIGHT: Face view (freeze-etched) of a single disc. The fracture reveals a paracrystalline array of particles, termed *quantasomes*. (Left courtesy of W. M. Laetsch; Right courtesy of R. B. Park and Lawrence-Berkeley Lab., Univ. of Calif.)

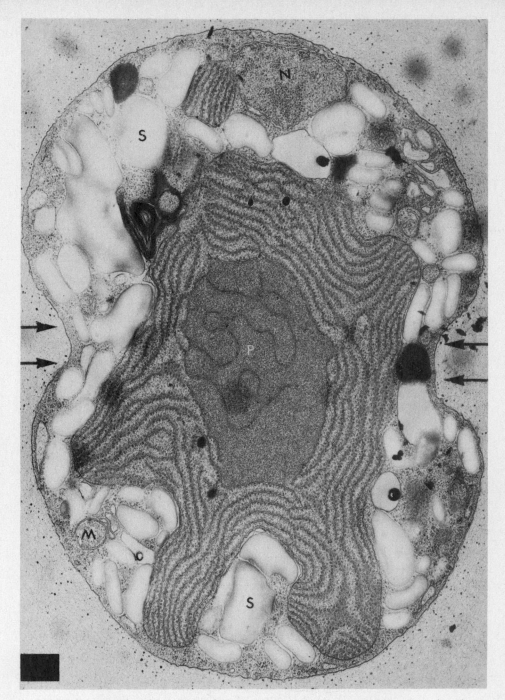

FIGURE 5–15. The Chloroplasts of Algae. A single, large, central chloroplast often dominates the cytoplasm of these cells. A prominant pyrenoid (P), which serves as the site of starch synthesis, is characteristic, as is the lack of grana. The spheres on the chloroplast lamallae are called phycobilosomes. This particular cell was about to divide (arrows). One of the two daughter nuclei (N) is evident. Starch granule (S). Mitochondrion (M). [Courtesy of E. Gantt and S. F. Conti, *J. Cell Biol.*, **26**:365 (1965).]

5–6 THE PHOTOSYNTHETIC LIGHT REACTION

We have seen that the chemical oxidation of glucose to carbon dioxide and water is accompanied by a $\Delta G°$ of -686 kcal/mole. It follows, therefore, that the same amount of energy is needed to synthesize glucose from carbon dioxide and water. In fact, the input of free energy must be even more than 686 kcal/mole if the equilibrium for the overall reaction is to lie significantly on the side of glucose formation. The energy comes from light, for the chloroplast is an energy transducer. It can convert light energy to chemical energy, much as a solar battery might use light to run a transistor radio. Light is an electromagnetic radiation, different from radio, radar, infrared rays (heat rays), X rays, and gamma rays only in the frequency with which the source is oscillating. Visible and ultraviolet light would fall between infrared and X rays in this list, which is ordered to reflect progressively faster frequencies or shorter wavelengths. The frequency of oscillation of the source, ν, is related to the wavelength, λ, of the emitted radiation by the following equation:

$$\lambda = \frac{c}{\nu} \qquad (5\text{–}17)$$

where c is the speed of light, 3×10^{10} cm/sec *in vacuo*. The frequency of the source is measured by the rate at which electrical and magnetic fields of the radiation rise and collapse as radiation passes a given point; wavelength is the distance a wavetrain travels in one complete cycle, or the distance between conjugate points in the oscillating field.

THE PHOTOELECTRIC EFFECT. Einstein suggested in 1905 that light and other electromagnetic radiations travel in discrete packets called *photons*, and that when light interacts with matter, it does so by annihilating complete photons, never a part of one. This hypothesis explained P. Lenard's 1899 observation that electrons ejected from a polished metal plate by light have velocities, and hence energies, that are independent of the intensity of the light. According to Einstein's *photoelectric theory*, it takes one photon to eject one electron. Thus, an increase in the intensity of light, and hence the flux of photons, only increases the *number* of electrons affected, not their *velocities*. On the other hand, changing the wavelength of light does change the velocities of ejected electrons, implying that the energy of a photon must be related to its wavelength. That relationship was found by Max Planck in 1900 to be:

$$E = h\nu = \frac{hc}{\lambda} \qquad (5\text{–}18)$$

where h, called *Planck's constant*, has a value of 1.585×10^{-34} cal-sec. Avogadro's number (6.023×10^{23}, or a "mole") of photons is called an *einstein*. At $\lambda = 700$ nm (7000 Å or 7×10^{-5} cm, a dark red to the eye), one einstein of photons has an energy of about 41 kcal (see Fig. 5–16), calculated as follows:

$$N_{av}E = \frac{N_{av}hc}{\lambda} = \frac{(6.023 \times 10^{23})(1.585 \times 10^{-34})(3 \times 10^{10})}{7 \times 10^{-5}} = 40.8 \text{ kcal}$$

When a molecule absorbs a photon of light, it is absorbing a *quantum* of energy. Several things can happen to that energy: (1) It can be dissipated in molecular motion, manifest as heat. (2) It can be re-emitted as a new photon of light at a longer wavelength, with the shift representing losses to other processes. If re-emission occurs very quickly, it is called *fluorescence*; if there is a long lag (milliseconds to seconds) between absorption and re-emission, the process is called *phosphorescence*. Or (3) the energy of light can cause a chemical change in the compound that absorbs it. It is this latter possibility that interests us, since it is one of the things that can happen when a molecule of chlorophyll absorbs a photon.

When chlorophyll absorbs a photon of light, an electron may be ejected. That electron may be transferred to a molecule of $NADP^+$ and, along with a second electron, cause the reduction of $NADP^-$ to NADPH. Oxidized chlorophyll, of course, then needs an electron donor to resume its original state. In the blue–green algae

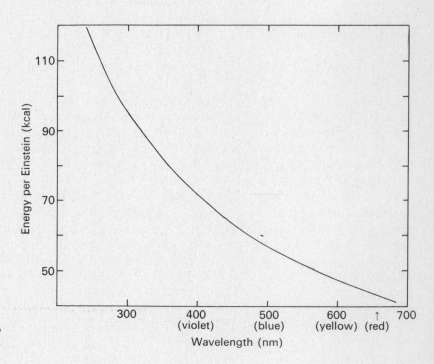

FIGURE 5–16. The Relationship Between Wavelength and Energy.

and higher plants, but not in photosynthetic bacteria, the electron donor is water, which thus becomes oxidized to oxygen.

LIGHT CAPTURE IN CHLOROPLASTS.   According to the above statements, electrons ejected from chlorophyll cause the reduction of $NADP^+$. Ejection is the result of light absorption and the photoelectric effect. We might expect, therefore, that the *action spectrum* of photosynthesis—that is, the relative effectiveness of various wavelengths of light—should follow the absorption spectrum of pure chlorophyll. In fact, this is not the case, as one can see from the graphs in Fig. 5–17. There are two reasons for this discrepancy: (1) chlorophyll *in vivo* does not have the same absorption spectrum as chlorophyll *in vitro*; and (2) other pigments in the quantasomes are capable of transferring energy from absorbed light to chlorophyll, thus considerably broadening the action spectrum toward the middle wavelengths of visible light.

The transfer of energy from pigment to pigment is important to the efficiency of photosynthesis, because it means that a photon of light from almost any part of the visible spectrum (about 400 nm to 700 nm) may be absorbed by a chloroplast and its energy utilized in photosynthesis. Not all of the energy of the shorter-wavelength photons is available, however, for a photon that is absorbed by a pigment other than chlorophyll will be transferred to a pigment with a longer-wavelength absorption maximum in a process closely akin to fluorescence, but without the production of light. The transfer of an absorbed photon continues from pigment to pigment until the one with the longest-wavelength (lowest energy) absorption maximum is reached. If that pigment cannot do something useful with the energy, it will then be dissipated internally or lost by true fluorescence. In the chloroplast, this longest-wavelength absorber is a special fraction of the chlorophyll—the only fraction that is capable of ejecting electrons.

Thus, energy from light absorbed at a wide range of wavelengths is funneled, in about 10^{-9} sec, through the pigments to a long-wavelength "trap" that utilizes it. According to this scheme, a low energy photon absorbed by the trap molecule itself should be as effective as a higher energy photon absorbed by another pigment and transferred to it. The efficiency of photosynthesis, therefore, depends on the wavelength of light used to support the process.

But what is the nature of this energy trap? According to the action spectrum, it must absorb light at or very near 700 nm, a wavelength that is much longer than the usual absorption maximum of chlorophyll. And yet the trap molecules are chlorophyll, distinguished from the bulk chlorophyll only by a difference in environment.

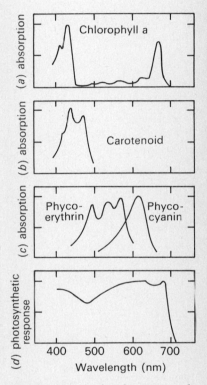

FIGURE 5–17. Light Absorption and the Photosynthetic Response. Chloroplasts contain a variety of pigments, each of which absorbs light maximally in a different region of the visible spectrum. (See (a)–(c) for examples.) Since the energy of captured photons can be transferred from pigment to pigment, photosynthesis can be sustained by light of any visible wavelength. Note the two absorption peaks of chlorophyll in (a). The short-wavelength peak is known as the Soret band, and is found also in heme compounds. Note also the sharp drop in photosynthetic response in the far red, at 700 nm.

Since the chlorophyll molecules in chloroplasts are not all in the same environment, we find that they absorb light over a rather wide range. Chlorophyll *a*, which has an *in vivo* absorption maximum at 675 nm, actually has significant absorption from about 660 to around 700 nm. This observation led Bessel Kok to suggest that the various fractions of chlorophyll constitute a cooperative unit, part of a *pigment system*. Only a few (about 0.3%) of the chlorophyll *a* molecules have an absorption maximum at 700 nm; it is these molecules that constitute the long-wavelength energy trap, designated *P700*.

$R = -CH_3$ in chlorophyll *a*

$R = -CHO$ in chlorophyll *b*

THE DUAL PIGMENT SYSTEM. The organization of the chloroplast chromophores, as it has been developed so far, fails to explain an interesting observation made at the University of Illinois by Robert Emerson and his colleagues in 1956. They found that while light of 700 nm is very inefficient in producing photosynthesis in chloroplasts by itself, the addition of a second beam of light at a shorter wavelength (they used 650 nm) greatly improves the overall efficiency. In other words, two beams, one at 650 and one at 700 nm, cause more oxygen production per photon than the sum of the two used individually. But if all absorbed energy is funnelled to $P700$, then one beam should have the same "quantum efficiency" as the other—i.e., it should produce as much oxygen for a given number of photons absorbed. The dual light, or "Emerson effect," led to the suggestion that at least two different traps exist, each with a pigment system feeding energy into it. The second trap is sometimes designated $P690$ or as a_{II}.

The pigments in chloroplasts thus seem to be divided into two parts, called *pigment system I* (PS I) and *pigment system II* (PS II), each with a long-wavelength trap into which absorbed energy is funneled. That the functions of the two pigment systems are coupled chemically rather than through an energy transfer was made clear by experiments in which the two beams of light used for the Emerson effect were flashed one at a time, with several seconds between them. That such a time lapse still allows the shorter-wavelength light to reinforce the longer-wavelength beam is interpreted to mean that a chemical intermediate is produced; if the coupling between the pigment systems were based only on energy transfer, it would die out a billion times faster than it does. As it turns out, the chemical reaction by which the two pigment systems are coupled involves the reduction and oxidation of a whole set of components similar to the electron transport chain of mitochondria. The set includes cytochromes, plastoquinones, and plastocyanin, all of which are proteins that contain metal ions.

Apparently electrons removed from water are transferred to the electron transport chain by PS II, and from the chain to $NADP^+$ by PS I (see Fig. 5–18). This latter step involves as an intermediate ferredoxin, which is the substance that actually reduces $NADP^+$ to NADPH. The oxidation of water by PS II is less well understood than the other reactions, and there is good reason to believe that not all the redox intermediates have been identified.

PHOTOPHOSPHORYLATION. The scheme outlined above explains the production of reduced nicotinamide nucleotides, using water as an electron donor and light as a source of energy. The nucleotides reduce carbon dioxide to carbohydrates; the latter are used to support the synthesis of other cellular constituents, including the chloroplasts themselves. The process of producing carbohydrate requires ATP as well as NADPH. One might assume that all the ATP needed could be produced by mitochondria in the same cell, using reduced nicotinamide nucleotides as an energy source.

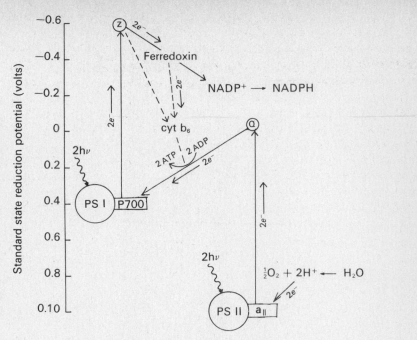

FIGURE 5-18. The Dual Pigment System. The shorter-wavelength trap, sometimes designated *P690* or a_{II}, captures light energy and uses it to oxidize water. The electrons are passed to an electron transport chain at some acceptor, Q, from which they cyclically oxidize and reduce plastoquinone, cytochrome *b*, cytochrome *f*, and plastocyanin before being passed to PS I. An additional photon input, trapped by *P700*, "boosts" the electrons to a level from which NADP$^+$ can be reduced. Cyclic electron flow (dotted lines), which utilizes PS I only, provides a way of using light to make ATP without evolving oxygen or reducing NADP$^+$.

However, measurements of the rate of carbon dioxide fixation show that mitochondria could not possibly provide ATP fast enough, implying that ATP must also be produced as a by-product of photosynthesis itself. This assumption was confirmed by measurements made on isolated chloroplasts.

ATP is produced by chloroplasts through electron transport. The theories advanced in section 5-2 to explain that process in mitochondria are also used to explain it in chloroplasts, although the potential drop from beginning to end of the chloroplast electron transport chain is only about 0.6 volt, equivalent to about 12 kcal per "mole" of electrons. This potential difference is about half of the 1.14 volt drop in mitochondria. *Photophosphorylation,* if it has the same efficiency as oxidative phosphorylation, might produce one molecule of ATP per electron transported. Whether in fact the process is that efficient is not yet known.

The synthesis of ATP by electron transport is an attractive prospect because it utilizes what would otherwise be wasted energy: the 0.6 volt drop from PS II to PS I as the electrons make their way from water to NADP$^+$. But apparently a less efficient mechanism is also available, a process called cyclic photophosphorylation (see Fig. 5-18). In this scheme, ATP is synthesized without NADPH production and without oxygen evolution. The ejected electrons are passed from PS I to an intermediate (possibly ferredoxin), which transfers them back to the electron transport chain, and so back to PS I again. Each cycle would produce a maximum of one ATP per photon of light absorbed by PS I. Although the reaction has been shown to occur under special conditions *in vitro*, its physiological significance is not yet known. However, it does represent a mechanism whereby the cell could correct an ATP deficit when the NADPH supply is already adequate.

THE STOICHIOMETRY AND EFFICIENCY OF PHOTOSYNTHESIS. The implication so far has been that one photon can cause the ejection of one electron from either pigment system. This scheme is still a little controversial, but we will assume for the following discussion that it is entirely correct. In addition, we will assume that every electron traversing the electron transport chain produces one ATP from ADP and P_i. The light reaction can then be summarized as follows, where $h\nu$ represents a photon:

1. Light reaction II, catalyzed by PS II, is

$$H_2O \xrightarrow{2h\nu} \tfrac{1}{2}O_2 + 2H^+ + 2e^- \tag{5-19}$$

2. Light reaction I, catalyzed by PS I, is

$$2e^- + H^+ + NADP^+ \xrightarrow{2h\nu} NADPH \tag{5-20}$$

3. The total reaction, including ATP synthesis, is

$$H_2O + NADP^+ + 2ADP + 2P_i \xrightarrow{4h\nu}$$
$$\tfrac{1}{2}O_2 + H^+ + NADPH + 2ATP \tag{5-21}$$

In other words, eight photons produce one molecule of oxygen (O_2) plus two molecules of NADPH, with as many as 4ATP as a by-product of electron transport.

The reduction of one $NADP^+$ at the expense of water has a $\Delta G°$ of about +52 kcal/mole, which is about the same as the standard state free energy for the reduction of NAD^+. Since light at 700 nm has an energy of about 41 kcal/einstein (Avogadro's number of photons), the four einsteins that it takes to produce one mole of NADPH thus represent about 164 kcal. If, in addition to NADPH, a full two moles of ATP are produced, requiring about 16 kcal, the overall process of photosynthesis can be said to conserve 68 kcal (52 + 16) of the 164 available, or a little over 40%. The figure is less if ordinary daylight is assumed, since higher energy photons are presumably no more effective than photons with longer wavelengths. In either case, however, the photosynthetic light reaction is a very efficient process.

5-7 THE PHOTOSYNTHETIC DARK REACTION
The "light reaction" just described accounts for the capture of radiant energy and explains how this energy is made available for

biosynthesis in the form of NADPH and ATP, evolving oxygen in the process. The biosynthetic reaction most often associated with photosynthesis is, of course, the synthesis of glucose (or glucose polymers such as starch) from carbon dioxide and water. Since glucose synthesis can occur without light as long as NADPH and ATP are available, it is termed the "dark reaction." Thus, when illuminated chloroplasts are transferred to the dark, carbon dioxide fixation continues to occur for some time, and can be prolonged even further by adding NADPH and ATP. In addition, the light reaction is centered in the material of the quantasomes, while the dark reaction is enzymatically catalyzed within the stroma (see Fig. 5–19).

CARBON DIOXIDE FIXATION. When radioactively labelled carbon dioxide ($^{14}CO_2$) is made available to illuminated chloroplasts, radioactive carbon appears very quickly in carbohydrate, and later in other chemical fractions of the organelle. In the late 1940s, Melvin Calvin and his coworkers at the University of California at Berkeley traced the path of this fixed radioactive carbon in algae. They used a brief exposure to the isotope followed by rapid extraction of the chloroplast contents. The first stable product in which ^{14}C could be found was 3-phosphoglycerate (3-PGA), an intermediate in glycolysis from which any other chemical constituent of the cell can be made. (Glucose, for example, is readily synthesized from 3-PGA by a reversal of the first six steps of glycolysis.)

Closer examination of labelled PGA revealed that radioactivity appears first in its carboxyl carbon, suggesting that the fixation of carbon dioxide occurs by a carboxylation. One might assume that the carbon dioxide acceptor is a two-carbon compound since PGA contains three carbons and CO_2 contains one. However, the primary carbon acceptor was soon identified as the five-carbon sugar derivative, ribulose 1,5-diphosphate (RDP). The result of car-

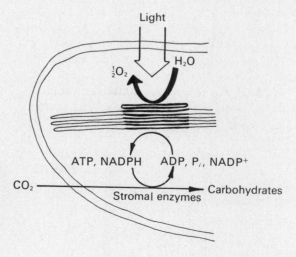

FIGURE 5–19. Localization of the Light and Dark Reactions. The light reaction is catalyzed by chloroplast lamellae, especially in the grana. The dark reaction, on the other hand, is catalyzed by enzymes of the stroma.

boxylation is a six-carbon addition compound that is immediately hydrolyzed to two molecules of PGA, thus partially offsetting an energetically unfavorable reaction, carboxylation, with a favorable one, hydrolysis:

$$
\begin{array}{ccc}
\text{H}_2\text{C}-\text{OPO}_3{}^{2-} & \text{H}_2\text{C}-\text{OPO}_3{}^{2-} & \text{H}_2\text{C}-\text{CPO}_3{}^{2-} \\
| & | & | \\
\text{C}=\text{O} & {}^{-}\text{OOC}^*-\text{C}-\text{OH} & \text{HC}-\text{OH} \\
| & | & {}^*\text{COO}^- \\
\text{HC}-\text{OH} \xrightarrow[\text{carboxylation}]{{}^*\text{CO}_2} & \text{C}=\text{O} \xrightarrow[\text{hydrolysis}]{\text{H}_2\text{O}} & \\
| & | & \text{COO}^- \\
\text{HC}-\text{OH} & \text{HC}-\text{OH} & | \\
| & | & \text{HC}-\text{OH} \\
\text{H}_2\text{C}-\text{OPO}_3{}^{2-} & \text{H}_2\text{C}-\text{OPO}_3{}^{2-} & | \\
& & \text{H}_2\text{C}-\text{OPO}_3{}^{2-}
\end{array}
$$

RDP   3-PGA

The enzyme that catalyzes the carboxylation of RDP is present in large quantities in green plants—up to about 15% of the soluble protein of spinach, for example. It is called *ribulose diphosphate carboxylase* or *ribulose diphosphate carboxy-dismutase*. Some of its subunits seem to be specified by nuclear genes and some by chloroplast DNA, a situation analogous to the cytochrome oxidase of mitochondria and several other proteins.

To prove their contention that RDP is the carbon acceptor, Calvin and his colleagues followed the change in RDP concentration when the supply of CO_2 was interrupted. The concentration of RDP immediately increased, while the concentration of PGA decreased, representing the expected build-up of substrate and removal of product from the blocked reaction.

THE CALVIN CYCLE. It is obvious that some of the 3-PGA formed by the carboxylation of RDP must be used to regenerate RDP, since otherwise fixation would stop for lack of acceptor. Specifically, only the fraction of carbons corresponding to CO_2 addition, one carbon out of six, can be used for other purposes; the remaining $\frac{5}{6}$ are needed for RDP regeneration.

The cyclic process whereby RDP is regenerated is called the *Calvin cycle*, or the *Calvin-Benson-Bassham cycle*, and is represented in Fig. 5-20. As you can see, 3-PGA is phosphorylated by ATP to 1,3-diphosphoglycerate, and then reduced to glyceraldehyde-3-phosphate (GAP). The reduction is accomplished by an NADPH-linked enzyme. Regeneration of RDP starts with GAP and proceeds through 4-, 5-, 6-, and 7-carbon intermediates.

The net carbon fixation reaction, neglecting RDP, is

$$6CO_2 + 18ATP + 12NADPH + 12H^+ \longrightarrow$$
$$18P_i + 18ADP + 12NADP^+ + \text{glucose} \qquad (5-22)$$

FIGURE 5–20. The Calvin Cycle in Spinach Chloroplasts. Also called the C3 pathway.

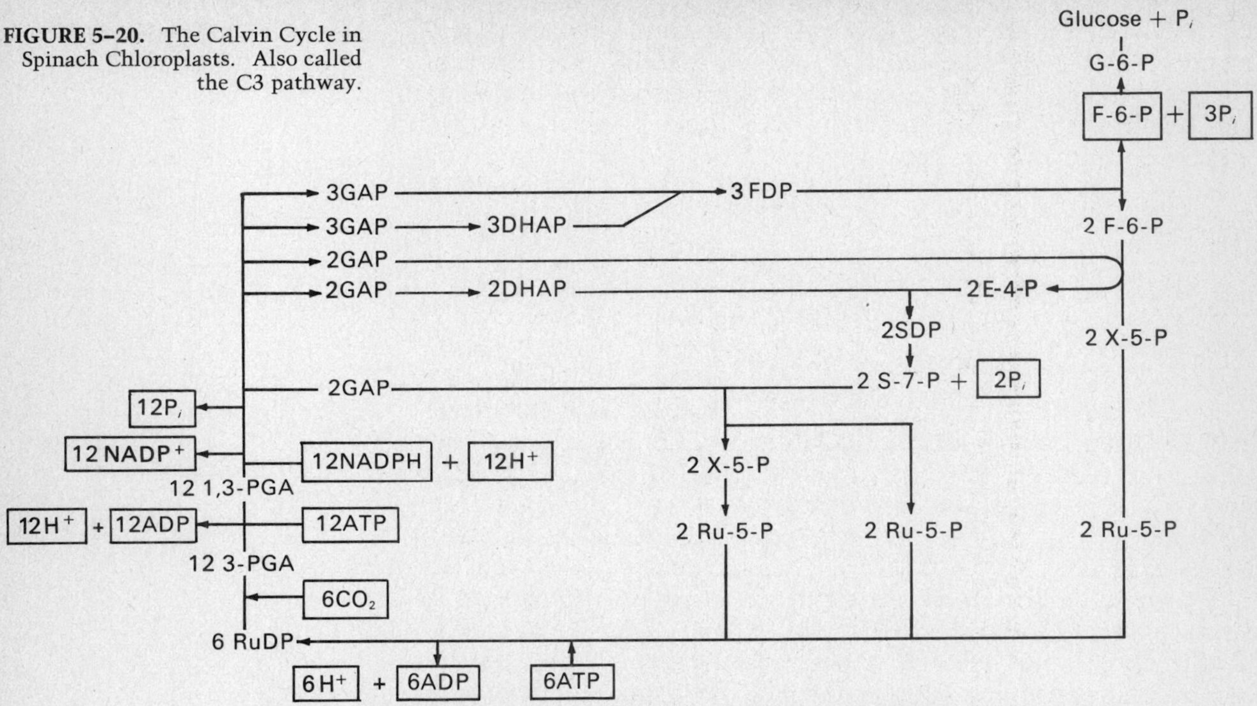

Glucose + P_i

G-6-P

F-6-P + 3P_i

3GAP ⟶ 3 FDP ⟶ 2 F-6-P

3GAP ⟶ 3DHAP

2GAP

2GAP ⟶ 2DHAP ⟶ 2E-4-P ⟵ 2 X-5-P

2SDP

2GAP ⟶ 2 S-7-P + 2P_i

12P_i ⟵

12 NADP$^+$ ⟵ 12NADPH + 12H$^+$

2 X-5-P

12 1,3-PGA

12H$^+$ + 12ADP ⟵ 12ATP

2 Ru-5-P   2 Ru-5-P   2 Ru-5-P

12 3-PGA

6CO$_2$

6 RuDP ⟵

6H$^+$ + 6ADP   6ATP

6CO$_2$ + 12NADPH + 18ATP^{4-} ⟶ F-6-P^{2-} + 12NADP$^+$ + 18ADP^{3-} + 17P^{2-} + 6H$^+$

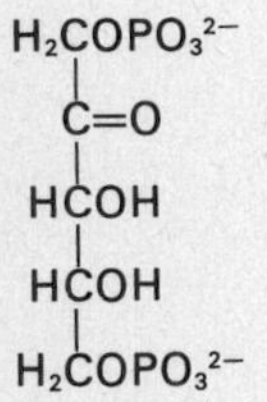

H$_2$COPO$_3^{2-}$
|
C=O
|
HCOH
|
HCOH
|
H$_2$COPO$_3^{2-}$

RuDP
ribulose—1,5—
diphosphate
(Ru-5-P = ribulose-
5-phosphate)

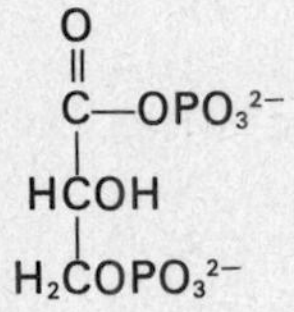

O
‖
C—OPO$_3^{2-}$
|
HCOH
|
H$_2$COPO$_3^{2-}$

1,3-PGA
1,3-diphospho-
glycerate
(3-PGA = 3-
phosphoglycerate)

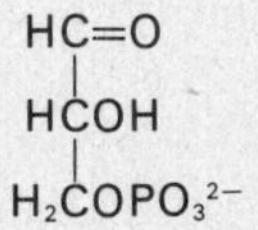

HC=O
|
HCOH
|
H$_2$COPO$_3^{2-}$

GAP
glyceraldehyde
-3-phosphate

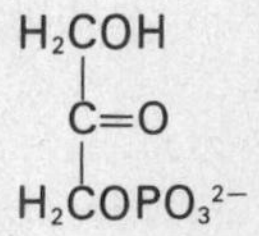

H$_2$COH
|
C=O
|
H$_2$COPO$_3^{2-}$

DHAP
dihydroxy-
acetone-3-
phosphate

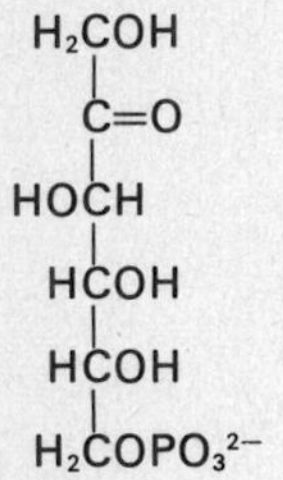

H$_2$COPO$_3^{2-}$
|
C=O
|
HOCH
|
HCOH
|
HCOH
|
H$_2$COPO$_3^{2-}$

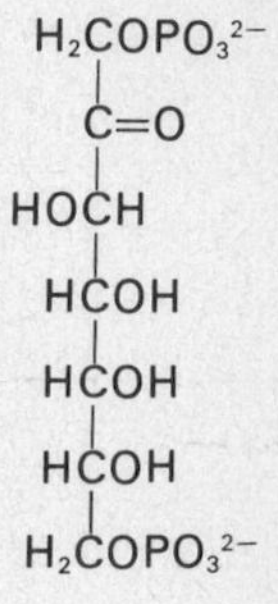

SDP
sedoheptulose
-1-7-diphosphate
(S-7-P = sedoheptulose-
7-phosphate)

H$_2$COH
|
C=O
|
HOCH
|
HCOH
|
HCOH
|
H$_2$COPO$_3^{2-}$

F-6-P
fructose-6-
phosphate

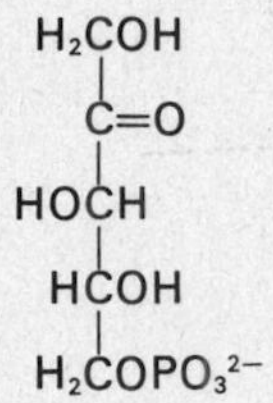

H$_2$COH
|
C=O
|
HOCH
|
HCOH
|
H$_2$COPO$_3^{2-}$

X-5-P
xyulose-5-
phosphate

HC=O
|
HCOH
|
HCOH
|
H$_2$COPO$_3^{2-}$

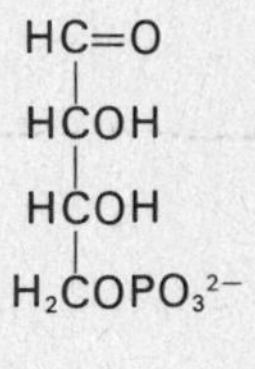

E-4-P
erythrose-4-
phosphate

If one considers 8 kcal/mole as the energy of hydrolysis of ATP and 52 kcal/mole as the "reducing power" of NADPH (see equation (5-1)), glucose synthesis can be said to require a total of 768 kcal/mole. Since the complete oxidation of glucose would yield 686 kcal/mole, the process of glucose synthesis via the above reactions is very favorable, with an overall $\Delta G° = -82$ kcal/mole. Clearly, equilibrium lies strongly on the side of glucose formation.

In accordance with the stoichiometry already discussed, the light reaction, equation (5-21), requires at least 48 photons to produce a molecule of glucose. Taking the most optimistic estimate of ATP production—one ATP per electron transported between pigment systems—we are left with 6 ATP after the 12 NADPH and 18 ATP are consumed. If this is, in fact, the *in vivo* situation, it represents an ATP surplus that would be available for other biosynthetic reactions. In addition, mitochondria continue to oxidize carbohydrates even in periods of illumination, a process called *photorespiration*, thus generating still more ATP. Mitochondria use much less oxygen than illuminated chloroplasts produce, however, except in times of darkness, when mitochondrial activity must provide all the energy necessary to run the cell. A photosynthetic plant cell is thus a little like a storage battery: it stores excess energy as carbohydrate during periods of illumination, and utilizes it when solar energy is not available.

Experiments with cell fractions indicate that carbohydrate synthesis can take place entirely within the chloroplasts. In fact, protein synthesis and a certain amount of lipid synthesis—functions that we normally associate with other parts of the cytoplasm—can also take place in these organelles. As mentioned earlier, a portion of the protein synthesis in chloroplasts (e.g., some of the enzymes responsible for chlorophyll synthesis) is under the control of local (chloroplast) genes, though the cell nucleus remains by far the most important container of genetic information.

ALTERNATE PATHWAYS OF CARBON FIXATION. The Calvin cycle just described is not the only carbon-fixing reaction sequence. In 1966 two Australian scientists, M. D. Hatch and C. R. Slack, suggested an alternate pathway for carbon fixation in corn and certain other hot-weather, dry-climate plants that contain dimorphic chloroplasts (see Fig. 5-21). In recent versions of their scheme (see Fig. 5-22), phosphoenolpyruvate (PEP) is carboxylated by CO_2 to form oxalacetate (OAA) in a dimorphic chloroplast. The OAA is then transferred to a cell with the usual type of chloroplast, where OAA releases its new carbon as CO_2 again, leaving pyruvate. After transfer back to the original cell, pyruvate is phosphorylated to PEP at the expense of an ATP-to-AMP hydrolysis in a reaction that has no counterpart in animal cells. The regenerated PEP is then ready to accept another molecule of carbon dioxide.

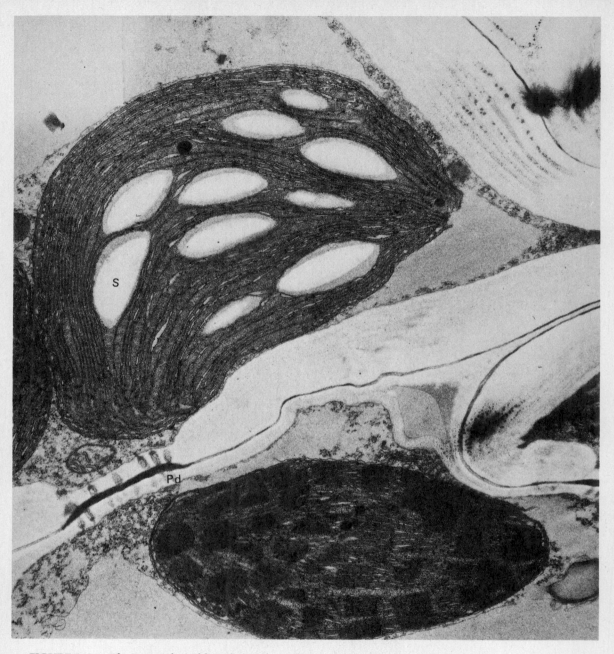

FIGURE 5–21. The Dimorphic Chloroplast. Plants that utilize the Hatch and Slack pathway also have dimorphic chloroplasts, notable for their abundance of starch and lack of grana. One is shown here in a bundle-sheath cell of sugar cane (top). The cell is connected via plasmodesmata (Pd) to a mesophyll cell (bottom) containing the usual chloroplast configuration. Starch (S). [Courtesy of W. M. Laetsch, *Sci. Prog.* (Oxford), **57:**323 (1969).]

The effect of the Hatch and Slack pathway, which is also called the C4 pathway since OAA is a four-carbon molecule, is to accumulate CO_2. Plants that have it grow rapidly and require relatively less water than other plants. Some of them can use this

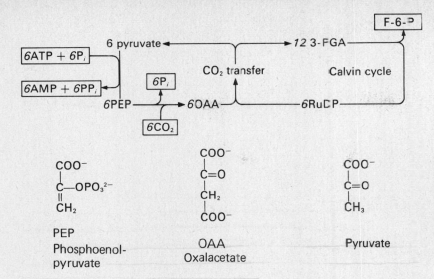

FIGURE 5-22. The Hatch and Slack Pathway for Carbon Fixation. Also called the C4 pathway. The net reaction is the same as for the Calvin cycle, except for the input of $6\ ATP \rightarrow 6\ AMP$, which is the equivalent of an additional 12 ATP consumed in the production of each molecule of glucose.

pathway to fix CO_2 at night and produce carbohydrate from it during the day.[2]

Still other mechanisms for carbon dioxide fixation have been suggested. A number of decarboxylations are known to take place in cells, almost any of which are candidates for reversal, leading to the fixation of CO_2 instead of to its evolution. Whether any of these play an important role in photosynthetic carbon fixation however, is still an open question.

5-8 PHOTOSYNTHESIS IN PROCARYOTIC CELLS

Photosynthesis is also found in the blue-green algae and in some bacteria. Blue-green algae have a photosynthetic arrangement similar to that of higher plants, including dual pigment systems, each with an energy sink, and the capacity to use water as an electron donor, thus evolving oxygen. Blue-green algae contain cytoplasmic lamellae with which most of the photophosphorylation system is associated (Fig. 5-23). In fact, isolated particles of these lamellae are capable of photophosphorylation.

Photosynthetic bacteria, on the other hand, are quite different from either the blue-green algae or higher plants. Although they do contain cytoplasmic membranes (lamellae) with which photophosphorylation is associated, there is no oxygen evolution and

2. It has been known for many years that succulent plants, such as cacti, open the stomata of their leaves only at night, when water loss is at a minimum, and that they accumulate malate during this period. Such plants are said to have Crassulacean acid metabolism (CAM—from the genus *Crassula*), often described as a third pathway for carbon fixation. It is now clear, however, that these plants also utilize the C4 pathway: NADPH and phosphoenolpyruvate (PEP) are supplied by starch breakdown at night, with the NADPH being used by malate dehydrogenase (see Table 5.2) to reduce oxalacetate reversibly and temporarily to malate.

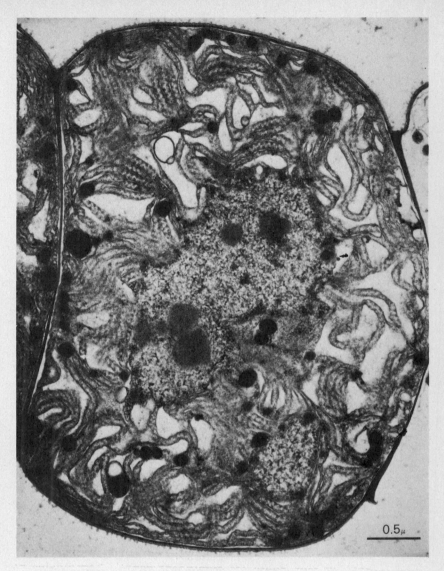

FIGURE 5–23. Blue-Green Alga. Note the photosynthetic lamellae inside this freshwater procaryote, *Fremyella diplosiphon.* The spheres on the membranes appear similar to those on the photosynthetic membranes within the chloroplasts of some eucaryotic algae (see Fig. 5–15). The DNA is the lighter, granular material in the center. [Courtesy of E. Gantt and S. F. Conti, *J. of Bact.,* **97:**1486 (1969).]

0.5μ

no Emerson enhancement. It appears, therefore, that only one photosystem is used. Some of the pigments of this system are unique to bacteria, including a special bacterial chlorophyll. The single photosystem appears to have an energy sink, *P870*, that absorbs well into the red. It represents approximately 2% of the chlorophyll molecules.

That photosynthetic bacteria do not evolve oxygen during photosynthesis means that an electron donor other than water must be used. Some use carbon monoxide, methane, or other organic compounds. Others use inorganic electron donors such as hydrogen gas or hydrogen sulfide. The latter is utilized by the *purple sulfur bacteria,* producing elemental sulfur:

$$H_2S \longrightarrow S + 2H^+ + 2e^-$$

Purple sulfur bacteria can also "fix" nitrogen from the atmosphere, reducing it to ammonia. They can therefore derive all their major atoms — C, H, O, and N — from air.[3] Bacterial photosynthesis is represented by the general reaction

$$CO_2 + 2H_2A \xrightarrow{hv} (CH_2O) + 2A + H_2O \qquad (5-23)$$

where H_2A is oxidized to A and (CH_2O) represents the newly added carbohydrate unit. Photosynthesis in plants and blue-green algae is the special case where H_2A is water and $2A$ is O_2.

Thus, there are substantial differences between bacteria and blue-green algae in their pattern of photosynthesis. Both organisms are considered to be more primitive than eucaryotic plants or animals. In fact, it has been proposed that the chloroplasts of eucaryotes are descendants of the blue-green algae, whereas mitochondria evolved from aerobic bacteria.

5-9 ORIGIN OF MITOCHONDRIA AND CHLOROPLASTS

It is believed by many biologists that both mitochondria and the plastids are descended from procaryotic cells that took up a symbiotic residence in some ancient eucaryote. This hypothesis is based largely on striking similarities between mitochondria, chloroplasts, and procaryotes (see Fig. 5-24.)

Both mitochondria and the plastids are semiautonomous; both reproduce by fission (see Fig. 5-25) at a rate that is not necessarily synchronized to that of their "host"; both types of organelles contain genes that direct the synthesis of at least some of their own components; and both can utilize a protein-synthesizing system contained within the organelle itself. Moreover, although the genetic content of a mitochondrion may be as little as 1% of that of a typical procaryotic cell like *Escherichia coli*, the amount of genetic material in chloroplasts approaches that of the smaller procaryotes. In fact, there are significant genetic homologies between blue-green algae and some eucaryotic plastids. There is no question of complete genetic autonomy, however, for there is also clear evidence for nuclear control over portions of each of the organelles.

Unlike most eucaryotic organelles, mitochondria and the plastids have a double membrane. The inner membrane of mitochondria and chloroplasts has numerous folds projecting into

---

3. The fixation of nitrogen is extraordinarily difficult because of the very strong, 225 kcal/mole, triple bond in N_2. The process is limited to only a few strains of microorganisms. All life depends on these, for other organisms are forever utilizing nitrogen compounds, releasing N_2 through oxidation.

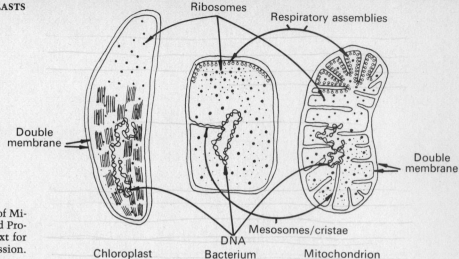

FIGURE 5–24. A Comparison of Mitochondria, Chloroplasts, and Procaryotic Cells. See text for discussion.

their interior. Thus, the cristae of mitochondria are formed in a way that is clearly reminiscent of bacterial mesosomes. Furthermore, the membranous interior of photosynthetic procaryotes contains light-capturing pigments, just as quantasomes are embedded in the thylakoids of chloroplasts. An additional similarity is that electron transport is associated only with membrane-bound com-

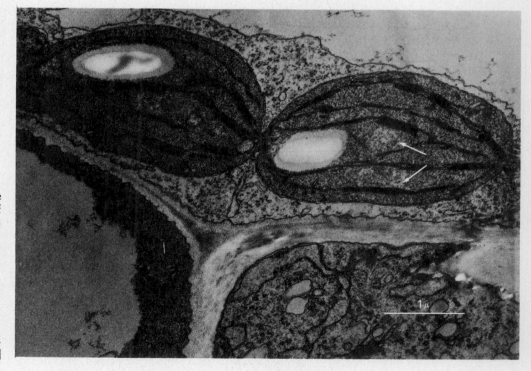

FIGURE 5–25a. The Replication of Chloroplasts. From a tobacco leaf. Note the prominent starch granule in each daughter plastid, and the light areas within the stroma that are probably DNA. [Courtesy of D. A. Stetler and W. M. Laetsch, *Amer. J. Botany*, **56:**260 (1969).]

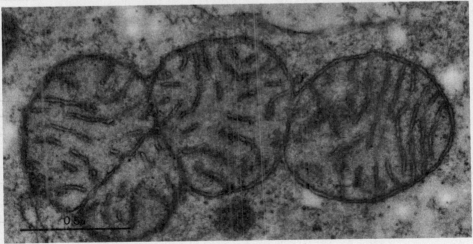

FIGURE 5–25b. The Replication of Mitochondria. From the fat body of an insect, *Calpodes ethlius*. Note that the partitions (arrow) are constructed like cristae—that is, they are invaginations of the inner membrane. [Courtesy of W. J. Larsen, *J. Cell Biol.*, **47**:373 (1971).]

ponents of chloroplasts, mitochondria, and procaryotes. The mushroomlike stalks seen in electron micrographs of mitochondrial cristae have also been reported on the plasma membrane of some aerobic bacteria.

There are also marked similarities in other chemical and physical characteristics, including the following: (1) Studies of the primary structures of cytochrome c and some other proteins from procaryotes indicate that they are remarkably similar to the corresponding proteins found in the organelles of eucaryotes. (2) Mitochondria, chloroplasts, and procaryotes are all similar in size, though chloroplasts tend to be somewhat the larger. (3) The genetic material of mitochondria, chloroplasts, and procaryotes is in the form of naked strands of DNA, found in closed circular form and attached to membranes rather than in the more complicated protein–DNA chromosomal structure of eucaryotic nuclei (see Fig. 5-26). (4) In all three structures being compared, the ribosomes, which are the sites of protein synthesis, are unmistakably smaller than and different from eucaryotic ribosomes found outside the organelles. For example, procaryotic, mitochondrial, and chloroplast ribosomes respond in the same way to certain antibiotics: chloramphenicol inhibits protein synthesis on these smaller ribosomes, but not on eucaryotic ribosomes found outside the organelles[4]; and cycloheximide inhibits protein synthesis on eucaryotic ribosomes from the cytosol, but not on ribosomes found in mitochondria, chloroplasts, or procaryotic cells.

These similarities and others more technical in nature have led to the postulate that both mitochondria and chloroplasts are the evolutionary remnants of procaryotic cells. Photosynthetic activity probably developed before oxidative phosphorylation, since the former produces oxygen required by the latter. Therefore, it has also been suggested that chloroplasts evolved from some ancient procaryote, while mitochondria may have evolved from chloroplasts.

In support of the procaryotic origin for mitochondria and chloroplasts, it should be noted that intracellular symbiosis (endosymbiosis) is a well-known phenomenon. For example, certain species of paramecia (which are single-celled animals) harbor gram-negative bacteria. In addition, there are a very large number of symbiotic relationships involving simple aquatic animals and procaryotic blue-green algae. Margit Nass of the University of Pennsylvania was able in 1969 to establish a similar relationship artificially. She introduced chloroplasts into cultured mammalian cells, where they were able to continue to function for some time,

---

4. The sensitivity of mitochondria to chloramphenicol may be the reason for the occasional, severe side effects observed when this common antibiotic is used to control bacterial infections in man.

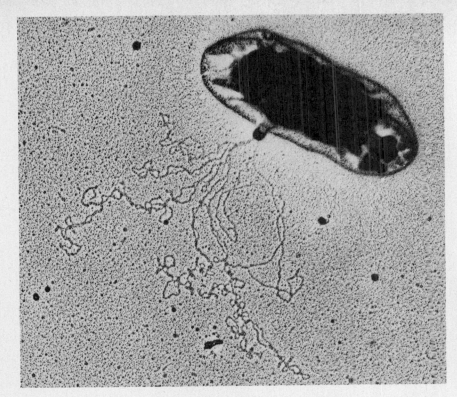

FIGURE 5–26a. Bacterial DNA, Partially Extruded. Note the supercoiling. The total contour length of the complete molecule is about a millimeter (1000 μ). [Courtesy of M. M. K. Nass, from *Biological Ultrastructure*, P. J. Harris, ed., Corvallis: Ore. State Univ. Press, 1971, p. 41.]

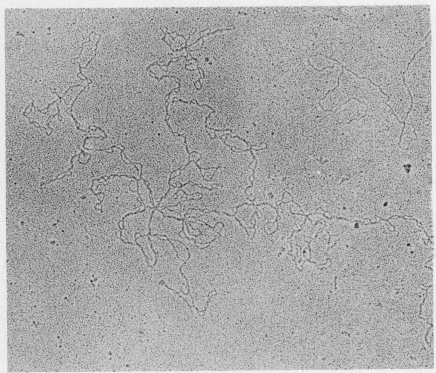

FIGURE 5–26b. Chloroplast DNA. From spinach. The contour length is 42 μ, enough to specify at least a hundred proteins. [Courtesy of J. E. Manning, D. R. Wolstenholme, and O. C. Richards, *J. Cell Biol.*, **53:**594 (1972).]

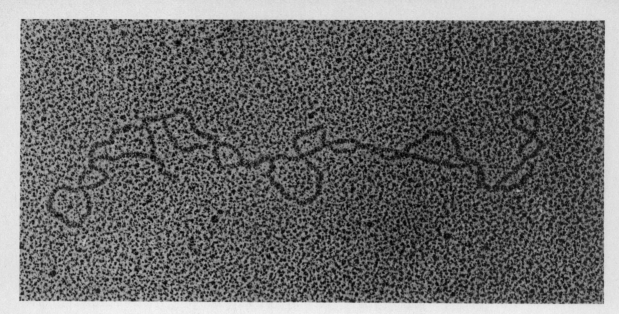

FIGURE 5–26c. Mitochondrial DNA. From an Ehrlich ascites tumor of the mouse. Contour length is 4.9 μ, enough to specify perhaps a dozen proteins. (Courtesy of D. R. Wolstenholme.)

leading to the facetious suggestion that future astronauts receive chloroplast implants in their skin, thus permitting them to recycle their respired carbon dioxide and so reduce food requirements.

The advantages of an endosymbiosis involving photosynthetic procaryotes, then, are that the photosynthetic cell confers on its host an ability to capture light energy, secreting nutritive products of that capture into the host cytoplasm; the photosynthetic cell, in return, secures a sheltered and rather constant environment in which to grow and reproduce. However, in such an environment, the procaryote may have little use for a number of functions that are necessary for survival in the free state. Since the loss of these functions would confer no disadvantage, and perhaps a positive advantage in avoiding duplication, one can understand how a very old symbiotic relationship might have evolved to the present system of a eucaryotic cell with semiautonomous organelles.

SUMMARY

5–1 Mitochondria use molecular oxygen as an electron acceptor in the oxidation of coenzymes (e.g., NADH), fatty acids, carbohydrates, and amino acids. The process produces water and releases a large quantity of energy, a portion of which is captured by the synthesis of ATP.

5–2 Mitochondrial electron transport uses molecules which, through reversible oxidations and reductions, pass electrons from one component to another, terminating with the reduction of oxygen to water. Electron transport thus utilizes (via NADH or FADH$_2$) some of the energy from oxidation to produce ATP.

5–3 The mitochondrion houses numerous oxidative pathways that feed electrons to the electron transport chain. Central to these pathways is the citric acid cycle or Krebs cycle, which begins with the condensation of acetate (as acetyl-SCoA) and oxalacetate to form citrate, a six-carbon tricarboxylic acid. With the citric acid cycle and electron transport, 38 ATP are obtained from each mole (180 g) of glucose, a nineteen-fold increase over the yield of 2 ATP obtained from glucose catabolism via anaerobic glycolysis.

5–4 The mitochondrion is a compartmentalized organelle, with the inner space (matrix), outer space, and two membranes each contributing in a unique way to its overall function. The inner membrane contains many enzymes as an integral part of its structure. In addition, the highly selective inner membrane maintains within the matrix high concentrations of important substances. The outer membrane, on the other hand, is relatively permeable and more closely akin to other cell membranes in structure.

5–5 Chloroplasts capture light, the energy of which can be used by them to synthesize carbohydrates. The photosynthesis of carbohydrates can be divided into two parts, a light reaction and a dark reaction. The light reaction in plants is a photochemical process involving the use of light to oxidize water to oxygen, reducing $NADP^+$ to NADPH. It also produces ATP via electron transport. The light reaction is catalyzed by elements of the internal membranes, specifically the grana. The dark reaction, on the other hand, is an enzymatic pathway within the stroma that uses ATP and NADPH to fix carbon dioxide, reducing it to carbohydrate.

5–6 Chlorophyll exhibits the photoelectron effect, which is actually a light-driven oxidation. The corresponding reduction is $NADP^+ \rightarrow NADPH$. In order to restore chlorophyll prior to its next oxidation, electrons are transferred to it from water, creating molecular oxygen. These light-dependent reactions seem to be divided into two parts via two pigment systems, or photosystems. Photosystem II uses light energy to oxidize water and transfers the electrons to the first member of an electron transport chain. The chain terminates at photosystem I (PS I), which uses the electrons, "boosted" by energy from additional light capture, to reduce $NADP^+$ to NADPH. Electron transport itself results in the production of ATP (photophosphorylation).

5–7 The fixation of carbon dioxide is catalyzed by ribulose diphosphate carboxylase, an enzyme that adds carbon dioxide (as a carboxyl group) to ribulose diphosphate (RDP) and then cleaves the addition product to two molecules of 3-PGA (phosphoglyceric acid). The major fraction ($\frac{5}{6}$) of the PGA must be used to regenerate RDP via the Calvin cycle, so that more molecules of CO_2 can be captured. The other $\frac{1}{6}$ of the PGA is available for conversion to carbohydrate, lipid, protein, or nucleic acid.

The Hatch and Slack, or C4, pathway is an alternate route for carbon fixation. It utilizes the carboxylation of phosphoenolpyruvate (PEP), creating oxalacetate. The latter transfers its new carbon to RDP.

5–8 Photosynthesis is also found in certain strains of bacteria and in all blue-green algae. The blue-green algae have a photosynthetic pathway that is nearly identical to that of chloroplasts. Bacteria, on the other hand, use one photosystem rather than two. They also have a different form of chlorophyll, and they do not use water as a source of electrons. Hence, bacterial photosynthesis does not result in oxygen evolution.

5–9 Mitochondria and chloroplasts have many things in common with procaryotes. This observation led to the suggestion that mitochondria and chloroplasts are evolutionary descendants of procaryotic symbionts living in some ancient eucaryote. In particular, mitochondria and chloroplasts (1) are self-replicating, dividing by binary fission as do procaryotes; (2) contain genes responsible for synthesizing some of their own proteins; (3) have ribosomes similar to those of procaryotes, but unlike other eucaryotic ribosomes; and (4) are similar in size to procaryotic cells.

STUDY GUIDE

5–1 (a) Define: outer space, cristae, intracristal space, matrix. (b) Mitochondria are found arranged about the base of a sperm cell flagellum. What is significant about this placement?

5–2 (a) Define: electron transport, respiratory chain, oxidative phosphorylation, substrate-level phosphorylation, cytochrome oxidase. (b) Summarize the three schemes presented here for coupling electron transport to ADP phosphorylation. (c) If direct coupling were used to drive ATP synthesis with electron transport, which reactions in the electron transport chain would be the most likely candidates? (d) What is a "P/O ratio," and what is its probable value for NADH? For $FADH_2$? Why are these values not certain?

5-3 **(a)** Define and draw the general reaction sequence for an oxidative decarboxylation. **(b)** One way of measuring the caloric value of food is to calculate the amount of ATP produced in its aerobic degradation. Then, at 8 Calories (8 kcal or kilogram calories) per mole of ATP, one can assign a numerical value to the food. With this scheme, find the ATP-capturable caloric content of 100 g (3.5 oz.) of glucose. [Ans: 169 Calories/100 g]

5-4 **(a)** List the four "compartments" into which mitochondrial reactions can be placed, and outline the function of each. **(b)** How do the inner and outer membranes differ in composition and permeability, and what are the advantages of the permeability differences?

5-5 What is the difference between the photosynthetic light reaction and dark reaction?

5-6 **(a)** Define: photon, a quantum of energy, an einstein, action spectrum, fluorescence. **(b)** What is the photoelectric effect? How would you increase the velocity of ejected electrons? How would you increase their number? **(c)** What are the possible fates (as discussed here) of an absorbed photon? Which fate is most applicable to the chloroplast? **(d)** What is "Emerson enhancement"? To what do we attribute it? **(e)** If absorbed light is utilized only by P700, what is the percent of energy lost in transferring it to P700 if the absorbed light is (1) yellow $(\lambda = 600 \text{ nm})$, (2) blue $(\lambda = 500 \text{ nm})$, or (3) violet $(\lambda = 400 \text{ nm})$? [Ans: 14%; 29%; 43%]

5-7 **(a)** What is the Calvin cycle, and what is its purpose? **(b)** Based on the stoichiometry suggested here, show why eight photons are needed per CO_2 molecule "fixed" and reduced to the oxidation level of glucose. **(c)** Write the reaction (according to the Calvin cycle) for the net production of one molecule of 3-PGA, regenerating any RDP used in the process. **(d)** What is the relationship of the Hatch and Slack pathway to Calvin's pathway for carbon fixation?

5-8 **(a)** The dual light effect (Emerson enhancement) is not observed with photosynthetic bacteria. Why? **(b)** What other differences are there between photosynthesis in bacteria and the mechanism used by blue-green algae or higher plants?

5-9 **(a)** What features do mitochondria and chloroplasts have in common, and which of these are shared with procaryotes but not found elsewhere in eucaryotes? **(b)** What conceivable advantage would be conferred by an endosymbiotic blue-green alga in an animal cell?

REFERENCES

GENERAL

GIBBS, MARTIN, ed., *Structure and Function of Chloroplasts.* New York: Springer-Verlag, 1972. Nine comprehensive reviews.

KROGMANN, D. W., *The Biochemistry of Green Plants.* Englewood Cliffs, N. J.: Prentice-Hall, Inc., 1973. A short text.

TANDLER, BERNARD, and C. L. HOPPEL, *Mitochondria.* New York: Academic Press, 1972. (Paperback.) Excellent micrographs.

MEMBRANES OF THE MITOCHONDRIA AND CHLOROPLASTS

BRANTON, D., and R. PARK, "Subunits in Chloroplast Lamellae." *J. Ultrastructure Research,* **19**:283 (1967).

HESLOP-HARRISON, J., "Structural Features of the Chloroplast." *Science Progress* (Oxford), **54**:519 (1966).

MACLENNAN, DAVID H., "Molecular Architecture of the Mitochondrion." *Current Topics in Membranes and Transport,* **1**:177 (1970). Mitochondrial membranes.

OELZE, J., and G. DREWS, "Membranes of Photosynthetic Bacteria." *Biochim. Biophys. Acta,* **265**:209 (1972). For comparison with chloroplasts.

PACKER, LESTER, "Functional Organization of Intramembrane Particles of Mitochondrial Inner Membranes." *J. Bioenergetics,* **3**:115 (1972).

RACKER, EFRAIM, "The Membrane of the Mitochondrion." *Scientific American,* February 1968. (Offprint 1101.)

SMOLY, J. M., B. KUYLENSTIERNA, and L. ERNSTER, "Topological and Functional Organization of the Mitochondrion." *Proc. Nat. Acad. Sci.* (U. S.), **66**:125 (1970).

THE CITRIC ACID CYCLE AND RELATED REACTIONS

KREBS, H. A., "The History of the Tricarboxylic Acid Cycle." *Perspectives in Biol. & Med.,* **14**:154 (1970).

————, "The Citric Acid Cycle." In *Nobel Lectures, Physiology or Medicine, 1942–1962.* Amsterdam: Elsevier Publ. Co., 1964, p. 399. Nobel Lecture, 1953.

LEHNINGER, A. L., *Biochemistry.* New York: Worth Publ., 1970. The function of mitochondria and chloroplast gets extensive consideration in Chaps. 16–21.

LIPMANN, FRITZ, "Development of the Acetylation Problem: A Personal Account." In *Nobel Lectures, Physiology or Medicine, 1942–1962.* Amsterdam: Elsevier Publ. Co., 1964, p. 413. Nobel Lecture, 1953.

WARBURG, OTTO H., "The Oxygen Transferring Ferment of Respiration." In *Nobel Lectures, Physiology or Medicine, 1922–1941*. Amsterdam: Elsevier Publ. Co., 1965, p. 254. Nobel Lecture, 1931.

ELECTRON TRANSPORT AND PHOSPHORYLATION

AZZONE, G. F., "Oxidative Phosphorylation, A History of Unsuccessful Attempts: Is It Only an Experimental Problem?" *J. Bioenergetics*, 3:95 (1972).

GREEN, D. E., "The Electromechanical Model for Energy Coupling in Mitochondria." *Biochimica Biophysica Acta*, 346(BR2):27 (1974). A synthesis of chemiosmotic and conformational models.

GREEN, D. E., and JOHN H. YOUNG, "Energy Transduction in Membrane Systems." *American Scientist*, 59:92 (1971).

MITCHELL, P., "Chemiosmotic Coupling in Energy Transduction: A Logical Development of Biochemical Knowledge." *J. Bioenergetics*, 3:5 (1972).

_____, "A Chemiosmotic Molecular Mechanism for Proton-Translocating Adenosine Triphosphatases." *FEBS Letters*, 43:189 (1974).

RACKER, EFRAIM, "Inner Mitochondrial Membranes: Basic and Applied Aspects." *Hospital Practice*, 9(2):87 (February 1974). Support for the chemiosmotic theory.

SENIOR, A. E., "The Structure of Mitochondrial ATPase." *Biochim. Biophys. Acta*, 301(BR1):249 (1974). The inner membrane spheres.

SLATER, E. C., "The Mechanism of Energy Conservation in the Mitochondrial Respiratory Chain." *Harvey Lectures*, Ser. 66 (1970–1971). New York: Academic Press, 1970, p. 19.

VANDERKOOI, G., and D. E. GREEN, "New Insights into Biological Membrane Structure." *Bioscience*, 21:409 (1971).

PHOTOSYNTHESIS

BASSHAM, J. A., "The Path of Carbon in Photosynthesis." *Scientific American*, June 1962. (Offprint 122.)

_____, "The Control of Photosynthetic Carbon Metabolism." *Science*, 172:526 (1971).

BJORKMAN, OLLE, and J. BERRY, "High-Efficiency Photosynthesis." *Scientific American*, October 1973. (Offprint 1281.) The Hatch and Slack pathway.

DEVLIN, R. M. and A. V. BARKER, *Photosynthesis*. New York: Van Nostrand Reinhold Co., 1971.

FOGG, G. E., *Photosynthesis* (2d ed). Amsterdam: Elsevier Publ. Co., 1973. A brief, general introduction.

GABRIEL, M. L., and S. FOGEL, eds., *Great Experiments in Biology*. Englewood Cliffs, N.J.: Prentice-Hall, 1955. Includes excerpts from the writings of Priestley, Ingen-Housz, de Saussure, R. Hill, Buchner, and Warburg, among others.

GIBBS, MARTIN, "The Inhibition of Photosynthesis by Oxygen." *American Scientist*, 58:634 (1970).

LAETSCH, W. M. "Relationship Between Chloroplast Structure and Photosynthetic Carbon-Fixation Pathways." *Science Progress* (Oxford), 57:323 (1969). The dimorphic chloroplast.

LEVINE, R. P., "The Mechanism of Photosynthesis." *Scientific American*, December 1969. (Offprint 1163.)

RABINOWITCH, E. I., and GOVINDJEE, "The Role of Chlorophyll in Photosynthesis." *Scientific American*, July 1965. (Offprint 1016.)

_____, *Photosynthesis*. New York: John Wiley & Sons, 1969. (Paperback.)

DEVELOPMENT AND AUTONOMY OF MITOCHONDRIA AND CHLOROPLASTS

COHEN, SEYMOUR S., "Are/Were Mitochondria and Chloroplasts Microorganisms?" *American Scientist*, 58:281 (1970).

FLAVELL, RICHARD, "Mitochondria and Chloroplasts as Descendants of Prokaryotes." *Biochem. Genetics*, 6:275 (1972).

GOODENOUGH, U. W., and R. P. LEVINE, "The Genetic Activity of Mitochondria and Chloroplasts." *Scientific American*, November 1970. (Offprint 1203.)

KROON, A. M., and C. SACCONE, eds., *The Biogenesis of Mitochondria*. New York: Academic Press, Inc., 1974. Reviews plus original work.

MAHLER, H. R., "Biogenetic Autonomy of Mitochondria." *CRC Critical Revs. in Biochem.*, 1:381 (1973). Speculative.

MARGULIS, LYNN, "The Origin of Plant and Animal Cells." *American Scientist*, 59:230 (1971). Model for the evolution of eucaryotes through a serial symbiosis.

NASS, MARGIT M. K., "DNA Threads, Circles, and Chains." *Science I.*, 5A:46 (1969). Mitochondrial DNA.

_____, "Uptake of Isolated Chloroplasts by Mammalian Cells." *Science*, 165:1128 (1969).

CHAPTER 6

Membranes and the Regulation of Transport

6–1   MEMBRANE STRUCTURE   231
The Lipid Bilayer
The Danielli-Davson Model
The Unit Membrane
The Fluid Mosaic Model
Membrane Diversity

6–2   THE MEMBRANE AS A PASSIVE BARRIER   241
Plasmolysis
Oil/Water Partition Coefficients

6–3   MEMBRANE TRANSPORT   246
Permeases
The Advantages of Proteins as Transmembrane Carriers
Non-Protein Carriers

6–4   METABOLICALLY COUPLED TRANSPORT   251
Group Translocation
Translocation of Ions
The Sodium Pump
Na^+-Coupled Transport

6–5   ENDOCYTOSIS   259

SUMMARY   263

STUDY GUIDE   264

REFERENCES   264

The membranes of a cell are an integral part of its metabolic machinery. The plasma membrane, for instance, helps to determine the composition of the cytoplasm by controlling which materials get in and out and the rate at which they do so. It thus permits the selective uptake of nutrients and secretion of waste products. Membranes found elsewhere in the cell, such as those around the nucleus, mitochondria, chloroplasts, lysosomes, peroxisomes, and so forth, also represent highly selective barriers to the passage of materials. Though most membranes are quite similar in structure, their properties vary considerably. We shall proceed by first outlining those structural features that seem to be common to most membranes, then go on to describe various modifications that have evolved to provide specialized functions.

6-1 MEMBRANE STRUCTURE

When cells are grown together in culture, there will be an occasional merging of two cells to form one. The frequency with which this fusion occurs can be greatly increased by the addition of inactivated *Sendai virus* (a myxovirus, named after a city in Japan), which adheres strongly to the cell surface. Its "sticky" nature glues cells together, providing more opportunity for fusion between neighboring cells (see Fig. 6-1 and Fig. 6-2). Two cells of different strains form a *somatic hybrid* or *heterokaryon*. (A somatic cell is any cell other than a reproductive cell. The name heterokaryon refers to the two unlike nuclei.)

When two cells with distinctively different surface features fuse, the way in which their membranes merge can sometimes be followed. When this experiment is carried out, it is found that the surface features of two fused cells quickly spread themselves around a heterokaryon, producing a hybrid membrane. This process leaves one with the impression that membranes behave more like a fluid than like a solid. One gets the same impression from experiments designed to show where new material is added to a membrane when it grows: except during division, when different parts of the cell are stressed differently, new material seems to be incorporated at a great many different sites. And finally, their fluid nature accounts for the ability of most membranes to seal

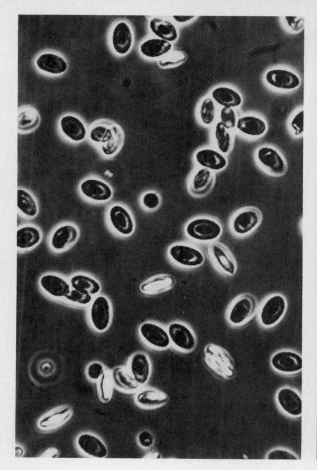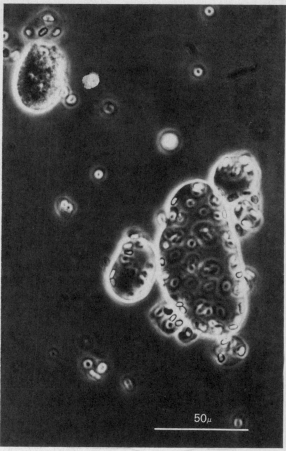

FIGURE 6–1. Cell Fusion. Induced by the presence of inactivated Sendai virus. These phase-contrast photomicrographs are of chicken erythrocytes. LEFT: Before fusion. RIGHT: After fusion. Note the many nuclei in the fused cell. [Courtesy of Z. Toister and A. Loyter, *Biochem. Biophys. Res. Commun.,* **41**:1523 (1970).]

small punctures. If a fine needle is used to pierce a cell, the hole closes when the needle is withdrawn.

All of these observations lead us to the conclusion that the typical membrane is not a rigid structure composed of stiffly connected subunits, but a fluid in which the components can move about. The basis for this structure is apparently a lipid film containing protein plus small amounts of nucleic acid and carbohydrate (see Table 6.1).

THE LIPID BILAYER. The first indications that lipids might be an important constituent of biological membranes came in the last years of the nineteenth century, primarily from the work of E. Overton. In 1895 he published a report on the permeability properties of various membranes, in which he noted how easily lipid-soluble substances penetrate them compared with their relative impenetrability to water-soluble substances. On the basis of "like dissolves like," Overton concluded that the surface permeability barrier of cells must be predominantly lipid. As such, it would readily be penetrated only by lipid-soluble substances.

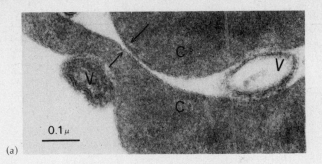

(a)

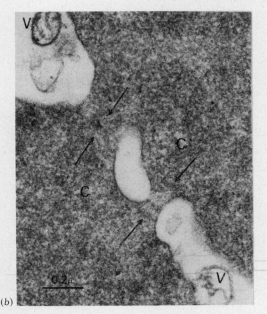

(b)

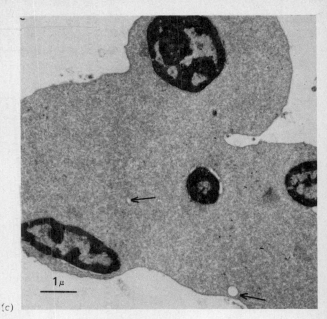

(c)

FIGURE 6–2. The Mechanism of Cell Fusion. Electron micrograph study of Sendai virus-induced fusion of chicken erythrocytes. **(a)** Two cells (C) held in close proximity by attachment to a common virus particle (V). **(b)** The formation of cytoplasmic bridges (arrows) between adjacent cells. **(c)** A fused cell with four nuclei. Note the small vacuoles (arrows), presumably the residue of those formed in the fusion process and visible also in part b. [Courtesy of Z. Toister and A. Loyter, *J. Biol. Chem.*, **248**:422 (1973).]

The most common membrane lipids are the phospholipids—molecules of glycerol containing a negatively charged phosphate ester at one hydroxyl and long chain fatty acid esters at the other two hydroxyls (see Chap. 3). Phospholipids have both a hydrophilic end in the form of a phosphate, often with other polar residues at-

TABLE 6.1   The composition of some membranes[a]

	LIPID	PROTEIN	$\dfrac{[\text{CHOLESTEROL}]}{[\text{POLAR LIPID}]}$
Myelin	80%	20%	0.7–1.2
Chloroplast lamellae	50	50	0
Erythrocytes	20–40	60–80	1
Mitochondrion			
Outer membrane	45	55	0.03–0.09
Inner membrane	25	75	0.02–0.04
Bacteria	20–30	70–80	0

[a] Most data taken from E. D. Korn, *Annual Reviews of Biochemistry*, **38**:253 (1969).

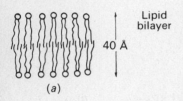

Lipid
bilayer

40 Å

(a)

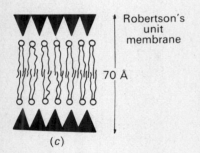

Danielli-Davson
model

Lipid interior

(b)

Robertson's
unit
membrane

70 Å

(c)

FIGURE 6–3. The Early Membrane Models. **(a)** Gorter and Grendel suggested a simple lipid bilayer. **(b)** Later, Danielli and Davson extended the model to include a coating of protein. The interior in their model was an unspecified "lipoid" layer. **(c)** The unit membrane, first proposed by Robertson in the 1950s, explained the trilaminate appearance of many membranes in the electron microscope by including an extended layer of protein on either side of the bilayer.

tached to it, and a hydrophobic end provided by fatty acid "tails." The physical chemist I. Langmuir demonstrated that molecules of this nature arrange themselves into monomolecular films at air–water interfaces. He proposed that the molecules in such films are all oriented with their polar ends into the water and their hydrophobic ends away from the water, a behavior that was discussed in Chap. 2, when micelles were described. Furthermore, Langmuir devised a physical technique for measuring the expanse of such films by gathering the molecules together with a barrier until a resistance is met, indicating a close-packing.

Capitalizing on Langmuir's work, the Dutch scientists E. Gorter and F. Grendel of the University of Leiden (1925) set out to measure the total film size produced by the lipids extracted from human erythrocytes (red blood cells). Knowledge of the physical area covered by the lipids, they reasoned, might enable them to produce a model for the way in which lipids are arranged in membranes. Human erythrocytes were a logical choice for these experiments, since the cells are easy to obtain and were known to be extremely simple. Since human erythrocytes have no nuclei and little or no internal membrane, one might presume that extracted lipids all come from the surface membrane.

Gorter and Grendel extracted erythrocytes with acetone and measured the monomolecular film formed from the extraction product with Langmuir's device, called a *Langmuir trough*. They found a total film area of about 200 square microns per cell, about twice their estimate for the surface area of the erythrocyte. Thus, they concluded that there is just enough lipid in each cell to cover it twice, and that the lipids are associated in a bilayer (see Fig. 6–3a). As it turns out, both area measurements were in error: the extraction process had only removed about three-quarters of the lipid, and the erythrocyte area is closer to 150 square microns. (Their measurements were based on light microscopy and the assumption that erythrocytes are discs when, in fact, they are biconcave.) By a happy coincidence, the two errors approximately cancelled and their conclusion—that the erythrocyte membrane is composed of a lipid bilayer—was quite correct.

The bilayer proposal by itself did not explain all of the known facts. For example, erythrocytes do not always behave as if they have a lipid exterior. For one thing, their surface tension, which is a measure of the tenacity with which molecules at the surface of a liquid cling to their own kind, is far too low. Fats and oils in an aqueous environment have very large surface tensions because of hydrophobic bonding; lipids extracted from erythrocytes do not.

THE DANIELLI-DAVSON MODEL. J. F. Danielli and E. N. Harvey suggested that the anomalously low surface tension of erythrocyte lipids is the result of contamination by protein, which would naturally seek the surface of a lipid droplet and thereby change its character. The hydrophilic behavior of intact cells could also be explained by this assumption, which led Danielli and Hugh Davson to propose the first complete membrane model in 1935.

Danielli, then at Princeton University, envisioned a membrane with a lipoid center, coated on either side with protein (see Fig. 6-3b). The phospholipids in his model are oriented in two monomolecular layers, with their hydrophobic tails toward the inside of the structure and their hydrophilic phosphates on the surface, contacting the layers of protein. The fundamentals of this structure are still accepted today, though with some important modifications.

THE UNIT MEMBRANE. In the middle 1950s, J. D. Robertson, at University College, London, found a way to verify the basic features of the Danielli model with the electron microscope. He devised staining techniques that resolve most membranes into two distinguishable lines on micrographs, whereas both electron and light microscopists had hitherto seen only single lines (see Fig. 6-4). It was later demonstrated by others that partial extraction of the lipids with acetone or enzymes leaves the double lines virtually intact, suggesting that the stain was revealing protein layers on the two surfaces of the membrane. That the original trilaminate structure remained intact after extraction of lipids suggested an important structural role for the associated protein.

FIGURE 6-4. The Trilaminate Structure of Membranes. It is this appearance that gave rise to the unit membrane model. **(a)** Red blood cell membrane. Cytoplasm (Cy). **(b)** A vacuole from a mussel. See also Fig. 1-18a. [(a) Courtesy of J. D. Robertson, Duke University; (b) courtesy of N. B. Gilula and P. Satir, *J. Cell. Biol.*, **51:**869 (1971).]

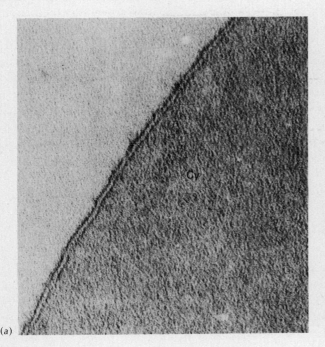

(a)

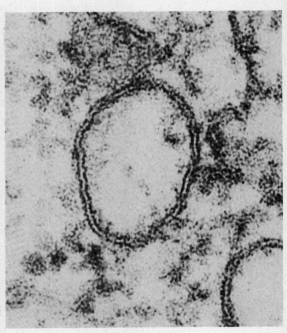

(b)

This idea represented a departure from the concepts presented in Danielli and Davson's paper, as they had dismissed the protein portions rather lightly in assuming that the important part of the membrane, both structurally and biologically, is the lipid barrier.

Robertson's original model has rather thin protein layers on the two surfaces (see Fig. 6–3c), more consistent with an extended protein structure than with the compact globular proteins suggested by Danielli. Furthermore, Robertson's model is not necessarily symmetrical—while the internal surface in the model is coated with protein, the outer surface could be either a glycoprotein (protein with attached carbohydrate) or polysaccharide. These features, along with the single lipid bilayer leaflet 40 to 65 Å thick, identify the *unit membrane* model, a name that implies a homogeneity in structure.

The Fluid Mosaic Model. One gets quite a different impression of membrane structure with another electron microscopical technique, a procedure using freeze-fracture followed by freeze-etching. In this process, the sample is prepared for electron microscopy by freezing, cutting with a microtome, and then subliming part of the surface ice. Structures containing relatively large amounts of water sink away from the split surface, leaving the more resistant portions standing out in relief, as in a metal etching (see Fig. 6–5). The combination of only partial dehydration with a lack of stain eliminates many of the objections of conventional sample preparation. Freeze-etching does not provide exactly the same kind of information as do the older techniques, since it produces a topo-

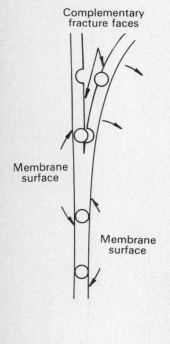

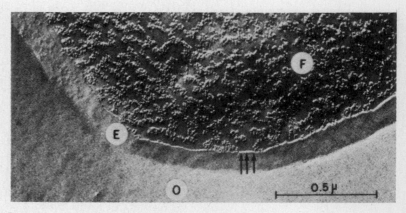

(a)   (b)

FIGURE 6–5. The Freeze-Fracture and Freeze-Etching of Membranes. **(a)** Cutting a solidly frozen specimen tends to fracture membranes through their relatively weak lipid interior as diagrammed here. Proteins embedded in the membrane are left exposed. **(b)** A fractured red blood cell. The fractured surface (F) reveals numerous particles, apparently within the membrane before fracture. The edge of the fracture is marked by triple arrows. Adjacent to this edge is a thin region (E) where the true exterior surface of the membrane has been exposed by removal of water in the etching process. Note the slightly granular appearance. The region marked O is that portion of the outside of the cell that was still covered by ice when the specimen was coated with metal. See also Fig. 1–38. [(b) Courtesy of P. Pinto da Silva and D. Branton, *J. Cell Biol.*, **45**:598 (1970).]

graphical map. But it is unique in often providing a look at membrane interiors, because attempts to cut a solidly frozen specimen cause membranes to fracture down their center as the knife strikes them at a shallow angle.

The freeze-etching and freeze-fracture techniques give membranes a pebbly appearance. These bumps are presumed to be proteins. The intimate association between lipid and protein suggested by freeze-etching satisfactorily explains several puzzling observations. For example, isolated membrane proteins are often very insoluble in aqueous solution under mild conditions unless detergents are present, implying that they have a hydrophobic exterior unable to interact with the surrounding water. And yet, if membrane proteins exist in an ionic environment *in vivo*—facing the polar ends of lipid molecules on one side and the surrounding aqueous phase on the other—then the surface of the protein should be ionic. The lipophilic surface characteristic of these proteins is however, consistent with the idea that lipids are a filler between protein molecules (see Fig. 6–6a).

The presence of protein within membranes, rather than merely on their surface, not only explains the hydrophobic interaction between protein and lipid, but also helps to explain certain aspects of membrane permeability, as we shall see. In addition, it provides enough room to allow the proteins to assume a globular form like that of most enzymes, which frequently have diameters of 40 Å or more. Various physical measurements carried out on membrane proteins suggest that they are indeed globular, and therefore inconsistent with the space allotted them in the unit membrane. One would thus interpret the trilaminate appearance of membranes in thin section as being due to the fact that only those portions of the proteins rising above the surface of the membranes would be stained.

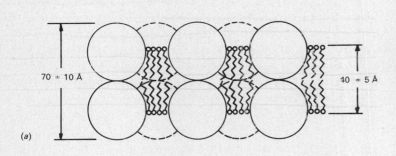

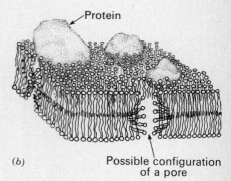

Protein

Possible configuration of a pore

70 ± 10 Å

40 ± 5 Å

(a)

(b)

FIGURE 6–6. The Protein Crystal and Fluid Mosaic Models. The size and solubility properties of membrane proteins suggest that they are not confined to the surface, but are interspersed among the lipids. **(a)** The protein crystal model is a regular arrangement of protein and bilayer. **(b)** A more irregular structure is known as the fluid mosaic model, here including pores. [Compare with Fig. 6–5.] [[a] Courtesy of G. Vanderkooi and D. E. Green, *Proc. Nat. Acad. Sci.* (U. S.), **66**:615 (1970).]

237

In addition, the relatively recent experiments described at the beginning of this chapter suggest that membrane proteins are not rigidly held in place, but are relatively free to move about laterally. This concept of membrane structure has been termed the *fluid mosaic model* by S. J. Singer, who has been one of its most effective advocates (Fig. 6–6b). Though discussion and research on membrane structure continues at a high level of activity, the basic concepts of the fluid mosaic model have become widely accepted as a satisfactory description of most biological membranes.

MEMBRANE DIVERSITY. In the preceding discussion of membrane models, it was tacitly assumed that the structure of most, if not all, membranes can be described by minor variations of a common model. Whether or not that assumption is valid, it tends to ignore the wide variation in chemical composition (see Table 6.1) and physical appearance of membranes from various sources. For example, one can readily distinguish at least three classes of membranes in a typical eucaryotic cell. These are: (1) the inner membranes of mitochondria and chloroplasts; (2) *exoplasmic membranes*, such as the plasma membrane itself; and (3) *endoplasmic membranes*, such as those found in the endoplasmic reticulum.

The inner membranes of mitochondria and chloroplasts were described in Chap. 5. They are relatively thin (50–60 Å), have a very high protein to lipid ratio, and they have a unique appearance in electron micrographs: rather than the expected trilaminate structure, both appear to be composed of a relatively regular array of subunits—respiratory assemblies in the case of mitochondria, and quantasomes in chloroplast lamellae (see Figs. 5–5 and 5–14, respectively). The type and abundance of protein may well be responsible for these properties, which can be accommodated by a fluid mosaic model having sparse areas of bilayer broken up by numerous multi-protein respiratory or photosynthetic assemblies. However, it has also been proposed that these membranes are composed of true subunits, with little or no lipid bilayer.

Exoplasmic and endoplasmic membranes typically have a trilaminate appearance in high-resolution micrographs of thin sections. Exoplasmic membranes include the plasma membrane, the outer cisternae of the Golgi bodies, and some secretory vesicles. The "outer cisternae" of the Golgi bodies refers to the oldest cisternae (new ones are added at the other side—see Fig. 1–14b); they give rise to secretory vesicles that bud from a Golgi cisterna, sometimes fuse with vesicles from the endoplasmic reticulum, then migrate to and fuse with the plasma membrane. (The latter step is similar to the sequence seen in Fig. 6–7, in which the contents of a vacuole are discharged to the extracellular space around an erythrocyte.)

Note that when a vesicle buds from a membrane, the outside–inside relationship of the membrane is preserved—that is, the surface

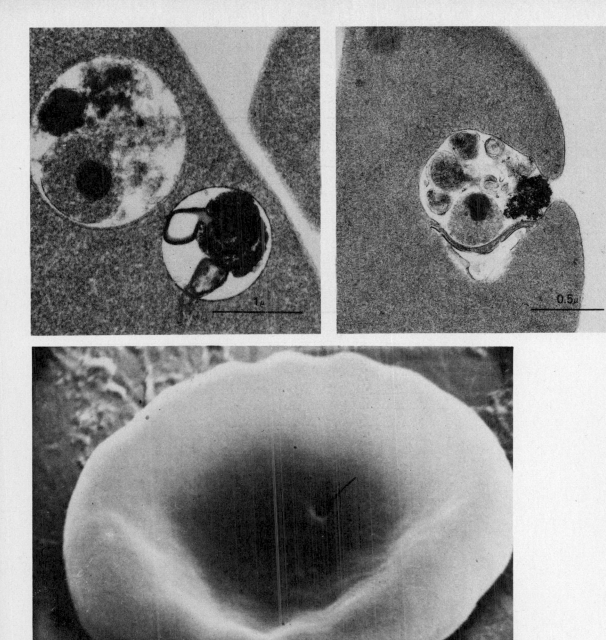

FIGURE 6–7. Fusion of a Vacuole and Plasma Membrane. Note that fusion enlarges the plasma membrane, and that the inside surface of the vacuole becomes a part of the outside surface of the plasma membrane. Any substances contained within the vacuole will be discharged to the extracellular space. TOP: Fusion of a vacuole with the plasma membrane of a human erythrocyte. BOTTOM: Scanning electron micrograph of a similar cell, revealing a pit (arrow) presumably left temporarily by the opening of a vacuole. [Courtesy of B. Schnitzer, D. L. Rucknagel, and H. H. Spencer, *Science*, **173**:251 (1971). Copyright by the American Association for the Advancement of Science.]

that was on the inside of the source membrane lies on the inside of the vesicle. (This is seen in Fig. 1–31, which shows a virus budding from the surface of a cell.) In contrast, when a vesicle opens onto the surface, as in Fig. 6–7, the inside of the vesicle becomes a portion of the outside of the cell. The former process, where the original relationship is preserved, is called *exotropy* (turning out), whereas the alternate process is called *esotropy* (turning in). The distinction is important because membranes are typically asymmetrical, with distinctly different surfaces. The protein components, which are relatively free to move longitudinally, do not ordinarily rotate or tumble, but keep their same side out. In the case of the plasma membrane, for example, only the outside surface contains an abundance of glycoproteins, which are proteins having carbohydrate covalently attached to some of their amino acid residues. (The carbohydrate portion may be composed of a number of different sugars, a common one being D-mannose and its derivatives, the sialic acids and neuraminic acids.) Glycoproteins are more familiar to us as the major constituent of mucus secretions, which contain a type of glycoprotein known as a *mucoprotein*. This polysaccharide coat on a cell, called its *glycocalyx*, contains patches that provide for specific cellular recognition by viruses and by other cells. Glycoproteins, including those found in the cell membrane, receive their carbohydrate portion from the Golgi body.

Endoplasmic membranes include those associated with the nuclear envelope, endoplasmic reticulum, inner cisternae of the Golgi bodies, and outer membranes of the chloroplasts and mitochondria. Endoplasmic membranes tend to have a thinner lipid bilayer than exoplasmic membranes (possibly due to more overlap in the fatty acid side chains), and to have less cholesterol, less carbohydrate, and more protein. A great many enzymes are associated with endoplasmic membranes—for example, those responsible for the oxidation and consequent detoxification of drugs in the liver. In spite of the differences between endoplasmic and exoplasmic membranes, their basic similarity is emphasized by reports of occasional continuities between the two, as in the cells shown in Fig. 6–8a and b.

Endoplasmic membranes not only have a close structural relationship to one another, but there have been repeated observations of interconnections among them: between endoplasmic reticulum and the outer membrane of mitochondria, between endoplasmic reticulum and the nuclear envelope, and elsewhere. The connection between the endoplasmic reticulum and nuclear envelope is particularly significant, as it appears that at least some of the endoplasmic reticulum is derived from the nuclear envelope. This proposal arose from the study of *annulate lamellae*, which have a structure very similar to that of the nuclear envelope, including the presence of pores or annuli (see Fig. 6–8c), and which are thought to mature to become typical cisternae of the endoplasmic reticulum.

6-2 THE MEMBRANE AS A PASSIVE BARRIER

The cytoplasm of the cell has a composition very different from that of the fluid surrounding it. This fact is *prima facie* evidence for a selectively permeable membrane. Though it is clear that molecules do get into and out of a cell, it is just as clear that the passage of substances across the cell membrane is a regulated—and regulatory—function. The first experimental verification of the selective nature of membranes came at about the turn of the century, primarily through the work of several plant physiologists including W. Pfeffer, E. Overton, and others.

PLASMOLYSIS. Early experiments on membrane permeability utilized the phenomenon of *plasmolysis* to demonstrate the selective permeability of the membrane. To plasmolyze a cell, it is suspended in a medium that is hypertonic—i.e., one that contains a higher total solute concentration than the cytoplasm. Water will then leave the cell in an attempt to equalize its concentration on the two sides. If water is the only molecular species capable of making that transfer, the volume of cytoplasm will shrink. With a microscope, one can watch such a cell quickly shrivel down to a small fraction of its former size. In the case of plant and bacterial cells, which have a rigid cell wall outside the plasma membrane, one can see the membrane and wall separate as the volume of the cytoplasm gets smaller (Fig. 6-9). It was through such experiments that the existence of two structures, a wall and a separate membrane, first became clear.

Overton found that cells of certain plant root-hairs could be plasmolyzed by 7.1% sucrose solutions, although 7.0% sucrose (0.21 M) had no such effect. The latter concentration, then, is *isotonic* or *iso-osmolar*—i.e., the total concentration of dissolved molecules and ions is the same as that of the cytoplasm. At a sucrose concentration of 7.5%, a complete and uniform plasmolysis was achieved in ten seconds or less, and the cells remained in their shrunken condition for 24 hours or more. Apparently the cells are not even slightly permeable to sucrose; otherwise, they would slowly regain their former volume (*de-plasmolize*) as sucrose enters them.

However, when Overton used solutions containing 7% sucrose plus about 3% methyl or ethyl alcohol, no plasmolysis was observed. Overton correctly decided that the failure to plasmolyze in relatively concentrated alcoholic solutions is because alcohol can distribute itself equally on the two sides of the plasma membrane

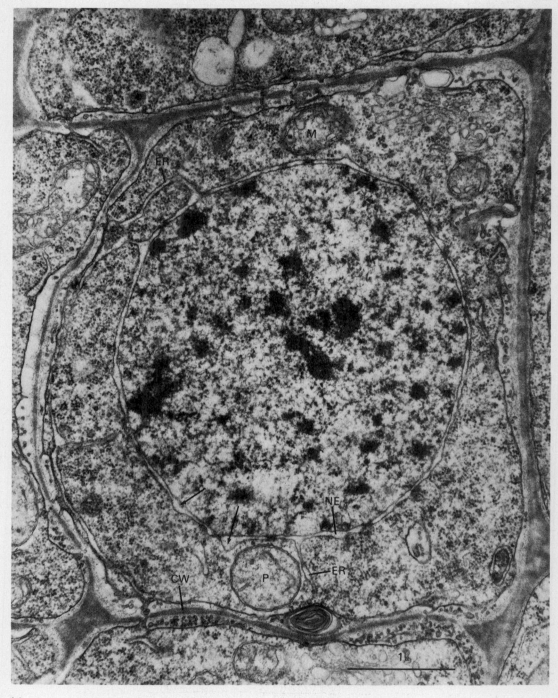

(a)

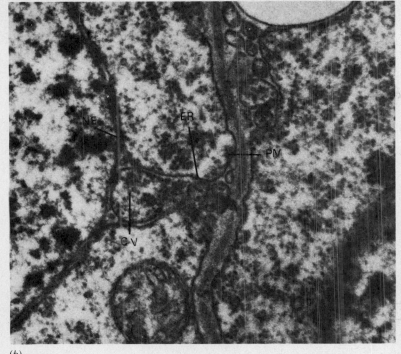

(b)

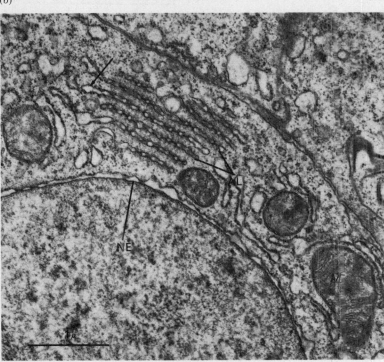

(c)

FIGURE 6–8. The Continuity of Cellular Membranes. **(a)** Androgonial cell from a liverwort (*Basia pusilla*) showing continuity of the nuclear envelope (NE) and endoplasmic reticulum (ER). Arrows indicate possible blebbing of the nuclear envelope. Plastid (P). Mitochondrion (M). Cell wall (CW). **(b)** Detail from another cell, clearly showing the continuity of the NE, ER, and plasma membrane (PM). One can also see a cytoplasmic vesicle (CV) enclosed by the outer nuclear membrane and ER. **(c)** Annulate Lamellae (AL). Their construction is like that of the nuclear envelope from which they are thought to be derived by blebbing. In this photo of chick embryo liver after two days of organ culture, one can see annulate lamellae apparently maturing to become typical rough endoplasmic reticulum (arrow). Note also the ribosomes attached to the outer nuclear membrane. [(a) and (b) Courtesy of Z. B. Carothers, *J. Cell Biol.*, **52**:273 (1972). (c) Courtesy of C. A. Benzo and A. M. Nemeth, *J. Cell Biol.*, **48**:235 (1971).]

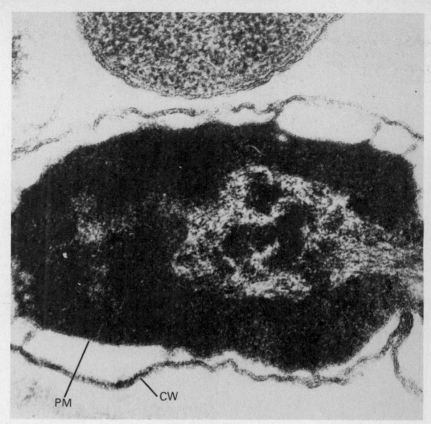

FIGURE 6–9. A Plasmolysed Cell. An *E. coli* cell plasmolysed with 20% sucrose. (Almost all cells are impermeable to sucrose.) Note connections between the plasma membrane (PM) and the cell wall (CW). A portion of a normal, unplasmolysed cell is seen at the top. The light area in the plasmolysed cytoplasm is DNA. [Courtesy of M. E. Bayer, *J. Gen Microbiol.*, **53**:395 (1968). Reprinted by permission.]

in less time than plasmolysis requires—in other words, in something less than ten seconds. The membrane, so impermeable to sucrose, must be extremely permeable to these simple alcohols.

In an attempt to better define the permeability characteristics of the plasma membrane, Overton repeated his plasmolysis experiments with a wide range of solutes, taking care to use relatively nontoxic materials in order to keep the cells from dying. He found that a number of substances, including most lipids, can penetrate cells readily. Others, such as sucrose, cannot get in at all. A third group was found to penetrate the cells only at a very slow rate: glycerol, the hexoses, and amino acids, for example, cause a rapid plasmolysis followed by a slow water regain, indicating a slow penetration of solute. When the chemical nature of the various solutes is examined, it becomes clear that in most cases there is a direct correlation between ease of entry and the lipid solubility of the substance. Though there are some notable exceptions to this rule, water itself being one of them, the cell seems to present a lipid barrier to many substances. This observation, of course, is consistent with the membrane models presented earlier and was the basis for proposing them.

OIL/WATER PARTITION COEFFICIENTS. To put the plasmolysis experiments on a more quantitative basis, later investigators measured and compared the *permeability coefficients* of various solutes with the *oil/water partition coefficients* of the same materials. Permeability coefficients are easily measured by determining the rate of entry of radioactively labelled solute into the cytoplasm at various external concentrations. An oil/water partition coefficient, on the other hand, is measured by shaking the solute in an oil-water mixture and then letting the phases separate. The partition coefficient is the concentration found in the oil phase at equilibrium, divided by the concentration found in the aqueous phase. The relationship between these two coefficients is shown in Fig. 6–10.

It is clear from these experiments that some molecules are able to enter a cell by simple diffusion. Diffusion, however, represents a movement from a region of higher concentration to a region of lower concentration. Therefore, it is possible for diffusion to maintain a constant flow into a cell only if the molecules are chemically altered immediately upon entry, thus keeping their internal concentration lower than their external concentration. Similarly, waste products may leave the cell by simple diffusion as long as they are washed away from the cell's surface at a rate fast enough to maintain a favorable concentration gradient for efflux—in other words, as long as the external concentration of the waste product remains less than its cytoplasmic concentration.

Diffusion is an attractive mechanism for getting things in and out of a cell because it does not require any specialized apparatus,

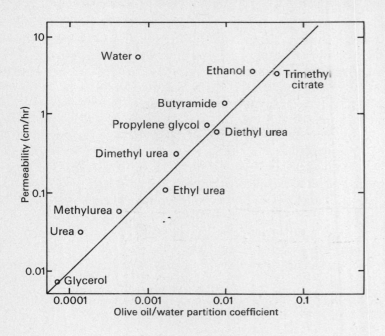

FIGURE 6–10. Permeability to Solutes as a Function of Oil/Water Partition Coefficient. For many substances, there is a direct relationship between cell permeability and lipid solubility. For others, however, the relationship fails completely. Note that water is in the latter category. [Data replotted from R. Collander, *Physiol. Plantarum*, **2**:300 (1949).]

nor does it require any expenditure of energy on the part of the cell. Even maintaining a favorable gradient for entry through chemical modification is not necessarily expensive, since modification can be the first step necessary to the utilization of the material.

Free diffusion, however, does not explain how a great many substances that are virtually insoluble in lipids, including certain ions and most natural metabolites and waste products, nevertheless get into and out of cells at relatively rapid rates. It has been suggested, therefore, that these substances penetrate the lipid portion of the membranes through pores consisting of 4 to 5 Å hydrophilic channels such as those shown in Fig. 6–6b. The problem with this idea is that it does not explain the very great selectivity exercised by cells. However, a few substances, including water, may be regularly moved through pores—very small pores would explain the high permeability to water without altering the selectivity to most other substances because water is the smallest biologically important molecule.[1]

Thus, although diffusion studies do confirm the existence of a lipid barrier, in keeping with our models, diffusion remains an incomplete explanation for membrane permeability. For a cell to live, many molecules must be assisted in their passage across the membrane barrier. We refer to this process as *membrane transport*.

6–3 MEMBRANE TRANSPORT

Specific transport mechanisms are required to account for the degree of selectivity exercised by cells, and to allow the passage of essential lipid-insoluble molecules (e.g., sugars) at rates fast enough to maintain life. Influx and efflux by diffusion through the lipid barrier of a membrane, or through pores therein, cannot provide the degree of sophistication actually attained by cells in regulating transmembrane movement. It is now well established that many substances gain entry to the cytoplasm, or are removed from it, by a reversible combination with specific molecules designed to facilitate translocation across the membrane. Once on the other side, the transported material leaves its *carrier*, which is then free to assist in the passage of the next molecule much as a ferry assists automobiles and passengers to cross a river. The process, if it does not require energy, is called *facilitated diffusion* (see Fig. 6–11).

PERMEASES. Carrier-assisted transport exhibits a very great selectivity, so that one molecule may enter readily whereas a nearly identical molecule is totally excluded. For example, glucose and

---

1. A water molecule is roughly spherical, with a diameter of less than 3 Å. Because of their shell of hydration, all ions are larger.

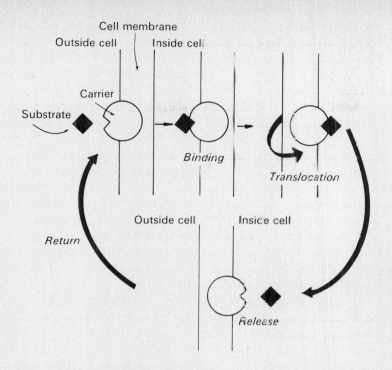

FIGURE 6–11. Facilitated Diffusion. In this schematic, the substance is bound to the carrier outside the cell, transported inside by movements of the carrier, then released. Since binding and release are straightforward equilibria, transport can only occur "down" an electrochemical gradient.

2-deoxyglucose are apt to compete with each other for entry to a cell, although neither affects the transport of galactose.

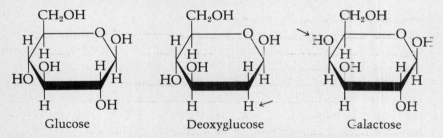

Since galactose and glucose are isomers, differing only in the position of the hydroxyl at carbon 4, there must be a highly specific receptor for glucose and a different one for galactose.

This ability to make extremely fine distinctions between closely related molecules is usually associated with the binding sites of enzymes and certain other proteins, because only proteins have the molecular variety and structural flexibility necessary to provide a wide range of specificities. Accordingly, the transport of certain molecules is attributed to a class of proteins called *permeases*, though it should be emphasized that not all membrane carriers are proteins. A number of different permeases are present in the cell, each type assisting in the transport of a different class of substances. The suffix "-ase," normally associated with enzymes, is used in order to emphasize the similarity between these transport molecules and true enzymes: (1) permeases accelerate transport; (2) they provide it with great selectivity; and (3) they are themselves unchanged in the process, being recycled after each assisted entry or exit. In other words, permeases seem to catalyze the entry and exit of other molecules.

The analogy between permeases and enzymes appears to break down at one very important point: enzymes, being true catalysts, do not change the equilibrium position of a reaction but merely the rate at which equilibrium is approached; permeases, on the other hand, may greatly alter an equilibrium. One might expect that entry of a substance, whether assisted or not, should be possible only so long as the substance's external concentration is greater than its internal concentration. But, in fact, many substances may be accumulated to internal levels many times greater than their concentration in the surrounding medium, even when the substances cannot be metabolized or chemically altered in any way. This phenomenon, called *active transport*, involves the coupling by a permease of transport to a more favorable reaction, such as the hydrolysis of ATP, just as an enzyme can couple two reactions. This system will be treated in more detail later.

THE ADVANTAGES OF PROTEINS AS TRANSMEMBRANE CARRIERS. On the surface, proteins might not seem the best choice for carriers, since proteins are relatively large and seemingly "expensive" to make. The advantages that they bring to the transport role are as follows:

1. Proteins are specific in their capacity to bind other molecules. They may be given exactly the right shape, and precisely the right distribution of positive and negative charges, of hydrophobic and hydrophilic regions, and of reactive groups of various sorts to allow them to make fine distinctions between those molecules that will be bound and those that will not. This specificity can be of great importance to the cell, for a close regulation of metabolic activity can only be achieved if the entry and exit of molecules are effectively controlled. Thus, it is possible for the cell to provide a mechanism for a particular molecule to get in and out while in the presence of much larger quantities of other kinds of molecules. Furthermore, the rate of transport of a given substance is easily controlled by regulating the number of specific carriers.
2. The capacity of proteins to alter their binding affinity for substrate has some important applications when this function is present in a transport molecule. In particular, it offers the same possibilities for control (allosteric control) as seen in cooperative systems.
3. Proteins, as catalysts, also make active transport possible by allowing an obligate coupling between two reactions—the energetically unfavorable transport, and an energetically favorable reaction such as the hydrolysis of ATP.

And finally, it should be pointed out that the use of proteins for transport is actually a conservative choice, for they are the direct products of genes. When small organic molecules are chosen as carriers, their synthesis is apt to require not just one, but several proteins in the form of enzymes. That is, the production of a protein permease may require only one gene, whereas establishing a biosynthetic pathway to produce any other kind of carrier might require several enzymes, and therefore several genes.

A number of proteins have been isolated that probably serve as carriers for membrane transport, although it is difficult to prove that they actually have this function *in vivo* because the appropriate assay (measure of biological activity) does not exist *in vitro*. One can measure their capacity to bind the substance to be transported, but of course a protein does not have to be a membrane carrier to have a specific affinity for a small molecule. Any enzyme, and a host of other proteins, also fit that description.

However, some of these isolated proteins have been shown to be necessary for transport, even if it should turn out that they themselves are not permeases. For example, when bacteria are plasmolyzed in sucrose and then quickly diluted with water, their rapid re-expansion causes some proteins to be released and some transport capacity to be lost. That some of the released proteins are carriers was inferred from the observations that: (1) mutant cells lacking the capacity to transport a particular substance (called *cryptic mutants*) also fail to release the corresponding binding protein when osmotically shocked; (2) in cases where transport can be induced (i.e., when the capacity is developed only after exposure to the substance), uninduced cells fail to release the binding protein; (3) the binding constant between a substance and the free proteins has been shown in several cases to be virtually identical with the binding constant for transport of the same substance in normal cells; and (4) there are reports that incubation of the shocked cells with solutions of some of the released binding proteins at least partially restores lost transport capacity.

NON-PROTEIN CARRIERS. Only a few small molecular weight carriers have been isolated. This is not surprising, as they would be present in only tiny quantities in the cell and would therefore be difficult to identify. Much of the work on small organic transport molecules has centered on the ionophorous (ion-carrying) antibiotics. These are generally macrocyclic (ringlike) compounds, produced by microorganisms and capable of sequestering inorganic ions. By virtue of its lipid solubility, the ion–antibiotic complex can diffuse through a lipid barrier that would be quite impermeable to the ion itself.

One of the more widely studied ionophorous antibiotics is *valinomycin* (see Fig. 6–12). It is a 36-atom ring of alternating hydroxyl acids and amino acids.

The ionophorous antibiotics are capable of increasing the perme-

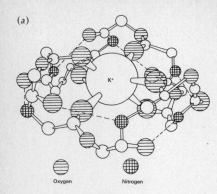

Oxygen   Nitrogen

FIGURE 6–12. Valinomycin. **(a)** Three-dimensional representation based on X-ray crystallography. Dotted lines indicate possible hydrogen bonds. The side chains, which would point either straight up or straight down, are omitted. Note that K⁺ is held in the 4.5 Å diameter pore by bonds to six oxygens **(b)** Chemical structure of valinomycin. [Adapted from M. Pinkerton, L. K. Steinrauf, and P. Dawkins, *Biochem. Biophys. Res. Commun.*, **35**:512 (1969).]

(b)

Picrate

ability of both natural and artificial lipid membranes to small inorganic ions. They often exhibit remarkable selectivity in this role. For instance, valinomycin has about a 10,000-to-1 preference for K⁺ over Na⁺, an observation that is not easily explained merely on the basis of ionic size. Valinomycin is a powerful uncoupler of mitochondrial oxidative phosphorylation, presumably by increasing K⁺ permeability and thus upsetting transmembrane potentials. This is also the basis for its antibiotic action.[2]

The biological action of the ionophorous antibiotics has been studied with model systems such as simple chloroform barriers. In one such experiment, chloroform was placed at the bottom of a U-tube, with a KCl solution above it in one arm and water above it in the other (see Fig. 6–13). Valinomycin, which is capable of transporting a cation–anion pair, was added to the chloroform. Although potassium is the cation favored by valinomycin, K⁺Cl⁻ is not a suitable ion pair for the antibiotic. However, picrate (Pc⁻ or 2,4,6-trinitrophenol) was found to be transportable with potassium. Thus, when potassium picrate was added to the KCl solution on one side, K⁺Pc⁻ ion pairs were transported across the chloroform by diffusion and concentrated on the other side until the following equilibrium was satisfied:

---

2. A compound is an antibiotic if it stops the growth of bacteria. This does not necessarily mean that it is medically useful, as it may also kill people.

$$[K^+]^L[Pc^-]^L = [K^+]^R[Pc^-]^R \qquad (6\text{--}1)$$

Here L and R represent the left and right arms of the U-tube, respectively. This equilibrium relationship was obeyed regardless of the starting concentrations of potassium picrate and potassium chloride on the two sides. The Cl^- is not picked up by valinomycin, and is thereby effectively ignored when the equilibrium is established.

Because valinomycin itself is uncharged, it must move both a cation and an anion to maintain neutrality. (Charged species are generally less soluble in lipids.) Other ionophorous antibiotics (e.g., schizokinen, Fig. 6–14) are charged and may therefore transport a single ion, promoting an ion exchange (e.g., Na^+ for K^+, H^+ for K^+, etc.) or creating an electrostatic gradient, in which case the equilibrium position will be altered. The principle is that the carrier alone, or the complex of carrier and ion(s), is lipid-soluble and can therefore diffuse across the lipid barrier. An ion or ion pair is picked up on one side according to the equilibrium constant between it and the carrier, and dropped on the other side according to the equilibrium prevailing there.

Since the transport of K^+ by valinomycin is driven only by a concentration gradient, the process is facilitated diffusion. The name indicates that the carrier improves permeability to a substance without altering the direction in which the substance would tend to move on its own. Facilitated diffusion, usually with proteins acting as carriers, is widely employed by living cells. However, cells are also capable of transporting material against concentration and voltage gradients by coupling transport to an energetically favorable reaction, generally the hydrolysis of ATP. The process is then referred to as active transport.

6–4 METABOLICALLY COUPLED TRANSPORT

A cell can accumulate a substance to internal concentrations far in excess of the expected equilibrium concentration in three ways: (1) by removing the transported molecules from solution once they are inside the cell; (2) by chemically altering the molecules during or

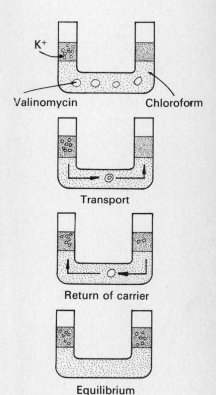

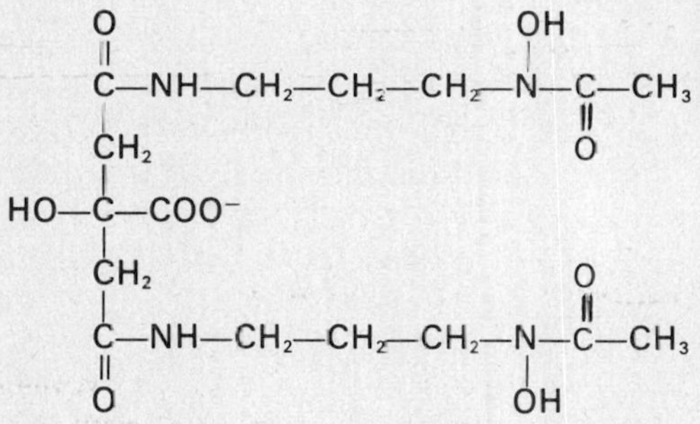

FIGURE 6–13. A Model System for Facilitated Diffusion. Valinomycin can transport a K^+/anion pair through the chloroform barrier until an equilibrium described by equation (6–1) is attained.

FIGURE 6–14. Schizokinen. This molecule is thought to be the natural iron-transport carrier of *Bacillus megaterium*.

after transport; or (3) by coupling transport directly to a second reaction that is energetically favorable.

The first of these processes may be accomplished by binding the transported molecule to a receptor site or by causing its precipitation. These are the suspected mechanisms in a large number of cases, but their existence is difficult to prove. One instance where precipitation is known to operate is in the accumulation of Ca^{2+} by cisternae of the endoplasmic reticulum of muscle, where the ions are stored as a calcium phosphate gel. Even in this instance, however, there is a net excess of free ion, indicating that an active transport mechanism is also at work.

The second type of accumulation process, chemical alteration, is the traditional explanation for the uptake of many simple sugars, for they are immediately phosphorylated. The phosphosugars cannot be transported back out, so the combination of facilitated diffusion followed by phosphorylation ensures their accumulation. When the phosphorylation is an integral part of the transport step, the mechanism is given the special name of *group translocation*, which is actually an example of the third type of accumulation process mentioned above.

GROUP TRANSLOCATION. A system for sugar transport in many bacteria, discovered in the mid-1960s, consists of the following reaction:

$$\text{sugar} + \text{PEP} \longrightarrow \text{sugar-P} + \text{pyruvate} \qquad (6\text{--}2)$$

In other words, phosphoenolpyruvate (PEP) is used as the phosphate donor. The standard state free energy change for this reaction is about -10 kcal/mole because of the properties of PEP, making the reaction virtually irreversible. The regeneration of PEP can take place in several ways, including the catabolism of glucose in the glycolytic pathway. However, another mechanism is provided through phosphoenolpyruvate synthetase:

$$\text{pyruvate} + P_i + \text{ATP} \longrightarrow \text{PEP} + \text{AMP} + PP_i \qquad (6\text{--}3)$$

This reaction is followed by the hydrolysis of pyrophosphate ($PP_i \rightarrow 2P_i$), pulling reaction (6–3) to the right. Note that equation (6–3) is also part of Hatch and Slack's alternate pathway of carbon fixation, described in Chap. 5. The responsible enzyme is found only in bacteria and in plants.

The group translocation of sugars requires at least three proteins: a small (under 10^4 amu) protein called HPr, plus two enzymes. Transport occurs in two steps, the sum of which is equation (6–2):

$$\text{PEP} + \text{HPr} \xrightarrow{E_1} \text{pyruvate} + \text{P—HPr} \qquad (6\text{--}4)$$

$$\text{P—HPr} + \text{sugar} \xrightarrow{E_2} \text{sugar—P} + \text{HPr} \qquad (6\text{--}5)$$

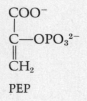

$$\begin{array}{l} COO^- \\ | \\ C\text{—}OPO_3^{2-} \\ \| \\ CH_2 \end{array}$$

PEP

There appears to be a specific enzyme (E_2) for each sugar or class of sugars utilizing this mechanism. In fact, E_2 may also be the membrane carrier itself.

Group translocation can accumulate sugars until a chemical equilibrium is reached in which the unfavorable free energy change for further influx is just balanced by the favorable free energy change of the coupled reactions.

From Chap. 2, we can predict that the Gibbs free energy change for transport of a substance, S, should be given by an equation of the following form

$$\Delta G = - RT \ln \frac{[S]_{ext}}{[S]_{int}} \qquad (6-6)$$

where R is the gas constant and T is the absolute temperature. When $[S]_{ext}$ is greater than $[S]_{int}$, the free energy change for entry will be favorable, as expected (i.e., $\Delta G < 0$). At 298 °K (25 °C), the value is

$$\Delta G = - 1360 \log \frac{[S]_{ext}}{[S]_{int}} \text{ cal} \qquad (6-7)$$

Since equation (6-2) has a $\Delta G° \approx -10$ kcal/mole, and since $\log 10^{-7} = -7$, a sugar could in theory be concentrated to internal concentrations that are more than 10^7 times greater than its external concentration.

TRANSLOCATION OF IONS. When the substance to be transported is charged, the existence of electrical as well as concentration gradients must be considered. Nearly all cells have a voltage gradient across their plasma membrane—typically, 50-100 mV (millivolt or 10^{-3} volt), negative inside with respect to the outside. Because of this potential, and because many transported substances are ions, the complications conferred by electrical considerations cannot be ignored.

The free energy change due to the translocation of an ion is, under standard conditions (25 °C, 1 atm, etc.)

$$\Delta G = z \left(\frac{F}{4.18} \right) (E_{int} - E_{ext}) \text{ cal} \qquad (6-8)$$

Here the charge on the ion is given by z (+1 for Na^+, −1 for Cl^-), F is the Faraday constant, and E_{int} and E_{ext} are the electrical potentials (voltages) inside and outside, respectively. When the Faraday constant is given in its usual units, 96,500 coulombs, and the electrical potential is in volts, 4.18 converts the answer to calories, as shown. We could also write equation (6-8) in a simpler form,

$$\Delta G = z(23,068) \Delta E \text{ cal} \qquad (6-9)$$

where ΔE is the membrane potential (often designated $\mathscr{E}$), arbitrarily taken as zero on the outside. Note that a negative membrane potential (negative on the inside with respect to the outside) results in a negative ΔG for entry of a cation—i.e., spontaneous influx.

The total free energy change for entry of an ion will be the sum of the contributions made by its concentration gradient, equation (6–6), plus that due to the voltage difference, equation (6–8). When the two influences are equal and opposite, equation (6–6) can be combined with equation (6–8) to get the *Nernst* equation:

$$RT \ln \frac{[S]_{ext}}{[S]_{int}} = \left(\frac{zF}{4.18} \right) (E_{int} - E_{ext}) \qquad (6\text{–}10)$$

An ion obeying this equation is in *electrochemical equilibrium* and so shows no tendency to move in either direction. Thus, the measured concentration ratio of an ion at equilibrium can be used to predict the potential difference that just balances this concentration:

$$\Delta E = \frac{RT}{(zF/4.18)} \ln \frac{[S]_{ext}}{[S]_{int}} \qquad (6\text{–}11)$$

or

$$\Delta E = \frac{0.058}{z} \log \frac{[S]_{ext}}{[S]_{int}} \text{ volts at } 25\,°C \qquad (6\text{–}12)$$

Likewise, measuring the voltage permits one to predict the counterbalancing concentration ratio at equilibrium:

$$\frac{[S]_{ext}}{[S]_{int}} = e^{(zF\Delta E)/(4.18RT)} \qquad (6\text{–}13)$$

Since $\log 10 = 1$, a ten-fold concentration ratio of a monovalent ion can be offset by a 58 mV (0.058 V) electrical gradient. Similarly, a hundred-fold ratio is equivalent to a 116 mV gradient ($\log 100 = 2$), and so on.

The membrane potentials found in nearly all cells are due to a net excess of negatively charged ions (nucleic acids, proteins, and so forth) in the protoplasm. The concentration of cations needed to equalize these negative charges is kept artificially low by the selective permeability of the membrane and by the active extrusion of some of the cations that do get in. The active transport mechanisms used to translocate ions against their electrochemical gradients are referred to as pumps. The best known among them is the so-called sodium pump.

THE SODIUM PUMP. The sodium pump was discovered by A. L. Hodgkin and R. D. Keynes in 1955, and was associated with ATP hydrolysis *in vitro* by J. C. Skou in 1957. It was the first documented instance of active transport. The sodium pump (see Fig. 6–15) is a mechanism for maintaining a low internal sodium concentration in the face of an unfavorable electrochemical gradient and constant Na⁺ influx due to leakage of various types. It functions by coupling the hydrolysis of ATP with the removal of Na⁺ from the cytoplasm.

Three varieties of sodium pump can be identified: those that exchange Na⁺ and K⁺ on a one-for-one basis; those that translocate Na⁺ along with an anion such as Cl⁻; and those that transport Na⁺ alone. The first two leave the membrane potential unchanged; the third type of pump clearly changes the potential of the membrane, and for this reason it is referred to as an *electrogenic sodium pump*.

One of the most widely studied sodium pumps is found in mammalian erythrocytes. When erythrocytes are rapidly transferred to a medium of low solute concentration, water will enter the cells, causing them to swell and perhaps to lyse. (This process, called *hemolysis*, is the opposite of the unfortunately named plasmolysis.) The small holes in the surface membrane caused by hemolysis seal quickly, but they permit hemoglobin to escape and allow external fluid to enter, a process that is sometimes referred to as "reverse hemolysis." Reverse hemolysis permits certain characteristics of the erythrocyte's sodium pump to be studied.

Reverse hemolysis was used to demonstrate that the erythrocyte sodium pump is activated by a rise in the internal concentration of Na⁺ or by a rise in the external concentration of K⁺. The vectorial nature of ion translocation and the inside-vs.-outside differences in

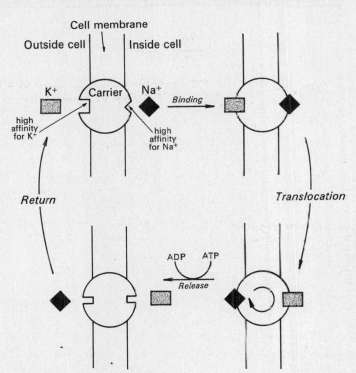

FIGURE 6–15. The Sodium Pump. In this hypothetical model, the carrier can assume two configurations, one with high affinities for the ions and one with low affinities. The transition is fostered by the hydrolysis of ATP, and lasts until the carrier has returned to its initial position—i.e., with its K⁺ binding site on the outside and its Na⁺ site on the inside. As a result, Na⁺ is pumped out and K⁺ is pumped in.

response to Na^+, K^+, and ATP, led to the suggestion that the enzyme is located within the cell membrane, with binding sites exposed on both sides. The examples isolated so far are consistent with that model, since **they** are relatively large (several hundred thousand amu) lipoprotein complexes with binding sites for Na^+, K^+, and ATP.

One of the most important functions of the sodium pumps (including indirectly the pumps that exchange one ion for another) is osmotic regulation. That is, they maintain the concentration of dissolved substances at a level such that the cells neither swell nor plasmolyze—especially important in animal cells, since animal cells lack a rigid cell wall to contain the cytoplasm. Consider, for example, that isotonic (iso-osmolar) saline is 0.15 M NaCl. Since pure water has a concentration of 55.55 M, the active translocation of one Na^+ and Cl^- pair causes the passive translocation of 370 water molecules (55.55/0.15) in order to maintain iso-osmotic conditions. Water itself is not actively transported, since this much more efficient system is available.

NA$^+$-COUPLED TRANSPORT. Because of the activity of sodium pumps, the typical cytoplasmic concentration of Na^+ is at least ten times less than its extracellular concentration—e.g., 9.2 and 120 mM, respectively, in frog muscle. This imbalance is in contrast to K^+, which is apt to be much more common in the cytoplasm than outside the cell (140 and 2.5 mM in frog muscle). Aside from osmotic regulation, one consequence of the Na^+ gradient is the transmembrane potential mentioned earlier. By applying equation (6–8) or (6–9), one can predict that the Gibbs free energy change for the entry of any cation into frog muscle, since the latter has a transmembrane potential of -90 mV, is -2.1 kcal/mole. And from equation (6–6) or (6–7), the concentration-driven Gibbs free energy change for Na^+ entry is -1.5 kcal/mole. Hence, Na^+ influx is driven by a total Gibbs free energy difference of about -3.6 kcal/mole. In contrast, the same analysis of K^+ reveals that it is very near (but not exactly at) electrochemical equilibrium.

The large negative free energy change for sodium entry has been exploited by cells in numerous ways. The electrical excitability of nerve and muscle membranes, for example, relies on the sodium gradient (see Chap. 7). But a more widely used device (among animal cells, where the sodium pump is important) is the Na^+-coupled entry of metabolites.

Robert K. Crane investigated glucose accumulation of the intestinal epithelial cells (the "brush border cells") of the hamster, and concluded that glucose accumulation against its concentration gradient is accomplished by an obligate coupling of glucose and Na^+ transport. The carrier for this process must have binding sites for both Na$^+$ and glucose, and must exist in two conformational states—one with a low affinity for glucose and one with a high affinity for glucose.

At the surface of the membrane, where the Na$^+$ concentration is high relative to its concentration inside the cell, the carrier accepts a Na$^+$ ion, which causes it (the carrier) to assume the conformation with high affinity for glucose. Once both substrates are bound, the carrier rotates or translates to expose its binding sites to the cytoplasm. Because of the low internal Na$^+$ concentration, the transported sodium ion is released, causing the protein to relax to its conformation having a low affinity for glucose. Hence glucose is also released, freeing the carrier to return for another load (see Fig. 6-16).

This model for Na$^+$-coupled transport results from the observations that glucose transport is Na$^+$-dependent and that the rate of its transport increases with an increasing extracellular concentration of Na$^+$. Furthermore, a number of monovalent cations, including K$^+$, compete for the sodium binding site but inhibit the transport of glucose. Competition from K$^+$ could help force the internal release of glucose because the cytoplasmic K$^+$ concentration is relatively high compared to the external K$^+$ concentration (the opposite of the Na$^+$ gradient). Such a model also predicts that if the transmembrane gradient of Na$^+$ were reversed, the cell should be able to transport glucose out against a concentration gradient as

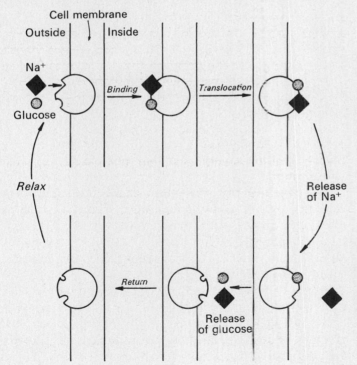

FIGURE 6-16. Sodium-coupled Transport. The favorable electrochemical gradient for Na$^+$ entry can be utilized to transport a second substance against the latter's own gradient. In this model, binding Na$^+$ increases the affinity of the carrier for glucose. Both substances are then translocated and released together inside the cell.

easily as it normally transports glucose inward. Crane performed this experiment and reported that the mechanism is, indeed, reversible.

Many other substances (e.g., some amino acids) are now known to be transported by a Na^+-coupled mechanism, and the phenome-

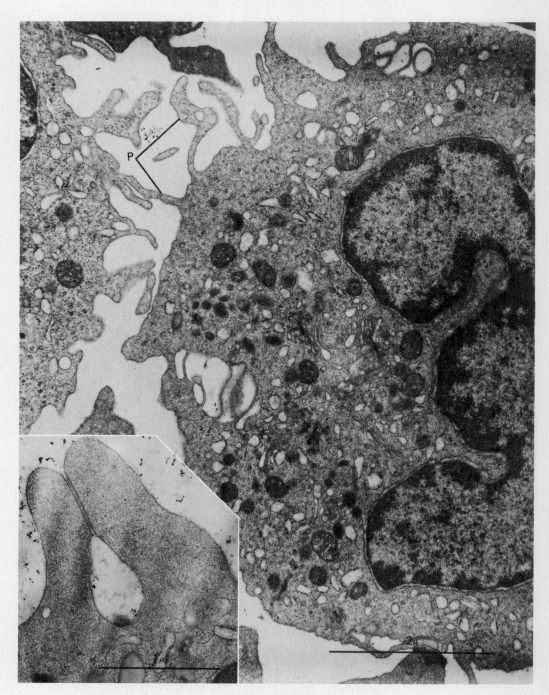

(a)

non has been observed in many cell types. It depends on a carrier with variable and controllable affinity for the substance that must be moved against its concentration gradient. Cooperative behavior of this type is normally associated only with proteins, which emphasizes again the advantages of utilizing proteins for transmembrane carriers, even when a catalytic activity is not a part of the transport process.

6-5 ENDOCYTOSIS

Most cell membranes can enclose external material in a vacuole without any break in the continuity of the surface. The process, called *endocytosis*, is a kind of active transport (see Fig. 6-17). If the material is particulate, the term *phagocytosis* may also be used; endocytosis of liquids is called *pinocytosis*. The reverse sequence, whereby materials are discharged from the cell, is an *exocytosis* (see Fig. 6-7). Exocytosis is the mechanism commonly

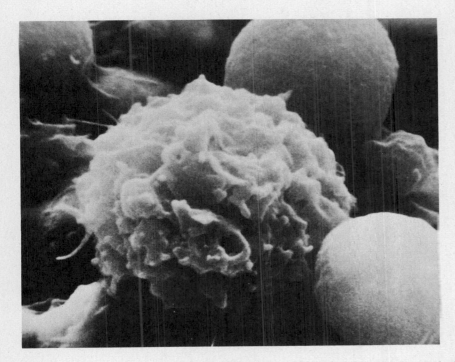

(b)

FIGURE 6-17. Endocytosis. **(a)** The undulating surface membrane of these human leukemic white blood cells demonstrates the mechanism of endocytosis. Note the numerous pseudopodia (P) and the newly formed and nearly formed vacuoles. The indented nucleus is characteristic of the white blood cells known as monocytes. INSET: Detail of another cell, showing the membranes of two pseudopodia about to fuse to form a vacuole. (The zipper-like junction is not observed in specimens from other sources.) **(b)** Scanning electron micrograph of a normal human monocyte, giving a three-dimensional view of the undulating surface membrane. The smaller, smooth cell at lower right is a type B lymphocyte, indistinguishable by other forms of microscopy from the type T lymphocyte seen in Fig. 1-38. Lymphocytes exhibit only minimal phagocytic activity, as one might guess from this micrograph. [(a) Courtesy of F. T. Sanel. Large photo from *Science*, **168:**1458 (1970). Copyright by the A.A.A.S. (b) Courtesy of Aaron Polliack. Procedure described in *J. Exp. Med.*, **138:**607 (1973).]

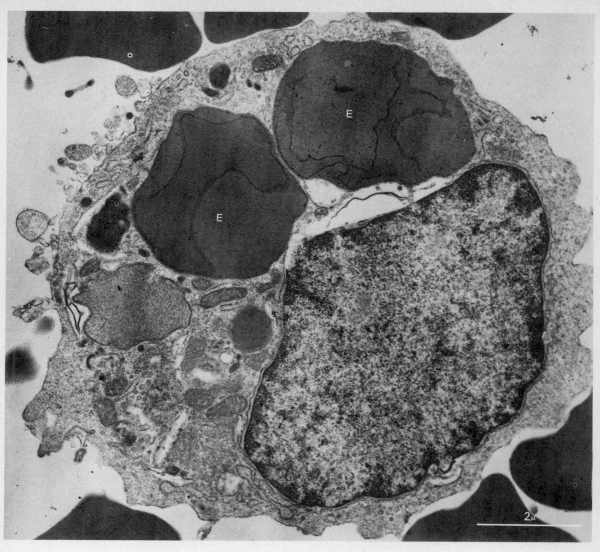

(a)

FIGURE 6–18. Phagocytosis and Digestive Vacuoles. **(a)**
Mouse macrophage with ingested erythrocytes (E). **(b)**
Portion of a guinea pig macrophage, containing ingested
bacteria (B) in phagocytic vacuoles or phagosomes (PV).
Lysosomes (L) are seen in the Golgi region (G). Centriole
(Ce). INSET: Fusion of a lysosome and bacteria-con-
taining phagosome to form a digestive vacuole. **(c)** Por-
tion of a rabbit macrophage, stained for the lysosomal en-
zyme aryl sulfatase. Note the spread of the enzyme from
the lysosome. In **(d)**, from a guinea pig, the enzyme
nearly surrounds a trapped bacterium. [(a) Courtesy of
F. Gudat, T. Harris, S. Harris, K. Hummeler, *J. Exp.
Med.*, **132**:448 (1970); (b)-(d) courtesy of B. Nichols,
D. Bainton, and M. Farquhar, *J. Cell Biol.*, **50**:498 (1971).]

employed in secretion, as in the discharge of insulin by pancreatic cells.

The most comprehensive studies of endocytosis have been carried out on amebas of the genera *Amoeba* and *Chaos*. For example, it has been found that proteins and dyes which most effectively stimulate phagocytosis in *Amoeba* are positively charged at the pH employed in the experiments. Various salts of sodium, potassium, calcium, magnesium, lithium, and ammonium were also found to stimulate endocytosis—in fact, Mg^{2+} or Ca^{2+} is essential in some systems. In addition, the acidic and basic amino acids were found to be effective, but not those with uncharged side chains. Sugars and nucleic acids were relatively ineffective as inducers, though they could be included in vacuoles formed in response to other stimuli.

Due to the prevalence of positively charged inducers, it has been suggested that simple neutralization of the negatively charged surface membrane can initiate vacuolarization at a point. Such a mechanism does not suggest much selectivity; yet there are a few

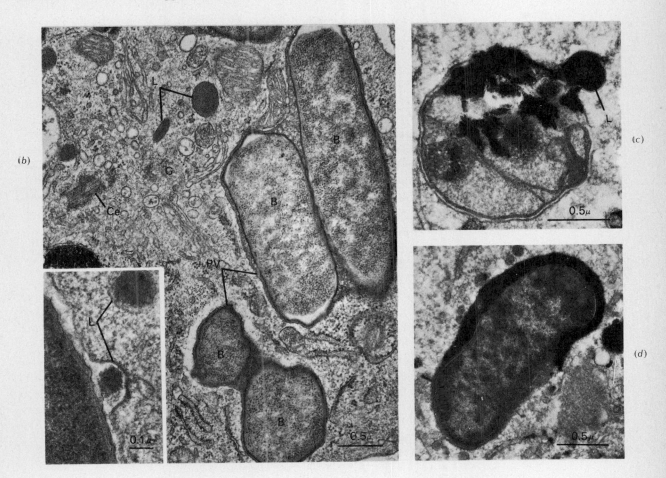

experiments that seem to ascribe considerable selectivity to endocytosis, at least in mammalian cells. Selectivity could be the result of specific binding to the cell surface, perhaps at some kind of protein receptor site. For example, mammalian cells in tissue culture reportedly show a marked ability to discriminate in favor of poly-D-lysine over poly-L-lysine, or arginine-rich histones (a basic protein from the nucleus) over lysine-rich histones. In both cases a factor other than charge must be involved.

The vacuoles produced through endocytosis are still bounded by the plasma membrane; hence, one might presume that they retain their former permeability properties. In actual fact, some change is noticed. For example, the plasma membrane of *Amoeba* seems to be impermeable to glucose; yet when the sugar is trapped in an endocytotic vacuole, it rapidly escapes into the cytoplasm accompanied by a visible change in the structure of the vacuolar membrane.

Endocytosis, as one would expect, requires energy. Poisoning the cell stops the process, whereas the addition of ATP sometimes stimulates it. Endocytosis is thought to result from the activity of microfilaments situated just inside the cell membrane. The proteins of these filaments bear a strong resemblance to actin, which with myosin (also found in ameba) converts ATP to mechanical energy during muscle contraction (see Chap. 7).

The usual fate of an endocytotic vacuole, which may be called a *phagosome* or *pinosome*, is fusion with a lysosome to form a *secondary lysosome* or *digestive vacuole* (Fig. 6–18). Lysosomes are probably pinched off from the Golgi apparatus or endoplasmic reticulum. The digestive enzymes of the lysosome hydrolyze ingested material to small organic products that escape (via membrane carriers?) into the cytoplasm. They also aid in the normal turnover of cellular constituents. For example, mitochondria may have a half-life as short as ten days due to digestion ("autophagocytosis") by lysosomes. Residual material may be removed from the cell by exocytosis, thus eliminating the vacuole and completing the cycle.

Endocytosis and exocytosis are extremely important mechanisms for entry to or exit from the cell. These mechanisms act in addition to the membrane carriers discussed earlier. The prevalence of endocytosis was long underestimated because of a failure to see smaller vacuoles with the light microscope—in *micropinocytosis*, some vacuoles are only 300 Å in diameter. The electron microscope, however, reveals the tiniest vacuoles, and often shows pits in plasma membranes representing the formation of new endocytotic vacuoles or the discharge of exocytotic (secretory) vacuoles (see Fig. 6–7).

The existence of endocytosis and the various transport mechanisms clearly vindicate the position taken at the beginning of this chapter, that the membrane is not just a barrier, but is an active, dynamic part of a cell's metabolic apparatus. Its structure and function give it a critical role in regulating cellular processes.

SUMMARY

6-1 Membranes behave more like a fluid than a layer of rigidly connected subunits. The basis for this behavior seems to be a bilayer of phospholipids with which various other lipids and a substantial number of proteins associate through noncovalent bonds.

Several models have been used to explain the distribution of lipid and protein within a membrane. The unit membrane concept is that of an essentially continuous lipid bilayer coated with protein. It is based on the trilaminate appearance of many membranes (e.g., plasma membranes) when viewed in the electron microscope. More recently, it has been suggested that membrane protein and lipid are in a more intimate contact. This explanation accounts better for the amount of protein and for the size of the individual protein molecules in membranes. The fluid mosaic model is the most flexible version of this idea, in that it allows proteins to move about longitudinally within the lipid bilayer. It accounts for the properties of most, or perhaps all, membranes, including the three general classes found in a typical eucaryotic cell: the inner membranes of mitochondria and chloroplasts, which have also been described as consisting of true subunits; exoplasmic membranes, such as the plasma membrane itself; and endoplasmic membranes, which are thinner, have more protein but less carbohydrate (less glycoprotein) than exoplasmic membranes.

6-2 The permeability of cellular membranes was first investigated by plasmolysis, and later by the measurement of permeability coefficients. The ability of many substances to penetrate cells is roughly proportional to their oil/water partition coefficients, implying that entry to the cytoplasm occurs by simple diffusion across a lipid barrier. A great many other substances, however, get in and out much more easily than their partition coefficient would predict. The possibility of membrane pores has been offered as an explanation. Such a scheme would explain the ease with which water and perhaps a few other small molecules get in and out, but it does not suggest the kind of selectivity actually observed with most substances.

6-3 Substances may enter a cell in one of three ways: (1) by simple diffusion across the membrane (or, at least in the case of water, perhaps through pores in the membrane), (2) by facilitated diffusion, in which a carrier combines with the substance on one side of the membrane and then releases it on the other side; and (3) by active transport, which also uses carriers, but which couples transport to the utilization of energy. The latter can support the cytoplasmic accumulation of free substrate to levels far greater than its external concentration. Both facilitated diffusion and active transport show considerable specificity because of the binding properties of the carriers involved, some of which are proteins called permeases.

6-4 Accumulation of a substance to an internal concentration that exceeds its external concentration can be accomplished (1) by providing internal receptors that bind or precipitate the substance, (2) by chemically altering the transported substance, or (3) by coupling influx to an energetically favorable reaction. Accumulation schemes that require metabolic activity (generally because they require ATP) are called active transport. The amount of energy required to transport a substance will depend on the material's concentration gradient and (for ions) the electrical gradient across the membrane.

Many cell types eject Na^+ from their cytoplasm by coupling the process to the hydrolysis of ATP. The responsible enzyme (permease) is called a sodium pump. Although the purpose of the pump is presumably to regulate cell volume, the artificially low Na^+ concentration thus created may be exploited by cells to allow the accumulation of other substances through a Na^+-coupled transport.

6-5 Endocytosis (phagocytosis or pinocytosis) and exocytosis also represent mechanisms for the energy-dependent uptake or discharge of materials. The material is surrounded by a membrane and pinched off as a vacuole to bring it into the cell, or discharged from the cell by fusion of its vacuole with the plasma membrane.

6-1 (a) What is a "lipid bilayer"? What role does it play in our concept of membrane structure, and why is it assigned to this role? (b) What are the major differences and similarities among the membrane models known as the Danielli-Davson model, the unit membrane, and the fluid mosiac model? What observations prompted the formulation of each? (c) Contrast the properties of the various classes of cellular membranes.

6-2 (a) What is plasmolysis and how can it be used to define the permeability of a membrane? (b) What kinds of substances have cellular permeability coefficients proportional to their oil/water partition coefficients? What kinds of substances do not?

6-3 (a) What is a permease, and in what ways can it affect transport? (b) What are the general features of cell permeation by: (1) diffusion, (2) facilitated diffusion, and (3) active transport?

6-4 (a) Lactose can be accumulated by *E. coli* to an internal concentration that is about 2,000 times its external concentration. What is the standard state free energy change for this process at 26 °C? [Ans: −4.5 kcal/mole] (b) The acidity of your stomach, which is about pH 2, is a result of proton pumps in the surrounding cells. Assuming that the pH of cell protoplasm is 7, what is the standard state free energy change for this translocation? Is one ATP hydrolysis per H^+ transported likely to be adequate? (Neglect voltage gradients and assume normal body temperature of 37 °C.) [Ans: 7.1 kcal/mole] (c) What membrane potential would the H^+ gradient used in the above problem balance, neglecting all other factors? [Ans: −310 mV] (d) Define: diffusion potential, sodium–potassium exchange pump, electrogenic sodium pump. (e) Define group translocation.

6-5 (a) Is endocytosis more akin to facilitated diffusion or to active transport? Explain. (b) Why is there reason to believe that endocytosis can exercise some discrimination over the material to be ingested?

REFERENCES

MEMBRANE STRUCTURE (See references in Chap. 5 on mitochondrial and chloroplast membrane structure and in Chap. 1 for the electron microscopy of membranes.)

BANGHAM, A. D., "Models of Cell Membranes." *Hospital Practice*, March 1973, p. 78.

BRANTON, D., and R. B. PARK, *Papers on Biological Membrane Structure.* Boston: Little, Brown & Co., 1968. (Paperback.) An excellent introduction, plus reprinted papers by Overton, Gorter and Grendel, Danielli and Davson, Robertson, Green, and others.

CAPALDI, R. A., "A Dynamic Model of Cell Membranes." *Scientific American*, March 1974, p. 26.

CHAPMAN, DENNIS, "Lipid Dynamics in Cell Membranes." *Hospital Practice*, February 1973, p. 79.

DANIELLI, J. F., "The Bilayer Hypothesis of Membrane Structure." *Hospital Practice*, June 1973, p. 63.

EPHRUSSI, B., and M. C. WEISS, "Hybrid Somatic Cells." *Scientific American*, April 1969. (Offprint 1137.)

FINEAN, J. B., "The Development of Ideas on Membrane Structure." *Sub-Cellular Biochem.*, 1:363 (1972).

FOX, C. FRED, "The Structure of Cell Membranes." *Scientific American*, February 1972. (Offprint 1241.)

GREEN, D. E., and R. F. BRUCKER, "The Molecular Principles of Biological Membrane Construction and Function." *Bio-Science*, 22:13 (1972).

HARRIS, HENRY, *Cell Fusion.* Cambridge, Mass.: Harvard Univ. Press, 1970. (Paperback.)

HUGHES, R. C., "Glycoproteins as Components of Cellular Membranes." *Progress in Biophysics and Molec. Biol.*, 26:189 (1973).

KENT, P. W., ed., *Membrane Mediated Information.* New York: American Elsevier Publ. Co., 1973. In two volumes. The first contains reviews on the biochemical properties of membranes, and the second on structure.

LUCY, J. A., "The Fusion of Cell Membranes." *Hospital Practice*, September 1973, p. 93.

NYSTROM, R. A., *Membrane Physiology.* Englewood Cliffs, N. J.: Prentice Hall, 1973. Chapters two and three discuss structure and permeation.

POSTE, G., "Mechanisms of Virus-Induced Cell Fusion." *International Review of Cytology*, 33:157 (1972).

ROBERTSON, J. D., "The Unit Membrane and the Danielli-Davson Model." In *Intracellular Transport* (Symp. Int. Soc. Cell Biol., 5), edited by K. B. Warren. New York: Academic Press, 1966, p. 1.

SHARON, N., "Glycoproteins." *Scientific American*, May 1974, p. 78.

SINGER, S. J., "Architecture and Topography of Biologic Membranes." *Hospital Practice*, May 1973, p. 81. The fluid mosaic model.

SINGER, S. J., and G. L. NICOLSON, "The Fluid Mosaic Model of the Structure of Cell Membranes." *Science*, 175:720 (1972).

TANFORD, CHARLES, *The Hydrophobic Effect. Formation of Micelles and Biological Membranes.* New York: Wiley-Interscience, 1973.

VANDERKOOI, G., and DAVID E. GREEN, "New Insights into Biological Membrane Structure." *BioScience*, 21:409 (1971).

ZINGSHEIM, H. P., "Membrane Structure and Electron Microscopy. The Significance of Physical Problems and Techniques (Freeze-Etching)." *Biochim. Biophys. Acta*, 265:339 (1972). A review.

MEMBRANE TRANSPORT, GENERAL CONSIDERATIONS

BERLIN, RICHARD D., "Specificities of Transport Systems and Enzymes." *Science*, 168:1539 (1970).

BITTAR, E. EDWARD, ed., *Membranes and Ion Transport* (three volumes). New York: John Wiley & Sons, 1970-71. Written for the novice.

DAMADIAN, R. V., "Cation Transport and Bacteria." *CRC Crit. Revs. Micro.*, 2:377 (1973). Points out possible weaknesses in generally accepted concepts of ion pumps and active transport.

DAVIES, M., *Functions of Biological Membranes* (Wiley Outlines). New York: John Wiley and Sons, Inc., 1973.

HEPPEL, LEON A., "Selective Release of Enzymes from Bacteria." *Science*, 156:1451 (1967). Enzyme and permease release from bacteria by osmotic shock.

HOKIN, L. E., *Metabolic Transport* (*Metabolic Pathways*, 3rd ed., Vol. 6, D. Greenberg, ed.). New York: Academic Press, 1972. Reviews on nearly all aspects of membrane transport.

PARDEE, A. B., "Membrane Transport Proteins." *Science*, 162:632 (1968).

STEIN, W. D., *The Movement of Molecules Across Cell Membranes*. New York: Academic Press, 1967.

ACTIVE TRANSPORT AND RELATED EVENTS

AZZONE, G. F., and S. MASSARI, "Active Transport and Binding in Mitochondria." *Biochim. Biophys. Acta*, 301(BR1):195 (1974).

CHOKE, HO COY, "Genetical Studies on Active Transport." *Science Progress* (Oxford), 59:75 (1971).

CRANE, R. K., "Structural and Functional Organization of an Epithelial Cell Brush Border." In *Intracellular Transport* (Int. Symposium Cell Biol., 5), edited by K. B. Warren. New York: Academic Press, 1966, p. 71.

GORDON, S., and Z. A. COHN, "The Macrophage." *Int. Rev. Cytology*, 36:171 (1973). Phagocytosis.

JAIN, M. K., A. STRICKHOLM, and E. H. CORDES, "Reconstitution of an ATP-mediated Active Transport System Across Black Lipid Membranes." *Nature*, 222:871 (1969).

JAMIESON, JAMES D., "Membranes and Secretion." *Hospital Practice*, December 1973, p. 71.

KIMMICH, G. A., "Coupling Between Sodium and Sugar Transport in Small Intestine." *Biochimica Biophysica Acta*, 300:31 (1973).

KORNBERG, H. L., "Carbohydrate Transport by Microorganisms." *Proc. Royal Soc. London B*, 183:105 (1973).

MacROBBIE, E. A. C., "The Active Transport of Ions in Plant Cells." *Quart. Revs. Biophys.*, 3:251 (1970).

SATIR, B., C. SCHOOLEY, and P. SATIR, "Membrane Reorganization during Secretion in Tetrahymena." *Nature*, 235:53 (1972).

SIEKEVITZ, P., "Dynamics of Intracellular Membranes." *Hospital Practice*, November 1973, p. 91. Emphasis on endocytosis and exocytosis.

STOSSEL, T. P., "Phagocytosis." *New Eng. J. Med.*, 290:717, 774, and 833 (1974). A three-part review of mechanisms and some medical implications.

TRUMP, B. F., "The Network of Intracellular Membranes." *Hospital Practice*, October 1973, p. 111. Emphasis on endocytosis and exocytosis.

WHITTAM, R., "Enzymic and Energetic Aspects of the Sodium Pump." In *Essays in Cell Metabolism*, edited by W. Bartley, H. L. Kornberg, and J. R. Quayle. London: John Wiley & Sons, 1970, p. 235.

CHAPTER 7

Excitability and Contractility

7-1 THE NEURON 267

7-2 THE NERVE IMPULSE 273
Membrane Potentials
The Spike Potential
Initiation of Spike Potentials

7-3 PROPAGATION OF THE NERVE IMPULSE 277
Chemical Conduction
Conduction by Local Currents
The Cable Effect and Myelinated Neurons
Transmission Between Neurons
Neuromuscular Transmission

7-4 MUSCLE CELLS AND THE MECHANISM
OF CONTRACTION 285
Myofibrils
Actin Filaments
Myosin Filaments
The Sliding Filament Theory

7-5 CONTRACTION–RELAXATION CONTROL 295
The Sarcotubular System
Excitation–Contraction Coupling
Graded Responses
Control in Smooth Muscle

7-6 NEUROMUSCULAR INTERACTION 302
The Neuromuscular Junction
Red and White Muscles

7-7 CONTRACTION IN NON-MUSCLE SYSTEMS 307
Protein Conformation and Mechanical Movement
Microtubules and Movement
Actin and Myosin in Non-Muscle Systems

SUMMARY 312
STUDY GUIDE 314
REFERENCES 314

It was pointed out in Chap. 6 that a voltage gradient exists across most biomembranes, and that this gradient is particularly important to the specialized function of nerve and muscle cells. These are the most thoroughly studied examples of excitable cells. By "excitable" we mean that the membrane of these cells is capable of generating and conducting an electrical impulse in the form of a temporary reversal of its membrane potential.

In the resting state, excitable cells, like nearly all other cells, are electrically negative on the inside with respect to the outside. However, if one momentarily reduces this charge difference beyond a fixed threshold level, a sudden reversal of polarity occurs, rendering the inside positive with respect to the outside. The polarity reversal, which may last only a millisecond or so, is called a *spike potential*. The spike potential is one of several kinds of active membrane responses involving changes in the electrical properties of membranes, and known collectively as *action potentials*.

A spike potential can be triggered at almost any point on the plasma membrane of an excitable cell. Once initiated, it spreads out in all directions. Thus, a stimulation at one end of such a cell is soon sensed at the other end. In addition, there are mechanisms for transmitting an action potential from one cell to another, either through a chemical intermediate or through direct electrical coupling. This interaction permits nerve cells to communicate with other nerve cells; it allows them to actuate muscles or other cells with excitable membranes; and it provides a mechanism to receive signals from the sense organs, all of which also contain excitable cells.

We shall proceed by describing the neuron, the mechanism of intracellular and intercellular propagation of action potentials, the structure and function of muscles, and finally, the way in which the function of muscle cells is controlled by action potentials. The chapter closes by pointing out the similarities between muscle contraction and the contractile activities of other types of cells.

7-1 THE NEURON

The nervous network of higher animals is divided into two parts, the central nervous system and the peripheral nervous system. The central nervous system, which includes the brain and spinal cord,

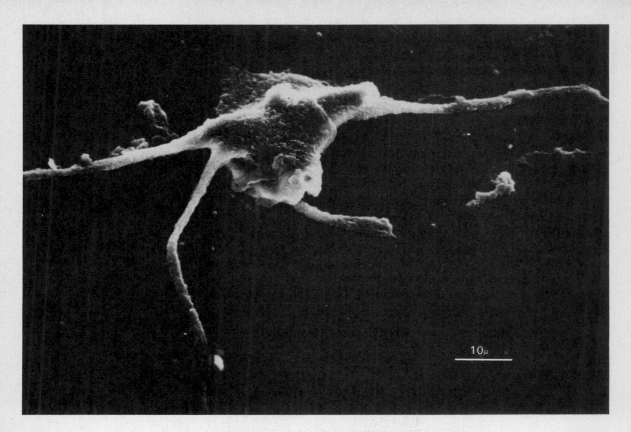

FIGURE 7-1. The Neuron. Scanning electron micrograph of a hypoglossal neuron from a rabbit. (Hypoglossal neurons innervate the region at the base of the tongue.) This cell should be viewed as an example only, not as a "typical" neuron, for neurons are so diverse that none can be so construed. [Courtesy of A. Hamberger, H. Hansson, and J. Sjöstrand, *J. Cell Biol.*, **47**:319 (1970).]

can integrate and process signals received from the peripheral nervous system, as well as return signals to it. The function of the peripheral system is to provide a communication link between the central nervous system and other parts of the body. This communication requires components with two capacities, motor and sensory. Motor function refers to the ability of nerves to affect the activity of target cells (e.g., the contraction of a muscle), while sensory capacity provides knowledge of the outside world in the form of touch, taste, smell, sight, and hearing.

The basic unit of both the central and the peripheral nervous system is an excitable cell called the *neuron* (see Figs. 7-1 and 7-2). The main body of the neuron, containing the nucleus, is called its *perikaryon, cyton,* or *soma.* Usually there are a number of short processes, called *dendrites,* radiating from the perikaryon. In addition, there will be a single longer process, sometimes branched, called an *axon.*[1] Axons contain numerous tubules and

---

1. Unfortunately, "axon" and "dendrite" are defined in two ways. In the functional definition, the impulse always passes from dendrites to the axon. But when the dendrites and axon are defined on the basis of their structure, we may find the opposite direction of travel. Although the two definitions are compatible in most cases, some of the sensory neurons in the peripheral system are notable exceptions.

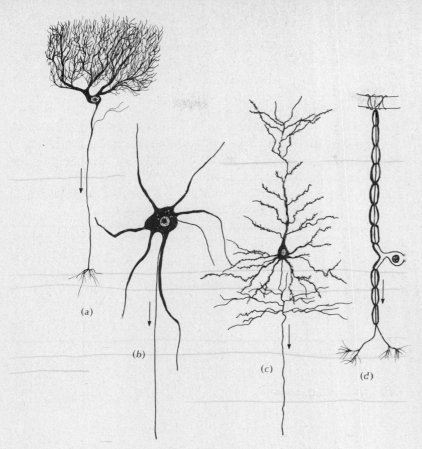

(a)

(b)

(c)

(d)

FIGURE 7–2. Neuronal Variety. Arrows show direction of impulse along axons. **(a)–(c)** Diagrammatic representations of some neurons found in the brain. Note the axonal branching in (c). **(d)** A sensory neuron. A single long branch (dendrite) conducts an impulse from a sensory cell or specialized ending. The cell body is located in a ganglion near the spinal cord. A shorter branched process delivers the impulse to the central nervous system.

filaments running their length, including the 250 Å *microtubule*, which may aid in maintaining structure and possibly participate in axonal flow of materials, a 100 Å *neurofilament*, and the smaller, approximately 60 Å, *microfilament*. Unlike microfilaments, neurofilaments have a microtubule-type construction. However, whereas the walls of microtubules have 13 strands, leaving an open core of 150 Å, neurofilaments have walls of four strands with a 30 Å core.

In most neurons, the impulse travels from the cell body toward the tip of the axon, from where it may be passed to another neuron via the latter's dendrites or directly to its soma at raised structures called *synaptic knobs* (visible in Fig. 7–1). Alternatively, the signal may be passed to a muscle cell or other excitable cell type. The junction between two neurons, or sometimes between any two excitable cells, is called a *synapse*. The connection between a neuron and a muscle cell is also referred to as a *neuromuscular* junction or *myoneural junction*. (The prefixes "myo-" and "sarco-" are used to designate muscle.)

Although processes of individual neurons are usually only a few

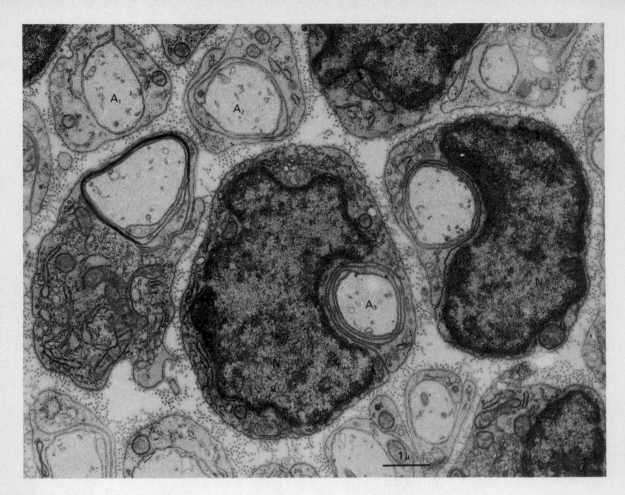

FIGURE 7–3. The Schwann Cell. From the sciatic nerve of a newborn rat. Schwann cell nucleus (N). The Schwann cell around axon A₁ makes a single turn, the one around A₂ makes about 2⅛ turns, and the one around A₃ makes 3⅜ turns. [Courtesy of Henry de F. Webster. Similar to *J. Cell Biol.*, **48**:348 (1971).]

microns in diameter, they may be several feet long in larger animals—extending, for example, from a perikaryon near the base of the spine to the big toe. A process is usually surrounded and protected by a sheath of cells. In the case of peripheral nerves, these adjacent cells are called *Schwann cells* (see Fig. 7–3); in the central nervous system they are called *glial cells*. The nerve cell process and its sheath together constitute a *nerve fiber*. Nerve fibers are commonly found in bundles, held together by connective tissue. It is these bundles that are identified as the anatomical unit called a *nerve* (see Fig. 7–4). It is not unusual to find a thousand fibers, each 10 to 20 μ in diameter, within a mammalian nerve. Finally, several nerves may be joined together as a *nerve trunk*.

The longer nerve cell processes of vertebrates are often myelinated. *Myelin* is the name given to the glistening, white, fatty covering about a nerve process that is now known to be a tight spiral wrapping of the associated sheath cells (see Fig. 7–5). When the individual processes within a nerve are myelinated, the nerve

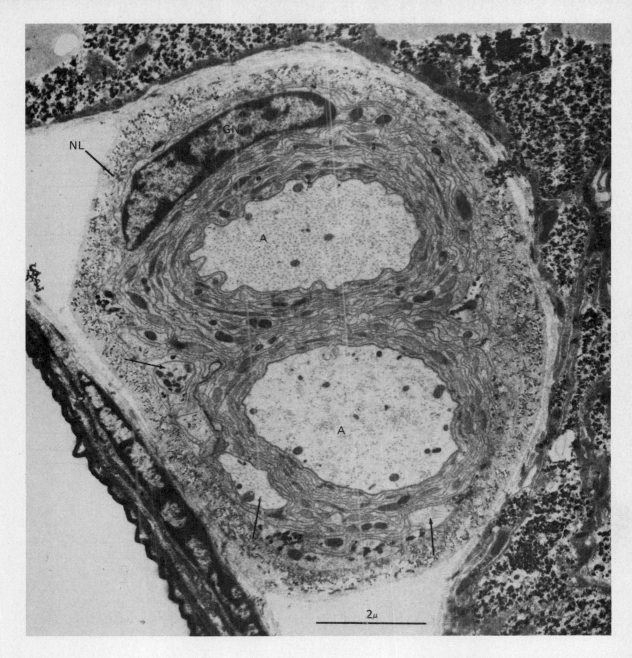

itself is said to be myelinated or *medullated*. Myelinated nerves are found both in the central and peripheral nervous systems of vertebrates. They comprise the white matter of the central nervous system; duller, unmyelinated nerves and their associated cells form the gray matter.

Each individual Schwann cell of a myelinated peripheral neuron covers perhaps a millimeter of the process, with a space of about a

FIGURE 7–4. The Nerve. A nerve may contain thousands of axons, plus associated cells. The photo shows the rather simple salivary duct nerve of a cockroach, containing two large axons (A), several smaller ones (arrows), several layers of glial cell wrapping (glial nucleus, GN), and the outer connective tissue covering (NL). [Courtesy of A. T. Whitehead, *J. Morph.*, **135**:483 (1971).]

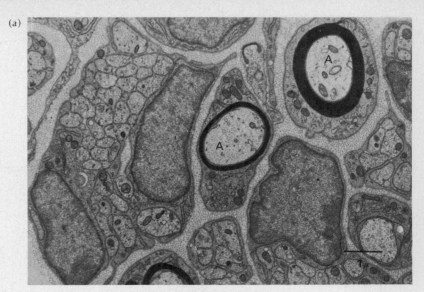

(a)

FIGURE 7–5. Myelin. **(a)** Part of the sciatic nerve of a week-old rat. Myelination has occurred mostly since birth (compare with Fig. 7–3), producing a heavy layer around two of the axons shown (A). Each of the other Schwann cells in the micrograph enwraps several unmyelinated axons. **(b)** Myelinated axon from a newt, showing the distinct layers of the myelin sheath. Microtubules (Mt, three in the circle) and mitochondria (M) are visible in the axoplasm. Neurofilaments (Nf). [(a) courtesy of T. L. Lentz, *J. Cell Biol.*, **52**:719 (1972).]

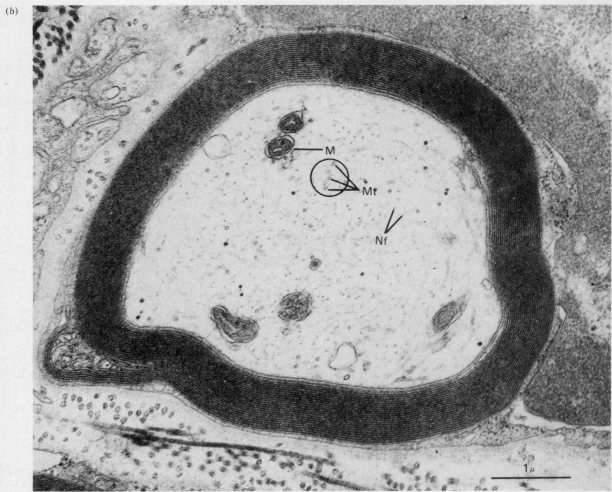

(b)

micron between it and the next Schwann cell. These gaps where the process is relatively exposed are called *nodes of Ranvier* (see Fig. 7–6). In general, motor nerves that actuate skeletal muscles are heavily myelinated, those from sensory organs have lighter myelin, and those to visceral, involuntary muscles have very light myelin or none at all. As we shall see, the presence of myelin leads to a greatly increased velocity of nerve impulse conduction.

Invertebrates do not have myelin. Rapid conduction of impulses is achieved in invertebrates by the use of processes with large diameters. For example, an English zoologist, J. Z. Young, pointed out in 1933 that the swim reflex of the Atlantic squid, *Loligo*, is controlled by neurons with axon diameters that are often as much as 0.5 mm (500 μ). This was a fortunate observation, for the large size of these axons makes it possible to do experiments that would be technically very difficult or impossible with smaller cells. In addition, the squid's giant axon can remain excitable for hours after being dissected from the animal, providing only that it is tied at both ends to prevent loss of cytoplasm (called *axoplasm* in this case). In other words, an isolated giant axon can be stimulated to produce and conduct an impulse even though the axon is nothing more than an isolated cell fragment. Most of what is known about the excitable properties of nerve cells has been learned through studies on this system.

7–2   THE NERVE IMPULSE

The voltage gradient that exists across the neuronal membrane is made possible by the selective permeability of the membrane and by the existence of specialized "pumps" that regulate the ionic content of the cytoplasm. It is changes in this gradient that are identified as the nerve impulse.

MEMBRANE POTENTIALS.   A typical cell has a high internal potassium ion and low internal sodium ion concentration, in spite of the situation being reversed in the surrounding fluid. As an example, consider the squid's giant axon: its Na^+, K^+, and Cl^- contents are, respectively, about 0.05 M, 0.40 M, and 0.05 M. If the axon is suspended in sea water, with concentrations of 0.46 M, 0.01 M, and 0.54 M for these three ions, a potential difference of -60 millivolts will be measured across the axon membrane—negative on the inside with respect to the outside, according to our earlier convention. Application of the Nernst equation, which may be written in an abbreviated form as

$$\Delta E(\text{millivolts}) = \frac{58}{z} \log \frac{c_{\text{out}}}{c_{\text{in}}} \qquad (7\text{–}1)$$

to the ion concentrations just given shows that the equilibrium

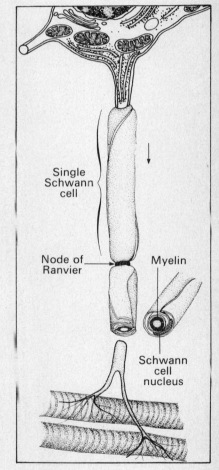

FIGURE 7–6. A Motor Neuron. Part of the innervated muscle is shown. Note the node of Ranvier. The cutaway shows a myelin layer plus a few loose turns of the responsible Schwann cell.

Single Schwann cell

Node of Ranvier

Myelin

Schwann cell nucleus

273

voltage in each case (that is, the membrane potential necessary to balance the observed concentration ratios of the ions) is as follows:

for Cl⁻ $(z = -1)$ $-58 \log 0.54/0.05 = -58 \times 1.033 = -60$ mV

 for K⁺ $(z = +1)$ $+58 \log 0.01/0.40 = +58 \times -1.6 = -93$ mV

for Na⁺ $(z = +1)$ $+58 \log 0.46/0.05 = +58 \times 0.98 = +57$ mV

Thus, Cl⁻ is in or very near equilibrium with the actual membrane potential of −60 mV. On the other hand, there is a slight tendency for K⁺ to flow outward if unopposed, and a strong tendency for Na⁺ to flow inward. The influx of Na⁺ is driven both by a voltage difference and by a concentration gradient. In fact, the free energy change for Na⁺ influx is better than −2.5 kcal/mole.

Sodium and potassium gradients are maintained by the "sodium pump." This conclusion was first suggested by the use of radioactive ions, which reveal a constant low flux of both Na⁺ and K⁺ across the membrane. Such a flow indicates a need for metabolically obtained energy to drive the pumps. Even in the absence of metabolic activity, however, it would be many hours before either ion reached a final electrochemical equilibrium, for the axonal membrane is only slightly permeable to K⁺ and even less permeable to Na⁺.

THE SPIKE POTENTIAL. In view of the ionic imbalance just outlined, consider what would happen if a portion of a neuron's membrane were suddenly to become permeable to sodium by the opening of some kind of channel or "gate." The inrush of ions would offset the negative potential in the region of the leak, and because of the greater concentration of Na⁺ outside, would actually push the potential toward the Na⁺ equilibrium point of +57 mV. If, before that point is reached, the Na⁺ channel were closed again, diffusion and the Na⁺ pump would eventually restore the original ionic concentration on both sides of the membrane, and with it the original −60 mV potential. However, if a channel for K⁺ were to open at about the time the Na⁺ gate is closed, K⁺ would leave the axoplasm, driving the potential rapidly toward the K⁺ equilibrium value of −93 mV and thus causing a quick return to the *resting potential*. At that point, the original permeability might be reinstated.

The situation just described is a relatively accurate description of the way in which the excitable membranes of nerves and most muscles respond to the proper stimulus. That stimulus may be in the form of an electrically induced partial depolarization of perhaps 20 mV or more (e.g., a change from −60 to −40 mV), or it may be in the form of a chemical. In either case, the result is to open a channel for sodium and trigger the events as they were outlined. The resulting reversal of polarization and its restoration constitute

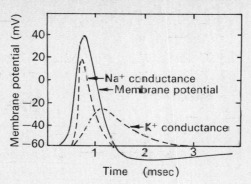

FIGURE 7–7. The Spike Potential. Curves show relative changes in ion permeabilities and voltage as a spike potential passes a pair of recording electrodes in a nerve fiber. Note the temporary hyperpolarization caused by K⁺ efflux in the absence of Na⁺ influx. The refractory period, during which a second stimulation would produce no response, is roughly 2 msec for this fiber. [Original data from A. L. Hodgkin, *Proc. Roy. Soc.* (London) *B*, **148**:1 (1958).]

the *spike potential* (see Fig. 7–7), which may be propagated as a nerve impulse.[2]

A spike potential, then, is an electrical response of a membrane to certain kinds of stimulation. In particular, the spike potential consists of a swing in the voltage from some negative resting potential to zero and then through a positive overshoot, eventually to return to the original resting potential. The depolarization and positive overshoot are due to a sudden increase in the otherwise slow sodium influx; the return to the resting potential is greatly accelerated by a corresponding, but delayed, increase in potassium efflux, which begins just before the sodium ion movement reaches its maximum.

If, as suggested, the size of the spike potential is due almost exclusively to changes in sodium ion permeability, then altering the concentration of sodium ion in the bathing fluid should affect the magnitude of the spike. This experiment was executed at the Plymouth, England marine biological laboratories in 1947 by A. L. Hodgkin and Bernhard Katz. They found that both the rate of voltage change and the magnitude of the spike potential vary with the external sodium ion concentration in accordance with predictions. The resting potential, on the other hand, is almost independent of Na⁺ concentration in the bathing fluid. Rather, it depends strongly on the K⁺ concentration, becoming more negative as the external K⁺ is increased. This, too, is consistent with the model, which includes a greater permeability to potassium than to sodium.

If only chloride is in passive equilibrium across the membrane, then the resting potential can be calculated from the distribution of K⁺ and Na⁺ as follows:

---

2. The explanation of action potentials in terms of changes in ionic permeabilities is by no means a new concept. As long ago as 1902, Julius Bernstein of the University of Halle, in Germany, attributed the action potential to a transient change in sodium permeability. He erred mainly in failing to predict the positive overshoot as the Na⁺ equilibrium potential is approached, and in failing to recognize the importance of K⁺ efflux to the quick restoration of a resting potential.

$$\Delta E = 0.058 \log \frac{[\text{K}^+]_{\text{out}} + b[\text{Na}^+]_{\text{out}}}{[\text{K}^+]_{\text{in}} + b[\text{Na}^+]_{\text{in}}} \text{ volts} \qquad (7\text{--}2)$$

Here b is the ratio of sodium to potassium permeability. With the sodium and potassium concentrations already given, the observed resting potential would be achieved with a value of about 1/15 for b—not so very different from the measured ratio of ion permeabilities in the squid's giant axon. The effect of the slow inward leakage of sodium ion, then, is to moderate the resting potential from −93 mV (the value that would be reached by K$^+$ alone) to the measured potential of −60 mV.

Thus, by the early 1950s, it was clear that the spike potential results from changes in permeability to Na$^+$ and K$^+$. Permeability to sodium increases first and normalizes first, but these changes are overlapped and followed by corresponding alterations in potassium ion permeability. The result is a return to the resting potential in a few milliseconds.

INITIATION OF SPIKE POTENTIALS. A spike potential may be initiated by partial depolarization or by chemical stimulation. Each plays an important role in nervous function and integration.

The detailed study of permeability changes as a function of membrane potential was made possible by a device called the *voltage clamp*, developed by K. S. Cole in the 1940s. It provides a way of holding the membrane potential at a preset value so that conductances can be measured without triggering the explosive changes associated with the spike potential. In the case of the squid axon, electrodes are inserted into the cut end of the axon and into the surrounding fluid, one set to supply an external voltage and a second set to monitor the total membrane potential.

With the voltage clamp, it was found that a depolarization of only 10 mV increases the sodium permeability about eightfold. However, when the membrane potential is set to zero with the voltage clamp, and held there, a transient inward flow of sodium ions is followed within a couple of milliseconds by an outward flow of potassium ions. The latter continues as long as the membrane is held in a depolarized state. In other words, whereas the induced increase in sodium permeability is self-correcting, potassium permeability follows the membrane potential (see Fig. 7–8). Since K$^+$ efflux normally drives the potential negative again following a spike potential, it appears that potassium permeability is controlled by negative feedback.

It is difficult to explain the molecular bases for these changes in ionic permeability, because we do not know enough about the membrane's components or how they interact. However, changes in the way in which the membrane scatters light during passage of a spike potential, along with measurements of other kinds, suggest

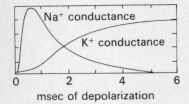

msec of depolarization

FIGURE 7–8. Ion Conductances in a Voltage-Clamped Neuron. The neuron was subjected to a constant depolarization via the device known as a voltage clamp. Note that Na$^+$ entry is transitory, while potassium conductance varies with membrane potential. (Original data by A. L. Hodgkin and A. F. Huxley.)

the presence of a considerable change in the conformation of membrane proteins during this period. Such ideas are not hard to reconcile with our experience, for there are many cases in which small conformational changes in proteins are known to be associated with large changes in biological activity. Since conformational shifts can be induced by a wide variety of stimuli, we can imagine that certain membrane proteins respond to changes in the membrane potential. Such changes could be responsible for controlling sodium and/or potassium permeability.

The initiation of a spike potential through chemical, rather than electrical, stimulation puts us into more familiar territory, for the existence of conformational shifts in proteins due to the binding of small molecules was discussed in Chaps. 3 and 4. Several examples of chemical stimulation will be discussed later, in the sections on intercellular transmission.

Whether a spike potential is initiated chemically or electrically, there is a period immediately after firing when neither type of stimulation can trigger a new spike potential. This time interval, which is known as a *refractory period*, lasts until the resting potential is reattained. The maximum rate at which spike potentials can be generated depends on the duration of the refractory period for a particular nerve. Furthermore, as the frequency of stimulation is increased, there comes a point when it can no longer be distinguished as a string of separate stimuli, for it will have the same effect as a continuous stimulation. This phenomenon, which is most noticeable to us in the perception of sights and sounds, is known as *flicker fusion*.

7-3 PROPAGATION OF THE NERVE IMPULSE

We shall now consider ways in which spike potentials can be *conducted* by an excitable membrane, and ways in which they may be *transmitted* from cell to cell. These two processes, conduction and transmission, are generally regarded as being fundamentally different.

Three mechanisms have been suggested for the conduction of spike potentials (see Fig. 7-9). They are: (1) chemical conduction, wherein a substance that is released by a spike potential at one point on the membrane diffuses to nearby points to start spike potentials there; (2) direct electrical stimulation by local currents, which supposes that a spike potential in any given area will depolarize an adjacent area to its threshold value; and (3) the cable effect, which is the way an electrical impulse is carried by a wire. Each mechanism has its own advantages and disadvantages, and each has apparently been adopted by Nature to serve different functions.

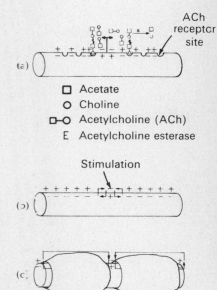

FIGURE 7-9. Propagation of the Impulse. Three mechanisms are diagrammed: **(a)** chemical conduction through the cyclical release and hydrolysis of a substance such as acetylcholine; **(b)** conduction by local currents, in which an action potential depolarizes adjacent areas to push them past their threshold potential; **(c)** the cable effect, or saltatory propagation, thought to function in myelinated axons.

CHEMICAL CONDUCTION. Chemical stimulation is the easiest to eliminate as a generalized scheme for nerve impulse conduction, although it is used for impulse transmission between nerve cells and from nerves to muscles. Chemical stimulation not only requires the maintenance of sodium and potassium gradients, as do the other mechanisms, but it also requires either the destruction and resynthesis or the reabsorption of the chemical transmitter (see Fig. 7–9a). These events involve the expenditure of metabolic energy. The application of metabolic poisons to isolated axons indicates that impulses can be conducted in the absence of metabolism as long as the sodium and potassium gradients remain, and that they decline along with the gradients. Such observation would seem to be inconsistent with chemical conduction.

The independence of impulse conduction from the metabolic machinery of the nerve cell axon was directly demonstrated in the early 1960s by two groups working independently at the marine biological stations in Plymouth, England, and Woods Hole, Massachusetts. They removed the axoplasm from isolated squid axons and replaced it with solutions of various composition—a process called *perfusion*. As long as the perfusing fluid was chosen properly, including the requirement that it be high in K^+ and low in Na^+, the axon was capable of conducting hundreds of thousands of spike potentials before failing. Only a thin layer of axoplasm adheres to the membrane of the perfused axon, clearly showing that the bulk of the axoplasm is not involved in conduction. Hence, conduction appears to be independent of the metabolic activities of the cell.

Similar conclusions are reached in studies using synthetic membranes. For example, in 1967 P. Mueller and D. O. Rudin reported the generation of spike potentials in a synthetic membrane consisting of a lipid bilayer, a crude bacterial protein extract, and protamine (a basic protein that is associated with the DNA of sperm). Partial depolarization from an applied electrical source triggered voltage changes that resemble those seen in cellular systems (see Fig. 7–10). Such arguments do not imply that chemicals are never used to propagate spike potentials, but they do imply that the mechanism is not an attractive way of explaining conduction in most excitable membranes.

CONDUCTION BY LOCAL CURRENTS. Impulse conduction by simple electrical stimulation is attractive as a general mechanism, for it requires no metabolic involvement except through the maintenance of ion gradients. It suggests only that a partial depolarization may result from a larger depolarization at a nearby point (see Fig. 7–9b). Such an effect is all that is necessary to ensure propagation, because once the threshold potential at that second point is

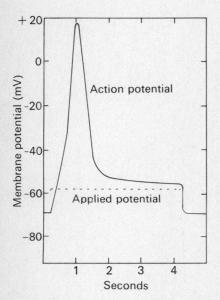

FIGURE 7–10. Spike Potentials in a Synthetic Membrane. The independence of the spike potential from metabolic function is emphasized by studies with synthetic membranes (lipid bilayers with added protein), for they can exhibit excitable properties remarkably like those of the living cell—even to reversible blockage by local anesthetics such as cocaine. A depolarization potential of about 11 mV was applied as indicated by the dotted line. The refractory period is about 10 seconds. [Data from P. Mueller and D. O. Rudin, *Nature*, **213**:603 (1967).]

reached, a new spike potential will be triggered, spreading the wave of depolarization across the surface.

The speed with which a spike potential spreads can be calculated from the dimensions of the axon, its permeability during various stages of the cycle, and the properties of the ions involved. According to equations derived by A. L. Hodgkin and A. F. Huxley in 1952, the speed of conduction should be proportional to the square root of the diameter of the axon, a relationship that has been verified in a wide variety of unmyelinated nerves.

THE CABLE EFFECT AND MYELINATED NEURONS. The third mechanism of impulse propagation is the cable effect, which is the way current travels in a wire. It requires a medium of low resistance, a requirement that is only marginally met by the nerve cell axon. Because of the electrical resistance of axoplasm, a spike potential propagated solely by the cable effect should be quickly distorted and dissipated. The distance over which it survives can be demonstrated with the local anesthetic procaine, which eliminates the spike potential wherever the drug is applied. To deaden an axon, at least a millimeter must be treated with procaine, implying that the spike potential can jump such distances, but no more. However, while the cable effect is thus eliminated as a general mode of conduction, it is still applicable to the special case of myelinated neurons.

A typical spacing between nodes of Ranvier in myelinated neurons is about a millimeter. Since spike potentials cannot be detected between nodes, where the nerve process is insulated by myelin, it has been proposed that the potential jumps from node to node via the cable effect. This mechanism is known as *saltatory conduction*. The impulse is renewed at each node as the propagated wave triggers a spike potential in it, much as a submarine cable sends telephone messages from one repeater station to the next (see Fig. 7–9c). One would expect the cable effect to conduct an impulse rapidly and, indeed, the fastest nerve fibers, which support velocities well in excess of 100 m/sec, are myelinated.

The length of time it takes an impulse to travel from node to node in a myelinated neuron is negligible compared to its regeneration time at a node. The conduction velocity of such a neuron will therefore depend on the distance between nodes. And that, in turn, is directly proportional to diameter. This relationship is derived from the way in which neurons and the Schwann cells surrounding them grow. As an animal gets larger, its neuronal processes both elongate and increase in diameter, one in proportion to the other. Since neither the number of neurons nor the number of Schwann cells increases during the growth period, Schwann cells must get larger along with the processes they cover. Thus, as an axon lengthens, each Schwann cell covers a longer stretch. The fact that length, diameter, and internodal distance all vary together provides a linear relationship between conduction velocity and axon diameter. In the case of the human ulnar nerve, for example, the internodal distance of various sized axons is about 100 times their diameter.

Since each spike potential must affect a finite length of the axon, myelination fails to confer any advantage in propagation to very small axons. Accordingly, one finds that in vertebrates, where myelination is used to increase the rate of conduction of larger axons, peripheral axons with diameters of less than about one micron are usually unmyelinated (see Fig. 7–11). In the central nervous system, however, axons as small as 0.2 μ may be myelinated. Where speed of conduction is vital, as with the sensory and motor neurons of a reflex arc, myelinated nerves with larger axons are found. Where a little delay in getting the message can be tolerated, as in the control of viscera, space seems to be conserved by using smaller fibers, often without myelination. Invertebrates, as noted earlier, rely on unmyelinated axons of extremely large diameters to provide rapid impulse conduction wherever that is vital. The squid's giant axon is in this group, since it is part of the motor nerve controlling the swim reflex.

TRANSMISSION BETWEEN NEURONS. Intercellular transmission of nerve impulses almost always relies on a chemical intermediate. The best understood of these intermediates is acetylcholine,

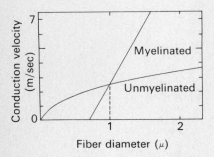

FIGURE 7–11. The Advantage of Myelination. An impulse can be propagated more rapidly in myelinated axons, presumably because it is conducted from node to node via the cable effect. In axons of very small diameter, with closely spaced nodes, that advantage is lost. [Data from W. Rushton, *J. Physiol.*, **115**:101 (1951).]

$$CH_3-\overset{\overset{\displaystyle O}{\|}}{C}-O-CH_2-CH_2-\overset{+}{N}(CH_3)_3$$

Acetylcholine

a substance that is very widespread through both the central and peripheral nervous systems, specifically in the *parasympathetic*, or *cholinergic*, branch of the autonomic nervous system. (The autonomic system controls the internal organs of vertebrates.) The other branch of the autonomic system is the *sympathetic*, or *adrenergic*, branch; the latter designation refers to the fact that noradrenaline and adrenaline (noradrenaline in humans) are the main neurotransmitters.[3]

Noradrenaline   Adrenaline

3. Adrenaline and noradrenaline are also referred to as epinephrine and norepinephrine, respectively. One of the precursors of noradrenaline is L-dopa, a drug that is being widely used to treat the neurological disorder known as Parkinson's disease.

Other neurotransmitters include the amino acid glutamate and its decarboxylated derivative, γ-amino butyric acid (GABA)—a system that has been studied most carefully in the crustacean myoneural junction, but that is also found in mammals.

$$
\begin{array}{ccc}
\text{COO}^- & & \text{COO}^- \\
| & & | \\
\text{CH}_2 & & \text{CH}_2 \\
| & & | \\
\text{CH}_2 & \longrightarrow & \text{CH}_2 \\
| & & | \\
\overset{+}{\text{H}_3\text{N}}\text{—CH—COO}^- & & \overset{+}{\text{H}_3\text{N}}\text{—CH}_3 + \text{CO}_2 \\
\text{Glutamate} & & \gamma\text{-Amino butyrate}
\end{array}
$$

Still other molecules have been implicated in various aspects of neurotransmission, though few details are known. Among the more common is serotonin, a tryptophan derivative that is synthesized mostly in the brain. In a few cases, however, transmission does not use any chemical intermediate at all, but depends on direct electrical connections between two cells by means of an intercellular conducting medium. And in still another variation, called the *electrotonic synapse* (see Fig. 7–12), there appears to be both electrical and chemical transmission. It should become clear from later discussions that electrical transmission, though speedy, does not provide for the flexibility inherent in chemical neurotransmission.

FIGURE 7–12. The Electrotonic Synapse. Ionic coupling, apparently via cytoplasmic connections, provides a rapid feedback to synchronize the firing of some communicating neurons. The micrograph shows two axons (A₁ and A₂) of a spiny boxfish in apposition to a common nerve cell body (an oculomotor neuron) at the bottom of the micrograph. The presynaptic and postsynaptic membranes share an area of junctional specialization, shown to better advantage in the inset. Mitochondria are seen in both axons. [Courtesy of M. Kriebel, M. Bennett, S. Waxman, and G. D. Pappas, *Science*, **166**:520 (1969). Copyright by the A.A.A.S.]

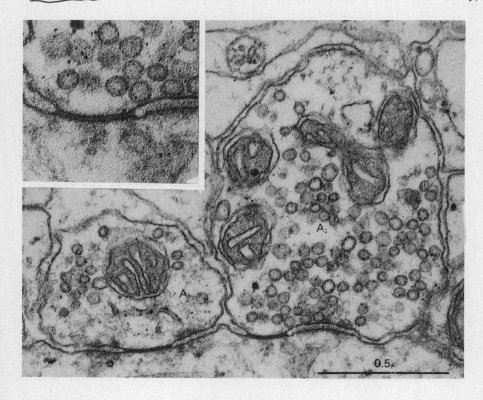

0.5

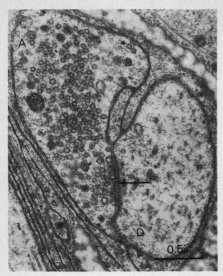

(a)

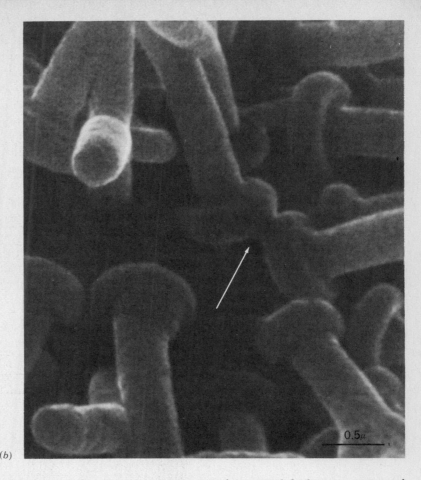

(b)

FIGURE 7–13. The Synapse. **(a)** From the cat (superior cervical ganglion). The axon (A) and dendrite (D) are enwrapped by Schwann cell cytoplasm. Note the numerous synaptic vesicles (smaller spheres) in the axon here and in Fig. 7–12. Synaptic cleft (arrow). **(b)** Scanning electron micrograph of neuronal processes from the mollusk, *Aplysia californica*. The knobs, or *boutons*, are presumed to be the termini of processes freed of their glial cells, though one cannot distinguish axons from dendrites in this view. Possible synaptic contacts are seen in several places (e.g., arrow). [(a) courtesy of J. J. Pysh and R. G. Wiley, *Science*, **176**:191 (1972), copyright by the A.A.A.S.; (b) courtesy of E. R. Lewis. Similar to *Science*, **165**:1140 (1969).]

The model for neurotransmission by acetylcholine consists of the following steps: (1) When a spike potential reaches the end of an axon, it causes acetylcholine to be released from storage vesicles found there (see Fig. 7–13). An essential intermediate is an increase in cytoplasmic Ca^{2+}. (2) Acetylcholine diffuses across the myoneural or synaptic gap. (3) At the postsynaptic membrane, the chemical encounters specific receptor sites to which it binds reversibly. (4) The appearance of acetylcholine at the postsynaptic receptors alters the ion permeability of the associated membrane, in some cases triggering a new spike potential. (5) The reversible binding of acetylcholine means that each molecule is also unbound and exposed for a time to hydrolysis by an enzyme, acetylcholine esterase, secreted by the adjacent neurons. (6) The hydrolysis products, acetate and choline, are inactive as neurotransmitters, but may diffuse back across the gap to be taken up again by the presynaptic neuron. (7) And finally, the amount of acetylcholine used in the process is replaced in the presynaptic neuron by transferring acetate from coenzyme A to choline and packaging the product into vesicles.

282

There is a good deal of support for this general mechanism, and a number of the details are now available. For example, from the work of Sir Bernard Katz and others, we have considerable information about the nature of acetylcholine release from its storage vesicles within the presynaptic neuron, especially from the *motor end plates* (terminal axons) of myoneural junctions. Acetylcholine vesicles are constantly being opened at a slow rate (e.g., one per second), but the number of acetylcholine molecules contained in each vesicle (a few thousand) is only enough to cause a transient change of about 0.5 mV in the potential of the postsynaptic membrane, not enough to trigger a spike potential. However, when an impulse reaches the region of acetylcholine vesicles in the presynaptic neuron, several hundred packets (a number that varies with the magnitude of the spike potential) may be released within a millisecond or so from each end-plate. Since the degree of depolarization in the target cell is proportional to the concentration of acetylcholine reaching it, a new spike potential may be produced there.

Katz, working in 1965 with R. Miledi, also measured the amount of time that it takes acetylcholine to transmit a nerve impulse. They placed a microelectrode at a neuromuscular gap, and found two electrical disturbances for each nerve impulse. The first signal marks the arriving nerve impulse itself. The second disturbance, 0.5 to 0.8 milliseconds later, is from the initiation of the new spike potential in the muscle cell membrane. This time lapse is consistent with the presence of a chemical transmitter, and is much too long to suggest any kind of direct, electrical transmission.

The events at the postsynaptic membrane have been studied in several ways. For example, the nature of the acetylcholine binding sites has been examined indirectly through the use of inhibitors. These inhibitors include the curare family (e.g., *d*-tubocurarine) and atropine, both of which compete with acetylcholine for the membrane receptor sites, and eserine, which inhibits acetylcholine esterase. However, more direct studies have also been carried out, such as those by the Argentine scientist Eduardo De Robertis, who reported the isolation of a lipoprotein from brain that may well be the acetylcholine receptor itself. The isolated molecules are highly elongated, about 15 Å by 150 Å, and aggregate to form tubule like structures which, he suggested, might be arranged transversely across the membrane. The channel through the center of the structure could be the site of Na^+ entry. In this model, the dimensions of the channel, or pore, might vary with the presence or absence of acetylcholine, as well as with other environmental changes such as depolarization.

De Robertis also isolated an acetylcholine receptor (another lipoprotein) from the electroplax (electric organ) of the eel, *Electrophorus*. This molecule has a site with very high binding affinity for acetylcholine $(K_{eq} = 10^8 \; M^{-1})$. When the preparation was added to an artificial lipid bilayer membrane, a significant increase in ion conductivity followed. Next, when acetylcholine was added, an additional, transient increase in conductivity was seen. This latter effect was blocked by d-tubocurarine, and was not observed at all with control membranes containing other lipoproteins. Such experiments give promise that eventually we shall understand how ionic permeabilities are altered in excitable cells.

NEUROMUSCULAR TRANSMISSION. Vertebrate skeletal and heart muscle cells (striated muscle) develop spike potentials through a Na^+–K^+ cycle of the type found in nerves. The spike potential of crustacean muscle (e.g., crab or barnacle) on the other hand, relies on Ca^{2+} to provide the depolarizing phase. Repolarization in both types of animal muscle is the result of K^+ efflux. (In a few excitable cell types, notably in plants, depolarization seems to be the result of anion efflux, commonly Cl^-, instead of cation influx.) However, the all-or-none response so common in neurons is found much less often in muscle, and some muscle cells do not propagate a spike potential at all.

As implied earlier, the basic mechanism by which impulses are transmitted between neurons is also responsible for neuromuscular transmission. There are some differences, however. For example, acetylcholine was discovered in 1921 by Otto Loewi as a chemical that mimics the effect of the vagus nerve (a parasympathetic nerve) on a frog's heart. But acetylcholine does not initiate spike potentials in the heart, for the rhythmic generation of these impulses originates within the heart itself. In fact, acetylcholine, like the vagus nerve, inhibits rather that stimulates spontaneous contraction of the heart. Instead of increasing permeability to Na^+, causing depolarization, acetylcholine increases the permeability of heart muscle cell membranes to K^+. This effect makes the initiation of spike potentials more difficult, as it decreases the value of b in equation (7–2), moving the resting potential closer to the more negative K^+ equilibrium value. Increasing the permeability to K^+, then, causes a *hyperpolarization*, inhibits the generation of action potentials, and slows the heart beat.

The parasympathetic nerves thus generate *inhibitory impulses* at the heart, though in other target cells the same transmitter, acetylcholine, is responsible for *excitatory impulses*. In the case of the heart, excitatory impulses come from sympathetic nerves, for the noradrenaline or adrenaline that they release increase the permeability to Na^+ in the membrane of heart muscle cells. The actual heart rate will therefore depend on the relative number of impulses arriving from both nervous systems. This integration of inhibitory and excitatory impulses is a common phenomenon, not only at the muscle cells associated with the internal organs, but in the central nervous system as well (see Fig. 7–14). It is one of several mechanisms by which the complex functions of the central nervous system are supported. Its application to the regulation of muscle contraction will be considered after the structure and contractile mechanism of muscle have been examined.

7–4 MUSCLE CELLS AND THE MECHANISM OF CONTRACTION

All muscle cells can be divided into two categories, *striated* and *smooth,* based on their appearance in the microscope (see Fig. 7–15). Striated muscles, as seen in the polarizing or phase contrast

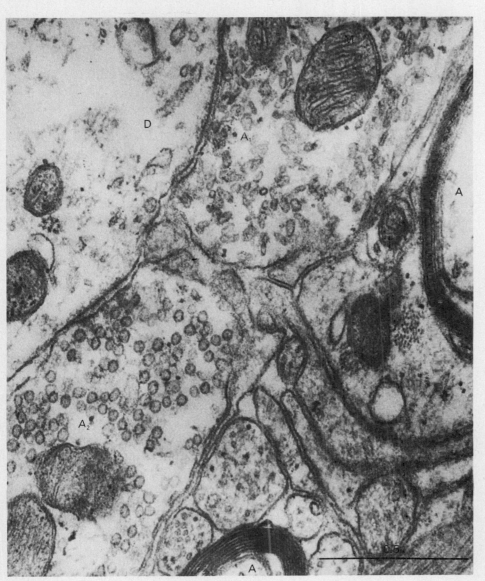

FIGURE 7–14. Synaptic Integration. More than one axon may contact the same dendrite. This micrograph, from the spinal cord of a monkey, shows two axons (A_1 and A_2) synapsing with a dendrite (D) of a common motorneuron. The two axons have differently shaped synaptic vesicles, leading to the hypothesis that they represent excitatory and inhibitory neurons. [Courtesy of D. Bodian, *Science,* **151:**1093 (1966). Copyright by the A.A.A.S.]

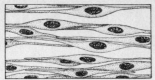

Smooth muscle

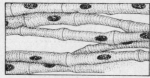

Cardiac muscle

Skeletal muscle

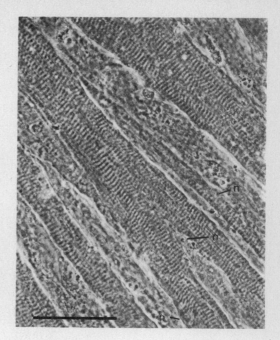

FIGURE 7–15. Muscle. Most muscle is striated, meaning that it has alternate light and dark bands when viewed in polarizing or phase contrast light microscopes (the micrograph shown here is phase contrast: nuclei (n); magnification bar = 50 μ). Although skeletal muscle and heart muscle are both striated, skeletal muscle cells are syncytial (many nuclei in a common cytoplasm, as in the micrograph), while cardiac muscle cells are mononucleated. Visceral involuntary muscle cells are also mononucleated, but they are smooth—i.e., without striations. [Micrograph courtesy of Y. Shimada, *J. Cell Biol.*, **48**:128 (1971).]

light microscope, are characterized by alternating light and dark bands. Skeletal muscle, which comprises about 40% of one's body weight, accounts for most of the striated muscle. Cardiac muscle is also striated, but unlike skeletal muscle it is involuntary and has but one nucleus per cell. Muscles associated with the other internal organs are smooth (nonstriated) and involuntary. Most of the discussion here will be devoted to striated muscle, for it is the more thoroughly studied.

MYOFIBRILS. Each muscle cell, or *muscle fiber*, has filaments (*myofilaments*) running its long axis.[4] In striated cells, but not in smooth muscle cells, the filaments are gathered together into bundles 1 to 2 μ in diameter called *myofibrils*. A nomenclature describing the appearance of myofibrils was developed from observations made originally with the polarizing light microscope (see Fig. 7–16). In this nomenclature, regularly spaced lines called *Z-lines* or *Z-discs* (from the German word *zwischen*, meaning between), divide the myofibril into *sarcomeres*. There is a light region at either end of the sarcomere called an *I-band*, because it appears isotropic (the same in every direction) when viewed with crossed polarizers. Between the two I-bands of each sarcomere is an *A-band* (anisotropic), which is generally much darker, except for

---

4. The prefixes "myo-" and "sarco-" are often used to define features of muscle cells that we now recognize to be the same for all cell types. Thus, the *sarcolemma* is the plasmalemma, or cytoplasmic membrane. A *sarcosome* is a mitochondrion, the *sarcoplasm* is the cytoplasm, and so forth.

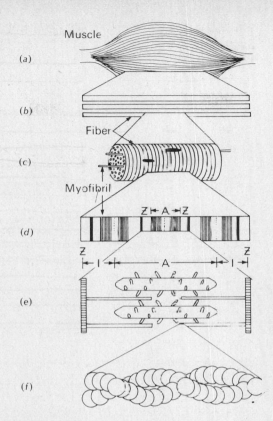

(a) Muscle

(b)

Fiber

(c)

Myofibril

(d) Z ⊢ A → Z

Z ⊢ I → ⊢ A → ⊢ I → Z

(e)

(f)

FIGURE 7–16. The Microanatomy of Skeletal Muscle. The muscle (part **a**) is composed of fibers shown in (**b**) and (**c**), each of which is a multi-nucleated cell. The fibers, in turn, contain parallel rows of myofibrils (part **d**). Each individual myofibril, as shown in (**e**), contains two classes of filaments, one of which (the actin filament) is illustrated in (**f**). By comparing (d) and (e), one can see how the striations arise.

a region of intermediate density at its center called the *H-zone* (from the German word *hell*, or clear). At the middle of the H-zone, which is also the middle of the sarcomere, is a thin, darker *M-line* (see Fig. 7–17a).

Electron micrographs taken at M.I.T. in the early 1950s by H. E. Huxley and J. Hanson made it clear that these striations are due to a pattern of overlapping myofilaments of two types. The I-bands contain only thin, actin filaments, which are attached to the Z-lines and extend toward the center of the sarcomere from either side. The A-bands contain actin filaments plus thicker, myosin filaments, stretching from one edge to the other and interdigitating with the thin filaments in the dark regions, which is what makes these regions dense. In cross-section, the filaments often appear to be hexagonally packed, with each thick filament surrounded in regions of overlap by six thin ones (see Fig. 7–17b). As we shall see, the fiber contracts by sliding these filaments past one another.

ACTIN FILAMENTS. Actin is a globular protein, about 55 Å in diameter, composed of one polypeptide chain with a mass of approximately 47,000 amu. It polymerizes in the presence of salts into long filaments, which are called *F-actin* to distinguish them from the depolymerized or globular form, *G-actin*.

FIGURE 7–17. The Myofibril. **(a)** Longitudinal section of a frog muscle cell, showing a line of mitochondria (M) adjacent to a row of myofibrils to which they supply ATP. Note that the myofibrils are in register, leading to cross-striations. In some invertebrates, they are out of register, causing oblique striations. The membranous structures are known as longitudinal vesicles (LV) and transverse tubules (TT). The latter are running nearly perpendicular to the plane of the section except at the periphery, where some angle toward the cell surface and open onto it. **(b)** Cross-section of a frog skeletal muscle cell, showing several myofibrils. Note the regular array of six thin filaments about each thick one. Sections of longitudinal vesicles (LV) can be seen between the myofibrils. The diameter of each myofibril is about 1 μ. [(a) courtesy of C. Franzini-Armstrong, *J. Cell Biol.*, **47**:488 (1970); (b) courtesy of H. E. Huxley.]

(a)

(b)

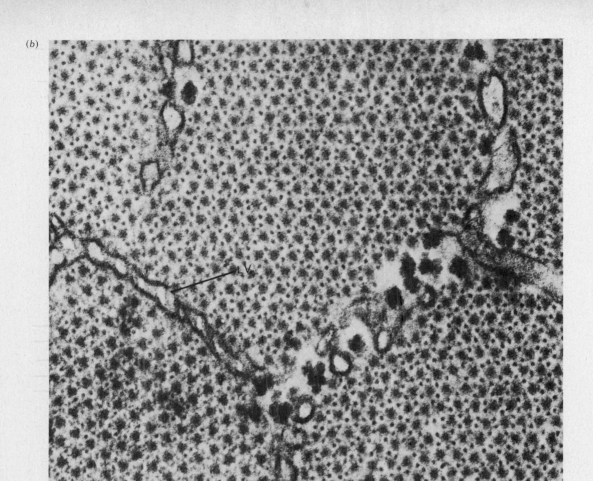

The filaments of F-actin form a right-handed double helix, like two strings of pop-beads wound about each other. The filaments have a pitch of 700 to 800 Å. At least two other proteins, called *tropomyosin* and *troponin*, are associated with actin filaments *in vivo* (see Fig. 7–18).

Tropomyosin is a protein of about 70,000 amu, in two polypeptide chains of 35,000 each, largely alpha helical in conformation and twisted about each other. The molecule has a length of about 400 Å, but a width of only 20 to 30 Å. According to a model of the actin filaments proposed by S. Ebashi and his colleagues of the University of Tokyo, the tropomyosin molecules are stretched end to end along the groove of the actin helix, with molecules of troponin acting as dividers (see Fig. 7–18b). Although troponin is roughly similar in mass to tropomyosin, it is globular, and therefore occupies much less space. The troponin–tropomyosin complex plays an important role in the contraction mechanism, to be discussed later.

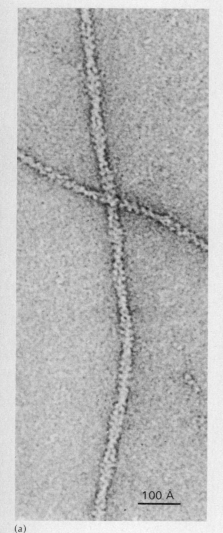

FIGURE 7–18. Actin. **(a)** Polymerized actin is called F-actin. Note its double-helical structure. **(b)** Model for the actin filament, showing the association of troponin and tropomyosin (see text). [(a) courtesy of H. E. Huxley, *J. Mol. Biol.*, 7:281 (1963).]

100 Å

(a)

MYOSIN FILAMENTS. Actin accounts for only about a quarter of the protein in myofibrils, and tropomyosin and troponin only a few percent each. More than half of the total protein in a myofibril is myosin, which is the major constituent of the thick filaments of most muscle. (Some invertebrate smooth muscles have filaments with diameters up to 1000 Å containing large quantities of a protein called paramyosin in addition to myosin.)

Myosin is an unusual protein, with a bilobed head and a rodlike 1300 Å (0.13 μ) tail that is perhaps 20 Å in diameter (see Fig. 7–19). It has a total mass of about 470,000 amu, including two polypeptide chains of about 210,000 amu each plus several (probably 3 or more) smaller chains. The major part of the two long polypeptide chains are in an alpha helical configuration and wound about each other in a coiled-coil arrangement like that of tropomyosin. The two-chain coil constitutes the long "tail" of the molecule and extends into the "head."

The head region of myosin, together with a portion (about 370 Å) of the tail, may be separated from the rest of the tail by the action of certain proteolytic enzymes. The resulting head fragment is called *heavy meromyosin;* the tail fragment is called *light meromyosin.* The enzymatic capacity of myosin, which consists of the ability to hydrolyze ATP to ADP and inorganic phosphate, is associated exclusively with heavy meromyosin. Thus, we know that the active site of the myosin ATPase is in the head region only.

Myosin is soluble in concentrated salt solutions (e.g., 0.6 M KCl), but polymerizes and precipitates when the salt concentration is lowered to about 0.16 M near pH 7. Polymerization of myosin starts with a tail-to-tail association that produces a bare central shaft with the heads at either end. Additional molecules are added at both ends, with the tails of the new molecules toward the center of the growing filament. Several chains formed in this way may aggregate by partially overlapping one another. In rabbit muscle preparations, the overlap places two heads (each bilobed) on one side of the filament opposite two heads on the other side, then repeats this pattern each 143 Å along the axis. Since there is a 120° rotation between one set of four heads and the next, a full revolution of the pattern is completed every 429 Å. The complete myosin filament is about 1.5 μ long, with a bare central shaft (the

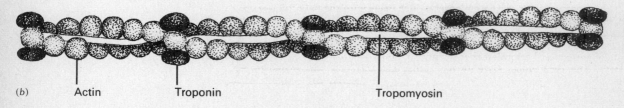

(b)   Actin   Troponin   Tropomyosin

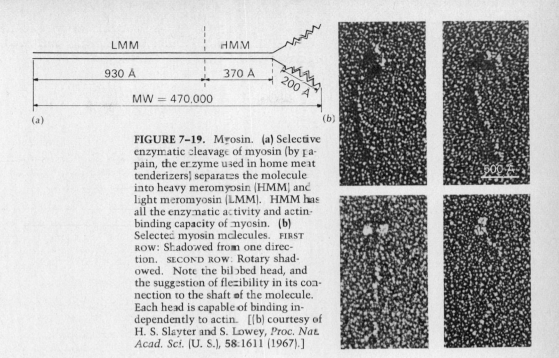

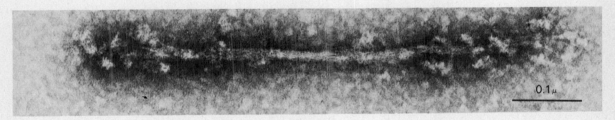

FIGURE 7–19. Myosin. **(a)** Selective enzymatic cleavage of myosin (by papain, the enzyme used in home meat tenderizers) separates the molecule into heavy meromyosin (HMM) and light meromyosin (LMM). HMM has all the enzymatic activity and actin-binding capacity of myosin. **(b)** Selected myosin molecules. FIRST ROW: Shadowed from one direction. SECOND ROW: Rotary shadowed. Note the bilobed head, and the suggestion of flexibility in its connection to the shaft of the molecule. Each head is capable of binding independently to actin. [(b) courtesy of H. S. Slayter and S. Lowey, *Proc. Nat. Acad. Sci.* (U. S.), **58**:1611 (1967).]

FIGURE 7–20. The Myosin Filament. Myosin polymerizes *in vitro* to form filaments that are symmetrical about a bare central shaft. Except in this bare region, projections, presumably due to myosin heads, are seen. [Courtesy of H. E. Huxley, *J. Mol. Biol.*, **7**:281 (1963).]

tail-to-tail region) of 0.15 to 0.2 μ where there are no heads (see Fig. 7–20).

THE SLIDING FILAMENT THEORY. The width of the A-bands in striated muscle remains unchanged during contraction, while the I- and H-bands get narrower or disappear entirely (see Fig. 7–21). This observation led to the *sliding filament theory*, which attributes contraction to the movement of the thick and thin filaments past each other.[5]

---

5. This hypothesis first arose from observations made with the light microscope by A. F. Huxley and R. Niedergerke, and was published at about the same time (1954) as the early electron microscopy of Hanson and H. E. Huxley. These two Huxleys, it should be noted, are not related. A. F. Huxley is from the prominent English family of authors and scientists that includes Aldous, Sir Julian, and Thomas Henry Huxley. The 1963 Nobel Prize for physiology and medicine was shared by A. F. Huxley, A. L. Hodgkin, and J. C. Eccles for their studies on the excitable membranes of nerve and muscle.

(a)

292

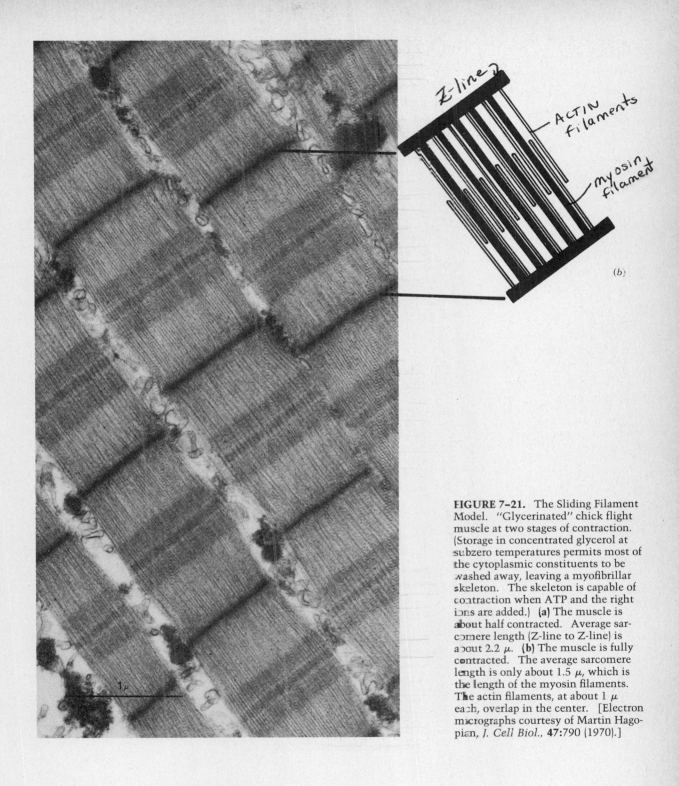

Z-line

ACTIN filaments

myosin filament

(b)

FIGURE 7–21. The Sliding Filament Model. "Glycerinated" chick flight muscle at two stages of contraction. (Storage in concentrated glycerol at subzero temperatures permits most of the cytoplasmic constituents to be washed away, leaving a myofibrillar skeleton. The skeleton is capable of contraction when ATP and the right ions are added.) **(a)** The muscle is about half contracted. Average sarcomere length (Z-line to Z-line) is about 2.2 μ. **(b)** The muscle is fully contracted. The average sarcomere length is only about 1.5 μ, which is the length of the myosin filaments. The actin filaments, at about 1 μ each, overlap in the center. [Electron micrographs courtesy of Martin Hagopian, *J. Cell Biol.*, **47**:790 (1970).]

1 μ

The significance of the bilateral symmetry of the myosin filaments should now be clear, for the two halves must pull actin in opposite directions. That is, the actin filaments from opposite ends of a sarcomere must be pulled toward each other along their longitudinal axis, and toward the center of the myosin filaments. Further examination reveals that the actin filaments, too, have a bipolar symmetry reflected about the Z-line. When H. E. Huxley investigated the binding of heavy meromyosoin (Fig. 7–19a) to actin, he found that all myosin heads align themselves in the same direction on a given actin filament, and that the direction is the same for all the filaments on one side of a Z-line.

The sliding filament model predicts, correctly, that contraction should stop soon after the Z'-lines meet the myosin filaments at each end of the sarcomere. In addition, if the motive force for contraction is derived from an interaction between thick (myosin) and thin (actin) filaments, then some overlap is always required, putting an upper limit on the sarcomere length corresponding to the combined length of a thick filament and two thin ones. In frog muscle, this limit is $1.5 \mu + (2 \times 1.1) \mu$ or about 3.7μ total.

The current view of the contraction process, then, is as follows: The actin and myosin filaments are parallel and connected by cross-bridges consisting of the enlarged ends of the myosin molecules (see Fig. 7–22). Each actin molecule is capable of binding one of these projections. Contraction occurs when myosin releases its contact with the adjacent actin, forms a new bridge with the next actin molecule in the chain, then reshortens to bring the newly contacted actin into the position occupied by the old one.

FIGURE 7–22. Myosin-Actin Interaction. Longitudinal section through a myofibril, showing cross-bridges between the actin and myosin filaments. The thickening in the middle of the myosin filaments is due to the *M-line protein*, now thought to be creatine kinase. Slender threads between myosin filaments are also sometimes seen in micrographs like this one. They may help hold the filaments in register. [Courtesy of H. E. Huxley, *J. Biochem. Biophys. Cytol.* (now *J. Cell Biol.*), **3**:631 (1957).]

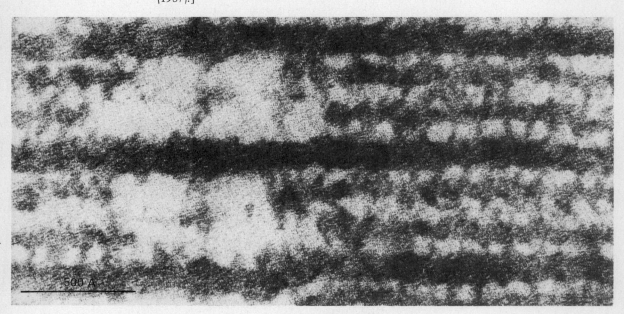

500 Å

Thus, the filaments move past each other in a ratchetlike manner. At any one time there must be enough stable bridges to prevent backsliding, but the force required to move the filaments would come from the alternate extension–contraction cycle of each of the myosin projections. Relaxation, in this scheme, requires that all the bridges be broken at once, leaving the filaments free to slide past each other without tension as they are pulled apart by the contraction of opposing muscles.

The energy needed for contraction comes from the breakdown of ATP. However, to help maintain ATP levels when demand is high, muscles also have large quantities of creatine phosphate, which can act as a donor, transferring its phosphate to ADP.

$$^-OOC—CH_2—N—C{=}NH + ADP$$

with CH$_3$ on the nitrogen and HNOPO$_3^{2-}$ below.

Creatine phosphate

creatine
kinase

$$^-OOC—CH_2—N—C{=}NH + ATP$$

with CH$_3$ on the nitrogen and $+NH_3$ below.

Creatine

Since the equilibrium for the transfer reaction (which is catalyzed by creatine kinase) lies strongly in favor of ATP, ATP levels stay approximately constant until nearly all of the creatine phosphate is gone. Thus, in effect, a considerable reserve of ATP exists, augmented by the additional mechanism shown in Fig. 7–23.

7-5 CONTRACTION-RELAXATION CONTROL

Contraction is normally initiated by nerves, through the action of neurotransmitters such as acetylcholine, adrenaline, and noradrenaline. When an impulse is received by a muscle cell, calcium ions are released from reservoirs within the cytoplasm and diffuse to the myofibrils. Relaxation occurs when these ions are no longer available to the myofibrils. Regulating the amount of free Ca^{2+} is the responsibility of specialized membranes within the cytoplasm and the enzymes they contain.

THE SARCOTUBULAR SYSTEM. A network of membranes called the *sarcotubular system* crisscrosses the cell both longitudinally and transversely. The longitudinal and transverse systems are not directly connected with each other, however, nor do they have the same origins or functions. The longitudinal component of the sar-

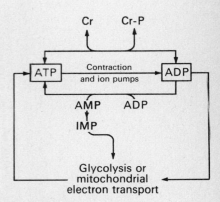

FIGURE 7–23. ATP Utilization and Regeneration in Muscle. In addition to ATP production through the usual metabolic channels, ATP maintenance is fostered through the creatine kinase equilibrium and through the adenylate kinase equilibrium (2 ADP ⇌ AMP + ATP). The latter is pulled to the right (toward ATP) by mass action when the adenine of AMP is deaminated to inosine, forming IMP. The IMP is reconverted to AMP during periods of rest.

295

cotubular system forms channels (the *longitudinal vesicles,* or LV)
that run parallel to the myofibrils. The LV contain periodic enlargements called *terminal cisternae.* This longitudinal network
is apparently derived from the endoplasmic reticulum of the immature muscle cell, or *myoblast,* and hence may be referred to as
the *sarcoplasmic reticulum.* (Unfortunately, the latter name also
is sometimes used interchangeably with sarcotubular system.)

The transverse component of the sarcotubular system, on the
other hand, consists of regularly spaced tubules, often about 400 Å
in diameter, that open onto the surface of the cell. This *T-system*
is, therefore, continuous with the sarcolemma from which it is
thought to be derived. At points where the longitudinal and transverse systems cross, one often sees a configuration called a *triad,*
composed of two terminal cisternae lying on either side of a transverse tubule (see Fig. 7–24). This close juxtaposition of the longi-

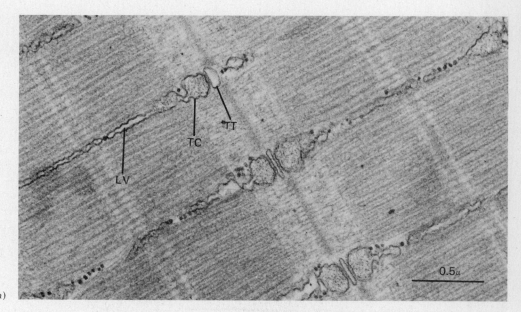

(a)

FIGURE 7–24. The Sarcotubular System. **(a)** From frog
muscle. Transverse tubules (TT) run perpendicular to the section, passing between terminal cisternae (TC) of adjacent longitudinal vesicles (LV). This arrangement is known as a triad.
(In some muscles, one sees dyads, pentads, etc.) Triads are also
visible in Fig. 7–17a. **(b)** Freeze-etched view of guinea pig heart
muscle. The cell is fractured, showing rows of T-system aper-
tures (arrows) on the surface, running parallel to the myofilaments (MF) beneath. A row of mitochondria (M) is deep to the
first row of sarcomeres as in 7–17a, a micrograph that also
shows tubules opening onto the surface. (The circled arrow indicates the direction of metal shadowing.) [(a) courtesy of
C. Franzini-Armstrong, *J. Cell Biol.,* **47:**488 (1970); (b) courtesy
of G. Rayns, F. O. Simpson, and W. S. Bertand, *Science,* **156:**656
(1970), copyright by the A.A.A.S.]

tudinal and transverse systems is the key to the coupling of excitation and contraction.

EXCITATION–CONTRACTION COUPLING. The longitudinal vesicles contain an active Ca^{2+} pump (Ca^{2+}-activated ATPase), which permits them to accumulate stores of the ion, much of which is present as a calcium phosphate gel. (The ATPase may comprise as much as 50% of the protein of rabbit sarcoplasmic reticulum.) When the sarcolemma is depolarized, an action potential is conducted into the interior of the cell by transverse tubules. As the impulse travels past terminal cisternae of the LV, Ca^{2+} is released and contraction ensues. Relaxation occurs when the released Ca^{2+} has been reaccumulated by the LV.

The first study in which the internal availability of Ca^{2+} was correlated with the development of tension following a normal ex-

(b)

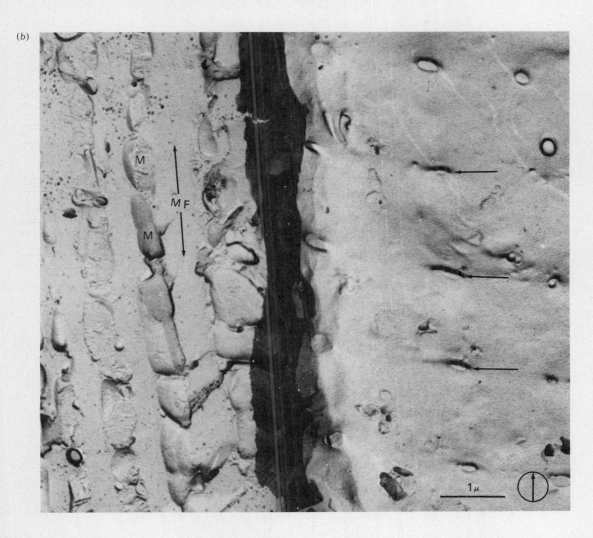

citation was described in a 1966 paper by F. F. Jöbsis and M. O'Conner of Duke University. They found that using dimethyl sulfoxide (DMSO) to increase membrane permeability could cause some toad muscles to take up significant quantities of Tyrian purple (murexide), a dye that changes color in the presence of calcium ions. The incorporated dye changed color immediately following an excitation, and then recovered slightly ahead of the development of peak tension.

A year later, E. B. Ridgway and C. C. Ashley made a similar observation with a different muscle and a different indicator. They used the protein aequorin, obtained from a jellyfish, to detect calcium in muscle fibers from giant barnacles. These fibers are typically 40 mm long by about 2 mm in diameter. (Vertebrate fibers rarely exceed 0.1 mm in diameter.) Aequorin produces a bluish luminescence, the intensity of which is directly related to the concentration of Ca^{2+} available to it. It was found that a barnacle muscle fiber that has been injected with aequorin glows briefly following excitation, but that the light fades again before maximum tension is reached (see Fig. 7–25).

The preceding studies reveal that a transient availability of Ca^{2+} is associated with excitation and the resulting contraction. Coupled with other observations, some of which have been mentioned, it is clear that contraction is a direct result of the appearance of free Ca^{2+} in the vicinity of the myofilaments, and that relaxation results from the removal of these ions. The slight delay between the appearance of Ca^{2+} and the development of maximum tension is caused by the damping effect of elastic elements both within and outside the muscle.

The availability of free Ca^{2+} at the myofibrils permits the utilization of ATP to power contraction. (Actually, it is the Mg^{2+} chelate of ATP that is used.) When both Ca^{2+} and ATP are present, the muscle cell contracts; when ATP is present without Ca^{2+}, the cell relaxes; and when ATP is absent, the cell enters a state of *rigor*,

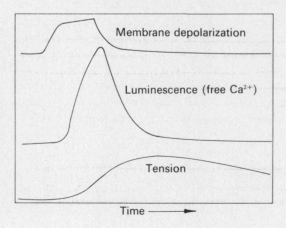

FIGURE 7–25. Excitation–Contraction Coupling. Electrical stimulation of a barnacle muscle fiber produces an action potential (top). (The sharp fall in depolarization marks the time when stimulation ceased.) The temporary appearance of free Ca^{2+} in the sarcoplasm is detected by luminescence from injected aequorin (middle). The fall in free Ca^{2+} is due to its binding by the myofibrils, resulting in tension in the fiber (bottom). Tension will last until Ca^{2+} has been reaccumulated by the sarcoplasmic reticulum. [Described by C. Ashley and E. Ridgway, in *J. Physiol.*, **209**:105 (1970).]

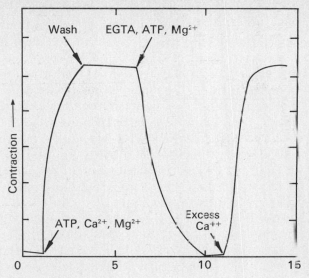

FIGURE 7–26. The Role of Ca^{2+} in Contraction. Follow the diagram from left to right: A solution of ATP, Ca^{2+}, and Mg^{2+} was added to washed, glycerinated fibers. Contraction followed. At the peak of the contraction, the fibers were washed, removing Mg^{2+} and ATP but not all the Ca^{2+}, some of which was bound to the fibrils. With ATP gone, the fibers enter a state of rigor (flat top on curve). New ATP and Mg^{2+} were then added along with EGTA, a chelating agent that ties up Ca^{2+}. With ATP but no Ca^{2+}, the fibers relax. Finally, addition of excess Ca^{2+} was used to saturate the EGTA, initiating a new contraction. [Experiment described by J. Bendall, in *Muscles, Molecules, and Movement.* New York: American Elsevier Publ. Co., Inc., 1969.]

with fixed cross-bridges, regardless of the presence or absence of Ca^{2+} (see Fig. 7–26).

The regulatory role of Ca^{2+} is made possible by troponin and tropomyosin, the proteins that were described earlier as being associated with the actin filaments. The cooperative interaction between actin and myosin needed to hydrolyze ATP and power contraction is prevented by troponin and tropomyosin in the absence of Ca^{2+}, but is allowed to occur when Ca^{2+} is bound to troponin.[6] It has been proposed that tropomyosin covers myosin acceptor sites on adjacent actins, and that Ca^{2+} causes a change in the shape of troponin which is propagated to the tropomyosin, causing the latter to uncover some or all of the sites.

GRADED RESPONSES. A contraction–relaxation cycle, then, is controlled by the amount of free calcium available for binding to the actin filaments at receptor sites located on troponin molecules. The variation in free Ca^{2+} near these sites is a result of release and reaccumulation of the ion by terminal cisternae of the longitudinal vesicles. Accumulation appears to be a continuous process, opposed by a rapid, transient release of Ca^{2+} in response to an excitation of the sarcolemma that is conducted inward by the transverse tubules. The rate of conduction in frog muscles is about 8 cm/sec, which should allow impulses to reach the center of a 0.1 mm diameter cell in much less than a millisecond. The result, ordinarily, is a single twitch.

---

6. At least some simple organisms achieve contraction-relaxation control by a different mechanism. The actin filaments of scallop muscles, for example, contain tropomyosin but not troponin, and hence have no Ca^{2+} receptors. Instead, Ca^{2+} sensitivity is conferred by one of the three light chains of the scallop myosin molecule.

To obtain a full contraction in striated muscles (also called *twitch muscles*), a string of impulses must arrive in rapid succession. The ability to add the effects of several impulses allows a graded response, so that many degrees of contraction can be achieved. Maximum tension is generated in a striated muscle only during *tetanus,* when the impulses are arriving too fast to permit any relaxation between twitches (see Fig. 7–27). As an example, twitches in frog sartorius muscle (from the leg) at 0 °C just fuse to form a tetanus when the impulse rate is 15 per second.

The ability of striated muscles to give a graded response to varying frequencies of incoming impulses is a consequence of Ca^{2+} mediation between excitation and contraction. Calcium release from the longitudinal vesicles of a muscle is directly related to the amount and duration of depolarization of the cell membranes. The amount of available Ca^{2+}, in turn, affects the degree of tension.

Most muscles are also capable of giving a graded response to subthreshold depolarizations, a fact that was exploited by A. F. Huxley and R. E. Taylor in 1958 to define the points at which a depolarization can be conducted into the interior. They found that a very small electric stimulus, which gives highly localized depolarizations insufficient to trigger a spike potential, can elicit a small twitch when applied at some points of the sarcolemma, but not at others. The sensitive points were found to be arranged in a regular pattern across the surface, and were later identified as the points at which the sarcolemma invaginates to form transverse tubules.

The response of a muscle is affected by signals from inhibitory neurons. In the case of the heart, acetylcholine (from the vagus nerve) produces hyperpolarization (a more negative potential) by increasing K^+ permeability, inhibiting the generation of spike potentials and hence the frequency of contraction. In other muscles, the same chemical increases both Na^+ and K^+ permeability, causing depolarization and triggering spike potentials.

The heart has a slower, more deliberate spike potential and contraction cycle than many other muscles, in keeping with the physiological role of that organ. However, even fast muscles have refractory periods of tens of milliseconds, which is very slow by neuronal standards. Although the rate of conduction of a depolarization is not so very different between nerves and muscles, events in muscle transpire more slowly than they do in neurons.

CONTROL IN SMOOTH MUSCLE. While striated muscle cells respond relatively slowly to excitation, smooth muscle cells are even slower. And, unlike striated muscles, they do not need a continuing input of nerve impulses to maintain a state of partial contraction. This ability to sustain a partially contracted state

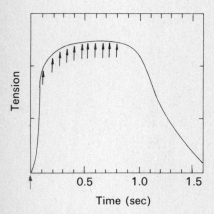

FIGURE 7–27. Tetanus. When excitations (arrows) fall too close together, twitches fuse into a tetanus, defined as a state of continued, complete contraction. (Continued partial contraction is a *tonus*.) The tension is greater than that produced in a single twitch, because there is time for elastic elements in the fiber to become maximally stretched. (Data from A. V. Hill.)

Tension

0.5   1.0   1.5

Time (sec)

with very little expenditure of energy is called *muscle tone*. The extreme example of a tonic contraction is found in the paramycsin *catch muscles* of mollusks and annelids, such as those that hold a clamshell shut.

One possible reason for the difference in contractile behavior between striated and smooth muscle is the paucity of endoplasmic reticulum in the latter cell type, which also has little or no T-system. (Smooth muscle cells also lack myofibrils, their seemingly less organized arrangement of actin and myosin filaments being responsible for the lack of striations—see Fig. 7-28.) Because of the absence of a sarcotubular system, smooth muscle cells

FIGURE 7-28. Smooth Muscle. This cross-section through a smooth muscle cell (from a rabbit vein) reveals a regular array of thick filaments and a not-so-regular array of thin ones. The filaments are not organized into myofibrils, nor is there any significant amount of endoplasmic reticulum. Mitochondrion (M). [Courtesy of R. Rice, G. McManus, C. Devine, A. Somlyo, *Nature/New Biology*, **231**:242 (1971).]

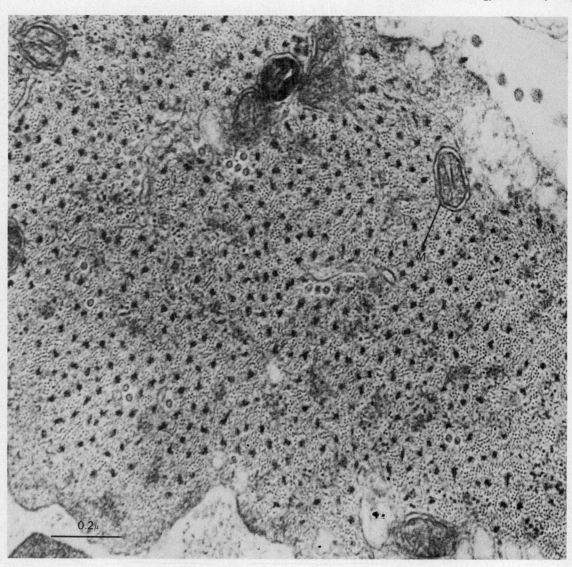

0.2μ

should have only a limited ability to release and recapture Ca^{2+}. In fact, it has been suggested that the plasmalemma of these cells serves as a source of calcium and as a mechanism for removing it. Calcium would have to be obtained from the fluid surrounding the cell and returned to it, a process that would certainly be slow enough to explain the relative sluggishness of smooth muscles.

7–6 NEUROMUSCULAR INTERACTION

One of the reasons for grouping the discussion of neurons and muscle cells into a single chapter is to provide an example of a close interaction between two cell types. You are already aware that most muscular contraction is initiated by impulses arriving from nerves. But innervation is also necessary for the maintenance of muscle cells, with denervation causing an *atrophy of disuse.* In addition, innervation is necessary to support the limited regeneration that may occur following injury, though the embryonic formation of muscles is independent of nerve supply. And finally, the choice of nerves that contact a given muscle cell may influence its biochemical and morphological characteristics.

We shall consider some of these interactions following a brief description of the neuromuscular junction itself.

THE NEUROMUSCULAR JUNCTION. A single motor neuron commonly controls a great many muscle fibers, known collectively as a *motor unit.* There may be many such units in a muscle, leading to a cumulative effect when a muscle is stimulated by more than one neuron. Synchrony is achieved by a rapid propagation of spike potentials along the muscle fibers and through their interior, the latter via transverse tubules. In some cases, however, especially in invertebrates, there are nerve endings at multiple points along each fiber, leading to a uniform contraction of the fiber without the propagation of spike potentials along the sarcolemma. This arrangement also facilitates a graded response of the fiber, with a contraction that varies in intensity with the frequency of incoming impulses.

There is no special structure associated with the neuromuscular junction in cardiac or smooth muscle. Axon endings, void of myelin, seem merely to pass near the surface of the muscle cells. The existence of a junction can be confirmed from electron micrographs, however, for synaptic vesicles containing the neurotransmitter can be seen within the axons at these points (see Fig. 7–29).

Striated muscles, on the other hand, normally contact nerves at identifiable points consisting of slightly elevated regions of the muscle surface, rich in nuclei and mitochondria. The "contact" between nerve and muscle is not complete, leaving a gap (actually a glycoprotein layer) of hundreds of angstroms to be crossed by the

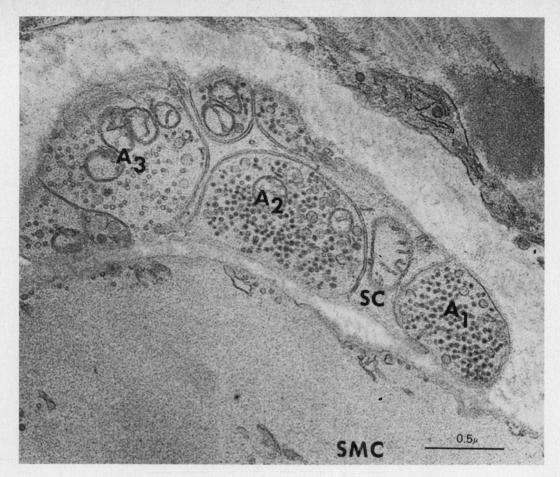

SMC

FIGURE 7–29. Neuromuscular Junction of Smooth Muscle. Three axons (A_1, A_2, A_3) pass near the sarcolemma of the smooth muscle cell (SMC). The axons, enclosed by a Schwann cell (SC), seem to be of two varieties, based on the lesser density of the synaptic vesicles in A_3. (A_3 is probably cholinergic, the others adrenergic.) The space between the axons and the sarcolemma is occupied by a fibrous basement membrane. [Courtesy of E. Nelson and M. Rennels, *Brain*, **93**:475 (1970).]

neurotransmitter. The junctions are distinguished by numerous folds in the sarcolemma called *synaptic troughs, junctional folds,* or *primary synaptic clefts*. These folds contain terminal branches of innervating axons, with their closely packed synaptic vesicles (see Fig. 7–30). When an impulse reaches the terminal part of the axon, the contents of some of these synaptic vesicles are emptied into the cleft between the two cells, causing in the muscle cell membrane an action potential that may lead to contraction.

Contraction is not the only way that a nerve can affect a muscle, however. It may also cause more subtle changes, such as the development of red or white characteristics.

RED AND WHITE MUSCLES. Striated muscle can be divided into three kinds of fibers with very different physical, physiological, and metabolic properties. The two extreme groups are called *red muscle* and *white muscle*, with an intermediate group that is sometimes designated as *pink*. Some characteristics of red and

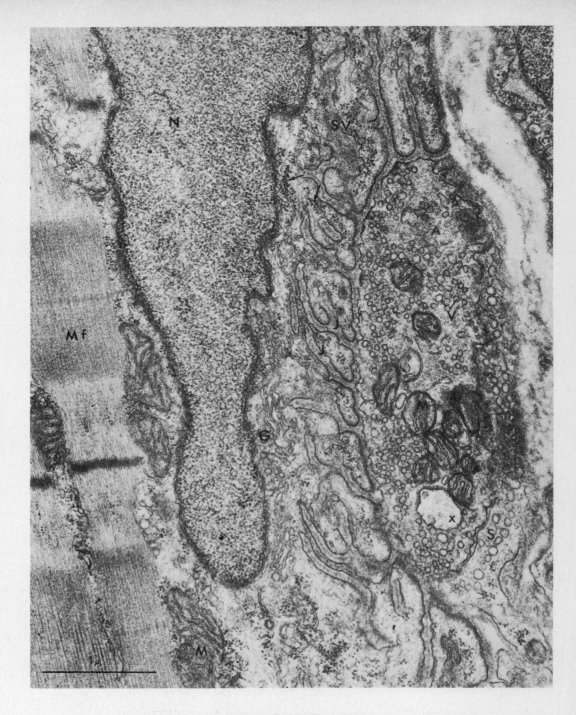

FIGURE 7–30. The Neuromuscular Junction. A red muscle fiber from a rat diaphragm, showing a single elliptical axonal ending (A) with synaptic vesicles (V) and a Schwann cell covering (S). Vacuole (x). Golgi apparatus (G). Rough endoplasmic reticulum (arrows). Myofibril (Mf). Sarcoplasmic vesicle (SV). A nucleus of the muscle fiber (N). [Courtesy of H. A. Padykula and G. F. Gauthier, *J. Cell Biol.*, **46:**27 (1970).]

DESIGNATION	RED	WHITE
Alternate designations[a]	slow, small, type I, type C, tonic, slow twitch	fast, large, type II, type A, phasic, tetanic, fast twitch
Fiber size	small	large
Vascularization	heavy	lighter
Endoplasmic reticulum	moderate	extensive
Mode of contraction	slow twitch	fast twitch
Stored fuel	fat	glycogen
Main source of ATP	fatty acid oxidation	glycolysis
Mitochondria	many	few
Myoglobin	much	little

[a] Intermediate fibers are also known as type II, type B, or pink.

white muscles are given in Table 7.1, which shows that the two types are clearly adapted for different purposes.

Red muscles, which have been described by W. Mommaerts as "pay-as-you-go-muscles," are highly vascularized, with relatively small surface-to-volume ratios, a great deal of myoglobin for oxygen storage and transfer, and numerous mitochondria. Their color is mainly due to myoglobin. These muscles, which are designed for relatively continuous use, oxidize fatty acids as a primary source of energy. White fibers, on the other hand, are larger, with less access to the circulatory system, relatively little myoglobin, and few mitochondria (see Fig. 7–31). Their main source of energy is glycolysis, which, because it is relatively inefficient, can sustain activity for only brief periods before rest is required to remove accumulated lactic acid and to obtain more carbohydrate. Mommaerts refers to these as "twitch-now-pay-later muscles."

The neuromuscular junctions and sarcotubular systems are decidedly different in red and white muscle cells. White fibers, for example, have the most extensive sarcotubular network, accounting in part for the relative rapidity of their contraction–relaxation cycle. Red muscles have a sparser network and hence respond more slowly to stimulation. (Smooth muscle, of course, has almost no sarcotubular system and is the slowest of all.)

Most skeletal muscles are a combination of red and white cells, with the overall classification of the muscle reflecting the predominant species. In addition, intermediate (pink) fibers may themselves be dominant in some muscles. Nevertheless, classifying a muscle as red or white is a useful description of its properties, and has the additional advantage of being familiar to most people through reference to the "white meat" of chicken or turkey.

White meat in domestic fowl is mostly the well-developed *pec-*

(a)

FIGURE 7–31. Red and White
Muscle. (a) A white muscle (bra-
chioradialis, from the arm of a rhesus
monkey), stained for succinate dehydrogenase, a mitochondrial enzyme
and therefore a marker for red fibers.
Note the predominance of large white
fibers, with a few smaller, red
(stained) ones. (b) Sartorius muscle
(from the thigh) of a rhesus monkey,
also stained for succinate dehydrogenase. This muscle is decidedly
mixed in its fiber type, and would be
classed as pink or red by visual
inspection alone. [(a) courtesy of
C. Beatty, G. Basinger, C. Dully, and
R. Bocek, *J. Histochem. Cytochem.*,
14:590 (1966); (b) courtesy of R. Bocek
and C. Beatty, *J. Histochem. Cytochem.*, **14**:549 (1966). Copyright by
the Williams & Wilkins Co.,
Baltimore.]

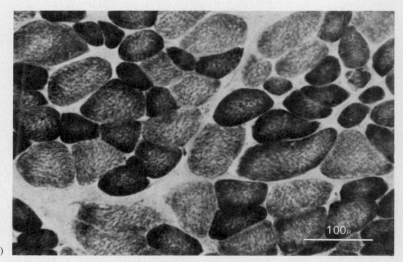

(b)

toralis major, or flight muscle. In spite of the size of this muscle,
these birds can only flutter along for a few yards before stopping to
rest, a pattern of usage that is characteristic of white muscles. On
the other hand, anyone who eats domestic fowl knows that their
leg muscles, which get almost continuous use in life, are predominantly red. When one turns to a migratory bird for comparison, he
finds the situation reversed: very red flight muscles to provide the
endurance for sustained flight, and little-used leg muscles that are
relatively white. The same distinction is found in mammals,

where intermittently used skeletal muscles are relatively white compared with the heart or other muscles designed for more continuous use.

The basis for differentiation into red and white fibers became clear around 1960, when it was found that the very considerable differences between these two types of muscle cells are due to the influence of nerves. This conclusion was reached from experiments in which nerves were switched between red and white muscles: cross-innervation turned white muscles into red ones and influenced red muscles to take on many of the properties of white muscles. It is not understood why one neuron influences a cell to become red while another influences it to become white, although the secretion of chemicals ("trophic" substances) by the neurons is often cited as a possible mechanism. We do know, however, that the influence includes an effect even on the structure of myosin (the pattern of light chains) and on the anatomy of the neuromuscular junction itself.

7-7 CONTRACTION IN NON-MUSCLE SYSTEMS

The conversion of chemical energy into mechanical motion is not a property unique to muscle cells. It must be responsible for cytoplasmic streaming, which refers to the ubiquitous movement of cytoplasm and organelles from one part of a cell to another—in some cells, the movement seems highly specific, in others it seems almost random. The conversion of chemical energy into mechanical movement must also provide cells with locomotion, whether they crawl along a surface like an ameba or get propelled through a liquid by the beat of their cilia or flagella.

In these cases and others, there is a net hydrolysis of ATP or an ATP-equivalent (e.g., GTP), the energy of which makes possible movement. It is convenient to classify the systems involved into three categories: (1) movement involving conformational changes in the protein subunits of a single filament, (2) movement due to the action of microtubules or microtubule-like structures, and (3) movement due to the interaction of actin and myosin in a mechanism akin to that of muscle. A possible fourth category will be introduced in Chap. 9 as being responsible for chromosome movement during cell division. These divisions are quite arbitrary, of course, since ultimately all movement must rely on enzymatic activity and conformational changes in proteins.

PROTEIN CONFORMATION AND MECHANICAL MOVEMENT. In Chap. 4, it was pointed out that the conformation (shape) of an enzyme may change during the course of its catalytic activity. In Chap. 5, this idea was used as the basis for the conformational coupling theory of oxidative phosphorylation in mitochondria. Then, in

307

Chap. 6 and earlier in this chapter, conformational changes driven by ATP hydrolysis were used to explain the sodium pump and cross-bridge movements between the actin and myosin filaments of muscle. The simplest application of these ideas to motile or contractile systems is probably the contraction of the tails of tadpole-shaped bacterial viruses when they penetrate and infect their host cells.

The T-even bacterial viruses described in Chap. 1 (see Figs. 1–25 and 1–26) have a tail composed of a helical arrangement of subunits (the outer sheath) surrounding a central core. After attachment to a bacterium, the outer sheath shortens, apparently driving the core through the wall of the host cell (Fig. 1–26). There is, at the same time, an injection of the viral DNA into the infected cell. The tail sheath is thought to be akin to a stretched spring, with potential energy stored in the conformation of the proteins of the sheath, but capable of only a single contraction because of a lack of metabolically derived energy to restretch the spring.

In other systems that use conformational transitions for mechanical movement, the high-energy conformation of the participating proteins would be reestablished by hydrolysis of a molecule such as ATP. This is thought to be the basis for the twisting of bacterial flagella, which rotate like corkscrews to drive their cells through the surrounding liquid. (Recall from Chap. 1 that bacterial flagella are constructed like a rope of individual polypeptide strands, in contrast to the complicated cylindrical construction of eucaryotic flagella and cilia.)

Other places where a similarly primitive type of motive force generation might be used is in the contraction of *myonemes* (M fibers) that run longitudinally through the stalks of some single-celled animals. These filaments, which are some 100–120 Å in diameter, contract to shorten the stalks. The cilia of the same organisms, however, have movements ascribed to a somewhat different mechanism.

MICROTUBULES AND MOVEMENT. Microtubules (see Figs. 1–18, 1–20, and 1–21) have been implicated in many types of movements. They are found, for example, running the length of neuronal processes, where they both provide for structural stability and (according to some workers) axonal transport, which is the flow of material from the cell body to the tip of the axon, sometimes at rates of 400 mm per day. (Both of these functions in nerves, it must be noted, have been alternately ascribed to neurofilaments, the construction of which is similar to but somewhat simpler than that of microtubules.) Microtubules have also been described as participating in cytoplasmic streaming and in ameboid locomotion of cells, though actin and myosin may be involved instead, as will be discussed later. And finally, microtubules or microtubule-like structures are certainly involved in the movements of chromosomes during mitosis and meiosis and in the beat of cilia and flagella.

Microtubules are composed of at least two types of subunits. The proteins are called *α and β tubulin.* At least some microtubules also contain a small amount of an additional protein called *dynein,* seen occasionally in high-resolution electron micrographs as periodic projections from the surface of the microtubules. Dynein, when isolated, has ATPase activity, and is therefore assumed to be associated with the conversion of chemical energy (ATP hydrolysis) into mechanical motion. Though no one really knows how cilia and flagella move, many workers suspect that the process includes the sliding of microtubules past one another, powered by dynein-catalyzed hydrolysis of ATP. Similarly, substances could be transported along cytoplasmic microtubules by passing them from one subunit to the next, again powered by dynein's enzymatic capacity, much as a cable car is pulled along its tracks.

ACTIN AND MYOSIN IN NON-MUSCLE SYSTEMS. The proteins actin and myosin were long thought to be unique to muscle cells, where they are the major constituents of the thin and thick filaments that slide past each other during contraction. It is now clear, however, that actin and myosin are produced in many types of cells, not only in animals but even in some single-celled organisms.

Non-muscle actin was discovered as a result of studies on microfilaments, for purified microfilament protein was found to be identical to muscle actin. The identification was readily made because the amino acid sequence and structure of actin has been highly conserved through evolution—even actin from an ameba is remarkably like that isolated from mammalian muscle. Further confidence in the identity of the microfilament protein came with the discovery that, like muscle actin, it is capable of interacting *in vitro* with muscle myosin to form threads, called *actomyosin,* that contract in the presence of Ca^{2+}, Mg^{2+}, and ATP. Contraction is accompanied by, and presumably powered by, the hydrolysis of ATP.

This ability of non-muscle actin to interact with myosin also permitted the demonstration of actin filaments *in vivo.* When heavy meromyosin (HMM—the head region and a portion of the tail of the myosin molecule, as explained in Fig. 7–19a) is added to solutions containing actin filaments *in vitro,* the HMM binds to the actin to form "arrowhead complexes." The name reflects that all the HMM on a given actin filament is aligned in the same way (see Fig. 7–32). This distinctive complex can also be induced in the cortical regions (that is, near the plasma membrane) of many cell types after the addition of HMM. The cortical region is also where one finds the greatest concentration of microfilaments, reinforcing the view that microfilaments and actin filaments are one and the same.

309

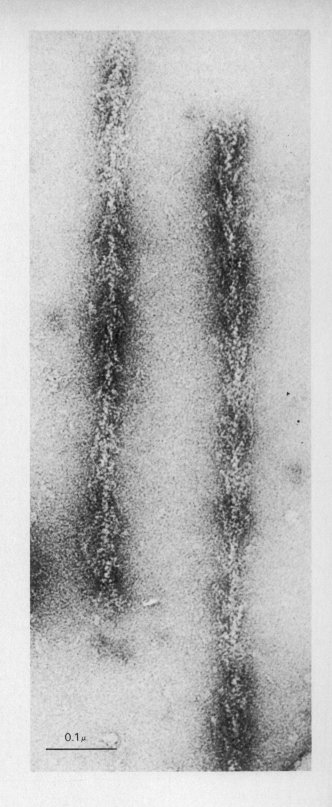

FIGURE 7–32. The Binding of Heavy Meromyosin (HMM) to Actin. The HMM fragments align themselves all in the same direction on a given actin filament. The two filaments shown here have opposite polarities. This distinctive arrowhead configuration is used to identify isolated proteins as being either actin or myosin. [Courtesy of H. E. Huxley, *J. Mol. Biol.*, **7**:281 (1963).]

0.1μ

The presence of native myosin molecules in a variety of non-muscle cell types has now also been confirmed. Unlike actin, there is considerable variation in the physical and chemical properties of the protein from one source to another. Identification is possible, however, by the ability of non-muscle myosin to form with either muscle or non-muscle actin the same distinctive arrowhead complexes described above. Whether myosin in non-muscle cells is polymerized into filaments or not is at this time still unclear. There have been a few reports of myosin-like filaments in non-muscle cells, but it has been only relatively recently that myosin filaments could be seen anywhere except in striated muscle—that is, not even in smooth muscle. Apparently, myosin filaments have a tendency to depolymerize during the routine preparation of specimens for electron microscopy. Hence, the absence of myosin filaments in micrographs of other cells is not adequate proof of their non-existence *in vivo*.

That myosin and actin may interact in non-muscle cells in ways similar to muscle cells is suggested by evidence for the presence of a tropomyosin-like protein in at least some cells. In fact, many types of cellular movements are known to be stimulated by Ca^{2+}, Mg^{2+}, and ATP, just as muscle requires those substances for contraction.

Of the various types of movements associated with cells, two of the more likely candidates for actomyosin involvement are the pinching of a cell into two daughter cells, and ameboid locomotion. The former will be discussed in Chap. 9. The latter refers to the kind of crawling, common to many cell types other than amebas, where a pseudopod ("false foot") forms on the cell surface, then enlarges as cytoplasm flows into it in sufficient quantity to provide a net movement in the direction of the pseudopod. The incremental motion is complete when the tail (*uroid*) of the cell is pulled an equal distance in the same direction (see Fig. 7-33). More detailed observations suggest that ectoplasm (the stiff and usually organelle-free outer layer of cytoplasm in the cortical region) is turned from a gel into a more soluble form at the uroid, flows forward through the interior of the cell, and is then returned to a gel at the leading edge. The material arriving at the front at any given moment will gradually get displaced back toward the uroid again as new material arrives.

It has long been recognized that the ectoplasm contains numerous microfilaments. The identification of these filaments as actin, plus the discovery of myosin in the same cells, implicates actomyosin in protoplasmic streaming and ameboid locomotion. For example, soluble myosin molecules might be swept along actin filaments, causing a net flow of the cytoplasm. Or filaments of actin and myosin might be pulled past each other. A full understanding of these phenomena must await further investigation.

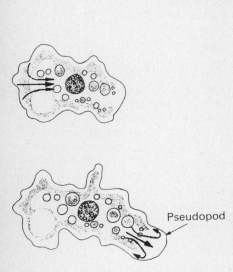

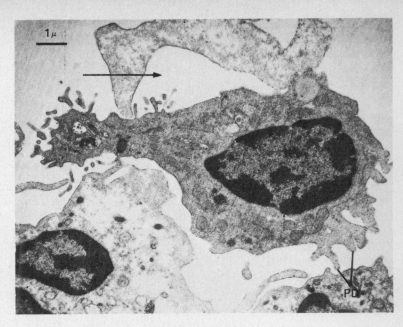

Pseudopod

1 μ

FIGURE 7–33. Ameboid Locomotion. Many cells can move along a surface by an ameboid-like crawling, diagrammed here. The micrograph is of a lymphocyte moving in the direction of the arrow. Pseudopod (PD). The lymphocyte is nearly clear of the two adjacent cells that seem almost to be pinching the uroid of the lymphocyte. [Micrograph courtesy of W. McFarland, *Science,* **163**:818 (1969). Copyright by the A.A.A.S.]

SUMMARY

7–1 The basic unit of any nervous system is the individual neuron, which consists of a cell body (the perikaryon) containing a nucleus and generally one long and several short processes called axons and dendrites, respectively. Glial cells (also called Schwann cells in the peripheral system) are often found wrapped about nerve processes, presumably to protect and insulate them from other cells. When Schwann cells are spirally wrapped in a tight sheath about an axon, the axon is said to be myelinated.

7–2 A neuron is capable of transmitting information in the form of a spike potential, which is a travelling wave of polarization reversal. A spike potential starts with an increase in permeability to Na^+, which enters in response to both its electrical and concentration gradients. As the membrane potential becomes positive, K^+ efflux begins. The Na^+ channel is then closed, while K^+ efflux continues until the resting potential is reachieved. Since the spike potential is driven by Na^+ entry, its magnitude varies with the external Na^+ concentration. The resting potential, on the other hand, is much more sensitive to the external K^+ concentration, reflecting the much greater permeability of the membrane to K^+ (see equation (7–2)).

Except during the refractory period after the last firing, a spike potential may be initiated by a slight (e.g., 20 mV) depolarization or by certain chemicals such as acetylcholine. In both cases, the stimulus acts by increasing Na^+ permeability—i.e., by opening a "Na^+ channel." The opening is temporary (self-correcting) even when the membrane potential is held constant by a voltage clamp. Permeability to K^+, on the other hand, varies with membrane potential.

7–3 There are three general mechanisms for the propagation of nerve impulses:

1. Nerve impulses can be propagated by chemicals. Although the use of metabolic poisons and isolated,

perfused axons indicates that chemicals are not commonly used for intracellular conduction, chemical propagation is the most common mechanism for intercellular transmission. In cholinergic nerves, for example, acetylcholine, released from storage vesicles within the presynaptic axon, diffuses across the synaptic gap to interact with receptor sites on the postsynaptic membrane. Other neurotransmitters are also known, among the most common of which are adrenaline and noradrenaline, collectively called adrenergic transmitters.

2. Nerve impulses may be propagated by direct electrical stimulation in the form of local currents that induce enough depolarization at an adjacent site to trigger a new spike potential there. This mode of propagation is commonly used for intracellular conduction in unmyelinated fibers.

3. The cable effect refers to the ability of spike potentials to be propagated directly to relatively distant points (saltatory conduction). This effect is apparently utilized by myelinated nerves to transmit action potentials from one node of Ranvier to the next. Since transmission is very rapid between nodes, myelinated nerves propagate impulses at great speeds. However, axons of a micron or less in diameter usually have nodes close enough together (because of their proportionately smaller Schwann cells) to eliminate the advantage gained by myelination. Hence, very small fibers are ordinarily unmyelinated.

7–4 All muscle may be classed as either striated or smooth, depending on its appearance in the polarizing light microscope. In general, voluntary muscles are striated while involuntary muscles (except for the heart) are smooth. Striations result from the organization of filaments (myofilaments) into discrete bundles, or myofibrils. Each myofibril is further divided into segments, called sarcomeres, by thin dark Z-lines (or Z-discs). Within each sarcomere one finds two kinds of filaments, thick and thin (myosin and actin). Thick filaments occupy only the A-zone in the middle of the sarcomere. Thin filaments extend into the A-zone from each Z-disc, interdigitating with the thick filaments.

According to the sliding filament theory, contraction is achieved by moving the two kinds of filaments past one another. This must be accomplished by a cyclical extension and contraction of the interfilament contacts, or bridges, which are known to be part of the myosin heads.

7–5 The sarcotubular system of striated muscle is divided into two components, longitudinal and transverse. The longitudinal component forms vessels and cisternae. The transverse system is a periodic set of tubules opening onto the surface of the cell. Where the two systems cross, one frequently sees two cisternae flanking a tubule, a configuration that is called a triad. Contraction is initiated by a depolarization of the sarcolemma that is conducted into the interior by transverse tubules. Depolarization causes Ca^{2+} to be released from terminal cisternae of the longitudinal vesicles. The free ions are then bound by troponin, which is associated with tropomyosin on the actin filaments. In the presence of Ca^{2+}, these proteins lose their capacity to prevent the cooperative actin–myosin interaction that causes ATP hydrolysis and contraction. Relaxation occurs when Ca^{2+} is removed by a calcium pump in the terminal cisternae.

7–6 The neuromuscular junction in striated cells (but not smooth muscle cells) is found at an elevated region of the muscle cell containing numerous junctional folds in the sarcolemma. The effect of nerves on muscles is normally to cause contraction, of course, but studies on the biochemical differentiation of red and white fibers reveal other effects as well.

The fully differentiated striated fiber can be classed as red, white, or pink. Red and white fibers are distinctly different from each other in color and size, and in their metabolism and contractile characteristics. This aspect of the differentiation of a muscle cell depends on which nerves innervate it, as one can demonstrate by switching nerves between a red and white motor unit.

7–7 All contractile activity depends on conformational changes in protein, powered in some way by hydrolysis of ATP or a similar "high energy" molecule. In bacterial flagella, conformational changes in the rope-like flagellar protein itself may be responsible for movement. Eucaryotic cilia and flagella, on the other hand, are thought to move by sliding microfilaments past one another, powered by dynein–catalyzed ATP hydrolysis. Microtubules have also been implicated in some types of cytoplasmic transport. However, the discovery of actin (microfilaments) and myosin in non-muscle cells suggests the presence also of muscle-type contractile activity, but whether or not actual myosin filaments are involved remains to be determined.

7–1 **(a)** What is the distinction between an axon and a dendrite? A Schwann cell and a glial cell? **(b)** Define perikaryon, myelin, node of Ranvier.

7–2 **(a)** According to one estimate, there are 13 Na^+ channels per square micron of axonal membrane. (Other estimates place the number at 100 or more.) At 10^{-12} moles per square cm of membrane, how many sodium ions will enter through each channel during a typical spike potential? [Ans: nearly 2,000] **(b)** How many spike potentials can be initiated in an isolated squid axon with a diameter of 0.5 mm before the internal Na^+ concentration will be doubled? (Assume that the Na^+ pump has been poisoned.) What is the corresponding figure for a human axon, if the only difference between the two is an axon diameter of 10μ in the latter? [Ans: about 150,000 and 3,000, respectively] **(c)** Describe the sequence of events comprising a spike potential. **(d)** The magnitude of the spike potential depends mostly on Na^+ concentration, whereas the magnitude of the resting potential depends mostly on K^+ concentration. Why? **(e)** What is the refractory period, and how is it related to flicker fusion?

7–3 **(a)** What are the three basic mechanisms of spike potential propagation? (Be careful to distinguish between the two varieties of electrical propagation.) **(b)** What kinds of experiments argue against chemical propagation as a general mechanism for intracellular conduction? **(c)** What is the significance of myelination? Why do very small fibers generally lack it? **(d)** What are the individual steps involved in cholinergic neurotransmission? **(e)** What are some of the noncholinergic neurotransmitters?

7–4 **(a)** Describe the sarcomere. **(b)** Myosin filaments have a bilateral symmetry about their midpoints; actin filaments are symmetrical about the Z-lines. What is the importance of these configurations to the contractile mechanism? **(c)** What will happen to the size of the I-, A-, and H-zones as a sar-comere shortens by sliding the two kinds of filaments past each other? **(d)** What is the role of creatine phosphate in ATP maintenance?

7–5 **(a)** Early attempts to find a relaxing factor—a substance that would terminate a contraction—resulted in the isolation of a material that did have such an effect when injected into a contracted muscle cell. The substance (called the Marsh factor or Marsh-Bendall factor) looks like membrane fragments in the electron microscope. What do you think it is, and how does it work? **(b)** How do the contractile properties of smooth muscle differ from those of striated cells, and what are some possible reasons for the differences? **(c)** Frog muscle undergoes a prolonged contracture when bathed in concentrated KCl solutions provided Ca^{2+} is also present, but not if Ca^{2+} is absent. Explain both the K^+-dependent and Ca^{2+}-dependent aspects of this observation. **(d)** It has been found that a portion of the Ca^{2+} in striated muscles can be readily exchanged for other ions during rest, and that still another fraction is available for exchange only when the muscle contracts. Explain.

7–6 **(a)** Describe the neuromuscular junction of striated muscle and of smooth muscle. **(b)** Frogs, fish, and rabbits are examples of animals in which most muscle tends to be very white. Dogs and ducks, on the other hand, tend to have many red fibers. Can you identify any behavioral patterns of these two groups of animals that might have led you to predict these differences?

7–7 **(a)** How does the structure and probable contractile mechanism of bacterial flagella differ from the corresponding properties of eucaryotic flagella? **(b)** In what motile functions other than ciliary and flagellar movements have microtubules been implicated? **(c)** What evidence is offered for the existence of actin and myosin in non-muscle cells? **(d)** In what types of contractile activity is non-muscle actomyosin possibly involved?

REFERENCES

GENERAL — NERVES AND MUSCLES

AIDLEY, D. J., *The Physiology of Excitable Cells.* London: Cambridge Univ. Press, 1971. (Paperback.)

BENDALL, J. R., *Muscles, Molecules and Movement. An Essay in the Contraction of Muscles.* New York: American Elsevier Publ. Co., 1968. A fine introductory text, concentrating on the energetics of contraction.

BILLINGS, SUSAN M., "Concepts of Nerve Fiber Development, 1839–1930." *J. of the History of Biol.,* **4:**275 (1971).

BLOOM, W., and D. W. FAWCETT, *A Textbook of Histology* (9th ed.). Philadelphia: W. B. Saunders, 1968. Contains excellent micrographs and discussion of nerves and muscles.

BRISKEY, E. J., R. G. CASSENS, and B. B. MARSH, ed., *Physiology and Biochemistry of Muscle as a Food, Vol. II.* Madison: Univ. of Wisconsin Press, 1970. An excellent series of review articles on the structure, function, and differentiation of muscle, as well as a few on its food aspects.

CARLSON, F. D., and D. R. WILKIE, *Muscle Physiology.* Englewood Cliffs, N. J.: Prentice-Hall, 1973. A small book, covering initiation, control, and energy conversion in muscle.

COLE, KENNETH S., *Membranes, Ions, and Impulses.* Berkeley: Univ. Calif. Press, 1968.

GOLGI, C., "The Neuron Doctrine—Theory and Facts." In *Nobel Lectures, Physiology or Medicine, 1901–1921.* Amsterdam: Elsevier Publ. Co., 1967, p. 189. Nobel Lecture, 1906.

KATZ, B., *Nerve, Muscle, and Synapse.* New York: McGraw-Hill, 1966. (Paperback.) An excellent introduction to neuronal function.

LOEWY, ARTHUR D., and S. RAMÓN Y CAJAL, "Methods of Neuroanatomical Research." *Perspectives in Biol. & Med.,* 15:7 (1971).

PEARSON, C. M., and F. K. MOSTOFI, eds., *The Striated Muscle.* Baltimore: Williams and Wilkins, 1973. A collection of reviews.

SCHMITT, F. O., *The Neurosciences: Second Study Program.* New York: The Rockefeller Univ. Press, 1971. A series of reviews, many of which are highly readable by the nonspecialist.

THE NEURON AND THE NERVE IMPULSE

ALBERS, R. W., G. J. SIEGEL, R. KATZMAN, and B. AGRANOFF, eds., *Basic Neurochemistry.* Boston: Little, Brown, 1972 (Paperback.)

BAKER, PETER F., "The Nerve Axon." *Scientific American,* March 1966. (Offprint 1038.)

COHEN, L. B., "Changes in Neuron Structure During Action Potential Propagation and Synaptic Transmission." *Physiol. Revs.,* 53:373 (1973). Evidence of conformational changes in membrane macromolecules.

DE ROBERTIS, E., *Cellular Dynamics of the Neuron,* New York: Academic Press, 1969.

ERLANGER, JOSEPH, "Mammalian Nerve Fibers." In *Nobel Lectures, Physiology or Medicine, 1942–1962.* Amsterdam: Elsevier Publ. Co., 1964, p. 34. Nobel Lecture, 1944.

GASSER, H. S., "Some Observations on the Responses of Single Nerve Fibers." In *Nobel Lectures, Physiology or Medicine, 1942–1962.* Amsterdam: Elsevier Publ. Co., 1964, p. 50. Nobel Lecture, 1944.

HODGKIN, A. L., "The Ionic Basis of Nervous Conduction." *Science,* 145:1148 (1964). Nobel Lecture, 1963.

HUXLEY, A. F., "Excitation and Conduction in Nerve: Quantitative Analysis." *Science,* 145:1154 (1964). Nobel Lecture, 1963.

KEYNES, R. D., "The Nerve Impulse and the Squid." *Scientific American,* December 1958. (Offprint 58.)

LERMAN, L., A. WATANABE, and I. TASAKI, "Intracellular Perfusion of Squid Giant Axons: Recent Findings and Interpretations." *Neurosciences Research,* 2:71 (1969). A review of the technique and some of the experiments it made possible.

STEVENS, LEONARD A., *Explorers of the Brain.* London: Angus and Robertson, 1973. Historical perspective and modern ideas of neuronal function and interactions.

USHERWOOD, PETER, *Nervous Systems.* London: Edward Arnold Publ., 1973. (Paperback.) Electrochemistry of the nerve impulse, including conduction and transmission.

INTERCELLULAR COMMUNICATION

AXELROD, JULIUS, "Noradrenalin: Fate and Control of Its Biosynthesis." *Science,* 173:598 (1971). Nobel Lecture, 1970.

———, "The Fate of Noradrenaline in the Sympathetic Neurone." In *The Harvey Lectures, 1971–1972 (Ser. 61).* New York: Academic Press, Inc., 1973.

———, "Neurotransmitters." *Scientific American,* June 1974, p. 58.

DE ROBERTIS, EDUARDO, "Molecular Biology of Synaptic Receptors." *Science,* 171:963 (1971).

ECCLES, JOHN C., "The Synapse." *Scientific American,* January 1965. (Offprint 1001.)

KATZ, B., "Quantal Mechanism of Neural Transmitter Release." *Science,* 173:123 (1971). Nobel lecture, 1970.

KITA, H., "Mechanisms for Neurotransmitter Release." *BioScience,* 24:13 (1974).

KRNJEVIĆ, K., "Chemical Nature of Synaptic Transmission in Invertebrates." *Physiol. Revs.,* 54:418 (1974).

LOEWI, OTTO, "On the Humoral Transmission of the Action of Heart Nerves." *Pflüger's Archiv.,* 189:239 (1921). Translated and reprinted in *Great Experiments in Biology,* ed. M. L. Gabriel and S. Fogel. Englewood Cliffs, N. J.: Prentice-Hall, 1955, p. 69.

———, "The Chemical Transmission of Nerve Action." In *Nobel Lectures, Physiology or Medicine, 1922–1941.* Amsterdam: Elsevier Publ. Co., 1965, p. 416. Nobel Lecture, 1936.

VON EULER, U. S., "Adrenergic Neurotransmitter Functions." *Science,* 173:202 (1971). Nobel Lecture, 1970.

WHITTAKER, V. P., "Membranes in Synaptic Function." *Hosp. Practice,* April 1974, p. 111.

STRUCTURE AND THE CONTRACTILE MECHANISM OF MUSCLE

BURNSTOCK, G., "Structure of Smooth Muscle and Its Innervation." In *Smooth Muscle,* ed. E. Bulbring, A. F. Brading, A. W. Jones, and T. Tomita. Baltimore: Williams and Wilkins, 1970, p. 1.

CLOSE R. I., "Dynamic Properties of Mammalian Skeletal Muscles." *Physiol. Revs.,* 52:129 (1972). Properties of red and white muscle.

HASELGROVE, J. C. and H. E. HUXLEY, "X-ray Evidence for Radial Crossbridge Movement and for the Sliding Filament Model in Actively Contracting Skeletal Muscle." *J. Mol. Biol.,* 77:549 (1973).

HUXLEY, H. E., "The Mechanism of Muscular Contraction." *Scientific American,* December 1965. (Offprint 1026.)

———, "The Mechanism of Muscular Contraction." *Science,* 164:1356 (1969).

KATSURA, I. and H. NODA, "Assembly of Myosin Molecules into the Structure of Thick Filaments of Muscle." *Adv. in Biophysics,* 5:177 (1973)

MARGARIA, RODOLFO, "The Sources of Muscular Energy." *Scientific American,* March 1972. (Offprint 1244.)

MURRAY, J. M. and A. WEBER, "The Cooperative Action of Muscle Proteins." *Scientific American*, February 1974, p. 58.

SMITH, DAVID S., *Muscle*. New York: Academic Press, 1972. (Paperback.) Primarily concerned with ultrastructure.

WHITE, D. C. S., and J. THORSON, "The Kinetics of Muscle Contraction." *Prog. in Biophysics and Molec. Biol.*, 27:175 (1973). Emphasis on crossbridge formation and release. The early part of the article will be of most interest.

EXCITATION–CONTRACTION COUPLING

EBASHI, S., M. ENDO, and I. OHTSUKI, "Control of Muscle Contraction." *Quart. Revs. Biophys.*, 2:351 (1969).

HOYLE, GRAHAM, "How is Muscle Turned On and Off?" *Scientific American*, April 1970. (Offprint 1175.)

PORTER, K. R. and C. FRANZINI-ARMSTRONG, "The Sarcoplasmic Reticulum." *Scientific American*, March 1965. (Offprint 1007.)

TAKEUCHI, A. AND N. TAKEUCHI, "Actions of Transmitter Substances on the Neuromuscular Junctions of Vertebrates and Invertebrates." *Adv. in Biophysics*, 3:45 (1972). Nice review.

ZACHAR, JOZEF, *Electrogenesis and Contractility in Skeletal Muscle Cells*. Baltimore: University Park Press, 1972. The nature of skeletal muscle action.

CONTRACTILITY IN NON–MUSCLE SYSTEMS

BLAKE, J. R. and M. A. SLEIGH, "Mechanics of Ciliary Motion." *Biol. Rev. (Cambridge)*, 49:85 (1974).

HATANO, S., "Contractile Proteins from the Myxomycete Plasmodium." *Adv. in Biophysics*, 5:143 (1973). Similarity to contractile proteins from muscle cells.

HOLWILL, M. E. J., "Some Physical Aspects of the Motility of Ciliated and Flagellated Microorganisms." *Science Prog. (Oxford)*, 61:63 (1974).

HUXLEY, H. E., "Muscular Contraction and Cell Motility." *Nature*, 243:445 (1973).

KOMNICK, H., W. STOCKEM, and K. E. WOHLFARTH-BOTTERMANN, "Cell Motility." *Int. Rev. Cytology*, 34:169 (1973). Cellular locomotion and cytoplasmic streaming.

POLLARD, T. D. and R. R. WEIHING, "Actin and Myosin and Cell Movement." *CRC Crit. Revs. Biochem.*, 2:1 (1974).

CHAPTER 8

Genes and Genetic Control

8-1 THE GENETIC CONCEPT   318
     Mendel's Observations
     Chromosomes and Genes
     The One Gene = One Enzyme Hypothesis
8-2 THE GENETIC MESSAGE   327
     Chemical Nature of the Gene
     Messenger RNA
     The Genetic Code
     Assigning Code Words
     Universal Nature of the Genetic Code
8-3 PROTEIN SYNTHESIS   342
     Transfer RNA
     Chain Elongation
     Ribosomes
     Initiation
     Termination
     Role of the Endoplasmic Reticulum and
     Golgi Apparatus
8-4 GENE REGULATION IN PROCARYOTES   353
     Induction and Repression
     Operon Theory
     The Regulation of RNA Polymerase
     Catabolite Repression
8-5 GENE REGULATION IN EUCARYOTES   361
     The Eucaryotic Chromosome
     The Histones in Gene Regulation
     Hormonal Control of Gene Expression
     Gene Amplification
     A Transcriptional Control Theory for Eucaryotes
     Translational Control
SUMMARY   376
STUDY GUIDE   378
REFERENCES   378

Genes could be considered the blueprints of a cell, because the information they contain is used—directly or indirectly—to construct other cellular components. In particular, the order of amino acids that uniquely defines an enzyme or other protein is determined by the sequence of nucleotides within a molecule of DNA. That sequence may be referred to as a gene.

In this chapter we consider the development of the genetic concept, the nature of the genetic message, and the mechanism whereby the genetic message is expressed. Included is a description of the way gene expression is regulated in procaryotic and in eucaryotic cells.

8-1 THE GENETIC CONCEPT

The earliest systematic observations of gene behavior date back to the mid-nineteenth century, when an Austrian monk named Gregor Mendel conducted a series of breeding experiments on garden peas. His observations helped define the gene in terms of a separately functioning unit responsible for a particular hereditary characteristic.

MENDEL'S OBSERVATIONS. Mendel identified seven characteristics that could be transmitted without change from one generation to the next, but that exist in more than one form. Some plants, for example, had long stems and some short. Generation after generation, inbred long-stemmed plants produced only long-stemmed progeny, while inbred short-stemmed plants produced only short-stemmed progeny. But when Mendel crossed a pure-breeding, long-stemmed plant with a pure-breeding, short-stemmed one, the offspring in the first generation (F_1, or *first filial generation*) resembled only the long-stemmed parent; however, the mating of two of these hybrid plants produced a generation (F_2, or *second filial generation*) consisting of both short- and long-stemmed plants. The ratio of long stems to short in the F_2 generation was about three to one. Each of the seven characteristics behaved this way when plants having alternate forms were crossed.

Mendel suggested that characteristics such as stem length are controlled by hereditary units (later called *genes*) which behave in the following way:

1. Genes affecting the same developmental process are always found in pairs in adult plants, with one member derived from each of the plant's two parents. The two genes of a given pair may or may not be identical. When they are different, and hence affect the same process in different ways, they are called *alleles* (or *allelomorphs*) of each other. The organism is then said to be *heterozygous* for the characteristic in question, to distinguish it from a *homozygous* organism in which the two paired genes are identical.

2. One allele may mask the effect of the other. For example, the allele for long stems is always *dominant*, meaning that it is expressed even in the presence of the allele for short stems, which is called *recessive*. Thus, a heterozygote always has a long-stemmed *phenotype* (appearance). However, one cannot tell by looking whether the *genotype*, or gene content, of a long-stemmed plant is homozygous or heterozygous.

3. A sex cell (gamete, or germ cell) contains only one member of each gene pair. That allelic pairs always separate during gamete formation, so that the two members of a pair end up in different gametes, is referred to as *Mendel's first law*, or the *law of independent segregation*.

4. Often, two allelic pairs controlling different characteristics seem to sort themselves during gamete formation such that a given allele of one pair has an equal chance of being found in a gamete with either allele of the second pair. Thus, if a plant contains both the dominant and recessive alleles for stem length (represented by L and ℓ, respectively) and the dominant and recessive alleles for seed shape (S for smooth, round seeds, and s for wrinkled seeds), it may produce gametes of four types: LS, Ls, ℓS, and ℓs. This principle is sometimes called *Mendel's second law*, or the *law of independent assortment*. It results in a reshuffling of dominant and recessive alleles and a corresponding mixing of characteristics in the next generation.

Let us apply these principles to a cross between a pure-breeding (i.e., homozygous), long-stemmed plant and a pure-breeding, short-stemmed one. Schematically, the cross could be represented as follows, keeping in mind that any male gamete usually can combine with any female gamete:

$$\male \; LL \times \ell\ell \; \female \qquad \text{first cross} \qquad \begin{cases} \male & \text{gametes} = L \;\&\; L \\ \female & \text{gametes} = \ell \;\&\; \ell \end{cases}$$

$$\overline{L\ell \quad L\ell \quad L\ell \quad L\ell} \qquad F_1 \text{ or hybrid generation}$$

$$\male \; L\ell \times L\ell \; \female \qquad \text{second cross} \qquad \begin{cases} \male & \text{gametes} = L \;\&\; \ell \\ \female & \text{gametes} = L \;\&\; \ell \end{cases}$$

$$\overline{LL \quad L\ell \quad L\ell \quad \ell\ell} \qquad F_2$$

Note that the F_1 plants are heterozygotes, each with a long-stemmed phenotype. In all, three-quarters of the plants in the second filial generation have long-stemmed phenotype, although heterozygotes ($L\ell$) outnumber long-stemmed homozygotes (LL) two to one within this group (see Fig. 8–1).

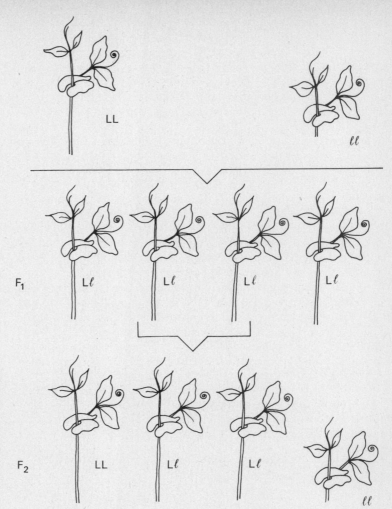

FIGURE 8–1. The Inheritance of Stem Length. The parents are each homozygous for stem length, but one has a long stem and the other has a short stem. The first filial generation (F₁) is entirely heterozygous, but with the long-stemmed phenotype. The second filial generation (F₂) has a 3:1 ratio of long-stemmed to short-stemmed phenotypes, and a 1:2:1 ratio of the genotypes LL, L*l*, and *ll*.

The more complicated cross between a homozygous long-stemmed plant with smooth seeds (LLSS) and one with short stems and wrinkled seeds (*ll*ss) produces gametes that are all LS from one parent and all *l*s from the other. The first filial generation, then, is entirely L*l*Ss. When two of these hybrids are crossed, however, there are four kinds of male gametes and four kinds of female gametes, yielding sixteen possible combinations containing nine genotypes and four phenotypes. The phenotypes are in a ratio of 9:3:3:1. This theoretical distribution, which can be predicted from Fig. 8–2, is remarkably close to Mendel's actual observations.

There are, as it turns out, many complications to the principles just outlined. For example, one member of an allelic pair is not necessarily dominant over the other—e.g., a pure-breeding red snapdragon and a pure-breeding white snapdragon produce heterozygotes that are pink. And the law of independent assortment

	gametes			
	LS	Ls	lS	ls
gametes LS	LLSS	LLSs	LlSS	LlSs
Ls	LLSs	LLss	LlSs	Llss
lS	LlSS	LlSs	llSS	llSs
ls	LlSs	Llss	llSs	llss

FIGURE 8–2. The Punnett Square. This scheme can be used to predict the offspring from any cross. In the present example, the gametes are derived from two doubly heterozygous parents, each LlSs. Independent assortment is assumed.

is actually violated with great regularity, as determined by the observation that certain genes usually sort together rather than independently. Such genes are said to be *linked*. Genes that do obey Mendel's second law are said to be unlinked or "in different linkage groups."

In spite of the exceptions, the principles developed by Mendel have been of great value in helping us to understand how hereditary characteristics are transmitted from one generation to the next in both animals and plants. It has also led to some unfortunate oversimplifications, such as the commonly held belief that eye color is controlled by a single gene, with the brown allele being dominant. While this does appear to be true most of the time, eye color is actually a more complicated trait, a fact that some blue-eyed parents first learn from their brown-eyed offspring.

CHROMOSOMES AND GENES. Although Mendel's observations were published in the mid-1800s, they were largely unappreciated and ignored until they were independently confirmed by several other workers at the end of the century. Then in 1903, Walter S. Sutton, a student of E. B. Wilson at Columbia University, authored a paper entitled "The Chromosomes in Heredity," pointing out the similarity between the behavior of genes and chromosomes. By that time cytologists had learned how to observe chromosomes during cell division, and some had suggested that chromosomes might be responsible for transmitting inherited traits from one generation to the next. Sutton's paper in defense of that hypothesis finally convinced a sizeable segment of the scientific community of its validity.

Consider the following parallel between genes and chromosomes:

1. Genes come in pairs. So do chromosomes.
2. Although the two members of each gene pair are normally present in the same cell, only one is found in a gamete, or germ cell. Similarly, paired, or *homologous*, chromosomes separate during gamete formation, specifically during a process that is called *meiosis* or *reductive division*. Thus, gametes contain half as many chromosomes as *somatic* (i.e., nongamete) cells of the same organism. In other words, gametes have a *haploid* chromosome number while somatic cells have a *diploid* number.
3. Except during gamete formation, genes must be accurately reproduced and transmitted to daughter cells in such a way that each cell contains

a set identical to its parent. Chromosomes, too, are very carefully replicated and separated during cell division, so that normally all descendants of a cell contain accurate copies of its chromosomes. This kind of cell division, where each daughter cell contains a chromosome set that is a replica of the parent set, is called *mitosis*.

4. Visible abnormalities in chromosomes can often be correlated with genetic changes. For example, the arms of two homologous chromosomes sometimes exchange ends. This exchange was first observed in 1909 by a Belgian investigator, F. A. Janssens. It was soon found to correlate with a genetic *crossover*, or interchange of allelic genes between homologous chromosomes. Thus, a plant that would otherwise produce AB and ab gametes might produce Ab and aB germ cells after a crossover, where A/a and B/b are dominant/recessive forms of any two characters found on the same chromosome.

TABLE 8.1. Early Development of the Gene Theory

1694	Camerarius publishes *De Sexu Plantarum Epistola*, in which it is demonstrated that plants are sexual.
1779	Lort reports the unusual pattern by which human color blindness is inherited.
ca. 1820	Nasse formulates the laws governing the inheritance of sex-linked characteristics.
1822	Goss and Seton independently note the segregation of recessive characteristics in peas.
1866	Mendel publishes his observations on inheritance in peas.
1875	Chromosomes are described by Strasburger.
1875	Hertwig proves that fertilization involves the fusion of two nuclei, contained in the egg and sperm, respectively.
1879–85	Flemming publishes his observations of chromosome "splitting" during cell division, with sister chromatids moving to opposite poles.
1883	Van Beneden notes that gametes contain only a haploid chromosome number.
1890	The re-establishment of a diploid chromosome number at fertilization by the joining of equal sets from the male and female gametes was noted by Boveri and by Guignard.
1902	Sutton points out the parallels between genes and chromosome behavior and suggests a chromosomal location for genes.
1906	Bateson and Punnett present data on linked genes, showing that their assortment during germ cell formation is non-random, in violation of Mendel's second law.
1908	Nilsson-Ehle provides a model for the inheritance of continuously variable (nondominant) characteristics.
1911	Morgan advances his gene theory. He proposed that genes are linearly arranged along chromosomes in a definite order. Thus, the number of linkage groups must be equal to the haploid number of chromosomes. Morgan then went on to prove this hypothesis by correlating the movement of chromosomes with the inheritance of sex-linked traits in the fruit fly, *Drosophila*.

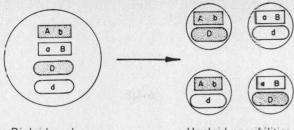

Diploid nucleus   Haploid possibilities

FIGURE 8–3. Linkage. Genes that reside on the same chromosome will be linked. That is, they will sort together during gamete formation in violation of Mendel's second law. If the haploid number of chromosomes is n ($n = 2$ in this example), there will be n linkage groups and 2^n different gametes. Note that the A/a and B/b loci are linked to one another, but that both sort independently of the D/d locus.

Taken together, such observations led to the conclusion that genes are carried by chromosomes, with the alleles for a given characteristic residing at corresponding points (or *loci*) on the two members of a homologous pair. (See Table 8.1.) Genes that are linked are carried by the same chromosome (see Fig. 8–3); except when crossover occurs, such genes will always violate the law of independent assortment (the second law). On the other hand, two genes that do sort independently must be part of nonhomologous chromosomes. The number of linkage groups, then, is the same as the number of homologous pairs, or the haploid chromosome number of the organism. This number, which is not necessarily related to the complexity of the organism, is 23 for humans (see Fig. 8–4), 7 for the garden pea, and 4 for the fruit fly, *Drosophila melanogaster*.

Additional evidence for the chromosomal location of genes comes from the observation that some genes behave as if their allele were present in females and absent in males—i.e., as if males were *hemizygous* for the gene. The basis for this behavior in animals is clear when the chromosomes are examined, for two of the chromosomes in the diploid sets of many animals have no physically homologous counterparts. These are the sex chromosomes, examples of which are the X and Y chromosomes. At the beginning of meiosis or mitosis in man and many other animals, the two strands (*chromatids*) of the X chromosome are joined near their center; the Y chromatids are much smaller and are joined near their ends, giving the Y chromosome a shape similar to chromosomes 13 through 15 in Fig. 8–4. It is this *heteromorphic* (dissimilar) pair that determines sex. In mammals, sex genes carried by the Y chromosome are dominant and masculine (with apologies to Women's Liberation). Thus, normal mammalian males are XY and females are XX (see Fig. 8–5). Aberrant individuals with XXY and XYY configurations are also males, but those with a single unpaired X are females.[1] (In chickens, on the other hand, it is the female that is heteromorphic.) This ability to correlate the dis-

---

1. Accidents involving distribution errors can also occur with *autosomes* (chromosomes other than the sex chromosomes), but generally with more drastic consequences. Mongolism, or Down's syndrome, for instance, is associated with an extra copy of chromosome #21. It is, therefore, a "trisomy 21."

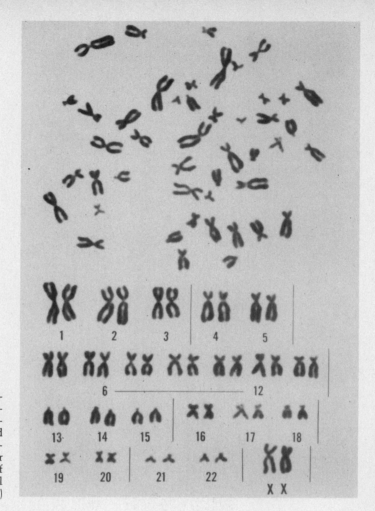

FIGURE 8-4. The Human Karyotype. Photomicrograph of the complete human chromosome set (female). The numerical labelling used here has ambiguities that can be resolved, if necessary, with other staining procedures. (Courtesy of Dr. Jacqueline Whang-Peng, National Cancer Inst.)

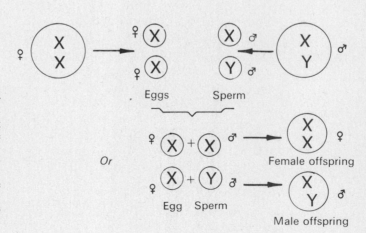

FIGURE 8-5. Determination of Sex. Gender, in humans and many other animals, is determined by the presence of a heteromorphic pair of sex chromosomes. A normal female mammal, for example, is XX and produces only X gametes (eggs) as diagrammed. The male is XY and produces both X and Y sperm. The gender of the offspring is then determined by which type of sperm gains entry to the egg.

tribution of heteromorphic chromosomes with the inheritance of sex is strong support for the chromosomal theory of genes.

Genes carried by the X and Y chromosomes will be distributed to progeny in a pattern that predictably follows the sex of the offspring—i.e., the genes are *sex-linked*. Hemophilia, for example, is caused by a recessive gene carried by the X chromosome. The normal allele from a second X chromosome prevents the expression of symptoms. Since the gene is missing in the abbreviated Y chromosome, all males who inherit the gene from their mothers will be affected, while a heterozygous female will be an unaffected carrier. (The only way a female can be a hemophiliac is for her father to be one, and then only if her mother is either a carrier or herself affected with the condition.) Red-green color blindness and several other conditions are transmitted in exactly the same way. These observations are all consistent with the chromosomal location of genes.

The location of genes on chromosomes thus explains several of Mendel's observations, including segregation (the separation of homologous chromosomes during gamete formation) and assortment. Independent assortment results from maternally derived and paternally derived homologues of a given pair being distributed to gametes without regard to the way in which other pairs are split Independent assortment, then, applies only to unlinked genes—i.e., to genes that reside on nonhomologous chromosomes. To understand the other aspects of gene behavior, such as dominance, we will have to understand how genes affect the properties of a cell.

THE ONE GENE = ONE ENZYME HYPOTHESIS. The connection between genes and proteins was first established experimentally in the early 1940s. The 1958 Nobel prize winners G. W. Beadle and E. L. Tatum, working at Stanford University, examined a number of mutations of the bread mold, *Neurospora crassa*, to discover the biochemical differences between mutant and normal strains. *Neurospora* was chosen for this work because certain of its metabolic pathways were known in some detail, and because the organism is haploid in the stage of its life cycle at which it is ordinarily observed. (Being haploid eliminates the complication caused by dominant and recessive genes, allowing all mutations to be phenotypically expressed.) Furthermore, the fungus can reproduce asexually by spore formation, allowing one to grow large quantities of genetically identical copies of an interesting specimen. In a cross, the four products of meiosis are neatly lined up in the ascus (spore sac), allowing easy genetic manipulation. And finally, native (wild-type) *Neurospora* can grow on a chemically defined, simple medium containing only biotin (a B vitamin), glucose, and inorganic salts. Glucose acts as the sole source of carbon for biosynthesis.

After exposure to X rays, a number of mutants of *Neurospora* were found that could no longer grow on this simple glucose medium. By the addition of various amino acids, nucleotide precursors, vitamins, and other substances, one at a time, Beadle and Tatum were able to determine the site in the metabolic scheme at which gene damage was manifest. Schematically, let us suppose that a desired nutrient "D" is manufactured in a three-step process from the starting material "A":

$$A \xrightarrow{E_1} B \xrightarrow{E_2} C \xrightarrow{E_3} D$$

Each step is catalyzed by its own enzyme; so if the cell grows on a simple glucose medium to which C or D is added but not on the same medium to which A or B is added, one concludes that the mutation has affected the activity of the second enzyme, E_2. Similarly, if D but not C will support growth, the defect would be in E_3.

By this sort of reasoning, Beadle and Tatum concluded that their mutants contained defective enzymes. It was logical to assume, then, that damaged genes were responsible for the presence of defective enzymes. Since each of the separately inherited mutations resulted in a defect in a different enzyme, the "one gene = one enzyme" hypothesis was born. In other words, it was supposed that the main (or only) function of genes is to control the synthesis of enzymes on a one-for-one basis.

Actually, the one gene = one enzyme hypothesis is true in only a limited sense. As knowledge of enzyme structure grew, it became clear that many enzymes are composed of more than one kind of polypeptide chain, and that mutations may affect one type of chain without affecting the others. In addition, proteins other than enzymes may also be affected by genetic mutations. And finally, it is now clear that not all genes make proteins. Thus, the original concept of one gene = one enzyme has evolved to the current position, which ascribes each polypeptide chain to a separate gene, while conceding that not all genes make polypeptide chains. In fact, the word "gene" no longer has a precise meaning, and so names with more restricted definitions are used in its stead. In particular, each polypeptide chain is said to be "coded" for by one *cistron*.

The reason for dominance can now be understood. The formation of a diploid organism by crossing a mutant and a normal haploid *Neurospora* strain, for instance, would result in the presence of two copies of each gene in the diploid cells, including one normal copy of the gene responsible for the missing (or inactive) enzyme of the mutant. This mutant gene would be recessive to the normal allele if the presence of the good copy is sufficient, as it is in most cases, to support the synthesis of enough enzyme to permit normal functioning of the cell.

The lack of phenotypic dominance is also straightforward. Pink flowers may be produced by crossing red and white parents because the offspring inherit equal numbers of "red" and "white" genes, diluting the red color to pink. Skin, hair, and eye color follow much the same scheme, except that more than one gene pair is involved. In these and most other characteristics, an average offspring is usually about midway between his two parents, reflecting a mixture of maternally derived and paternally derived alleles. Of course, there is enough opportunity for random variation in this mixture to permit chance deviations of considerable magnitude.

And finally, hemophilia and color blindness are sex-linked in humans because the proteins involved are specified by genes that occupy a site (a *genetic locus*) on the X chromosome, a site that is missing in the abbreviated Y chromosome. A male, being XY, carries only one gene for each of these proteins. If that gene is defective, he will have the associated disease. A female (XX) carrying one normal and one mutant allele will usually be an unaffected heterozygote since the normal gene produces the necessary protein in sufficient quantity to avoid symptoms.

8-2 THE GENETIC MESSAGE

So far, we have dealt with genes only in the abstract. To understand them more completely we will have to know their chemical structure.

CHEMICAL NATURE OF THE GENE. We know that genes are intimately associated with chromosomes which, in higher organisms, consist of nucleic acids (mostly DNA, but with some RNA) and proteins. Of these components, protein originally seemed the more likely candidate for a genetic role. Proteins contain more than twenty different kinds of amino acids and can assume a wide variety of physical conformations; nucleic acids, on the other hand, are chemically and physically much simpler. The assumption that genes are made of protein, though it was wrong, survived until the 1940s. The true situation was determined with the help of bacteria, starting with a series of experiments carried out in the 1920s by an English public health officer named F. Griffith.

Griffith was concerned with the epidemiology of pneumococcal pneumonia, a disease caused by a bacterium that is normally surrounded by a polysaccharide capsule. A mutant strain of the bacterium was known to be free of capsule and relatively nonvirulent.[2] Griffith found that nonencapsulated cells, when injected into a

---

2. Encapsulated cells are easily distinguished from nonencapsulated cells by the appearance of their colonies when grown on the surface of agar. Encapsulated (or "S") cells form large, smooth colonies. This appearance is due to their polysaccharide coat, which also makes them resistant to destruction by phagocytosis. The mutant "R" cells form smaller, rough colonies.

mouse along with heat-killed normal cells, caused a fatal infection. (The control experiments, where each preparation was used alone, were negative—i.e., there were no deaths.) Moreover, live encapsulated cells were isolated from the dead mice. Apparently, something from the heat-killed cells was capable of conferring on live cells the ability to make a polysaccharide capsule, and with it the ability to cause pneumonia. The process, which appears to change the genes of living cells, is called *transformation* (see Fig. 8–6).

The chemical nature of the transforming principle was determined in 1944 by O. T. Avery, C. M. MacLeod, and M. McCarty of the Rockefeller Institute in New York. They demonstrated that the only chemical component from one strain that is capable of transforming another strain is DNA (see Fig. 8–7). One of the suggested explanations for transformation was that DNA carries genetic information in a form that can be passed from cell to cell. However, because it ran counter to the popular belief that proteins should carry genetic information, this explanation did not receive universal acceptance. Some argued that the DNA was not pure, or that it was mutagenic rather than informational, or that bacteria are a special case—after all, they do not even have chromosomes in the eucaryotic sense.

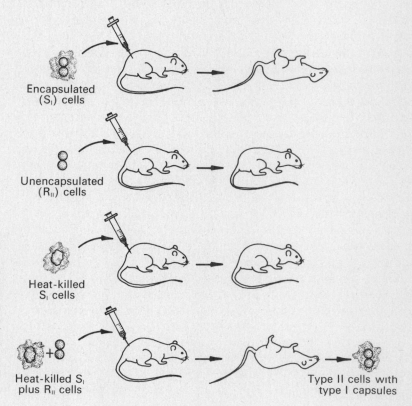

FIGURE 8–6. Bacterial Transformation. Transformation was discovered by F. Griffith, who noted that encapsulated (smooth or S) diplococci cause a fatal infection in mice, whereas nonencapsulated (rough or R) cells do not. Heat-killed S cells are likewise harmless, except when mixed with live R cells. In the latter case, a fatal infection can occur, and live cells having capsules characteristic of the S strain are found. [Described by F. Griffith in *J. Hygiene*, **27**:113 (1928).]

Encapsulated (S$_I$) cells

Unencapsulated (R$_{II}$) cells

Heat-killed S$_I$ cells

Heat-killed S$_I$ plus R$_{II}$ cells

Type II cells with type I capsules

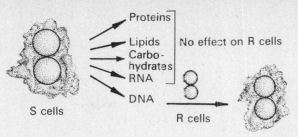

S cells

Prottins ⎤
Lipids ⎥ No effect on R cells
Carbo- ⎥
hydrates ⎥
RNA ⎦

DNA

R cells

Transformed R cells

FIGURE 8–7. The Transforming Principle. When encapsulated diplococci were chemically fractionated, and the transformation experiment performed *in vitro* with each fraction one at a time, O. T. Avery and his colleagues found that DNA is the agent responsible for transformation. [Described by O. T. Avery, C. M. MacLeod, and M. McCarty in *J. Exptl. Med.*, **79**:137 (1944).]

Rather than an isolated oddity, the bacterial transformation experiments were the first to indicate the true chemical nature of the gene. Though there are still some unanswered questions concerning the process, it is clear that transformation involves the transfer of genetic information from one cell to another via DNA isolated from the donor. A transformed cell incorporates the new genetic information into its own chromosome, where it is replicated as a normal part of the cell and thus passed along to progeny However, really wide acceptance of that position came only in 1952, from experiments by A. D. Hershey and Martha Chase at the Cold Spring Harbor (N. Y.) laboratory of the Carnegie Institute.

Hershey and Chase were studying the life cycle of bacteriophage T2 (see Chap. 1) with the newly available technique of isotope labelling. It was known that the virus brought to the infected cell new information, in the sense that the virus tells the cell to stop reproducing itself and to make new virus particles instead. Since T2 is composed almost entirely of DNA and protein, in nearly equal amounts, the objective was to be able to determine the fate of the DNA and protein during infection. When T2-infected cells were grown in the presence of ^{35}S (in the form of radioactive sulfate), T2 particles with labelled protein were produced. Similarly, when T2-infected cells were grown in the presence of ^{32}P in the form of radioactive phosphate, particles with labelled DNA were produced.

With the radioactive phage, Hershey and Chase were able to determine the fate of the protein and DNA during infection of a host (see Fig. 8–8). They found that nearly all the protein remains outside an infected cell and, in fact, can be sheared off by the violent agitation of a kitchen blender without terminating the infection. On the other hand, phage DNA was shown to enter the infected cell. Since the genes of a T2 particle are able to control the metabolic machinery of an infected cell, dedicating it to the task of turning out new T2 bacteriophage, it follows that if phage DNA and not its protein enters a host, the DNA must carry genetic information. Actually, a small percentage of the T2 protein does enter with the DNA, but the experiments were repeated with the same results using phages that do not have this DNA-associated

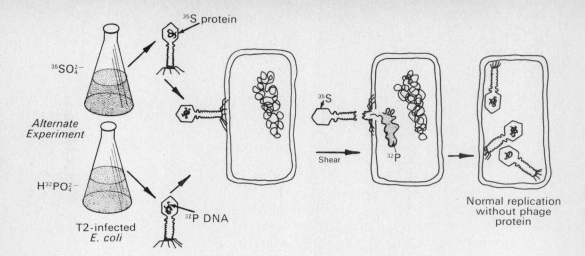

FIGURE 8–8. The Genetic Material of a Bacteriophage. Phage produced in the presence of ^{35}S will have radioactive protein. Phage produced on ^{32}P will have radioactive DNA. Hershey and Chase used this distinction to show that essentially no phage protein enters the host cell at infection. The phage genes, therefore, must be a part of its DNA. [Described by A. D. Hershey and M. Chase in *J. Gen. Physiol.*, **36**:39 (1952).]

"internal protein." And any lingering doubts about the nature of phage genes were clearly dispelled when it was found that naked, purified DNA from certain phages is by itself infective.

While DNA is clearly the genetic substance of some viruses, other viruses contain RNA instead of DNA. An example is tobacco mosaic virus (TMV), isolated and crystallized by Wendell Stanley in 1935 and later found to be 94% protein and 6% RNA. In it, RNA carries the genetic information necessary for replication. In a series of experiments in the mid-1950s, mostly from the laboratories of G. S. Schramm in Germany, and H. Fraenkel-Conrat of the University of California at Berkeley, it was found that purified RNA from TMV is infective; when RNA solutions were rubbed onto the surface of tobacco leaves, characteristic lesions containing TMV developed. The control experiment, using TMV protein, produced no infections.

Continuing the investigation of TMV infection, Fraenkel-Conrat showed that the protein and RNA components of the virus could be purified and then recombined to give intact, infectious particles. (This feat was "life in a test tube" to the newspapers.) A clear distinction between the roles of the RNA and protein was found in experiments using RNA from one strain of TMV and protein from another. When these hybrid particles were allowed to infect a plant, progeny viruses were always normal examples of the parental type from which RNA was obtained, not of the parental type that supplied the protein (see Fig. 8–9).

There is little doubt now about the nature of the genetic material: it is DNA in all cells and in those viruses that contain DNA, and RNA in those viruses that do not. That eucaryotic cells, like procaryotic cells, use DNA for their genes is indicated by the following observations: (1) Ultraviolet radiation causes genetic mu-

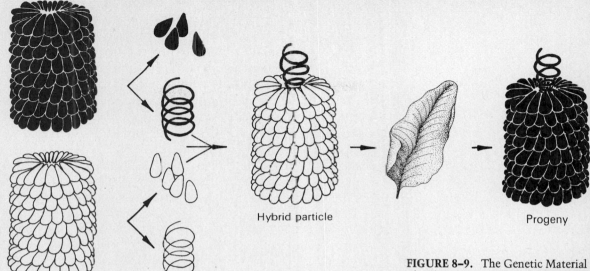

Two strains of TMV

Hybrid particle

Progeny

FIGURE 8–9. The Genetic Material of a Plant Virus. Most plant viruses, such as tobacco mosaic virus (TMV), contain RNA instead of DNA. In this experiment, RNA from one strain of TMV was reconstituted with protein from a second strain to produce an infective particle. Plants infected by such particles produced viruses corresponding to the strain that donated its RNA to the hybrid virus, demonstrating that RNA carries genetic information. [Described by H. Fraenkel–Conrat and B. Singer in *Biochim. Biophys. Acta,* **24**:540 (1957).]

tations, but only at wavelengths at which nucleic acids absorb. (2) Chemical agents that cause mutations can often be identified as reacting with DNA and not proteins. And (3) transformation of some animal cell lines has been achieved with DNA obtained from certain viruses. Later we will be able to point to similarities in gene function between eucaryotic and procaryotic cells, leaving virtually no room for doubt about the genetic role of DNA.

The questions that remain are: (1) how is DNA capable of storing genetic information, and (2) how is that information utilized to direct the synthesis of proteins? The discovery that an RNA molecule is an intermediary between DNA and protein synthesis helped to answer both of these questions.

MESSENGER RNA. It has been known since the mid-1950s that protein synthesis takes place in the cytoplasm, not in the nuclei of eucaryotic cells. More specifically, protein synthesis takes place at RNA–protein complexes called ribosomes, associated with the microsomal fraction of cell extracts (a fast-sedimenting fraction containing ribosomes bound to fragments of endoplasmic reticulum). Thus, if one feeds a cell radioactive amino acids or radioactive sulfur (which will become incorporated into amino acids, and hence into protein), the label appears first at the ribosomes and later in the cytoplasm as a part of free protein molecules.

Two French scientists, Francois Jacob and Jacques Monod (who shared the 1965 Nobel Prize) suggested a brilliant answer to the.

dilemma of how DNA in the nucleus could make protein in the cytoplasm. Clearly, they hypothesized, if the genetic information specifying an amino acid sequence resides in nuclear material, whereas the actual synthesis of protein takes place somewhere else, then there must be a carrier, or "messenger" molecule, responsible for the transfer of genetic information. The messenger was found to be a high molecular weight RNA that is not part of the ribosomes themselves, but becomes associated with them. This *messenger RNA*, or *mRNA*, is rapidly synthesized and, in most cases, almost as rapidly degraded, with a typical half-life in various bacteria of a minute or two. That messenger RNA is different from the two other classes of RNA, ribosomal and transfer RNA, became especially clear when it was shown that after infection by a virulent phage (one that causes immediate replication and subsequent lysis), no new ribosomes or tRNA's are made, although mRNA is produced rapidly.

Since mRNA carries genetic information from DNA to ribosomes, its base sequence should reflect the base sequence of phage DNA and not, for example, the sequence of bases in cellular DNA. E. Volkin and L. Astrachan, who in 1956 first described the rapid turnover of the RNA species later identified as messenger, also demonstrated that its base composition was similar to that of phage and not bacterial DNA (except, of course, that RNA has the base U where DNA would have T—see Chap. 3). The complementary nature of the actual base sequence, however, was elegantly demonstrated in 1961 by B. D. Hall and S. Spiegelman at the University of Illinois, who took advantage of the recent discovery that DNA could be denatured by heating in boiling water, a result of "melting" the hydrogen bonds that hold the strands together. If a solution of melted DNA is cooled rapidly, the DNA stays largely single-stranded; however, if a solution of denatured DNA is cooled very slowly, there is a specific reformation (*annealing*) of active double-stranded molecules. Hall and Spiegelman reasoned that if slow cooling were carried out in the presence of RNA having a base sequence that is complementary to a portion of the DNA (that is, matching U in the RNA with A in DNA, A in RNA with T in DNA, and so on for the G:C pairs) some specific DNA–RNA hybrids might form.

Hall and Spiegelman prepared DNA from stocks of phage T2, and then fractionated the contents of T2-infected cells to obtain some of the hypothetical messenger. Radioactive phosphorus (as $^{32}PO_4{}^{3-}$) was added after infection, so that newly synthesized RNA would be labeled. The DNA was heated, mixed with the labeled RNA and allowed to cool slowly. It was then placed in a centrifuge tube containing CsCl (a dense, high molecular weight salt) and spun to allow the salt to form a gradient of concentration, and hence of density (see Fig. 8–10). After centrifugation, the radioac-

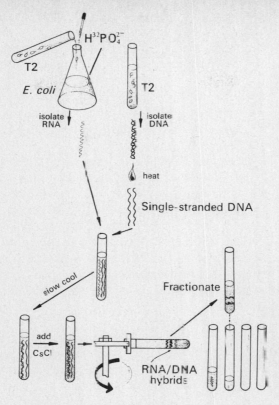

FIGURE 8-10. RNA–DNA Hybrids. Single-stranded DNA can be produced by heating double-stranded DNA. When RNA is added before slow cooling, hybrids will form if the DNA and RNA have complementary base sequences. The hybrids can then be isolated with a CsCl density gradient. (Experiment by B. D. Hall and S. Spiegelman, described in *Scientific American*, May, 1964, p. 48.)

tive label was found in two places: near the bottom of the tube, at a density corresponding to that of single-stranded RNA; and at a less dense position, in a band containing primarily DNA. When the experiments were repeated with DNA from other sources, including bacterial DNA, there was no association of RNA with the DNA band. Obviously, RNA that is synthesized immediately after phage infection reflects the base sequence of phage DNA. It is this RNA that directs the incorporation of amino acids into protein.

At about this same time, another group announced the purification of an enzyme that specifically polymerizes nucleotides into RNA that reflects the base sequence of an added DNA "primer." This step, which is called *transcription*, is catalyzed by a DNA-dependent RNA polymerase, or *transcriptase*. Thus, the flow of genetic information is from the linear sequence of nucleotides (or just "bases," since they distinguish one nucleotide from another) in DNA to the linear sequence of nucleotides in mRNA, and from there to the amino acid sequence of a polypeptide chain.

The transcription of information from double-stranded DNA raises the question of whether one strand of DNA alone serves as a template for the synthesis of RNA, or whether both strands are transcribed. The first experimental evidence that only one DNA strand is transcribed at a given gene was the discovery that the two strands of SP8 DNA (a phage that grows on *Bacillus subtilis*) are different enough in their overall base content, and hence densities, to be separable on CsCl gradients. The two strands can thus be tested independently for their ability to hybridize with the phage-specific messenger RNA produced after infection. Only one of the two strands of SP8 DNA forms hybrids, however, implying that only one of the two strands serves as a template in the transcription process.

Later work has shown that generally no one strand of DNA in a given molecule has a monopoly on transcription. Rather, some of the genes in a molecule of DNA are transcribed from one strand (e.g., the "Watson strand") while others are transcribed from the alternate strand (the "Crick strand"). In other words, the same gene is not ordinarily transcribed from both strands, an event that would lead to two messages bearing the same information in different (but complementary) languages. (Some mitochondrial DNA appears to be an exception, as both strands may be transcribed.)

Since the DNA-dependent RNA polymerase (transcriptase) always makes the 5′ end of an mRNA first, transcription of the two strands of DNA must proceed in opposite directions around the double-stranded molecule. Thus, the polymerase, after recognizing and fixing to some specified grouping of bases, proceeds to use one or the other DNA strand as a template for the polymerization of RNA. The direction of its movement depends on which strand is to be transcribed. The informational content of messenger RNA, as we shall see, gets translated into one or more polypeptide chains, beginning also at the 5′ end of the mRNA. Thus, it is possible for the process of protein synthesis (called *translation*) to begin even before the messenger is freed from DNA, at least in those systems where RNA synthesis and protein synthesis can take place in the same part of the cell—i.e., in procaryotic cells, but probably not in higher organisms.

The production of mRNA in eucaryotes is somewhat more complicated than in procaryotes. In the first place, most of the RNA made in the nucleus (up to 80% or more) never leaves it. This *heterogeneous nuclear RNA*, or HnRNA, is eventually degraded. In addition, most eucaryotic messengers (except those for the chromosomal proteins known as histones) have a 3′ terminal stretch of some 100 to 200 adenylic acid residues. This string of As could be transcribed from poly AT (that is, a string of A in one strand and a complementary string of T in the other) within the DNA. While there are reports of such regions in some eucaryotic cells, there is good evidence that As are added one at a time to the completed messenger. And finally, the observation that protein synthesis in eucaryotic cells seems to require the presence of nucleoli led to the suggestion that mRNA is processed in some way by nucleoli before transit to the cytoplasm, perhaps as some kind of complex with nucleoli-derived ribosomal precursors. Unfortunately, the significance of these observations and proposals is not yet clear.

Other aspects of transcription and its control will be considered later, but first we shall see how the discovery of mRNA allowed the genetic code to be deciphered.

THE GENETIC CODE. The first indication of how genes might specify proteins came in 1949 from the work of Linus Pauling and his colleagues at the California Institute of Technology. They demonstrated that sickle cell anemia — an inherited (genetic) condition affecting some 7 percent of Americans with African ancestry — is due to the presence of an abnormal protein called hemoglobin S (Hb S . Pauling was able to distinguish this protein from normal hemoglobin (Hb A) by its ionic charge, as reflected by its mobility in an electrostatic field.

In 1956, V. M. Ingram of Cambridge showed that normal and sickle cell hemoglobin differ by a single amino acid, with a valine substituting for glutamate (see Fig. 8–11).[3] Later, other abnormal hemoglobins were found. They too proved to be single amino acid replacements, as are many mutant proteins obtained from a wide variety of sources. Such proteins result from gene alterations, the simplest of which would be the substitution of one of the four nucleotides for another. To explain how such a substitution might lead to an amino acid replacement, we must seek an explanation for how four bases specify twenty amino acids in an unambiguous way.

Some of the earliest speculations on the relationship between a nucleotide sequence in DNA and the amino acid sequence of a protein specified by the DNA came from the Nobel prize-winning physicist and amateur cryptographer, George Gamow. He suggested that the problem was one of coding, pointing out that if every two base pairs along the DNA helix determine an amino acid, only sixteen different amino acids can be specified (four bases at each of two positions, hence $4 \times 4 = 16$); on the other hand, three bases allow for 64 ($4 \times 4 \times 4$) specifications, now called *codons* or *code words*. Sixteen codons is not enough, for it im-

---

3. Since glutamate is negatively charged at physiological pH, whereas valine is neutral, the substitution removes one charge and causes a reduction in the solubility of the protein. At low oxygen tension it loses still more charge, and may thus precipitate, causing the red cells to become crescent-shaped and to plug fine capillaries. People who are heterozygous for Hb S have both normal and abnormal hemoglobin in their erythrocytes and usually escape serious harm, though severe stress does pose significant risks. Homozygotes, on the other hand. suffer repeated sickle-cell crises and may die while still in their childhood. It should be noted that the evolutionary survival of this gene is apparently related to the ability of carriers to withstand malaria, a disease that is caused by a red cell parasite prevalent in the same geographical areas where sickle cell anemia is most common.

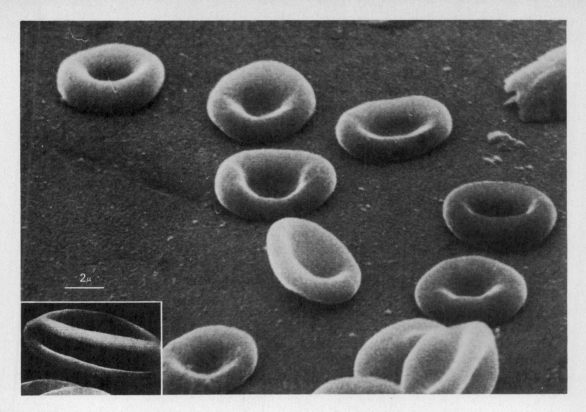

FIGURE 8–11. Normal and Sickle Cell Erythrocytes. Sickle cell anemia is caused by a single amino acid replacement in hemoglobin, resulting in its reduced solubility at low oxygen tensions. The cells may then "sickle" (inset) and plug fine capillaries to cause a sickle cell crisis. [Scanning electron micrograph of normal human erythrocytes courtesy of R. F. Baker, Univ. of Southern California. Inset courtesy of J. A. Clarke and A. J. Salsbury, *Nature*, **215**:402 (1967).]

plies a certain amount of ambiguity in the order of amino acids in a protein composed of twenty different amino acids. Sequence studies on insulin, completed in 1953 by F. Sanger, who also was later awarded the Nobel Prize, made it clear that no such ambiguity is present: all molecules of insulin obtained from the same species appeared to have the same amino acid sequence.

Gamow suggested, therefore, that the codon must consist of triplets (three bases). But three bases permits sixty-four possible codons, as we just pointed out, which seems an inefficient way to specify only twenty amino acids. For this and other reasons, it was further suggested that the triplets overlap one another. In the hypothetical base sequence

A-T-G-A-G-C-A-T-T

an overlap of one base per triplet would produce the code words ATG, GAG, GCA, and ATT. An overlap of two bases would produce the code words ATG, TGA, GAG, AGC, and so on. In contrast, a simple, nonoverlapping code would produce only the three code words, ATG, AGC, and ATT.

While overlap allows more amino acids to be specified with a given amount of DNA, it exacts a high price: an overlap of one means that only sixteen different amino acids can follow any specified amino acid, since the first base in the next code word is fixed by the last base of the current word; and an overlap of two bases means that only four different amino acids could follow. Although the data were not available through which such schemes could be ruled out when they were first presented, as more and more amino acid sequences became known, it also became clear that no such restrictions are present.

Thus, the most efficient way in which DNA can uniquely determine an amino acid sequence without placing prior restraints on the primary structure of the protein, is to use a nonoverlapping sequence of three bases per amino acid. Each group of three bases, according to this prediction, defines a code word. As it turns out, most amino acids are specified by more than one code word of three bases each, so that the genetic code is said to be *degenerate*.

ASSIGNING CODE WORDS. The year 1961 was a memorable one for molecular genetics. It saw the appearance of papers describing the *in vivo* characteristics of mRNA, the *in vitro* formation of RNA–DNA hybrids, and the purification of an enzyme (a transcriptase) responsible for DNA-dependent RNA synthesis. As we shall see later, it was also the year in which F. Jacob and J. Monod announced their famous operon model for the genetic regulation of protein synthesis. But perhaps the biggest bombshell came from the laboratory of Marshall Nirenberg at the National Institutes of Health in Bethesda, Maryland. In a series of Nobel Prize-winning papers coauthored with J. H. Matthaei, he reported the development of a cell-free system, obtained from the bacterium *E. coli*, that is capable of polymerizing amino acids.

The polymerization of amino acids by cell-free extracts of *E. coli* occurs only if high molecular weight RNA is added as a replacement for the missing mRNA. (Messenger RNA in bacterial cells is short-lived because of the presence of the enzyme ribonuclease in the cytoplasm.) At the International Congress of Biochemistry, held in Moscow in the fall of 1961, Nirenberg announced that synthetic polyuridylic acid (UUUUU . . .) stimulates cell-free systems to specifically incorporate phenylalanine into a polypeptide chain. This experiment not only defined the first code word (UUU = phe), but presented a system that could be used to define the other code words (see Fig. 8–12).

The first attempts to define the code involved correlating the statistical distribution of bases in a synthetic messenger with the uptake of amino acids by a cell-free system. This statistical approach was necessary because of the difficulty of synthesizing polynucleotide chains with known sequences. Some progress was made with these techniques, but by 1965 H. G. Khorana (also awarded the Nobel Prize) at the University of Wisconsin's Institute for Enzyme

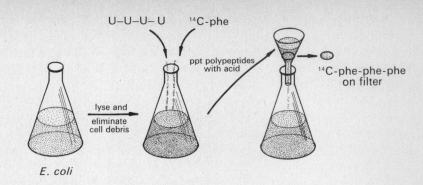

FIGURE 8-12. Breaking the Genetic Code. The first step was the demonstration that polyuridylic acid ("poly-U") causes the specific incorporation of phenylalanine into a polypeptide large enough to precipitate in acid. [Described by M. W. Nirenberg and J. H. Matthaei, in *Proc. Nat. Acad. Sci.* (U. S.), **47**:1588 (1961).]

Research had succeeded in producing synthetic messengers with strictly repeating base sequences. Such molecules stimulate a cell-free system to incorporate amino acids in strictly repeating sequences. For example, alternating uridylic and guanylic acid (UGUGUG . . .) causes the formation of alternating valine and cysteine (val-cys-val-cys . . .), implying that the polynucleotide is read in groups of odd numbers of bases. A triplet code would produce alternating codons (GUG and UGU in this example), while a code of two or four bases would produce a repeating codon—entirely UG or entirely GU for two bases per codon, UGUG or GUGU for four, etc. Thus, Khorana's alternating polymers demonstrated directly that the code is an odd number of bases, in agreement with earlier predictions.

Using a sophisticated combination of chemical and enzymatic techniques, Khorana was later able to synthesize messengers of repeating triplets and repeating groups of four, the results of which firmly established the triplet and degenerate nature of the code. Although codons cannot be assigned unambiguously from such polymers because of the difficulty of knowing which nucleotide starts the chain, the possibilities could be narrowed sufficiently to allow assignments when results obtained with several polymers were compared.

Most of the codon assignments, however, came from the ribosomal binding technique developed by Nirenberg and Leder in 1964. They found that when RNA triplets (synthetic codons) were added to a cell-free extract of *E. coli*, the triplets were bound by ribosomes along with an amino acid and another RNA, called tRNA. By using first one labeled amino acid and then another with a given triplet, they could determine the specificity of the codon. (The complex of amino acid, tRNA, codon, and ribosome sticks to nitrocellulose filters. When the proper radioactive amino acid was used for a given nucleotide triplet, filtering the reaction mixture left the filter "hot.") All possible triplets were tried, and on this basis most of the codons were assigned. The results of the binding experiments were not always unambiguous, but correla-

a allowed the codon assignments to be

nown to be a nonoverlapping sequence
(from mRNA) in a 5′ to 3′ direction.
ns specify an amino acid. The three
UAA, and UAG, which serve in most
that is, they stop the incorporation of
ine the end of a polypeptide chain.
in amino acid, these terminators were
odons.

se reveals an interesting pattern to its
s the terminal (3′) nucleotide in a codon
yrimidines, U or C, and still specify the
r cases, the third codon may be either of
without changing the specificity. This
robble hypothesis, in which the greatest
to the first two bases, leaving a certain
e third. The result is almost a "2½" base
he wobble lies with the geometry of the
properties of the tRNA molecules.

c Code[a]

	MIDDLE BASE				
BASE	U	C	A	G	3′ BASE
U	UUU ⎱ Phe UUC ⎰ UUA ⎱ Leu UUG ⎰	UCU ⎱ UCC ⎰ Ser UCA ⎱ UCG ⎰	UAU ⎱ Tyr UAC ⎰ UAA ochre UAG amber	UGU ⎱ Cys UGC ⎰ UGA opal UGG Try	U ⎱ pyrimidines C ⎰ A ⎱ purines G ⎰
C	CUU ⎱ CUC ⎰ Leu CUA ⎱ CUG ⎰	CCU ⎱ CCC ⎰ Pro CCA ⎱ CCG ⎰	CAU ⎱ His CAC ⎰ CAA ⎱ Gln CAG ⎰	CGU ⎱ CGC ⎰ Arg CGA ⎱ CGG ⎰	U C A G
A	AUU ⎱ AUC ⎰ Ile AUA ⎰ AUG* Met	ACU ⎱ ACC ⎰ Thr ACA ⎰ ACG ⎰	AAU ⎱ Asn AAC ⎰ AAA ⎱ Lys AAG ⎰	AGU ⎱ Ser AGC ⎰ AGA ⎱ Arg AGG ⎰	U C A G
G	GUU ⎱ GUC ⎰ Val GUA ⎰ GUG* ⎰	GCU ⎱ GCC ⎰ Ala GCA ⎰ GCG ⎰	GAU ⎱ Asp GAC ⎰ GAA ⎱ Glu GAG ⎰	GGU ⎱ GGC ⎰ Gly GGA ⎰ GGG ⎰	U C A G

[a] Note that in all cases but two (Try,Met), the third position may be occupied by either of the two purines or either of the two pyrimidines without changing the coding specificity. The terminator codons—UAA, UAG, and UGA—stop amino acid incorporation and free the growing polypeptide chain. (The names ochre, amber, and opal refer to the mutant bacterial strains in which the action of these terminators were first studied.) Chain initiation begins with AUG or GUG, marked with an asterisk (*), either of which can code (in procaryotes) for N-formylmethionine in addition to the amino acid shown for it.

TABLE 8.3. Amino Acid Sequence of the Bacteriophage MS2 Coat Protein (...) Sequence of its Gene. Note that 49 different codons are used to specify the 129 (...) An intercistronic region and the first few codons of the next gene are also shown (...) man, M. Ysebaert, and W. Fiers, *Nature,* **237:**82 (1972).]

Torn corner legend (partially visible):

> Compared with the Nucleotide amino acids of the coat protein. [Data from W. You, G. Haege-
>
> GUU· UGA· AGC· AUG·
> AAU· GGC· GGA· ACU·
> UAC· AAA· ...

. . .	(G)·	AUA·	GAG·	CCC·	UCA·	ACC·	GGA						
GCU· Ala 1	UCU· Ser	AAC· Asn	UUU· Phe	ACU· Thr 5	CAG· Gln	UUC· Phe	GUU· Val	CUC· Leu	GUC· Val 10	GAC· Asp			
GGC· Gly	GAC· Asp	GUG· Val	ACU· Thr	GUC· Val 20	GCC· Ala	CCA· Pro	AGC· Ser	AAC· Asn	UUC· Phe 25	GCU· Ala	AA_· Asn		
GAA· Glu	UGG· Trp	AUC· Ile	AGC· Ser	UCU· Ser 35	AAC· Asn	UCG· Ser	CGU· Arg	UCA· Ser	CAG· Gln 40	GCU· Ala	UAC· Tyr	AA_· Lys	
UGU· Cys	AGC· Ser	GUU· Val	CGU· Arg	CAG· Gln 50	AGC· Ser	UCU· Ser	GCG· Ala	CAG· Gln	AAU· Asn 55	CGC· Arg	AAA· Lys	UAC· Tyr	AC_· Thr
AAA· Lys	GUC· Val	GAG· Glu	GUG· Val	CCU· Pro 65	AAA· Lys	GUG· Val	GCA· Ala	ACC· Thr	CAG· Gln 70	ACU· Thr	GUU· Val	GGU· Gly	GGU· Gly	GU_· Val 75
GAG· Glu	CUU· Leu	CCU· Pro	GUA· Val	GCC· Ala 80	GCA· Ala	UGG· Trp	CGU· Arg	UCG· Ser	UAC· Tyr 85	UUA· Leu	AAU· Asn	AUG· Met	GAA· Glu	CUA· Leu 90
ACC· Thr	AUU· Ile	CCA· Pro	AUU· Ile	UUC· Phe 95	GCU· Ala	ACG· Thr	AAU· Asn	UCC· Ser	GAC· Asp 100	UGC· Cys	GAG· Glu	CUU· Leu	AUU· Ile	GUU· Val 105
AAG· Lys	GCA· Ala	AUG· Met	CAA· Gln	GGU· Gly 110	CUC· Leu	CUA· Leu	AAA· Lys	GAU· Asp	GGA· Gly 115	AAC· Asn	CCG· Pro	AUU· Ile	CCC· Pro	UCA· Ser 120
GCA· Ala	AUC· Ile	GCA· Ala	GCA· Ala	AAC· Asn 125	UCC· Ser	GGC· Gly	AUC· Ile	UAC· Tyr 129	UAA·	UAG·	ACG·	CCG·	GCC·	AUU·
CAA·	ACA·	UGA·	GGA·	UUA·	CCC·	AUG·	UCG· Ser 1	AAG· Lys	ACA· Thr	ACA· Thr	AAG· Lys 5	AAG· Lys	(U)

The codon assignments given in Table 8.2 have been verified in the most direct way possible. In 1972, W. Fiers and his associates at the State University of Ghent, Belgium, announced the complete nucleotide sequence of the gene that codes for the coat pro-

TABLE 8.4   Hemoglobin Variants and the Genetic Code

α-Chain position	30	57	58	68
Original amino acid	glutamate⁻	glycine	histidine	asparagine
Possible original codon	GAA	GGU	CAU	AAU
	↓	↓	↓	↓
Possible mutant codon	CAA	GAU	UAU	AAA
Amino acid replacement	glutamine	aspartate⁻	tyrosine	lysine
Name of mutant hemoglobin	Hb G Honolulu	Hb Norfolk	Hb M Boston	Hb G Philadelphia

β-Chain position	6	6	7	63
Original amino acid	glutamate⁻	glutamate⁻	glutamate⁻	histidine
Possible original codon	GAA	GAA	GAA	CAU
	↓	↓	↓	↓
Possible mutant codon	GUA	AAA	GGA	CGU
Amino acid replacement	valine	lysine⁺	glycine	arginine⁺
Name of mutant hemoglobin	Hb S	Hb C	Hb G San José	Hb Zürich

tein of the small RNA phage, MS2. Since the sequence of 129 amino acids in the protein was already known, a direct comparison between amino acid and nucleotide sequences was possible (see Table 8.3).

UNIVERSAL NATURE OF THE GENETIC CODE. The genetic code was determined through investigations on *E. coli*. There is every reason to believe, however, that the code is the same in all its essential aspects in every form of life.

For example, tobacco mosaic virus RNA can be used as a messenger to make TMV coat protein in cell-free extracts of *E. coli*. This would not be possible if the message were interpreted in a different way by bacteria and higher plants. In addition, most of the single amino acid replacements in human hemoglobin can be explained by single base changes in the *E. coli* code, as Table 8.4 makes clear. And in 1971, Carl Merril, Mark Geier, and John Petricciani of the National Institutes of Health announced the *in vivo* transcription and translation of a bacterial gene in human cells.

The human cells were derived from a patient with galactosemia, an inherited enzymatic defect that prevents the normal breakdown of galactose and may result in mental retardation or other disabilities. Merril and his coworkers used a defective version of phage "lambda" that was known to carry an accurate copy of a gene that makes the needed enzyme, picked up from the *E. coli* host cell in which the virus was formed. They found that when human cells, grown in tissue culture, ingested this phage by phagocytosis, the needed bacterial gene was both transcribed and translated. As a result, the cells were able to metabolize galactose in the normal way. This experiment demonstrated that a bacterial gene can be

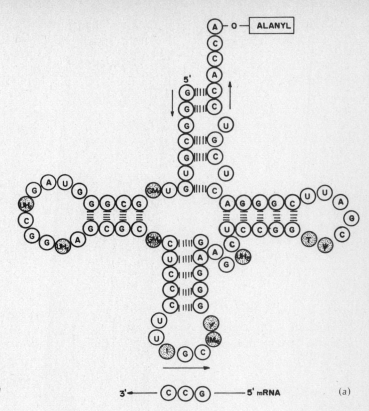

Unusual bases
ψ Pseudouridine
GCC Codon
IGC Anticodon

I = Inosine
UH₂ = Dihydrouridine
T = Ribothymidine
GMe = Methyl guanosine
GMe = Dimethyl guanosine
IMe = Methyl inosine

3' ←—— C C G ——— 5' mRNA   (a)

properly interpreted by the human protein-synthesizing apparatus, and thus demonstrated that a common genetic code is used.[4]

8–3 PROTEIN SYNTHESIS

We have seen that the informational content of DNA lies with its sequence of bases, and that the information may be transferred to messenger RNA by polymerizing the latter with a complementary sequence of bases. The next step is to polymerize amino acids in the proper sequence. That process, called translation, involves the ribosomes, several small proteins, messenger RNA, and an additional class of RNA known as *transfer RNA*, or *tRNA*.

TRANSFER RNA. The tRNA's are relatively small—about 80 nucleotides or 25,000 amu. They are smaller and more soluble in

---

4. This discovery also lends hope for the eventual therapy of a wide range of genetic disorders by packing desired genes in a virus, and allowing the virus to infect a person with the disorder. If the new DNA becomes a permanent part of the host—and there is reason to believe that it can—one such infection might produce a permanent cure.

(b)

FIGURE 8–13. Transfer RNA. (a) The first tRNA to have its sequence of bases determined was that for alanine, diagrammed here. Like the others deciphered since then, it contains "unusual" bases, caused by chemical modification after the tRNA is synthesized, and three hydrogen-bonded loops. *In vivo*, the tRNAs are probably folded into an "L," as in the molecular model of yeast phenylalanine tRNA shown. (b) In the model, deduced by X-ray crystallography, the anticodon loop is at the bottom, while the site for amino acid attachment is at the upper right. [Diagram adapted by permission of W. A. Benjamin, Inc. (Menlo Park, Calif.) from *Molecular Biology of the Gene*, by James D. Watson, copyright ©1970 by James D. Watson. Model courtesy of S. H. Kim, F. Suddath, G. Quigley, A. McPherson, J. Sussman, A. Wang, N. Seeman, and A. Rich, *Science*, **185**:435 (1974), copyright © 1974 by the American Association for the Advancement of Science.]

acid than the other RNA's, and hence they were originally referred to as *soluble RNA*, or sRNA. Like the other RNA's, messenger and ribosomal (mRNA and rRNA—the latter being permanently associated with ribosomes, in contrast with the temporary association between mRNA and ribosomes), tRNA is polymerized complementary to DNA genes. Although methyl groups (CH_3—) are also added to mRNA and rRNA after their synthesis, the tRNA's are unique in undergoing very extensive chemical modification. The tRNA specific for alanine, for example, has 10 unusual bases of 7 different varieties out of a total of 77 bases in the molecule (see Fig. 8–13). It was the first tRNA for which the sequence of bases was determined, Nobel Prize-winning work that was carried out under the direction of R. W. Holley in 1964.

Before an amino acid can be incorporated into a growing polypeptide chain, it must first be esterified to a suitable transfer RNA. That event is catalyzed by an enzyme, called an *aminoacyl-tRNA synthetase*, that is specific for the amino acid in question and drives the reaction by coupling it to the hydrolysis of ATP to AMP. The enzyme first produces an acid anhydride between AMP and the amino acid, using the carboxylate of the amino acid and the phosphate of AMP.

343

$$H_2N-CH-\overset{\overset{\displaystyle O}{\|}}{C}-OH + HO-\overset{\overset{\displaystyle O}{\|}}{\underset{\underset{\displaystyle O^-}{|}}{P}}-O-\overset{\overset{\displaystyle O}{\|}}{\underset{\underset{\displaystyle O^-}{|}}{P}}-O-\overset{\overset{\displaystyle O}{\|}}{\underset{\underset{\displaystyle O^-}{|}}{P}}-O-\text{Adenosine}$$
$$\underset{\displaystyle R}{|}$$

$$\downarrow$$

$$H_2N-CH-\overset{\overset{\displaystyle O}{\|}}{C}-O-\overset{\overset{\displaystyle O}{\|}}{\underset{\underset{\displaystyle O^-}{|}}{P}}-O-\text{Adenosine} + PP_i$$
$$\underset{\displaystyle R}{|}$$

$$\downarrow \text{(HO—)tRNA}$$

$$H_2N-CH-\overset{\overset{\displaystyle O}{\|}}{C}-O-tRNA + AMP$$
$$\underset{\displaystyle R}{|}$$

The complex of amino acid and AMP remains bound to the enzyme until a second reaction attaches the amino acid via an ester linkage to the 3' hydroxyl of the terminal nucleotide (which is always an adenosine) of a tRNA.

$$aa + ATP + E \longrightarrow (aa\text{-}AMP\text{-}E) + PP_i$$
$$(aa\text{-}AMP\text{-}E) + tRNA \longrightarrow aa\text{-}tRNA + E + AMP$$

Free pyrophosphate gets hydrolyzed to P_i by the enzyme pyrophosphatase, thus helping to drive the reaction to the right.

The function of the activating enzyme, the aminoacyl-tRNA synthetase, is to recognize a particular amino acid and tRNA and to couple them. It follows, of course, that there is at least one activating enzyme for each amino acid. An amino acid esterified to tRNA will be placed in the growing polypeptide chain at a point determined by the specificity of the tRNA alone. This function of tRNA was cleverly demonstrated by exposing a cysteine–tRNA complex to Raney nickel, a catalyst that removes the —SH of cysteine, thus converting it to alanine:

$$H_2N-CH-\overset{\overset{\displaystyle O}{\|}}{C}-O-tRNA^{Cys} \longrightarrow H_2N-CH-\overset{\overset{\displaystyle O}{\|}}{C}-O-tRNA^{Cys}$$
$$\underset{\underset{\underset{\displaystyle SH^2}{|}}{\displaystyle CH}}{|} \qquad\qquad\qquad\qquad\quad \underset{\displaystyle CH_3}{|}$$

The altered amino acid, bound to its tRNA, gets inserted into protein as if it were still cysteine. Hence, the coding identity of the complex must lie entirely with tRNA.

The amino acid–tRNA complex (called an *aminoacyl-tRNA* or

charged tRNA) is positioned at an mRNA by matching a three-nucleotide complementary *anticodon* of the tRNA with the corresponding codon of the messenger. In this way, the new amino acid is brought into the proper position with respect to the last amino acid in the growing chain. This is the step at which the coding ambiguity mentioned earlier as the wobble hypothesis is seen, for the third position in the codon is less reliably matched than the others. In particular, either of the two pyrimidines, U or C, might be paired with an anticodon G; and either of the two purines, A or G, might be paired with an anticodon U. (On the other hand, an anticodon C or A in the third position is thought to be rather specific for G or U, respectively, in the mRNA.) Still more flexibility is derived from the appearance of inosine in the third position of the anticodon, for this base, which is derived from adenine by changing its amino group to a carbonyl (NH_2 to $C{=}O$), may hydrogen bond to U, A, or C:

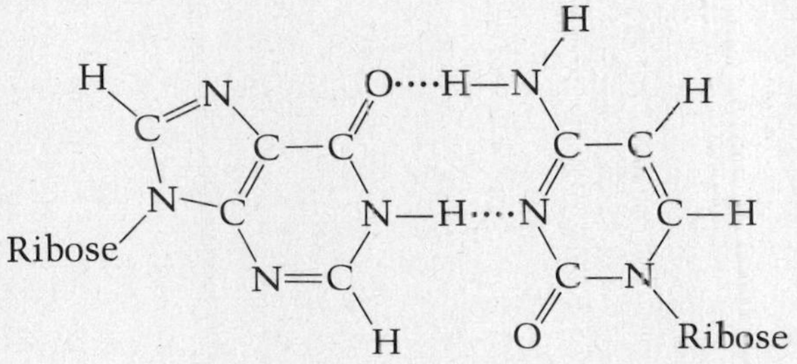

Inosine   Cytosine

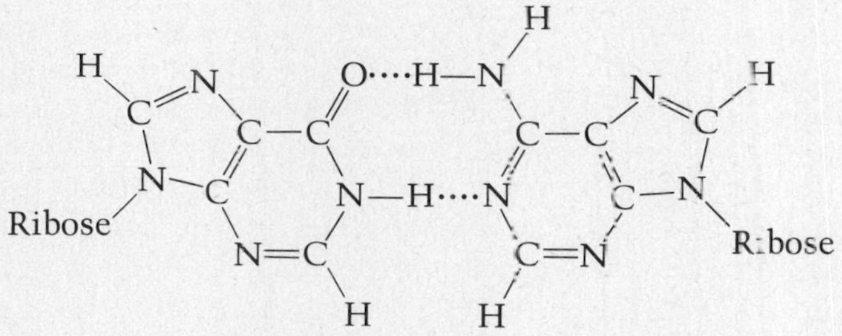

Inosine   Adenine

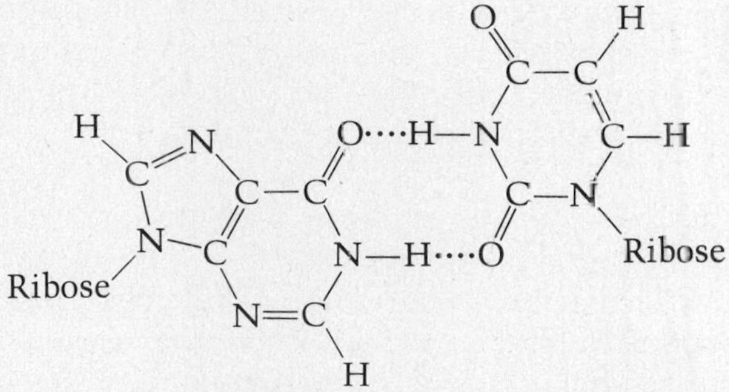

Inosine   Uracil

345

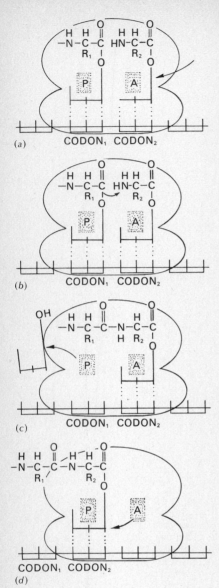

(a)

CODON₁ CODON₂

(b)

CODON₁ CODON₂

(c)

CODON₁ CODON₂

CODON₁ CODON₂

(d)

FIGURE 8–14. Polypeptide Chain Elongation. **(a)** A charged amino acid (tRNA-amino acid) is attracted to the acceptor, or aminoacyl (A) site. **(b)** The ester bond between the existing chain and the tRNA residing in the peptidyl (P) site is exchanged for a peptide bond to the new amino acid. **(c)** the tRNA at the P site is now free to leave. **(d)** Translocation of the ribosome moves the elongated chain from the A to the P site preparatory to starting another cycle.

CHAIN ELONGATION. Codon–anticodon pairing brings a new amino acid into the *A* (*aminoacyl*) *site* of the ribosome on which the mRNA rests. The existing polypeptide is attached to the *P* (*peptidyl*) *site* through an ester linkage to its own tRNA residing there (Fig. 8–14a). It is this ester bond that is replaced by a peptide bond, thus transferring the chain, now lengthened by one amino acid, to the newly arrived tRNA (Fig. 8–14b). The exchange, which is thermodynamically very favorable, is catalyzed by peptidyl transferase, an enzyme that is an integral part of the large subunit of the ribosome.

Lengthening the polypeptide chain is followed by a translocation that moves the polypeptide and attached tRNA from the A to the P site (Fig. 8–14d). Translocation thus frees the A site to make way for the next "charged" (i.e., aminoacyl) tRNA. Thus, the messenger RNA is moved past the ribosome three nucleotides at a time, so that a particular mRNA codon and its tRNA occupies each of two ribosomal positions in turn. The whole process of chain elongation occurs very rapidly, with growth rates up to 40 amino acids/sec in bacteria. The cellular rate of protein synthesis is enhanced further by the simultaneous presence of multiple ribosomes on the same mRNA, forming a complex that is sometimes called a *polysome* (see Fig. 8–15).

At least two non-ribosomal protein factors are needed for elongation. The first was formerly called transfer factor I, but is now designated EF-T in procaryotes and EF-1 in eucaryotes ("EF" for "elongation factor"). It is needed in binding the charged tRNA to the ribosome, and appears to consist of two distinct protein components, at least in procaryotes. The second factor, called EF-G in procaryotes and EF-2 in eucaryotes (formerly transfer factor II), is required for translocation of the message along the ribosome to the next codon. *In vitro* studies have shown that EF-G can function with mitochondrial ribosomes but not with eucaryotic cytoplasmic ribosomes. Only EF-2 is active with the latter.

Even though the equivalent of two ATP hydrolyses (i.e., an ATP to AMP) are needed just to charge the appropriate tRNA with an amino acid, still more energy is needed to complete chain elongation. The additional input of energy comes in the form of two GTP hydrolyses. One occurs during the binding of a charged tRNA to the ribosome and messenger, but appears to be needed not for the binding itself but for peptide bond formation. The second GTP hydrolysis is needed for translocation.

RIBOSOMES. The ribosome clearly plays an important role in protein synthesis, though the details of that role are by no means all known. The codon-by-codon reading of a messenger RNA implies the existence of a pointer to mark the three bases that constitute a codon. That function falls to the ribosome. A further indication

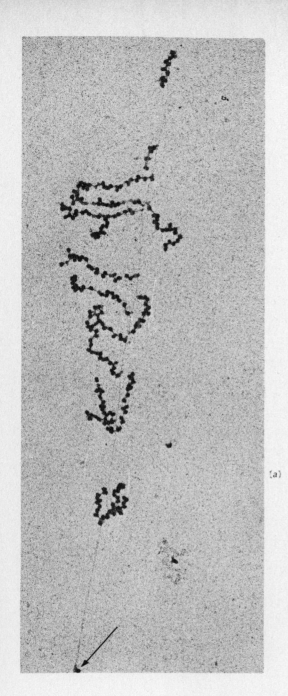

(a)

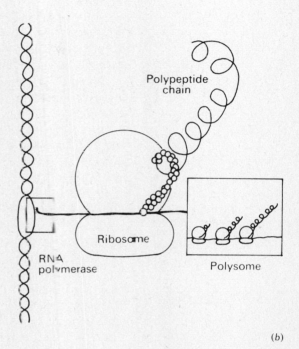

Polypeptide chain

RNA polymerase

Ribosome

Polysome

(b)

FIGURE 8–15. A Bacterial Gene in Action. **(a)** The production of messenger RNA by RNA polymerase (small dots joining side chains to the DNA fiber) is proceeding apace with translation, as evidenced by the numerous ribosomes attached to each mRNA. The longer chains (called polysomes) are further from the point of gene origin. Arrow indicates a presumed RNA polymerase molecule at or near the beginning of the gene. **(b)** A schematic interpretation of one polysome showing polypeptide chain elongation. [Micrograph courtesy of O. L. Miller, Jr., B. A. Hamkalo, and C. A. Thomas, Jr., *Science*, **169**:392 (1970). Copyright by the A.A.A.S]

of the importance of the ribosome in translation came with the discovery that streptomycin, neomycin, and certain other antibiotics cause mistakes to be made in the reading of mRNA by interfering with ribosomes. (Bacteria that have been selected for their resistance to one of these drugs have abnormal ribosomes which, though less efficient under ordinary conditions, protect the translation process from interference by the antibiotic.)

The bacterial ribosome, as it exists during translation, is roughly spherical, nearly 200 Å across, and has a mass of about 2.4×10^6 amu. It is referred to as a *70S ribosome*, a reflection of its sedimentation rate in a centrifuge. Ribosomes of mitochondria and chloroplasts are similar to bacterial ribosomes, but eucaryotic cytosol ribosomes are larger and heavier, with an 80S sedimentation rate. The 70S ribosome is composed of two major subunits, called 30S and 50S (40S and 60S in eucaryotic cytosol ribosomes), with the larger subunit about twice the mass of the smaller (see Fig. 8–16). The smaller, or 30S subunit, may be further broken down into a single 16S rRNA and some 20 different proteins. The larger 50S ribosome has two rRNA strands, one of 23S and one of 5S, plus 33 to 35 different proteins. (Eucaryotic ribosomes contain four RNA species: 28S, 18S, 7S, and 5S. The first three are cleaved from a common 45S nucleolar-derived precursor. The last is extranucleolar.) The ribosomal proteins, which together constitute about one-third the mass of the 70S ribosome, are translated from their own messenger RNA's in the same way as other proteins. Ribosomal RNA molecules have no obvious messenger activity, but are structural components of the ribosome and probably con-

FIGURE 8–16. 70S Ribosomes from a Bacterium, *E. coli.* Note that most particles are composed of two unequal subunits, designated 30S and 50S in bacteria. [Courtesy of H. E. Huxley and G. Zubay, *J. Mol. Biol.,* **2:**10 (1960).]

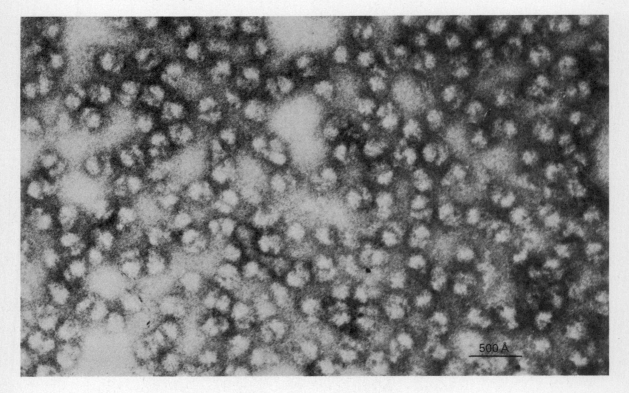

500 Å

tribute to the capacity of the ribosome to attract and properly orient mRNA and tRNA.

INITIATION. When they are not involved in translation, bacterial ribosomes exist as their free 30S and 50S subunits. The first step in translation, then, must involve the assembly of a 70S (or 80S) ribosome at a specified point on the mRNA (see Fig. 8–17). Since translation starts near the 5' end, there is always an assembly point in that region, but additional assembly sites may also exist further down the messenger if there is more than one cistron (which there usually isn't in eucaryotes).

The assembly of a 70S ribosome for translation begins with a complex between the 30S subunit, mRNA, and N-formyl-methionyl-tRNA. The 50S subunit is then added. GTP and at least three nonribosomal proteins, called *initiation factors*, must also be present. N-formylmethionyl-tRNA (*fmet-tRNA*) is a methionine-charged tRNA that has been formylated after its synthesis at the free amino group of methionine. Although there are two methionine transfer RNA's, only one will permit the enzyme-catalyzed formylation of its amino acid.

The existence of fmet-tRNA was discovered in 1964 by K. Marker and F. Sanger. The discovery was of considerable in-

$$H_2N—CH—C\overset{\displaystyle O}{\parallel}—O—tRNA$$
$$|$$
$$CH_2$$
$$|$$
$$CH_2$$
$$|$$
$$S$$
$$|$$
$$CH_3$$

Methionyl-tRNA (met-tRNA)

$$HC\overset{\displaystyle O}{\parallel}—NH—CH—C\overset{\displaystyle O}{\parallel}—O—tRNA$$
$$|$$
$$CH_2$$
$$|$$
$$CH_2$$
$$|$$
$$S$$
$$|$$
$$CH_3$$

N-formylmethionyl-tRNA (fmet-tRNA)

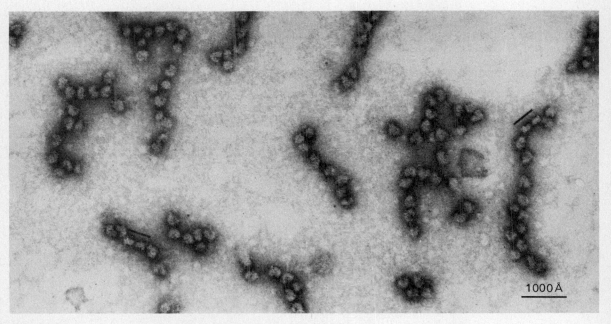

FIGURE 8–17. Rat Liver Polysomes. These eucaryotic cytosol ribosomes are 80S. Each bar underlines 37S (also called 40S) subunits, separated from 60S subunits and attached to mRNA. Initiation of transcription begins with configurations like these. [Courtesy of Y. Nonomura, G. Blobel, and D. Sabatini, *J. Mol. Biol.*, **60**:303 (1971).]

terest because the previous year J. P. Waller had pointed out the surprising fact that nearly 45% of all proteins in *E. coli* have methionine at their amino ends, even though methionine constitutes only about 2.5% of all the amino acids in these same proteins. It was immediately suggested that fmet-tRNA might initiate polypeptide chain synthesis, as the amino end is the only position that formylmethionine could occupy, its own amino group being blocked.

By 1966 several workers had demonstrated the role of fmet-tRNA in the initiation of proteins coded by RNA from the small phages f2 and R17. These RNA's act as messengers both *in vivo* and in cell-free extracts of *E. coli* (i.e., *in vitro*). When the phage coat protein made *in vitro* was analyzed, it was found to have *N*-formylmethionine at its amino end, followed by alanine and then serine. However, when the same protein is isolated from phage grown *in vivo*, the chains start with alanine, with serine as the next amino acid. Apparently, the chain is initiated by an *N*-formylmethionine that is later removed by some peptidase before the protein is incorporated into a maturing phage particle. In other proteins more than one amino acid may be removed, or the formylmethionine may merely be deformylated to allow the chain to begin with ordinary methionine.

Chain initiation in eucaryotes appears to be similar in most respects to the process in procaryotes. There are two tRNA's that can be charged with methionine, one of which actually permits formylation when tested with the bacterial formylating enzyme, although it does not seem to be formylated *in vivo*. It is this latter tRNA and its amino acid that initiate protein synthesis. In addition, cleavage of one or more amino acids from the amino end of the chain appears to be much more common in eucaryotes, since no one amino acid seems to be found there with inordinate frequency in the finished protein.

Formylmethionine-tRNA binds specifically to the normal methionine codon, AUG. *In vitro* experiments indicate that it will also bind to and initiate polypeptide synthesis (albeit less efficiently) at GUG, a codon normally associated with valine. Thus, the same two codons that specify ordinary methionine and valine when they occur in the interior of a polypeptide may also be responsible for chain initiation.

TERMINATION. The termination of polypeptide chain growth, like its initiation, takes place at specific codons—in this case UAA, UAG, and UGA. These are the "nonsense" codons that prematurely terminate translation when they appear by mutation. (Such mutants are referred to as "ochre," "amber," or "opal" mutations, for the three codons, respectively.) When a terminator codon is encountered at the A site of a ribosome, the polypeptide chain is usually released.

Recognition of the terminator codons comes not from tRNA, but from proteins called *release factors*. At least two of these exist in *E. coli*, one of which is involved in termination at UAA and UAG, and the other at UAA and UGA. When a ribosome encounters these codons in the presence of the appropriate factor, the growing polypeptide chain is freed from its tRNA.

In most cases, termination of a growing polypeptide chain takes place when the residue corresponding to the C-terminal amino acid of the mature protein has been added—i.e., at the end of a cistron, even when the messenger RNA contains several cistrons. The polypeptide, which may already be partially folded, is then released and spontaneously assumes its final secondary, tertiary and quaternary structures. In some eucaryotic systems, however, such as the translation of polio virus RNA in mammalian cells, a single large precursor polypeptide is apparently made from a polycistronic messenger and subsequently cleaved to yield the mature proteins.

ROLE OF THE ENDOPLASMIC RETICULUM AND THE GOLGI APPARATUS. A completed protein is generally released into the cytoplasm. The more hydrophilic proteins may remain in the cytosol, whereas hydrophobic proteins are usually attracted to and become part of membranes. However, some proteins are never released into the cytosol, but immediately become enclosed by the endoplasmic reticulum. Such proteins may later be found in lysosomes, peroxisomes, secretion vacuoles, and so on. It is thought that insertion of a protein into one of these membrane-enclosed organelles is made easier when protein synthesis occurs at the endoplasmic reticulum (ER), because it is the ER that is responsible for producing most membrane-enclosed organelles. Some of the ribosomes, or at least their larger subunits, seem to be firmly attached to the ER; others become attached to the ER when polysomes form. This attachment allows proteins to be polymerized at a site where packaging can take place.

Cells that secrete large quantities of protein have a larger fraction of ER-bound ribosomes than do cells that are not secretors. In other words, there is more rough ER in secretory cells, a reflection of the fact that protein destined for packaging by the ER is apparently very often synthesized by ER-bound ribosomes. There is, however, an unsolved problem of selectivity: Why would an mRNA carrying the message for a cytosol enzyme choose to associate with free ribosomes, whereas an mRNA carrying the message for a lysosomal, peroxisomal, or secretory protein associates preferentially with membrane-bound ribosomes? A further complication is that components of different membrane-enclosed organelles seem to originate at different places in the rough ER.

In any case, studies by G. E. Palade and others indicate that newly synthesized protein which is destined for secretion gets trapped in cisternae of the ER, and is then pinched off in small vesicles (Fig. 8–18). These vesicles fuse with condensing vacuoles, which appear to be saccules released by the Golgi apparatus as diagrammed. This process of accumulation is accompanied by a compacting (condensing) of the material, in part through the loss of water, after which the vacuole will be identified as a secretion granule.

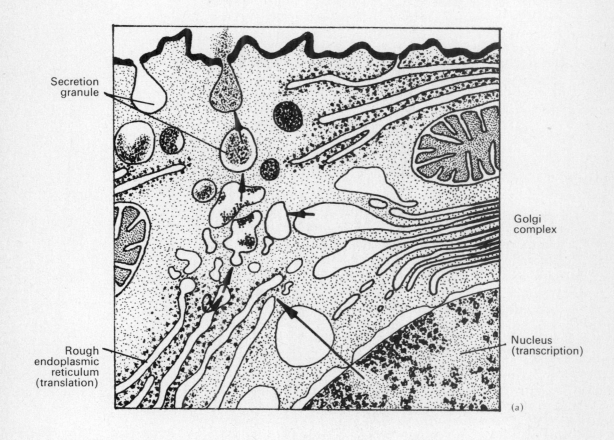

FIGURE 8–18. Compartmentation of Protein Synthesis. **(a)** Protein destined for packaging into membrane-enclosed vesicles (lysosomes, peroxisomes, secretion vacuoles) is usually synthesized on the rough endoplasmic reticulum and then packaged by it. The Golgi complex is involved in the concentration, packaging, and processing of protein that is to be secreted. **(b)** The electron micrograph shows a rabbit plasma cell engaged in the synthesis and secretion of antibodies. (Antibodies are glycoproteins.) Note the extensive rough endoplasmic reticulum (RER) with large, swollen cisternae (light areas) containing the synthesized antibodies. [Micrograph courtesy of F. Gudat, T. Harris, S. Harris, and K. Hummeler, *J. Exp. Med.*, **132**:448 (1970).]

Other types of movement to the Golgi apparatus also may occur. For example, lysosomal enzymes have been identified in small vesicles that have apparently been pinched off from the edge of Golgi saccules (see Fig. 1–14). One does not find many ribosomes in the immediate vicinity of the Golgi membranes, so it is presumed that these proteins are made elsewhere. They may, for example, have been collected by an earlier fusion of ER-derived vesicles with the Golgi apparatus. There is some indication that this collection may be for purposes of processing the protein—especially for the addition of carbohydrate, though the purpose is yet unknown. Why carbohydrate is found covalently attached in varying quantities to a great many different proteins, especially those that are secreted, is only one of the several mysteries about protein synthesis left to be solved.

8–4 GENE REGULATION IN PROCARYOTES

The main function of genes is to direct the synthesis of proteins. In this way, genes determine the chemical and physical properties of the cell. To maintain a balanced production of proteins, it is essential for cells to have ways of regulating the activity of their genes.

INDUCTION AND REPRESSION. When *E. coli* is grown on the sugar lactose as a carbon and energy source, each cell contains several

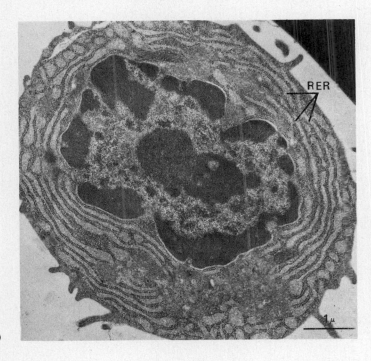

(b)

thousand copies of the enzyme, β-galactosidase, which hydrolyzes lactose to glucose and galactose.

β-Lactose

H_2O $\downarrow$ (β-galactosidase)

β-D-Galactose   + β-D-Glucose

Sister cells, genetically identical to the first culture but growing on glucose instead of lactose, will have only a few copies of this enzyme per cell. If lactose is added to this second growth medium, it will remain unused until all the glucose is consumed. There will then be a pause in the growth of the culture while cells begin making β-galactosidase and two companion proteins, galactoside permease (to facilitate lactose transport across the cell membrane) and a galactoside transacetylase. That having been accomplished, growth resumes. By the use of radioactive amino acids, one can readily show that the new enzyme activity appearing during this metabolic readjustment is due to new synthesis (i.e., *de novo* synthesis) of the enzymes in question, rather than to the activation of already existing proteins (see Fig. 8–19).

Enzymes, such as β-galactosidase, which do not appear until they are needed are called *inducible*. Inducible enzymes are synthesized in the presence of an appropriate substance, called an *inducer*, that is often a substrate of the enzyme itself, a substrate of another enzyme in the same pathway, or a chemical relative of one or the other.

In addition to inducible enzymes, there are *repressible* enzymes, the synthesis of which decreases rather than increases in response to the appearance of certain metabolites. A metabolite that lessens the rate of synthesis of such enzymes is called a *corepressor*. (The origin of the "co-" will be made clear in a moment.) For example, the enzymes of the histidine pathway are

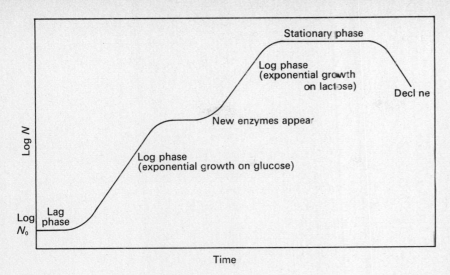

FIGURE 8-19. **FIGURE 8-19.** Biphasic Growth. A bacterial culture, started with N_0 cells, generally exhibits an initial lag phase, during which metabolic adjustments are made, followed by an exponential phase that terminates in a plateau, or stationary phase. In the example shown, both glucose and lactose were available. Thus, during the first plateau, cells adjust for the exhaustion of glucose by making the additional enzymes needed for lactose catabolism. This type of biphasic growth is also called *diauxie*.

repressible: when histidine is presented to a cell, the cell has less need for the enzymes that make histidine, and so the rate of synthesis of these enzymes is reduced.

A third group of enzymes, whose rate of synthesis is neither increased nor decreased by reasonable changes in the environment or food supply, are called *constitutive*. Even though their rate of synthesis does not change in response to external stimuli, there is always just the right amount of each—perhaps a few copies per cell, perhaps thousands of copies per cell. Whatever the number, it must be maintained in the face of loss due to degradation or dilution by growth. There must be a way, therefore, of synthesizing each protein at its own optimum rate. And in the case of inducible and repressible enzymes, that rate must be variable.

OPERON THEORY. Our knowledge of transcriptional control began in the 1950s with studies on the induction of β-galactosidase, especially by Francois Jacob and Jacques Monod, of the Institut Pasteur in Paris. As mentioned above, the appearance of this enzyme is coupled with the appearance of two others, galactoside permease and galactoside transacetylase. Experiments revealed that genes carrying the message for these three proteins, called *structural genes z, y,* and *a,* are controlled by two others, called the regulator and operator, or *i* and *o* genes, respectively. The four genes *o, z, y,* and *a* (operator, galactosidase, permease, and acetylase) were found to be next to each other, in that order. The *i* gene proved to be a short distance away from the others along the DNA, on the operator side (see Fig. 8-20, top diagram).

The functions of the two control genes, the regulator and the operator, were revealed by the properties of their mutants. Muta-

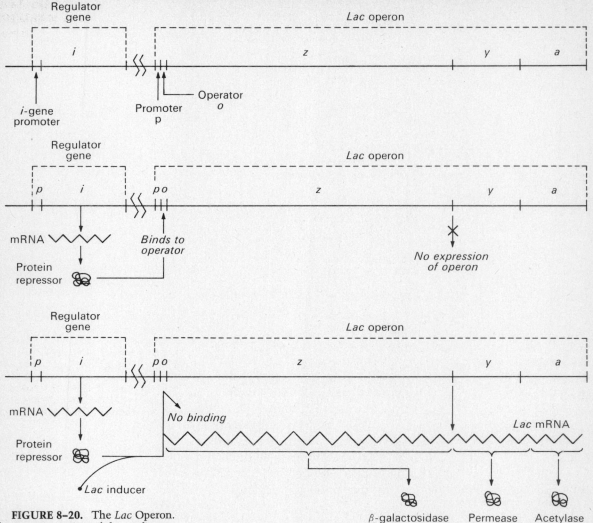

FIGURE 8-20. The *Lac* Operon. The operon consists of three adjacent structural genes, coding for proteins, plus an operator (only about 27 base pairs long) and promoter. The promoter is the site to which RNA polymerase first binds. The presence of the *i*-gene product prevents transcription of the operon unless an inducer is also present. The inducer is *allolactose*, which is lactose with its 1 → 4 linkage between monosaccharides replaced by a 1 → 6 linkage. Allolactose is produced from lactose by the enzyme β-galactosidase, the other function of which is to hydrolyze lactose to galactose and glucose. [For more on this point see A. Jobe and S. Bourgeois, *J. Mol. Biol.*, **69**:397 (1972).]

tion in either of them (the mutants are called i^- and o^c, respectively) results in constitutive synthesis of the three inducible proteins. That is, mutant cells continue to synthesize all three proteins regardless of the presence or absence of lactose. When partially diploid cells—i.e., cells containing some DNA from a second strain in addition to the DNA that is normally present—were examined, the i^+ and o^c genes appeared to be dominant over their alleles, i^- and o. This means that a cell containing both a wild-type and a mutant regulator gene continues to behave like a normal cell, but when both a normal and an o^c operator are present, there is constitutive synthesis of the three proteins (see Fig. 8-21). In addition, a second kind of mutation in the i gene was found, called i^s. The i^s mutation is dominant over the wild-

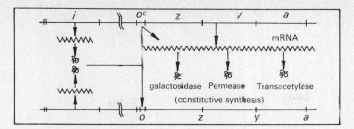

o^c is dominant over o^+

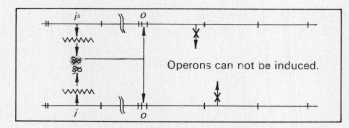

i^+ is dominant over i^-

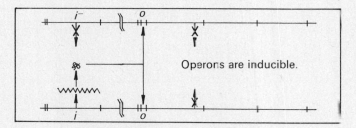

i^s is dominant over i^+

FIGURE 8–21. *Lac* Operon Mutations. The model of the *lac* operon, presented in Fig. 8–20, was deduced from observations made on three mutant strains, o^c, i^-, and i^s. Their behavior was observed in partially diploid cells that contained both a normal operon and a mutant version. (The superscript +, added for emphasis, always means the normal, or "wild-type," gene.)

type regulator and results in a cell that cannot be induced at all, in contrast to the i^- mutant, which is constitutive.

In 1961 Jacob and Monod proposed a model that offered a logical explanation for these observations. Their proposal, which had an enormous impact on molecular biology, helped to win for them the 1965 Nobel Prize, shared with André Lwoff. Jacob and Monod suggested that the i gene makes a molecule, called a *repressor*, that can bind to the operator gene (see Fig. 8–20). When it does so, synthesis of mRNA from the three structural genes is blocked. The presence of lactose (or a chemical derivative of it) blocks the repressor-operator interaction and results in induction. The i^- regulator gene, then, is recessive because the repressor from a wild-type allele in a partially diploid cell can bind to both operators. An o^c mutant is dominant because the structural genes that it controls will be actively transcribed regardless of the existence of repressors. And finally, the i^s mutant appears to make a repressor that always binds to a normal operator in spite of the presence of inducer (see Fig. 8–21).

In support of their model, Jacob and Monod offered the results of an experiment in which an i^+z^+ donor strain (male) was mated to an i^-z^- recipient (female) strain. Bacterial mating, or *conjugation*, involves the linear transfer of DNA from the donor to the recipient through a *conjugation tube* joining the two cells.[5] Transfer always starts at the same point on the chromosome (for a given strain) and proceeds at a more or less constant rate with time. It may thus be interrupted at any point by a vigorous stirring in a blender, producing partially diploid cells of the type just mentioned. In the present experiment, however, the recipient began making β-galactosidase as soon as it received the z^+ gene, even in the absence of lactose. About an hour later the i^+ gene was transferred. Thereafter, synthesis of β-galactosidase required the presence of lactose or other inducer. This response is precisely what one would expect if the i gene were releasing to the cytoplasm some product that is capable of inhibiting the activity of the structural genes.

The operator and adjacent genes influenced by it are known collectively as an *operon*. When the operon theory was first proposed in 1961, it was not clear how many genes were included in the lactose operon (*lac* operon for short), since the presence of transacetylase had not as yet been confirmed. The nature of the regulator molecule, the repressor, was also uncertain. But in 1966, Walter Gilbert and B. Müller-Hill of Harvard succeeded in isolating the repressor for the *lac* operon. They demonstrated that the *lac* repressor is a protein of four subunits (about 40,000 amu each), present in only a few copies per cell and quite ordinary in most respects. The repressor acts by binding directly to DNA at the operator site unless the appropriate inducer is also present. In that case, the inducer binds to the repressor and destroys its capacity to recognize the sequence of nucleotides defining the operator, presumably through a conformational change in the repressor protein. The binding of *lac* repressor to the operator prevents transcription of the *lac* operon by interfering with the activity of RNA polymerase (the transcriptase).[6]

Though it has been modified in several ways since its original in-

---

5. A male (donor strain) is one that contains a sex factor, or F factor. This factor is a segment of DNA that may exist either as a free, circular *episome* (extrachromosomal DNA) or integrated into the chromosome much like a provirus. In its free state, it may be transferred from one cell (a male or F+ cell) to another (a female, or F− cell) without transfer of chromosomal genes. An F− cell that receives the episome thereby becomes F+ —i.e., the cell is converted from female to male. Thus, masculinity in bacteria is a contagious condition, a situation that presumably causes them less confusion than it would cause us.

6. RNA polymerase attaches first to a site just at the edge of the *lac* operon, called the *promoter*, and would ordinarily pass over the operator region during its transcription of the operon. When repressor is absent, RNA polymerase synthesizes one messenger RNA for the entire operon, stopping only when it comes to a special sequence of nucleotides at the end of the operon. Recognition of the stop signal is provided through the help of a protein called a *rho factor* that probably associates with the polymerase.

ception, Jacob and Monod's operon theory has served as a useful model for the study of enzyme induction in both procaryotes and eucaryotes. Operon theory also explains the ability of certain viruses to remain in a provirus state. In fact, a repressor for phage lambda was isolated by Mark Ptashne of Harvard at about the same time as the *lac* repressor, and proved to control the lambda provirus genes in much the same way as the *lac* repressor controls the *lac* operon. While it prevents transcription of a prophage, the lambda repressor also prevents replication of superinfecting lambda particles, making the cell appear to be immune to them.

The *lac* operon is said to be under *negative control* of its regulator protein, as the protein must vacate the operator in order to get expression of the structural genes. Other operons are under *positive control*, meaning that the regulator protein must be present for expression of the structural genes. Either scheme can be used to explain repressible as well as inducible operons. Thus, the combination of regulator protein plus end product of a pathway might bind at the operator site to prevent transcription if the regulator exhibits negative control. Conversely, a positive regulator might be unable to bind in the presence of the end product.

It is also possible for the structural genes that are controlled by a given regulator gene to be scattered at several points about the chromosome, each cluster acting as an operon. These systems have been called *regulons* (Fig. 8-22) to distinguish them from the *lac*-type of gene control. And in still another variation of the basic operon model, the regulator protein appears to be the product of one of the structural genes of the operon itself (a system that has been named "autogenous regulation") rather than a product of a distant gene.

THE REGULATION OF RNA POLYMERASE. While the operon theory satisfactorily explains the regulation of certain genes, it is a relatively expensive mechanism. It requires, for each operon, an oper-

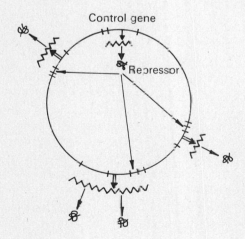

Control gene

Repressor

FIGURE 8-22. The Regulon. In the regulon, the product of a single regulator gene controls several operons at scattered sites around the chromosome. The arginine pathway of *E coli*, for example, has a total of nine structural genes distributed among six operon-type clusters, all of which are coordinately expressed by a single regulatory gene.

ator gene and a regulator gene, though the latter may be shared with other operons if their coordinate expression is desirable. The regulator gene must be transcribed and translated into a protein; hence, this mechanism not only presents a considerable genetic burden, but, if widely used, would clutter the cell with numerous proteins whose only function would be to control genes instead of carrying on the immediate task of metabolism. In addition, not all genes need to be turned off and on regularly, even though all do require a controlled rate of expression. It should not be surprising, then, that other mechanisms exist to control transcription. Again, much of our information comes from studies on *E. coli.*

Once synthesis of a particular mRNA begins, it seems to proceed at about the same rate for any gene in a given cell (about 55 nucleotides/sec in *E. coli* at 37 °C). So if two genes need to be transcribed at different rates to provide different amounts of their respective proteins, one should look for a way of controlling the initiation of mRNA synthesis. As an alternative to the operon mechanism, the promoter sites for two genes could be designed to have different affinities for RNA polymerase. To investigate that possibility, and to learn more about transcription, it was necessary to purify RNA polymerase. As so often happens, that task, initially viewed as relatively straightforward, presented some perplexing problems.

To purify an enzyme, one needs an assay for it. In the case of RNA polymerase, the appropriate assay is the enzyme's ability to polymerize RNA complementary to one or the other strand of a DNA template. However, a point in the purification of RNA polymerase was reached at which the enzyme remained almost completely active when tested with DNA from calf thymus, but lost much of its ability to transcribe *E. coli* DNA and almost all of its ability to transcribe DNA isolated from phage T4. The enzyme, which had hitherto been considered as quite nonspecific, was now capable of considerable discrimination in its choice of template. The purified enzyme was shown to contain four essential subunits of three varieties, designated $\alpha_2\beta\beta'$. The change in specificity was due to the removal of still another subunit, called a sigma (σ) factor. Addition of this protein to the larger structure restored the capacity of RNA polymerase to transcribe all the test templates.

Although the core enzyme (that is, the RNA polymerase without sigma factor) can transcribe most DNA, addition of the sigma protein increases its efficiency and alters its specificity. This change represents a mechanism by which transcription can be controlled, because in addition to the presence of promoter sites with varying affinities for the polymerase, a given site might be transcribed more or less often depending on the availability of a particular sigma factor. However, it does not appear at the present time that control by sigma factors is widely used.

CATABOLITE REPRESSION. Another example of transcriptional control through modification of a polymerase–promoter interaction is the phenomenon of catabolite repression. An example of catabolite repression (also called the *glucose effect*) was described earlier in our discussion as the ability of *E. coli* to exhaust the supply of glucose in its growth medium before allowing induction of the *lac* operon. Catabolite repression is more general than that, however, for the catabolism of a number of other sugars is similarly repressed in the presence of glucose. The mechanism involves the intracellular concentration of cyclic AMP (cAMP), which is obtained from ATP via the enzyme adenylate cyclase:

$$ATP \longrightarrow cAMP + PP_i$$

Adenosine 3′,5′-phosphate
cAMP

Cyclic AMP falls to low levels when glucose is being catabolized, but is at high levels when the cell is living on one of the other sugars. The effect of cAMP is mediated through a specific receptor protein (called CAP) which, in the presence of cAMP, can increase the affinity of RNA polymerase for the promoter regions of the various sugar operons, probably by binding directly to the promoters. Note that this capacity to bind to promoters distinguishes the mechanism of the cAMP receptor protein from that of *lac*-type regulator proteins, which bind instead to operators.

Thus, we see that procaryotes have a variety of mechanisms with which to modulate the activity of genes. Some variation of these mechanisms may also be found in eucaryotes, but the latter also have many features unique unto themselves, including control mechanisms about which relatively little is known.

8–5 GENE REGULATION IN EUCARYOTES

The considerable diversity in the properties of cells from different tissue types indicates the need for a way of turning some genes on while others are turned off. In addition, the cells of higher organisms, like those of unicellular life forms, must respond to changing conditions by altering their chemical composition. A number of human enzymes, for example, especially in the liver, are synthesized at a rate that varies with diet. A great many more are under the control of hormones. In fact, it is easier to control the intracellular concentration of enzymes in mammals than it is in bacteria, because bacteria lack an efficient mechanism for protein degradation, while the half-life of a typical mammalian enzyme may range from a few hours to many weeks, depending on the enzyme, the tissue, and conditions. Constant protein turnover makes the level of an enzyme in mammalian cells especially sensitive to the rate of its synthesis.

We can extrapolate the principles of gene regulation in procaryotes to the problem of gene regulation in eucaryotic systems. Some caution is required, however, for there are clearly certain differences in the way genes are organized and controlled in the two systems. One of the most notable differences between eucaryotes and procaryotes, for example, is the structure of their chromosomes.

THE EUCARYOTIC CHROMOSOME. The procaryotic chromosome is a simple double-stranded molecule of DNA. The eucaryotic chromosome, on the other hand, is a complicated structure containing a great deal of protein and some RNA along with the universal DNA core. Some of the chromosomal protein is acidic (negatively charged at neutral pH), but most of it consists of a small group of basic proteins known collectively as histones.

Histones are basic proteins because of a high proportion of the amino acids arginine and lysine, both of which have positively charged amino groups ($-NH_4^+$) at neutral pH. Histones have molecular weights in the range of 11,000 to 21,000, and together constitute about half of the total mass of a eucaryotic chromosome. They are generally divided into three groups, called *lysine-rich*, *slightly lysine-rich*, and *arginine-rich*. Each group contains only a few different proteins, found with very little variation in almost all eucaryotic cells.[7]

A chromatin fiber consists of a double helix of DNA plus protein, mostly histone. The exact arrangement is still being investigated. In one model, the DNA-histone complex, which would

---

7. There are two generally accepted nomenclatures for the histones. The "lysine-rich" histones are also called *fraction I* histones or *f1* histones. They apparently consist of a group of closely related proteins with only small differences in their amino acid sequences. The second group, called "slightly lysine-rich," contains at least three different proteins, designated IIb1, IIb2, and V in one scheme, and f2a2, f2b, and f2c in the other. Although histone V (f2c) seems to be limited to amphibian and bird red blood cells, the other two members of this group are widely distributed. The last group, "arginine-rich," is composed of at least two proteins. One of these is known as histone III or f3, and is unusual for its cysteine ($-SH$) content. The other is histone IV (f2a1), which seems to be found without variation in probably all animals and plants.

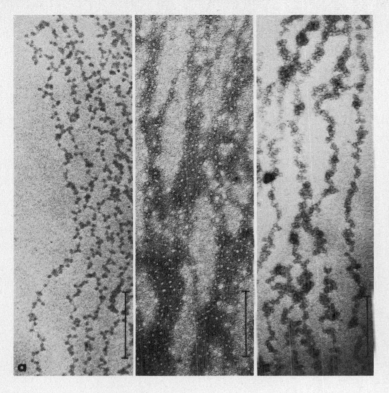

FIGURE 8-23. Chromatin. Micrographs like those shown here suggest that the DNA and histone of chromatin are tightly coiled or twisted together into discrete particles, separated from each other by strands of naked DNA. (a) Rat thymus chromatin, positively stained. (b) The same type of preparation, negatively stained. (c) Chicken erythrocyte chromatin, negatively stained. Note the clustering of particles in (c) and connecting strands of DNA between particles, especially visible in (b). Scale = 0.2 μ. Compare these micrographs with Fig. 8-24b, taken of chromatin prepared in a different way. [Courtesy of A. L. Olins and D. E. Olins, *Science*, **183**:330 (1974). Copyright © 1974 by the American Association for the Advancement of Science.]

have a diameter of 30-35 Å (DNA alone is about 20 Å), is coiled into a fiber some 80-100 Å in diameter and then this structure is sometimes coiled again into a 250 Å fiber. However, much of the recent attention of investigators has been directed toward models in which DNA and histone are twisted together into discrete packets, or *particles*, which are sometimes found in clusters but which are always separated from each other by strands of double-helical DNA (see Fig. 8-23).

Between cell divisions chromatin is normally dispersed, in which case it is referred to as *euchromatin*. However, some may remain in a more condensed, compact form, called *heterochromatin*. When these names were first coined, it was believed that euchromatin and heterochromatin were different entities, rather than merely different configurations of the same material. It is now clear, however, that most chromatin can undergo the transition between dispersed (euchromatic) and condensed (heterochromatic) states.

Prior to cell division, most chromatin condenses to form the discrete bodies that we know as chromosomes. According to some investigators, each chromatid (half of the recently condensed chromosome—see Fig. 8-24) is a single long fiber, with one DNA molecule as its core. Attempts at measuring the length of eucaryotic DNA to see whether any strands are long enough to account for the entire DNA content of a chromatid have been con-

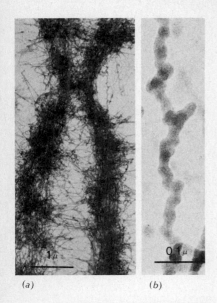

(a)   (b)

fronted with frustrating technical difficulties, and the results have not been highly reproducible. Nevertheless, estimates of more than 2 cm for single DNA molecules have been reported, a value that approaches the necessary length.

There are chromosomal configurations other than the common one shown in Fig. 8–24. For example, the chromatin of vertebrate oöcytes, which are precursors of female gametes or ova, have chromatin arranged in a characteristic form known as a *lampbrush chromosome*. This chromosome has a central axis consisting of a chromatin fiber with periodic stretches of dense folding and other stretches where one of the strands loops out from the axis (see Fig. 8–25). The pattern of looping and the structure of each loop are highly specific for a given species.

The *giant chromosomes* or *polytene chromosomes* of flies (*diptera*), such as those found in the salivary gland of *Drosophila melanogaster* (see Fig. 8–26), represent another unusual chromosomal configuration. These chromosomes, which remain visible

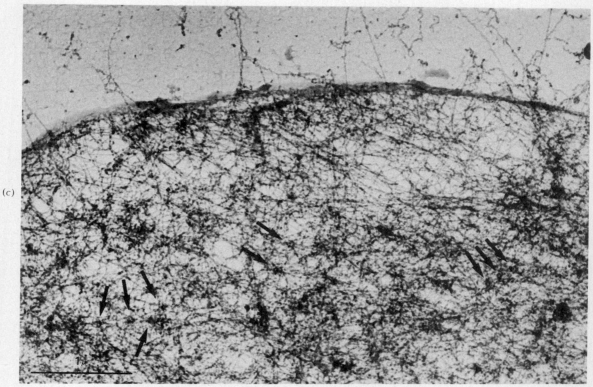

(c)

FIGURE 8–24. The Eucaryotic Chromosome. The eucaryotic chromosome is a tightly folded unit of chromatin. **(a)** Human chromosome from a white blood cell. **(b)** Individual chromatin fiber. Note that the micrograph suggests a supercoiled structure, unlike the micrographs of Fig. 8–23, which resemble beads on a string. Different preparative procedures were used. **(c)** The configuration of the chromatin in the nucleus between cell divisions, when it is not condensed into chromosomes. The arrows point to fragments of the nuclear envelope, resembling annuli, to which chromatin is still attached. Note that many fibers loop out from the nuclear envelope and then back to it. [Photos courtesy of F. Lampert. (a) and (b) from *Nature/New Biology*, **234**:187 (1971); (c) from *Humangenetik*, **13**:285 (1971).]

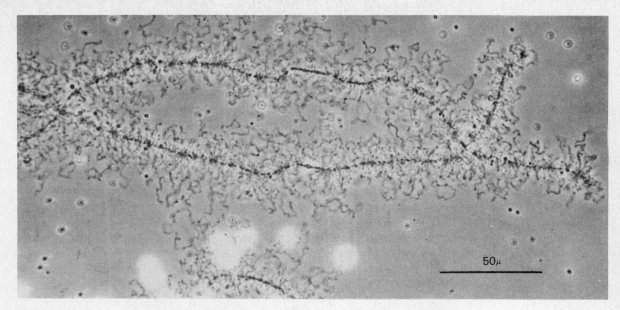

FIGURE 8–25. The Lampbrush Chromosome. An enormous amount of nuclear activity is necessary to produce the items needed for an amphibian egg. Apparently the oöcyte's lampbrush chromosome is designed for that purpose. The loops, which are actively being transcribed, are kinks in a continuous chromatin fiber that is coiled when between loops. (*Triturus viridescens* oöcyte, unfixed. Taken with phase-contrast optics and an electronic flash. Courtesy of J. G. Gall, Yale Univ.)

between replication cycles, may be several microns in diameter and hundreds of microns long. They are formed by parallel chromatids (in some cases over a thousand) aligned in perfect register, producing an identifiable pattern of light and dark bands that correspond to variations in the density (folding pattern) of the individual chromatids.

THE HISTONES IN GENE REGULATION. One can sometimes correlate the physical configuration of a chromosome, or section of a chromosome, with its involvement in transcription. In general, transcription is found only in the more dispersed states. In the lampbrush chromosome, for instance, one can show that RNA synthesis occurs only in the loops, not in the condensed regions. And RNA synthesis in polytene chromosomes takes place at "puffs," consisting of loosened areas of chromatin (Fig. 8–27). The puffing pattern of polytene chromosomes varies from one tissue to another, and within the same tissue at various developmental stages, reflecting a changing pattern of gene activation and repression.

More generally, one notes that a typical eucaryotic cell ceases to make RNA at about the time in the cell cycle when chromatin condenses into chromosomes—i.e., just prior to cell division. Chromatin that does not get dispersed at all may never be transcribed into RNA. For example, one of the two X chromosomes in female mammals is known to fall into this category; it forms a noticeable clump of heterochromatin, called a *Barr body* (see Fig.

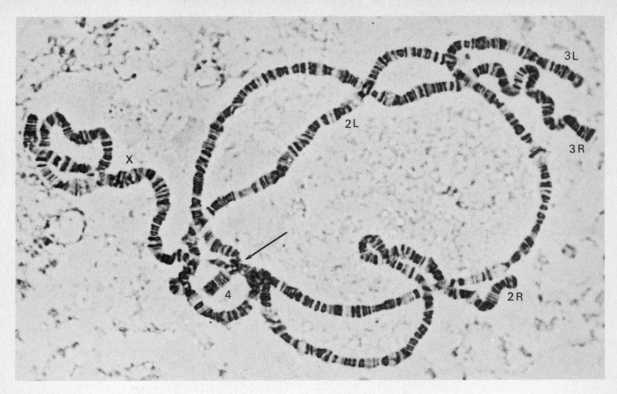

FIGURE 8–26. The Polytene Chromosome. The salivary gland chromosomes of *Drosophila* larvae, showing the four homologous pairs of polytene chromosomes. Radiating from the chromocenter (arrow) are five long arms and a short one. Four of the long arms represent the "left" and "right" ends of two autosomes (chromosomes 2 and 3). The fifth arm is the sex (X) chromosome, or chromosome 1. The short arm (about 10 μ) is the complete chromosome 4. (Courtesy of Paul A. Roberts, Oregon State Univ.)

8–28), in the nucleus when the rest of the chromatin is dispersed between cell divisions. X chromosome inactivation in certain strains of mice having a coat color mutation carried on the X chromosome results in heterozygotes having a mottled coat, consisting of a mixture of differently colored hairs. The pattern of

FIGURE 8–27. Chromosome Puffs. The correlation between chromatin condensation and transcription is supported by the observation of puffs in polytene chromosomes. They represent loosened areas, actively being transcribed. Here one can see a prominent puff near the end of chromosome 3L from *Drosophila melanogaster*. (Phase-contrast optics. Courtesy of J. G. Gall, Yale Univ.)

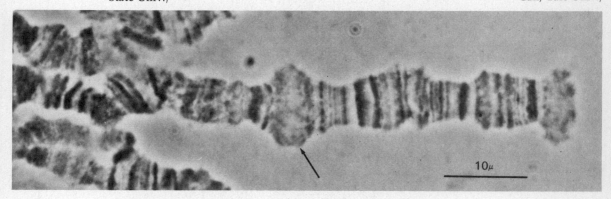

10μ

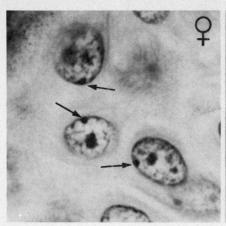
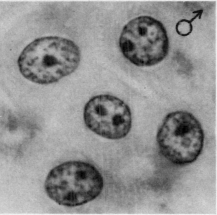

FIGURE 8-28. The Barr Body. The Barr body, or sex chromatin body (arrows), appears in cells having more than one X chromosome — e.g., in the cells of human females. It represents the inactivation (as heterochromatin) of one of the X chromosomes, randomly chosen in most mammals but always the paternally derived X in marsupials. The inactivation gives XX and XY cells nearly equivalent amounts of transcribable DNA, as the abbreviated Y chromosome carries relatively few genes. (Courtesy of Murray L. Barr, Univ. of Western Ontario.)

hairs reflects the activation of either the maternally derived or paternally derived X chromosome in a given cell. The original choice as to which chromosome will be active in a given cell is apparently made randomly at an early stage in embryonic development in most mammals. Once it has been made, however, all the cell's descendants inherit the same choice.

Because of the above observations, activation of a segment of DNA for purposes of transcription is thought to be associated with a loosening of the chromatin structure at that point. In other words, the chromatin must unfold. Transcription might therefore be controlled through a folding–unfolding sequence. Since the change is correlated with and possibly due to the presence of histones, these proteins are automatically implicated in the process of gene regulation. However, their lack of diversity from one cell type and species to another makes it difficult to understand how they could have the kind of gene specificity associated, for example, with the *lac* repressor in *E. coli*. Histones do, however, bind tightly to DNA, attracted by the negative charges on the phosphates of the polynucleotide (at least some appear to be stretched along the grooves of the helix) where they may interfere with RNA polymerase and thus prevent transcription.

As almost every cell in a eucaryotic organism contains the same set of genes, there is an obvious need for simultaneous and complete repression of all but a small fraction of genes within a cell of a given type. The histones provide a way of doing this requiring only that a mechanism be added to ensure the activation of genes that are needed. Several models have been proposed for this mechanism, all of which remain controversial. One, known as the *director hypothesis*, originated in 1965 with R. C. Huang and James Bonner of the California Institute of Technology. It involves a class of small RNA molecules, rich in an unusual base called dihydrouridine, that are covalently linked to a nonhistone protein. These molecules, according to the hypothesis, form larger aggregates with histones and provide the necessary specificity for histone binding.

A second model for conferring gene specificity on histones supposes that all genes are indiscriminately masked against transcription unless gene-specific acidic proteins are present. These proteins activate selected genes by removing or displacing histones from them, perhaps by binding to specific operator-type sites in a manner similar to that of the *lac* repressor. (The *lac* repressor, it should be remembered, is also an acidic protein.)

Chemically altering histones is another way in which they may be removed from selected genes. In fact, a number of different modifications of histones have been recorded. Though it is difficult to determine their physiological significance, any of them could affect DNA–histone interactions. The modifications include: (1) changes in the redox state of cysteine side chains, which can be in the oxidized (disulfide, —S—S—), or reduced (sulfhydryl, —SH HS—) condition; (2) the addition of methyl groups to lysine, arginine, or histidine (—NH—CH_3); (3) the addition of acetyl

$$\text{groups to lysine} \left(-NH-\overset{\overset{\displaystyle O}{\|}}{C}-CH_3 \right); \text{ and (4) the phosphorylation of}$$

serine (—CH_2—O—PO_3^{2-}) and several other amino acid residues. This latter mechanism, phosphorylation, has received the most attention. In certain cases it is known to be stimulated by a cAMP-activated protein kinase, which suggests a mechanism for hormonal involvement.

HORMONAL CONTROL OF GENE EXPRESSION. Hormones are substances that permit cells in one part of an organism to control functions of cells in distant parts. In higher animals, hormones are transported by the blood. Their effect on target cells may involve activation or inhibition of enzymes and/or they may alter the expression of genes. It is this latter function that is of primary interest here.

It is convenient to classify hormones according to whether or not they penetrate the plasma membrane of their target cells. The steroids and some other lipid hormones, by virtue of their solubility in the lipid bilayer of the plasma membrane, do gain entry to the cell; polypeptides, proteins, and other very water-soluble hormones (e.g., adrenaline) generally do not.

The soluble hormones appear to exert their actions via "second messengers," meaning that appearance of the hormone at the plasma membrane elicits the production of a second substance within the cytoplasm. Two such substances have been extensively studied: 3',5'-cyclic AMP (cAMP) and 3',5'-cyclic GMP (cGMP). The former was described earlier in this chapter as a regulator of bacterial genes; the latter substance differs from it only in having guanine instead of adenine as its organic base. The two cyclic nucleotides frequently have opposite physiological effects, consistent with the observation that hormones also frequently work in antagonistic pairs (e.g., insulin and glucagon).

In 1971, E. W. Sutherland, Jr. received the Nobel Prize for having discovered, in the late 1950s, that cAMP mediates the action of adrenaline by activating a protein kinase — that is, by activating an enzyme that catalyzes transfer of a phosphate from ATP to a protein. The original work was concerned with the conversion of glycogen to glucose. This conversion is accelerated when the cAMP-activated protein kinase phosphorylates a second enzyme, which in turn phosphorylates the ultimate target, the enzyme that actually catalyzes the breakdown of glycogen. This *enzyme cascade* amplifies the effect of minute quantities of adrenaline and provides several intermediate steps at which further controls can be implemented.

More recent work has shown that cAMP activates the first protein kinase by binding to a regulatory subunit of the enzyme. The regulatory subunit then dissociates from the catalytic subunit, activating the phosphorylating capacity of the latter. It is also clear now that cAMP-activated protein kinases are involved in much more than just the breakdown of glycogen. Among the additional targets are histones and the nuclear acidic proteins. Several different protein kinases are involved in histone phosphorylation, each specific for certain of the histones. As noted earlier, phosphorylation of histones can be expected to alter their ability to bind to DNA and thus to affect gene expression. In this way, an extra-cellular hormone might control the activity of genes.

The second class of hormones are those that are either taken up specifically by the cell or are soluble enough in the cell membrane to gain entry directly. The steroids are extensively studied examples of this class. Though in some cases, the cytoplasmic levels of cyclic nucleotides are affected by steroids, probably through adenylate and guanylate cyclases associated with the endoplasmic reticulum, the primary mode of action of the steroid hormones is through a soluble cytoplasmic receptor protein. The complex of hormone and receptor protein is able to pass directly into the nucleus where it can be found bound to chromatin, presumably affecting the expression of specific genes.

GENE AMPLIFICATION. Quite a different way of regulating the amount of a particular mRNA, and hence the rate of synthesis of the corresponding protein, is to control the number of genes from which that mRNA can be transcribed. If additional copies of a given gene were to be produced within a cell, there would also be more of that type of mRNA transcribed. The first clear example of this process in eucaryotes is the *gene amplification* by which a growing cell can produce ribosomes at a greatly accelerated rate. Gene amplification occurs in the nucleolus, where ribosomal RNA is made (see Fig. 8–29).

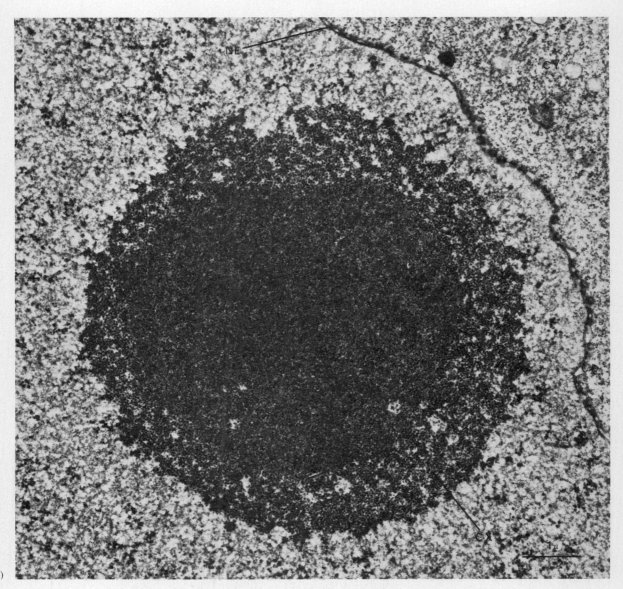

(a)

FIGURE 8–29. Gene Amplification. **(a)** Thin section through a nucleolus, composed of a dense fibrous core surrounded by a granular cortex (CX). The nuclear envelope (NE) is visible, with the cytoplasm (CY) beyond. **(b)** Amplified ribosomal genes being transcribed. The gradient of RNA lengths reflects movement of transcriptases from the points of RNA initiation. Newly made rRNA is coated immediately with protein in the first step toward ribosome synthesis. [*Triturus viridescens* oöcyte. Courtesy of O. L. Miller, Jr., and B. R. Beatty, Oak Ridge National Lab. (a) from *J. Cell Physiol.*, **74**, suppl. I:225 (1969); (b) from *Science*, **164**:955 (1969), copyright by the A.A.A.S.]

Ribosomal cistrons (specifically, the DNA coding for 28S, 18S, and 7S rRNA, all produced from a common 45S RNA precursor, probably by a specific transcriptase) are linearly repeated along a chromosome in a region known as the *nucleolar organizer*, often identifiable as a constriction in the chromosome. (There is usually

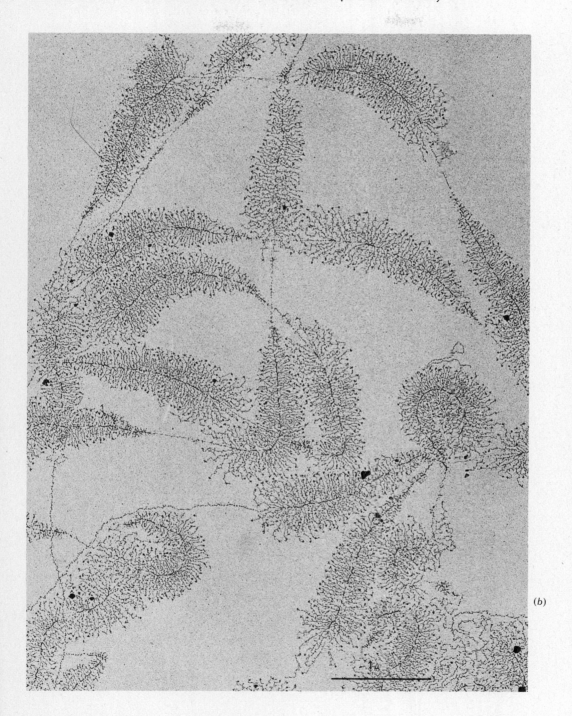

(b)

one nucleolar organizer per haploid set of chromosomes. A diploid cell, therefore, normally contains two nucleoli.) The nucleolar organizer of *Xenopus laevis* (a frog), for example, accounts for about 0.2% of the DNA and includes roughly 450 repeated genes (haploid complement) for ribosomal RNA, arranged in tandem. This redundancy within the chromosome provides an increased template for ribosome synthesis. However, still more ribosomal template becomes available when a nucleolus is formed, since the ribosomal genes in the chromosome are copied over and over again to form extrachromosomal circular units of repeated DNA. This DNA and its associated protein and RNA form the core of the nucleolus.

The most extreme case of gene amplification occurs during egg formation in vertebrates. In *Xenopus laevis*, the oöcyte (precursor of the ovum, or egg) contains literally thousands of nucleoli, containing altogether some 4,000 times as much ribosomal DNA (rDNA, or DNA that specifies ribosomal components) as is normally found in a haploid set of chromosomes. Gene amplification in *Xenopus* occurs during a period when large numbers of ribosomes are made for use after fertilization of the egg. This is also the period when the chromosome, in its lampbrush configuration, supports the synthesis of large numbers of *maternal messages*, which are inactive, stable, or "masked" mRNA's, destined to be translated only after fertilization.

Amplification apparently does not apply to most genes, but it could provide the same advantages for some others—i.e., a greatly increased template for transcription—as it provides for ribosomal genes in the oöcyte. It also helps explain why the DNA content per haploid or diploid cell varies so widely from one species to the next, seemingly without regard for the relative complexity of the organism. Frogs, salamanders, and corn, for instance, all contain far more DNA per cell than man, though we would not expect them to have a larger number of different genes.

Note that gene amplification is different in a very fundamental way from polyteny and polyploidy. Gene amplification, as it is presented here, is the proliferation of selected genes, not of whole chromosomes. The polytene chromosome, on the other hand, has multiple complete chromatids, while the polyploid nucleus has multiple copies of each chromosome. Although an increase in the number of chromosomes or chromatids may serve to support faster growth, it does not provide the selectivity associated with gene amplification.

A TRANSCRIPTIONAL CONTROL THEORY FOR EUCARYOTES. R. J. Britten and D. E. Kohne of the Carnegie Institute in Washington found that although most genes (exceptions being those for histones and those of the nucleolar organizer) have a single unique copy per haploid set of chromosomes, a large fraction of the DNA consists of sequences that are repeated as many as a million or more times per cell, clustered in so-called "g regions." For example, about 40% of calf thymus DNA consists of segments of some 400 nucleotide pairs, repeated from 10^4 to 10^6 times per nucleus. These repeated DNA's were formerly known as *satellite DNA*. Since most of the repeated segments of DNA appear to be too small to be structural genes, it has been suggested that redundant segments may be an integral part of the chromosome, involved in a complex transcriptional control mechanism that provides the basis for tissue-specific patterns of gene activation.

R. J. Britten and E. H. Davidson, in a 1969 theory proposed that most or all structural genes have multiple control genes analogous to operators (or perhaps promoters) associated with them. Each cell type in an organism would produce a small number of gene regulators, either proteins or (as originally suggested) RNA's, specific for that cell type. If a particular structural gene has a control element corresponding to one of these activators it will be "turned-on." Otherwise, histones prevent its transcription.

In a hypothetical case, a structural gene (a gene that specifies a protein) might have adjacent controlling genes recognized by tissue-specific activators A, B, E, and F, but not by C and D. The

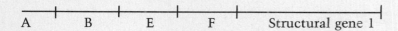

A   B   E   F   Structural gene 1

four control elements in this model permit the gene to be turned on in the four tissue types where the structural gene is needed—i.e., in tissue types where activators A, B, E, or F are present, but not in tissues containing C or D instead. A second structural gene might have three control elements of types A, C, and D, corresponding to tissues where that gene must be activated.

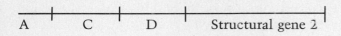

A   C   D   Structural gene 2

In other words, the two structural genes are recognized by an overlapping but nonidentical set of regulator molecules. Gene redundancy would be a natural consequence, since each regulator molecule would recognize numerous identical, or highly similar, control genes (A through F in this case) at scattered points about the chromosomes. The mechanism is thus analogous in some respects to the regulon concept of procaryotic gene control.

The above scheme, which must be considered speculative, is presented to emphasize some of the differences between procaryotic and eucaryotic systems. It does explain some otherwise perplexing facets of gene expression in eucaryotes, such as the phenomenon of mRNA processing. Eucaryotic genes, as noted earlier, seem to be transcribed in batches (a kind of oversized operon), producing RNA's with up to about 20,000 nucleotides; these are then broken down into smaller mRNA's before leaving the nucleus. The process is not a conservative one, however, for most of the RNA is degraded without ever leaving the nucleus. This degraded RNA could represent linkage regions between structural genes or gene groups, and might possibly consist of transcribed control elements. In fact, some of the RNA transcribed from control elements may be left on, for there are reports that some messenger has 5' ends (the end transcribed first) containing perhaps 50 or so nucleotides copied from repeated sequences of DNA.

TRANSLATIONAL CONTROL. Since the synthesis of a protein requires two steps, transcription and translation, it may be controlled at either level. In bacteria, where messenger RNA's have half-lives of minutes, one might suppose that translational control would be very inefficient. That conclusion is based on the assumption that mRNA can continue to be synthesized and degraded without being translated—an unnecessary assumption, as it turns out. There is, in fact, some indication that transcription and translation are (or can be) tightly coupled, so that interfering with translation also interferes with transcription. The opportunities for this kind of coupling are not so prevalent in eucaryotes, where transcription occurs in the nucleus and translation in the cytoplasm. On the other hand, the generally much longer (though highly variable) lifetimes of eucaryotic mRNA make translational control more important: merely turning off a gene does not stop the synthesis of its enzyme, as long as the appropriate mRNA is intact and available for translation.

In any case, there are numerous instances where translational control is the only satisfactory explanation for an experimental observation. For example, the addition of tryptophan to *E. coli* grown in a minimal medium results in an immediate cessation of tryptophan synthetase production. The rate of production of the enzyme, however, falls off much more rapidly than the rate of production of the corresponding mRNA, indicating that the addition of tryptophan affected both transcription and translation.

The translation of masked mRNA after fertilization of an egg is a second example of translational control. A third example is found in the formation of an umbrellalike cap on the single-celled alga, *Acetabularia:* although cap formation involves the synthesis of new enzymes, it can take place on schedule even when the nucleus of the plant has been removed at a much earlier developmental stage. The only satisfactory explanation, again, is activation of a stable mRNA.

The advantage of translational control is the speed with which protein synthesis can be turned on and off. Its disadvantage lies in the waste involved in making mRNA that is not used before it is degraded. The relative weight given to these two criteria obviously depends on how important immediate control of enzyme synthesis is in a given situation, keeping in mind that the regulation of already functioning enzymes, as discussed in Chapter 4, is very rapid.

The longer that an mRNA can survive intact, the more important it becomes to be able to control its rate of translation; and the more rapidly mRNA turns over, the more wasteful it becomes to avoid translating it without a feedback signal to also control mRNA synthesis. The lifetimes of mRNA in both eucaryotes and procaryotes are highly variable, but usually in a logical way. That is, enzymes whose quantity must be increased or decreased in response to external signals (e.g., diet) have shorter mRNA lifetimes than do enzymes whose synthesis is constitutive. The range of half-lives of most mRNA's in eucaryotes is from perhaps an hour to several days, although there are some notable exceptions. The extreme case of mRNA stability is found in an unfertilized egg, for masked mRNA may remain untranslated for very long periods, even years, until fertilization occurs. The analogous situation in plants is the formation of seeds and spores, both of which also contain stable messenger RNA.

One way of controlling the amount of translation of a given mRNA, then, is to regulate its lifetime. Unfortunately, it is not at all clear what makes some mRNA's more stable than others. One possibility is that lifetime is determined by secondary structure. It has been shown in several instances that complementary sequences within the same mRNA molecule result in a specific folding, consisting of loops and hairpins. While these local regions of hydrogen bonding must come apart as the ribosomes pass over them, they may make an mRNA more or less attractive to the ribonuclease that will eventually degrade it. An alternative scheme suggests that some mRNA's are translated a predetermined number of times. The number would be "ticketed" by the removal of one or more bases from the 5′ end each time translation is initiated until a region with high affinity for ribonuclease is exposed.[8] There are also reports that the poly A sequence at the 3′ end of eucaryotic mRNA shortens as the mRNA ages, though the significance of that observation is unknown.

The structure of the ribosomal binding site may also play a role in deciding how frequently a given mRNA will be translated within its lifetime. All we know so far, however, is that not all

---

8. The degradation of mRNA, in bacterial systems at least, always seems to start from the 5′ end, which is the end that was synthesized first. The enzyme responsible for degradation is called an exonuclease, because it shortens the chain one nucleotide at a time, starting from a free end.

ribosomal binding sites have the same sequence and that not all cistrons (even within the same mRNA, at least in procaryotes) are translated at the same rate.

The degeneracy of the code may provide still another method of ensuring that some mRNA's are translated more or less frequently than others. The various tRNA's for a given codon are usually present in quite different quantities, so it is conceivable that the use of a codon specifying a less abundant tRNA for a particular amino acid in preference to a more common tRNA could delay translation of the message. (Charged tRNA's have also been implicated as repressors in the transcriptional control of genes involved in amino acid synthesis.)

The intracellular level of a protein will be affected by the above mechanisms and by the rate at which the protein is either destroyed (turned over) or diluted by growth and cell division. Unfortunately, we know almost nothing about the mechanisms that regulate protein turnover, except that it does seem to be variable. For example, R. T. Schimke demonstrated in 1966 that when rats are starved, the amount of the enzyme arginase in their liver doubles in about six days, a change that is entirely due to a decreased rate of degradation. In contrast, most liver proteins are degraded more rapidly than normal under these same conditions. The increased stability of arginase and decreased stability of other enzymes in the starving rat may be related to the availability of substrates, for it is known that the *in vitro* digestion of enzymes by proteases is inhibited when the substrate of the enzyme is also present.

When two proteins differ in their rate of degradation, their level will also respond differently to gene regulators. Thus, Schimke reported that cortisone (a steroid hormone) stimulates the production by the liver of the two enzymes, tryptophan pyrrolase and arginase, to the same extent; yet the administration of cortisone has a much more profound effect on the level of tryptophan pyrrolase. The reason for this difference is that tryptophan pyrrolase is degraded more rapidly than the other enzyme, providing it a lower basal level and greater sensitivity to changes in its rate of synthesis.

The other factor affecting the level of enzyme in a cell—its rate of dilution during growth—will be considered in the next chapter, which deals with cell division and its regulation.

SUMMARY

8–1 The earliest systematic observations on the nature of genes were made by Gregor Mendel, using a eucaryotic organism (the pea). His description of gene behavior during gamete formation and mating paralleled descriptions of chromosome movement under the same conditions. When this fact was realized, it was proposed (by Sutton) that genes reside in or on chromosomes, a hypothesis that was amply supported by further observations made especially on sex chromosomes and sex-linked genetic traits.

Another of Mendel's concepts, the idea of dominance, has been shown to result usually from the fact that genes produce proteins. A recessive gene may produce an inactive or less active protein, or none at all. This condition is sometimes masked in a heterozygote by the presence in the cell of a gene producing the normal protein in sufficient quantity.

8-2 Genes consist of sequences of deoxyribonucleotides (DNA) except in the case of the RNA viruses. The genetic function of DNA was demonstrated first in bacterial transformation experiments and later in experiments showing that only viral nucleic acid, not viral protein, is needed to produce a normal infection.

The major task of genes is to make proteins. The nucleotide sequence of a strand of DNA is first transcribed to produce a complementary sequence of RNA nucleotides, preserving the informational content of DNA in the nucleotide sequence of mRNA. That information consists of an uninterrupted sequence of nucleotides, functionally grouped into code words or codons, each of which consists of three bases (nucleotides). Sixty-one of the 64 possible codons have been assigned to amino acids. The remaining 3 are terminator (or nonsense) codons, used to stop the incorporation of amino acids and to release the polypeptide chain.

8-3 The flow of information from gene to protein may be represented as

$$\text{DNA} \xrightarrow{\text{transcription}} \text{RNA} \xrightarrow{\text{translation}} \text{protein}$$

The second of these events, translation, includes the following steps:

1. Initiation, which includes assembly of a ribosome, if it is not already present, at a specific binding site on the mRNA, and insertion of methionyl-tRNA (formylated methionine in procaryotes) into a specific site, called the peptidyl (P) site of the ribosome, where it is complementary to an initiator codon of the mRNA.
2. Elongation of the peptide chain, which includes the hydrogen bonding of a charged tRNA to its mRNA codon at the acceptor (aminoacyl, or just A) site of the ribosome, followed by transfer of this amino acid to the C-terminal of the polypeptide at the P site (to methionine during initiation). The elongated chain, now bound by the new amino acid to the A site, is shifted (translocated) to the P site preparatory to another elongation step.
3. Termination of the polypeptide, which occurs when a terminator codon is encountered along with an appropriate releasing factor. The peptide chain is then released, and minus its initiating methionine,

folds into its final form. If destined for secretion or inclusion in another type of vesicle, it will be packaged by the endoplasmic reticulum.

8-4 The regulation of protein synthesis is accomplished through a variety of mechanisms, the best understood of which is the operon. The operon is a set of adjacent genes whose coordinate expression is under the control of a repressor (or regulator) protein, itself the product of a regulator gene. When the regulator protein attaches to the operator site of the operon, transcription may be either inhibited (negative control) or made possible (positive control). The ability of a regulator protein to bind to the operator may be influenced by small molecules acting either as corepressors or activators.

Some activated operons and some constitutive genes may be transcribed more often than others because their promoters have a higher affinity for RNA polymerase (transcriptase). The affinity of transcriptase for a promoter can be affected by sigma factors and by cAMP binding protein. The latter functions in catabolite repression of *E. coli*, where, in the presence of cyclic AMP, a cAMP receptor protein increases the normally low affinity of RNA polymerase for a variety of sugar operons. Since cAMP is at a low level during glucose catabolism, glucose will be utilized in preference to these other sugars.

8-5 There is a considerable difference in structure between procaryotic and eucaryotic chromosomes. DNA in eucaryotes is associated with roughly an equal mass of protein, mostly basic (arginine- and lysine-rich) histones. According to various models, the complex is either in a coiled-coil configuration or found in discrete particles separated by naked DNA, like beads on a string.

The conformation of chromatin correlates with its histone content. There is reason to believe that the more compact configurations do not participate in RNA transcription. Hence, it has been suggested that the presence of histones regulates genes through a folding/unfolding transformation. Activation might involve either the displacement of histones by small regulator molecules or the chemical modification of histones, for instance by cyclic AMP-activated protein kinases.

Hormones may affect transcription through changes in the cytoplasmic level of cyclic AMP or cyclic GMP. Those that get into the cytoplasm (e.g., the steroid hormones) may instead bind to a receptor protein and with it enter the nucleus to alter transcriptional rates.

Transcriptional rates may also be modified by gene amplification, in which chromosomal genes are reproduced in multiple extrachromosomal copies, thus increasing the total template available for transcription. This process is known to occur during the formation of nucleoli.

The rate of protein synthesis can be regulated at translation. There is good evidence for the existence of translational control in both procaryotes and eucaryotes, though the presence of long-lived, stable mRNA makes the mechanism more important in eucaryotes.

STUDY GUIDE

8-1 (a) What are the four principles of Mendelian genetics, as summarized here? (b) If one were to follow n human genes through several generations, what is the largest value of n that could be used without necessarily breaking the law of independent assortment? (c) Define: haploid, diploid, homologous chromosomes, hemizygous, somatic cell, sex-linked, and cistron. (d) What parallels exist between chromosome and gene behavior that would lead one to suspect a chromosomal location for genes? (e) What was the experimental foundation for the one gene = one enzyme hypothesis?

8-2 (a) Describe the first transformation experiments as they were carried out both *in vivo* and *in vitro*. (b) How do we know that DNA carries genetic information: (1) in bacteria? (2) in viruses of all types? and (3) in eucaryotic organisms? (c) What is messenger RNA, and what is its function? (d) Describe an experimental procedure that reveals whether one strand or both strands of a given gene are transcribed.

8-3 (a) What are the three classes of RNA, and what role does each play in translation? (b) How do we know that recognition of the proper amino acid for a tRNA is the function of a specific nonribosomal enzyme? (c) What is the basis for the "wobble hypothesis"? (d) At which two steps in translation is GTP involved? (e) What are the functions of the A and P ribosomal sites? (f) Summarize the events of initiation and termination. (g) What is the role of the endoplasmic reticulum and Golgi apparatus in protein synthesis?

8-4 (a) What distinguishes inducible, repressible, and constitutive genes from one another? (b) Describe the structure of the *lac* operon, and how it is controlled. (c) What is the difference between negative and positive control of an operon? What experiments can be performed to distinguish one from the other? (d) What is a σ factor and how does it function? (e) What is meant by catabolite repression? How does *E. coli* respond to catabolite repression by glucose?

8-5 (a) What are histones and what is their probable function? In what ways are they subject to chemical modification *in vivo*? (b) Describe the basic chromatin fiber. What is the difference between euchromatin and heterochromatin? (c) Describe the configuration of the polytene and lampbrush chromosomes. Where are they found? (d) What evidence exists for a correspondence between transcription and chromatin coiling? (e) How do hormones affect genes? (f) What is meant by gene amplification? What is its purpose? (g) What evidence exists for translational control?

REFERENCES

GENERAL

BROWN, DONALD D., "The Isolation of Genes." *Scientific American,* August 1973. (Offprint 1278.)

DuPRAW, E. J., *DNA and Chromosomes.* New York: Holt, Rinehart and Winston, 1970. (Paperback.)

STURTEVANT, A. H., *A History of Genetics.* New York: Harper & Row, 1965.

WATSON, J. D., *Molecular Biology of the Gene* (2d ed.). Menlo Park, Calif.: W. A. Benjamin Co., 1970.

REPRINT COLLECTIONS. Featuring original papers or excerpts thereof. Collections will be referred to later by their editor's initials.

ADELBERG, E. A., *Papers on Bacterial Genetics* (2d ed.). Boston: Little, Brown & Co., 1966. The introduction contains a clear discussion of mutation, transformation, and other processes.

CARLSON, ELOF A., *Gene Theory.* Belmont, Calif.: Dickenson Publ. Co., 1967. Excerpts from important papers, with annotation.

CARPENTER, BRUCE H., *Molecular and Cell Biology.* Belmont, Calif.: Dickenson Publ. Co., 1967. A collection of readings, consisting of excerpts from important papers.

GABRIEL, M. L., and S. FOGEL, eds., *Great Experiments in Biology.* Englewood Cliffs, N. J.: Prentice-Hall, 1955. Includes writings of Mendel, Sutton, Wilson, and Morgan.

HAHON, NICHOLAS, ed., *Selected Papers on Virology* Englewood Cliffs, N. J.: Prentice-Hall, 1964.

LEVINE, LOUIS, *Papers on Genetics: A Book of Readings.* St. Louis, Mo.: The C. V. Mosby Co., 1971.

LOOMIS, W. L., ed., *Papers on Regulation of Gene Activity During Development.* New York: Harper & Row, 1968.

PETERS, JAMES A., *Classic Papers in Genetics.* Englewood Cliffs, N. J.: Prentice-Hall, 1959. Reprints many of the earlier papers, with annotation. Included are works by Mendel, Sutton, Morgan, Beadle, and others.

STENT, GUNTHER S., *Papers on Bacterial Viruses* (2d ed.) Boston: Little, Brown & Co., 1965. The introduction presents a clear overview of the subject matter.

STERN, CURT, and E. R. SHERWOOD, eds., *The Origin of Genetics: A Mendel Source Book.* San Francisco: W. H. Freeman, 1966. Writings of Gregor Mendel, in translation and with annotation.

TAYLOR, J. HERBERT, *Selected Papers on Molecular Genetics.* New York: Academic Press, 1965.

VOELLER, BRUCE R., *The Chromosome Theory of Inheritance: Classic Papers in Development and Heredity.* New York: Appleton Century Crofts, 1968. Excerpts from historically important works.

ZUBAY, GEOFFREY L., *Papers in Biochemical Genetics.* New York: Holt, Rinehart & Winston, 1968. A second edition in 1973, edited by Zubay and J. Marmur, reprints mostly more recent papers.

THE GENETIC CONCEPT

BEADLE, GEORGE W., "The Genes of Men and Molds." *Scientific American*, September 1948. (Offprint 1.)

———, "Genes and Chemical Reactions in Neurospora." In *Nobel Lectures, Physiology or Medicine, 1942–1962.* Amsterdam: Elsevier Publ. Co., 1964, p. 587. Nobel Lecture, 1958.

CARLSON, ELOF AXEL, *The Gene: A Critical History.* Philadelphia: W. B. Saunders, 1966.

COLEMAN, WILLIAM, "Cell, Nucleus and Inheritance: An Historical Study." *Proc. American Philosophical Society*, **109**:125 (1965).

JUKES, T. H., "Possibilities for the Evolution of the Genetic Code from a Preceding Form." *Nature*, **246**:22 (1973).

MURAYAMA, M., "Sickle Cell Hemoglobin: Molecular Basis of Sickling Phenomenon. Theory and Therapy." *CRC Crit. Revs. Biochem.*, **1**:461 (1973).

TATUM, E. L., "A Case History in Biological Research." In *Nobel Lectures, Physiology or Medicine, 1942–1962.* Amsterdam: Elsevier Publ. Co., 1964, p. 602. Nobel Lecture, 1958, on the gene-enzyme relationship.

WATSON, J. D., and F. H. C. CRICK, "Genetical Implications of the Structure of Deoxyribonucleic Acid." *Nature*, **171**:964 (1953). Reprinted in JHT, EA, EC, and EA.

THE GENETIC MESSAGE

AARONSON, S. A., and G. J. TODARO, "Human Diploid Cell Transformation by DNA Extracted from the Tumor Virus SV40." *Science*, **166**:390 (1971). The first *in vitro* transformation of a human cell by purified DNA.

AVERY, O. T., C. M. MACLEOD, and M. MCCARTY, "Studies on the Chemical Nature of the Substance Inducing Transformation of Pneumococcal Types: Induction of Transformation by a Deoxyribonucleic Acid Fraction Isolated from Pneumococcus Type III.' *J. Exp. Med.*, **79**:137 (1944). Reprinted in JHT, EA, JP, and G & F.

BERNFIELD, MERTON R., and MARSHALL W. NIRENBERG, "RNA Codewords and Protein Synthesis." *Science*, **147**:479 (1965). Describes the trinucleotide binding technique developed in Nirenberg's lab.

CRICK, F. H. C., ' On the Genetic Code." In *Nobel Lectures, Physiology or Medicine, 1942–1962.* Amsterdam: Elsevier Publishing Co., 1964, p. 811. Nobel Lecture, 1962.

———, "The Genetic Code." *Scientific American*, October 1962. (Offprint 123.)

———, "The Genetic Code: III." *Scientific American*, October 1966. (Offprint 1052.)

DINA, D., I. MEZA, and M. CRIPPA, "Relative Positions of the 'Repetitive', 'Unique' and poly (A) Fragments of mRNA." *Nature*, **248**:486 (1974). In *Xenopus* embryo mRNA, poly (A) is on the 3' end, a transcribed segment of repeated DNA on the 5' end.

FRAENKEL-CONRAT, H., "Structure and Infectivity of Tobacco Mosaic Virus." In *The Harvey Lectures, 1957–1958* (Ser. 53) New York: Academic Press, 1959, p. 56.

GIERER, A., and G. S. SCHRAMM, "Infectivity of Ribonucleic Acid from Tobacco Mosaic Virus." *Nature*, **177**:702 (1956). Reprinted in GZ and NH.

HERSHEY, A. D., and M. CHASE, "Independent Functions of Viral Protein and Nucleic Acid in Growth of Bacteriophage." *J. General Physiology*, **36**:39 (1952). Reprinted in JHT, GS, and NH.

HURWITZ, J., and J. J FURTH, "Messenger RNA." *Scientific American*, February 1962. (Offprint 119.)

INGRAM, N. M., "How Do Genes Act?" *Scientific American*, January 1958. (Offprint 104.) Sickle-cell hemoglobin.

KHORANA, H. G., "Polynucleotide Synthesis and the Genetic Code.' In *The Harvey Lectures, 1966–1967* (Ser. 62). New York: Academic Press, 1968, p. 79.

———, "Synthetic Nucleic Acids and the Genetic Code." *J. Am. Med. Assoc.*, **206**:1978 (1968). Lasker Award, 1968.

LEDERBERG, JOSHUA, "A View of Genetics." In *Nobel Lectures, Physiology or Medicine, 1942–1962.* Amsterdam: Elsevier Publ. Co., 1964, p. 615. Nobel Lecture, 1958.

MERRIL, CARL R., M. R. GEIER, and J. C. PETRICCIANI, "Bacterial Virus Gene Expression in Human Cells," *Nature*, **233**:398 (1971).

NIRENBERG, M. W., "The Genetic Code: II." *Scientific American*, March 1963. (Offprint 153.)

OCHOA, SEVERO, "Enzymatic Synthesis of Ribonucleic Acid." In *Nobel Lectures, Physiology or Medicine, 1942–1962.* Amsterdam: Elsevier Publ. Co., 1964, p. 645. Transcriptase. Nobel Lecture, 1959.

SPIEGLEMAN, S., "Hybrid Nucleic Acids." *Scientific American*, May 1964. (Offprint 183.)

STEWART, P. R., and D. S. LETHAM, eds., *The Ribonucleic Acids.* New York: Springer-Verlag, 1973. Reviews on the synthesis and function of RNA.

YANOFSKY, CHARLES, "Gene Structure and Protein Structure." In *The Harvey Lectures, 1965–1966* (Ser. 61). New York: Academic Press, 1967, p. 145. Also see *Scientific American*, May 1967. (Offprint 1074.)

RIBOSOMES AND PROTEIN SYNTHESIS

COX, ROBERT, "The Ribosome—Decoder of Genetic Information." *Science J.*, **6**:56 (1970).

HOLLEY, R. W., "The Nucleotide Sequence of a Nucleic Acid." *Scientific American*, February 1966. (Offprint 1033.)

INGRAM, VERNON M., "On the Biosynthesis of Hemoglobin." In *The Harvey Lectures, 1965–1966* (Ser. 61). New York: Academic Press, 1967, p. 43.

KAJI, A., "Mechanism of Protein Synthesis and the Use of Inhibitors in the Study of Protein Synthesis." *Prog. Mol. and Subcellular Biol.*, **3**:85 (1973).

NANNINGA, N., "Structural Aspects of Ribosomes." *Int. Rev. Cytology*, **35**:135 (1973).

NOMURA, M., "Ribosomes." *Scientific American*, October 1969. (Offprint 1157.)

PACE, N. R., "The Structure and Synthesis of the Ribosomal Ribonucleic Acid of Prokaryotes." *Bact. Revs.*, **37**:562 (1973).

PALADE, GEORGE E., "A Small Particulate Component of the Cytoplasm." *J. Biophys. & Biochem. Cytol.* (now *J. Cell Biol.*), **1**:59 (1955). The ribosome. Reprinted in BC.

RICH, ALEXANDER, "Polyribosomes." *Scientific American*, December 1963. (Offprint 171.)

TATA, J. R., "Ribosomal Segregation as a Possible Function for the Attachment of Ribosomes to Membranes." *Subcellular Biochem.*, **1**:83 (1971).

WAINWRIGHT, S. D., *Control Mechanisms and Protein Synthesis*. New York: Columbia Univ. Press, 1972. A thorough inventory.

WATSON, J. D., "The Involvement of RNA in the Synthesis of Proteins." In *Nobel Lectures, Physiology or Medicine, 1942–1962*. Amsterdam: Elsevier Publ. Co., 1964, p. 785. Nobel Lecture, 1962.

GENE REGULATION IN PROCARYOTES

CLARKE, PATRICIA H., "Positive and Negative Control of Bacterial Gene Expression." *Science Progress* (Oxford), **60**:245 (1972).

GOLDBERGER, R. F., "Autogenous Regulation of Gene Expression." *Science*, **183**:810 (1974). Where the regulator protein is the product of one of the structural genes of the operon it controls.

GROS, F., "Control of Gene Expression in Prokaryotic Systems." *FEBS Letters*, **40** (Suppl.): S19 (1974).

JACOB, FRANCOIS, "Genetics of the Bacterial Cell." *Science*, **152**:1470 (1966). Nobel Lecture, 1965.

JACOB, F., and J. MONOD, "Genetic Regulatory Mechanisms in the Synthesis of Proteins." *J. Mol. Biol.*, **3**:318 (1961). Reprinted in JHT and GZ. The operon concept.

MARTIN, DUNCAN T. M., "The Operon Model for the Regulation of Enzyme Synthesis." *Science Progress* (Oxford), **57**:87 (1969).

MÜLLER-HILL, B., "The 'Lac' Repressor." *Science J.*, **5A**,48 (1969).

NIERLICH, D. P., "Regulation of Bacterial Growth." *Science*, **184**:1043 (1974). An overview of enzymatic and genetic regulation.

PASTAN, IRA, "Cyclic AMP." *Scientific American*, August 1972. (Offprint 1256.)

PTASHNE, M., and W. GILBERT, "Genetic Repressors." *Scientific American*, June 1970. (Offprint 1179.)

CHROMOSOME STRUCTURE

COMINGS, D. E., and T. A. OKADA, "Electron Microscopy of Chromosomes," In *Perspectives in Cytogenetics*, edited by E. Wright, B. Crandall, and L. Boyer. Springfield, Ill.: Charles C Thomas, 1972. A review of chromosome structure, with excellent micrographs.

KORNBERG, R. D., and J. O. THOMAS, "Chromatin Structure: Oligomers of the Histones." *Science*, **184**:865 (1974). Specific interaction among histones. It is followed by an article giving a possible model for chromatin structure based on these observations.

SIMPSON, R. T., "Structure and Function of Chromatin." *Adv. in Enzymology*, **38**:41 (1973).

STUBBLEFIELD, E., "The Structure of Mammalian Chromosomes." *Int. Rev. Cytology*, **35**:1 (1973).

THOMAS, C. A., JR., "The Genetic Organization of Chromosomes." *Annual Rev. of Genetics*, **5**:237 (1971).

WHITE, M. J. D., *The Chromosomes*. (6th ed.) London: Chapman and Hall, 1973. (Paperback.)

WISCHNITZER, S., "The Submicroscopic Morphology of the Interphase Nucleus." *Int. Rev. Cytology*, **34**:1 (1973). The nuclear envelope, nucleolus, and nuclear-cytoplasmic exchange.

GENE REGULATION AND PROTEIN TURNOVER IN EUCARYOTES

BIRNSTIEL, MAX, and MARGARET CHIPCHASE, "The Nucleolus: Pacemaker of the Cell." *Science Journal*, **6**:41 (1970).

BONNER, JAMES, M. E. DAHMUS, D. FAMBROUGH, R. C. HUANG, K. MARUSHIGE, and D. Y. TUAN, "The Biology of Isolated Chromatin." *Science*, **159**:47 (1968). Histones as gene regulators.

BRITTEN, R. J., and E. H. DAVIDSON, "Gene Regulation for Higher Cells: A Theory." *Science*, **165**:349 (1969).

BRITTEN, R. J., and D. E. KOHNE, "Repeated Segments of DNA." *Scientific American*, April 1970. (Offprint 1173.)

BROWN, D. D., and I. B. DAWID, "Specific Gene Amplification in Oöcytes." *Science*, **160**:272 (1968). Reprinted in WL.

DAVIDSON, E. H., and R. J. BRITTEN, "Organization, Transcription, and Regulation in the Animal Genome." *Quart. Rev. Biol.*, **48**:565 (1973).

FLAMM, W. G., "Highly Repetitive Sequences of DNA in Chromosomes." *Int. Rev. of Cytology*, **32**:2 (1972).

GALL, JOSEPH G., "Differential Synthesis of the Genes for Ribosomal RNA During Amphibian Oögenesis." *Proc. Nat. Acad. Sci.* (U. S.), **60**:553 (1968). Reprinted in WL.

GOLDBERG, A., E. HOWELL, J. LI., S. MARTEL, and W. PROUTY, "Physiological Significance of Protein Degradation in Animal and Bacterial Cells." *Federation Proceedings*, **33**:1112 (1974).

HALL, R. H., and J. J. MONAHAN, "Chromatin and Gene Regulation in Eukaryotic Cells at the Transcriptional Level." *CRC Crit. Revs. Biochem.*, **1**:67 (1974).

KENNEY, F., B. HAMKALO, G. FAVELUKES, and J. AUGUST, eds., *Gene Expression and its Regulation*. New York: Plenum Publ. Co., 1973. Current developments in the field.

KOLATA, G. B., "Repeated DNA: Molecular Genetics of Higher Organisms." *Science*, **182**:1009 (1973). A news summary of recent work.

LYON, MARY F., "The Activity of the Sex Chromosomes in Mammals." *Science Progress* (Oxford), **58**:117 (1970).

MACGREGOR, H. C., "The Nucleolus and Its Genes in Amphibian Oögenesis." *Biol. Revs.*, **47**:177 (1972).

MILLER, O. L., JR., "The Visualization of Genes in Action." *Scientific American*, March 1973. (Offprint 1267.)

MITTWOCH, URSULA, "Sex Differences in Cells." *Scientific American*, July 1963. (Offprint 161.) The Barr body.

RECHCIGL, M., JR., "Intracellular Protein Turnover and the Roles of Synthesis and Degradation in Regulation of Enzyme Levels." In *Enzyme Synthesis and Degradation in Mammalian Systems*, edited by M. Rechcigl. Baltimore, Md.: Univ. Park Press, 1971, p. 237.

SCHIMKE, R. T., "On the Roles of Synthesis and Degradation in Regulation of Enzyme Levels in Mammalian Tissues." *Current Topics in Cellular Regulation*, **1**:77 (1969).

———, "Control of Enzyme Levels in Mammalian Tissues." *Adv. Enzymology*, **37**:135 (1973).

SMITH, EMIL L., R. J. DeLANGE, and J. BONNER, "Chemistry and Biology of the Histones." *Physiological Revs.*, **50**:159 (1970).

STEIN, G., T. SPELSBERG, and L. KLEINSMITH, "Nonhistone Chromosomal Proteins and Gene Regulation." *Science*, **183**:817 (1974).

TOMPKINS, GORDON M., T. D. GELEHRTER, D. GRANNER, D. MARTIN, JR., H. H. SAMUELS, and E. B. THOMPSON, "Control of Specific Gene Expression in Higher Organisms." *Science*, **166**:1474 (1969).

WALKER, P., "Repetitive DNA in Higher Organisms." *Prog. Biophys. and Mol. Biol.*, **23**:145 (1971).

YUNIS, JORGE J., and W. G. YASMINEH, "Heterochromatin, Satellite DNA, and Cell Function." *Science*, **174**:1200 (1971).

HORMONES AND GENES

BITENSKY, M. W. and R. E. GORMAN, "Cellular Responses to Cyclic AMP." *Prog. Biophys. and Mol. Biol.*, **26**:409 (1973).

GOLDBERG, NELSON, "Cyclic Nucleotides and Cell Function." *Hospital Practice*, **9**(5):127 (May 1974). Cyclic AMP and GMP.

JENSEN, E., and E DeSOMBRE, "Estrogen–Receptor Interaction." *Science*, **182**:126 (1973).

KOLATA, G. B., "Cyclic GMP: Cellular Regulatory Agent?" *Science*, **182**:149 (1973). A news-style summary of recent work.

O'MALLEY, B. and A. MEANS, "Female Steroid Hormones and Target Cell Nuclei." *Science*, **183**:610 (1974).

RASMUSSEN, H., D GOODMAN, and A. TENENHOUS, "The Role of Cyclic AMP and Calcium in Cell Activation." *CRC Critical Revs. Biochem.*, **1**:95 (1972).

CHAPTER 9

Cell Division and Its Regulation

9–1 CELL DIVISION IN PROCARYOTES   384
The Replication of DNA
Replicases
Replication Models
Errors in Replication
Cytokinesis in Procaryotes

9–2 REPLICATION OF THE EUCARYOTIC NUCLEUS   395
The Replication of Chromatin
Mitosis
Meiosis

9–3 DIVISION OF THE EUCARYOTIC CELL   413
Synchrony Between Nuclear and Cytoplasmic Division
Cytokinesis

9–4 THE REGULATION OF CELL DIVISION   416
Regulation of Cell Division in Procaryotes
Initiation of the Eucaryotic Cell Cycle
Contact Inhibition
Chalones
Hormonal Control of Cell Growth and Division

SUMMARY   424
STUDY GUIDE   425
REFERENCES   425

Most cells reproduce by binary fission, so that one parent cell gives rise to two nearly identical daughters. There are some notable exceptions to this mode of replication, however. Yeasts, for example, can reproduce by budding, a process in which a small daughter cell is pinched off from the larger parent (see Fig. 9–1). In addition, many plants and some bacteria can produce and scatter spores, quite unlike the cells from which they were derived, that can germinate to perpetuate the species. In this chapter, we will be less concerned with these exceptions than with the more general mechanism of binary fission and its accompanying events, particularly the distribution of genetic material to daughter cells.

We will consider the mechanism and control of cell division in

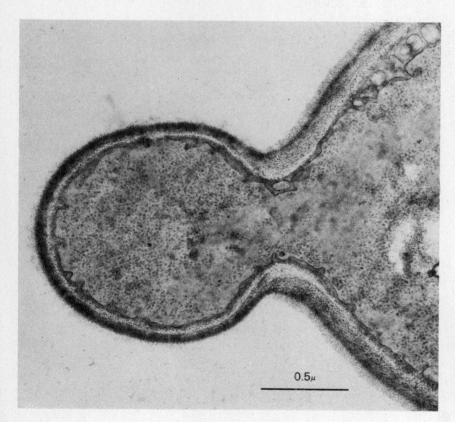

0.5µ

FIGURE 9–1. Reproduction by Budding. Micrograph of yeast (*Candida albicans*), showing formation of a daughter cell. Note the thinner cell wall in the bud. The granules are presumably ribosomes. (Yeasts are eucaryotic organisms.) [Courtesy of W. Diaczenko and A. Cassone, *J. Cell Biol.*, **52**:186 (1972).]

procaryotes and eucaryotes, beginning with procaryotes since, as usual, they operate in a simpler fashion.

9–1 CELL DIVISION IN PROCARYOTES

Cell division in procaryotes is a relatively simple process, reflecting the relative simplicity of the cells themselves. First the DNA is replicated, then the cytoplasm is divided (a process called cytokinesis) in a way that ensures an equal distribution of the replicated DNA to daughter cells.

THE REPLICATION OF DNA. Watson and Crick pointed out that genetic information could be conserved during cellular replication if each strand of the parent molecule were to serve as a template for the polymerization of a complementary strand in a daughter molecule. Thus, the sequence of paired bases would be reproduced.

If the two parent strands were to separate completely, so that each daughter molecule consisted of one new and one old strand, the mode of replication would be called *semiconservative*. On the other hand, one could imagine that a polymerization could occur with only a temporary and localized separation of the parent strands, leaving the parent molecule intact and producing a daughter molecule that is entirely new. This latter mechanism would be called *conservative* replication. An experiment capable of distinguishing between the two modes of replication was designed by M. Meselson and F. W. Stahl at the California Institute of Technology in 1958.

Meselson and Stahl utilized the fact that any solute not isodense with its solvent experiences a net force in a gravitational or centrifugal field. In a really strong centrifugal field, there may be a noticeable tendency for even small solutes to float or sink, depending on whether they are less or more dense than the surrounding solvent. Meselson and Stahl designed an experiment in which newly synthesized DNA would have a buoyant density different from its parent (or template), and thus be readily identified (see Fig. 9–2).

Cultures of *E. coli* were grown in a medium containing heavy nitrogen (^{15}N), washed, and then allowed to continue growing in a normal medium (^{14}N). At the time of transfer, isolated DNA was added to a dense (1.7 g/ml) solution of CsCl. The tube was spun in a centrifuge at speeds great enough to cause the CsCl to partially sediment, forming a density gradient ranging from about 1.75 g/ml at the bottom to 1.65 g/ml at the top. In this gradient, added DNA "bands" at the level at which it is isodense with the salt solution. After one generation of growth in the normal medium, isolated DNA formed a band that fell halfway between the isodense point for fully labelled ^{15}N-DNA and the position expected for normal,

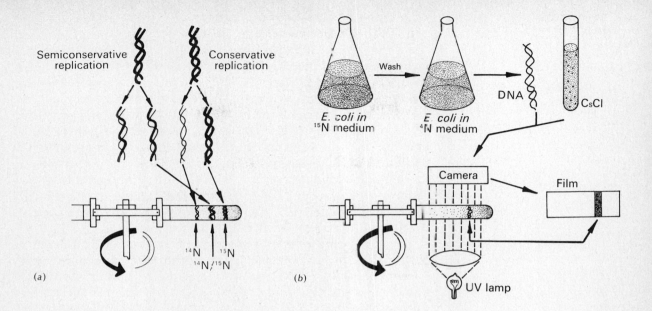

Semiconservative replication

Conservative replication

^{14}N ^{15}N
$^{14}N/^{15}N$

(a)

Wash

E. coli in ^{15}N medium

E. coli in ^{4}N medium

DNA

CsCl

Camera

Film

UV lamp

(b)

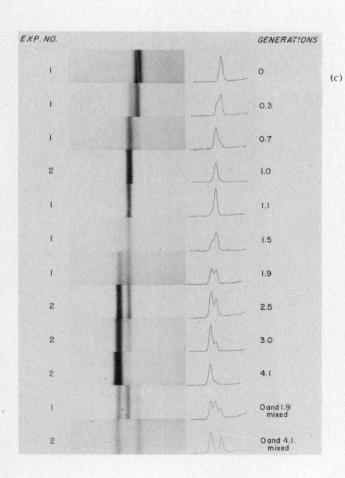

(c)

EXP. NO.	GENERATIONS
1	0
1	0.3
1	0.7
2	1.0
1	1.1
1	1.5
1	1.9
2	2.5
2	3.0
2	4.1
1	0 and 1.9 mixed
2	0 and 4.1 mixed

FIGURE 9–2. The Semiconservative Replication of DNA. **(a)** DNA labelled with heavy nitrogen (^{15}N, dark strands) is allowed to replicate once in the presence of the common isotope of nitrogen, ^{14}N. The diagram shows the results expected from conservative and semiconservative replication. **(b)** The experimental protocol used by Meselson and Stahl. **(c)** The results of different trials, showing the banding pattern revealed by the camera after various periods of growth on ^{14}N, plus a tracing of the films. (Dense CsCl is to the right.) [(c) courtesy of M. Meselson and F. W. Stahl, *Proc. Nat. Acad. Sci.* (U.S.), **44**, 671 (1958).]

light DNA. After one generation of growth, then, each DNA molecule was half original and half new. Heating the DNA to separate the strands and then recentrifuging them made it clear that hybrid molecules contained one parental (fully ^{15}N) and one newly synthesized (completely ^{14}N) polynucleotide strand. In other words, the DNA was replicated semiconservatively.

The semiconservative replication of DNA raises some perplexing problems (see Fig. 9–3). For example, since the length of the *E. coli* chromosome is about a millimeter, the 34 Å per turn pitch of the DNA helix means that about a third of a million twists are required to separate the two strands during replication. Since replication can occur in about 40 minutes, a twist rate of roughly 7,500 revolutions per minute is implied—truly a staggering number.

To get a better picture of the replicative process, an Australian scientist, John Cairns, used *autoradiography* to capture some *E. coli* chromosomes in the process of replication. First, he fed growing cells tritiated (^{3}H) thymidine. Tritium is a radioactive form of hydrogen that emits weak beta rays, or electrons; and thymidine, of course, is a DNA nucleoside. Next, the cells were placed in a cellulose casing and disrupted with detergent. When

FIGURE 9–3. The Replication Problem. This bacterium, an *Hemophilus influenzae* spheroplast, is sitting in the middle of DNA extruded from it—some 832 μ in a single molecule. Replicating a molecule so enormous, and then segregating the daughters into separate cells at division, would seem to present a formidible mechanical problem. [Courtesy of L. A. MacHattie. For details, see *J. Mol. Biol.*, **11**:648 (1965).]

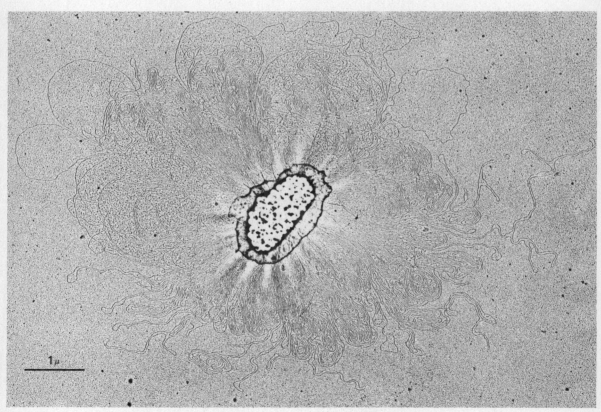

1μ

the solution was removed from the casing, some DNA molecules were left behind, adsorbed to the surface of the container. The casing was then transferred to a microscope slide and coated with a fine-grain photographic emulsion. As the tritium atoms decayed, the released electrons caused the conversion of silver halide in the emulsion to silver grains in much the same way as photons of light would do. After some weeks, the emulsion was developed to reveal a trail of silver grains corresponding to the positions of the decaying molecules (see Fig. 9–4a). Thus, the shape of the labelled DNA was visualized, although with rather poor resolution—the

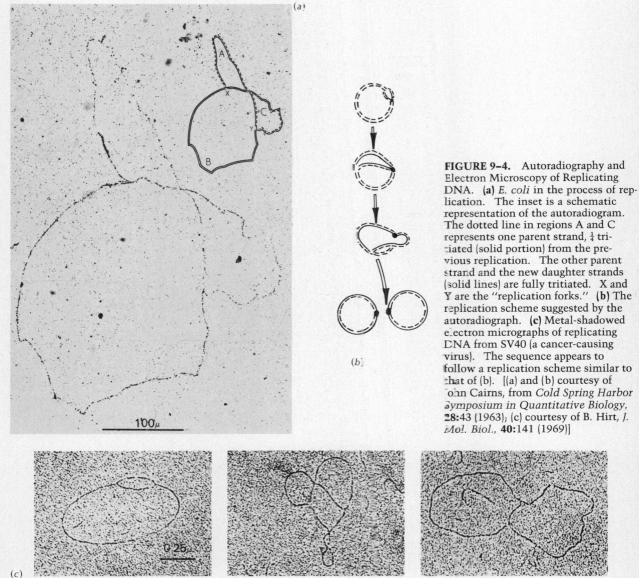

FIGURE 9–4. Autoradiography and Electron Microscopy of Replicating DNA. (a) *E. coli* in the process of replication. The inset is a schematic representation of the autoradiogram. The dotted line in regions A and C represents one parent strand, ½ tritiated (solid portion) from the previous replication. The other parent strand and the new daughter strands (solid lines) are fully tritiated. X and Y are the "replication forks." (b) The replication scheme suggested by the autoradiograph. (c) Metal-shadowed electron micrographs of replicating DNA from SV40 (a cancer-causing virus). The sequence appears to follow a replication scheme similar to that of (b). [(a) and (b) courtesy of John Cairns, from *Cold Spring Harbor Symposium in Quantitative Biology*, 28:43 (1963); (c) courtesy of B. Hirt, *J. Mol. Biol.*, 40:141 (1969)]

silver grains, after all, are much larger than the diameter of the DNA helix.

Autoradiography allowed Cairns to see the shape of the complete chromosome of *E. coli.* It turned out to be a closed circle (circumference 1100–1400 μ), even during replication. Seemingly, the new molecules are separated from the original in the same way that a closed zipper with its two ends sewed together would yield two circles when unzipped. The point of replication (the zipper pull) starts at a unique spot and always proceeds in the same direction around the complete circle.[1] A unique initiation point is common in procaryotic systems, but in some cases—phage lambda, for example—replication can go in both directions.

REPLICASES. An enzyme capable of catalyzing the replication of DNA (a *replicase*) was discovered in *E. coli* by Arthur Kornberg in 1955. It polymerizes a DNA strand complementary to an existing strand (the primer, or template), creating the double stranded molecule required by semiconservative replication. Kornberg's "DNA-dependent DNA polymerase," now called *polymerase I*, elongates a polynucleotide strand by adding a nucleotide to the 3' hydroxyl of the last ribose already present in the chain. The new nucleotide, initially a triphosphate, loses its terminal pyrophosphate (PP_i) upon addition to the chain. The pyrophosphate is then cleaved by a pyrophosphatase, making the overall reaction thermodynamically favorable. To add a guanine nucleotide to the chain, for example, involves two steps:

(1) $\quad$ (polynucleotide)$_n$ + GTP $\longrightarrow$ (polynucleotide)$_{n+1}$ + PP_i

(2) $\qquad\qquad\qquad\qquad PP_i \longrightarrow 2P_i$

The standard state free energy changes of the two steps are about 0.5 kcal/mole and −8 kcal/mole, respectively. Thus, when the steps are considered together, the combined reaction is very favorable.

But note that polymerase I adds only to the 3' end of an existing chain, while Cairns's data lead us to expect the simultaneous replication of two antiparallel chains at a single *replication fork*. The autoradiographs make it appear that one of the new strands is polymerized in the 5' → 3' direction, consistent with the mode of action of Kornberg's enzyme, while the other strand is being synthesized from its 3' to its 5' end. It was first assumed that another

---

1. Note that a unique initiation point means that genes near this point in growing cells will always be present in greater quantity than genes that are replicated last. This differential increase in template might be significant, and may account for the evolutionary choice of initiation point.

polymerase, undiscovered, catalyzes the $3' \rightarrow 5'$ mode, but all attempts to identify such an enzyme have failed.

The discovery in 1969 of an *E. coli* mutant that lacks polymerase I but grows nonetheless rekindled the search for additional replicases. Evidence for the attachment of DNA to the cell membrane led workers to look for enzymes in the membrane fraction of cell debris—a fraction that had earlier been routinely discarded—to discover there a second replicase, called *polymerase II*. However, the activity of polymerase II is similar in nature to that of the older replicase, in that it catalyzes nucleotide addition only at the 3' end of an existing strand. Still a third polymerase catalyzing $5' \rightarrow 3'$ growth is now assigned the role of major replicase; polymerases I and II may be largely (but not exclusively) concerned with the repair, rather than the replication, of DNA. It is known, for example, that mutants lacking polymerase I are unusually sensitive to ultraviolet light, a result of their reduced ability to cut out and repolymerize ultraviolet-damaged regions of their DNA.

REPLICATION MODELS. Thus, we have three problems to consider in explaining DNA replication: (1) how a high-speed twisting of a very long molecule might be initiated and sustained; (2) what mechanism might be responsible for lengthening the strand that grows in the $3' \rightarrow 5'$ direction; and (3) how initiation of DNA synthesis begins if replicases can only elongate existing strands.

There are several models that satisfactorily explain DNA replication, even with the above restrictions. The models depend on the existence of another pair of nucleases, one of which is capable of creating single-strand breaks (an endonuclease or "nickase") while the other, a DNA ligase, repairs them (see Fig. 9–5). Such breaks, temporarily created and repaired ahead of a replication point, allow unravelling to occur by swivelling about the P—O—P bonds in the adjacent chain without the necessity for twisting the whole molecule. (Unraveling may be aided by still another protein, capable of denaturing DNA by binding tightly only to single-stranded DNA.) The ligases also permit both strands to grow in the same direction: while one strand is elongated smoothly in the $5' \rightarrow 3'$ direction, the other may be elongated by back-filling, with short segments of DNA—each polymerized $5' \rightarrow 3'$—being joined to support a $3' \rightarrow 5'$ movement of the replication fork on that strand. Autoradiography and other techniques insensitive to very small changes in the size of DNA would overlook these pieces, called *Okazaki fragments* (after their discoverer), and see only a smooth elongation.

The discontinuous synthesis of DNA is supported by experiments with *E. coli* mutants having a DNA ligase that is unusually temperature sensitive. At slightly elevated temperatures, newly synthesized DNA in these mutants is found only in small

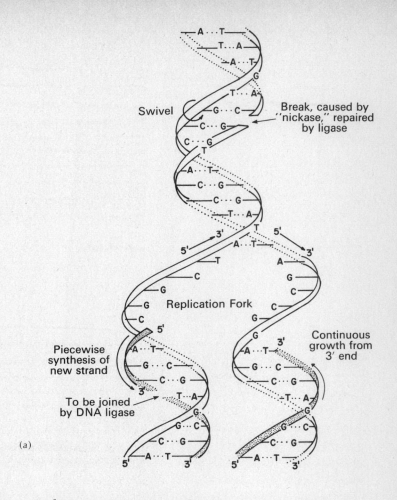

(a)

fragments. At lower temperatures, the fragments rapidly become incorporated into DNA of normal size.[2]

Finally, there is the problem of how a new segment of DNA is initiated with replicases that apparently add only to the 3' end of an existing strand. It appears now that initiation of replication relies on RNA primers of perhaps 50–100 nucleotides. The RNA, hydrogen bonded to a complementary region of one strand of DNA, would be elongated by the addition of deoxyribonucleotides to its 3' end. Thus, each new fragment of DNA, according to this model, should have a stretch of RNA at its 5' end. Presumably, the RNA is hydrolyzed by a ribonuclease and replaced by DNA before the new fragment is linked by a ligase to the fragment at its 5' end.

---

2.   This trick of obtaining temperature-sensitive mutants is used regularly in molecular biology. Cells are selected that exhibit the desired function at 20 °C, for example, but not at 37 °C. These mutants are a result of an amino acid substitution that makes the altered protein easily inactivated by heat.

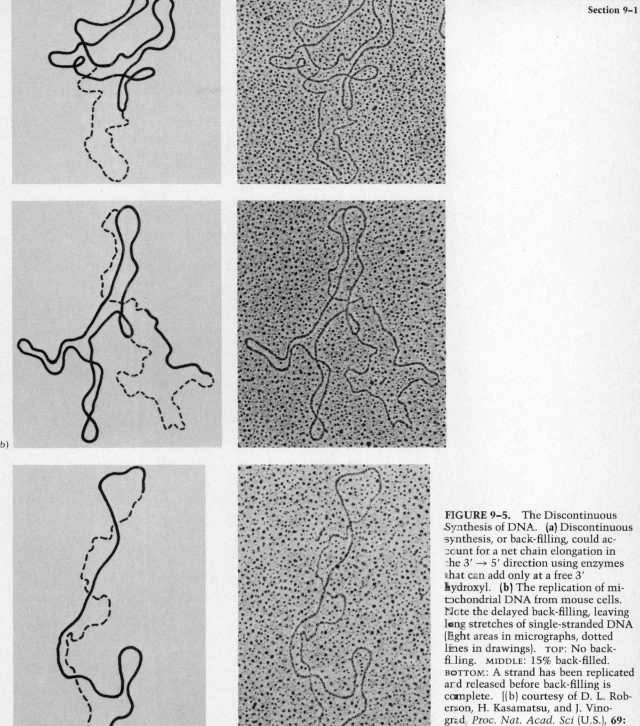

FIGURE 9–5. The Discontinuous Synthesis of DNA. **(a)** Discontinuous synthesis, or back-filling, could account for a net chain elongation in the 3′ → 5′ direction using enzymes that can add only at a free 3′ hydroxyl. **(b)** The replication of mitochondrial DNA from mouse cells. Note the delayed back-filling, leaving long stretches of single-stranded DNA (light areas in micrographs, dotted lines in drawings). TOP: No back-filling. MIDDLE: 15% back-filled. BOTTOM: A strand has been replicated and released before back-filling is complete. [(b) courtesy of D. L. Roberson, H. Kasamatsu, and J. Vinograd, *Proc. Nat. Acad. Sci* (U.S.), **69:** 737 (1972).]

ERRORS IN REPLICATION. The accurate transmission of genetic information requires an accurate replication of DNA. Mistakes in the process yield mutant daughter cells. It is not surprising, considering the complexity of the task, that such mistakes do occur from time to time. We can usually detect a new spontaneous mutation about once in a million gene replications. In phage T4, however, mutants in one segment of the chromosome (the "rII region") are detected with a frequency that approaches one in 10^4. Considering the number of nucleotides in that region, one concludes that the probability of making a mistake during replication is about 10^{-7} per nucleotide per replication.

There are several ways in which mistakes can occur. The wrong base might be inserted in the daughter strand because of the tendency of the bases to spend a very small amount of time in alternate tautomeric forms. In the case of thymine, for example, tautomerism could result in a satisfactory hydrogen bonding to guanine instead of to adenine.

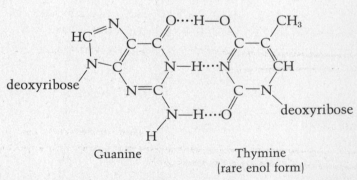

Adenine   Thymine
(normal keto form)

Guanine   Thymine
(rare enol form)

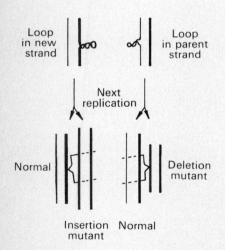

FIGURE 9-6. Insertions and Deletions. The dark strands represent newly synthesized DNA.

Another way in which replication errors can occur is for single-stranded loops to form in either the parent or daughter strand. A loop in the parent would cause bases to be missed, yielding a deletion mutant. A loop in the daughter would produce an insertion mutant. At the next replication, in either case, a fully mutant double-stranded molecule would be produced (see Fig. 9-6).

To these mechanisms for spontaneous mutation, one must add a long list of mutagenic agents, including radiation (ultraviolet radiation and higher-energy, or ionizing radiation) and a variety of chemicals. Some of these agents act by altering the bases, thus changing their hydrogen-bonding capacity. Other agents physically interfere with replication, in some cases by slipping in between bases (a process called *intercalation*) to distort the geometry of the molecule and increase the chance of adding or deleting bases in the replicating strand. Considering the complexity of the replication process and the variety of agents capable of interfering with it, we must marvel at the impressive reliability with which it occurs.

CYTOKINESIS IN PROCARYOTES. Cytokinesis in bacteria begins with an inward growth of the cytoplasmic membrane and wall, often at or near a folded area of the cytoplasmic membrane called a mesosome. The mesosome is also the site of attachment of the cell's DNA in many cases. Whether at a mesosome or not, the membrane attachment of DNA aids in its partitioning, for the daughter molecules move apart as membrane is added between them. Then, as cytokinesis proceeds, the membrane and wall material form a complete *transverse septum*, or cross-wall, that divides the cytoplasm into two approximately equal parts, each part with at least one chromosome (see Fig. 9–7). The daughters are then freed from one another by a cleavage of the septum. Should cleavage be incomplete, long chains of cells will develop, as happens frequently with certain types of bacteria.

In order for a cell to divide, both new wall and new membrane must be synthesized. Either of these events may be the immediate cause of division. To investigate which, one can grow bacteria without cell walls by exposing them to a substance, such as lysozyme or penicillin, that interferes with wall synthesis. Such cells, which are called *protoplasts* when completely free of wall material, sometimes cannot divide. The difficulty experienced by wall-less cells in trying to divide suggests that the formation of new cell wall material is necessary for cytokinesis, and that it may be the inward growth of the wall rather than the membrane that is really responsible for dividing the cell. The protoplasts of some bacteria, and all the bacteria of the genus *Mycoplasma*, which do not have walls, nevertheless manage to divide. However, the process in these cells is neither as regular nor as accurate as in normal, walled cells. In fact, mycoplasmas have been observed to pinch off bags of cytoplasm containing no chromosomes, an error that is obviously fatal to the offspring.

Implicit in the above description is the assumption of growth, for repeated cytokinesis without growth would yield an ever smaller cell size. With the exception of DNA, growth requires only that the proper structural proteins and enzymes be made in the right

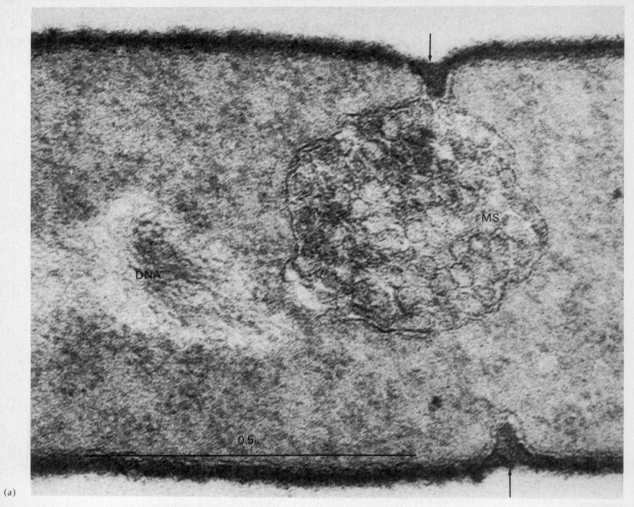

(a)

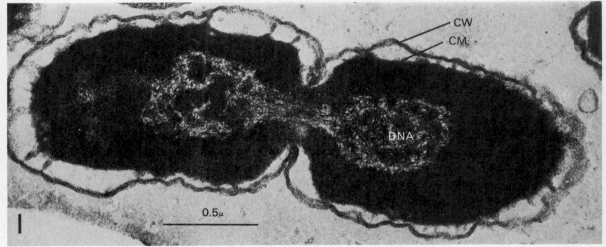

(b)

quantity. Hence, growth requires regulation of gene expression, as discussed in the last chapter, since enzymes, made under the direction of genes, are responsible for catalyzing the synthesis of other cellular components. For the most part, cellular components, once manufactured, seem to be self-assembling. For example, multisubunit proteins, ribosomes, and the cytoplasmic membrane itself form naturally from their constituent molecules. Assembly is driven by the formation of new chemical bonds—most of them weak bonds of the kind described in Chap. 2. Such interactions are, of course, a natural consequence of the arrangement of atoms in the molecules involved.

9–2 REPLICATION OF THE EUCARYOTIC NUCLEUS

Cell division in eucaryotes is complicated mostly by the multiplicity of chromosomes and by the necessity for their accurate replication and distribution to daughter cells. We will consider in this section the mechanisms that have evolved to overcome that complication.

THE REPLICATION OF CHROMATIN. The experiments of Meselson and Stahl demonstrated that DNA replication is semiconservative. It does not automatically follow that eucaryotic DNA, which is complexed with protein and RNA as chromatin, is similarly replicated. However, soon after Meselson and Stahl's experiments were published, workers were able to use similar techniques to show that DNA replication is almost certainly semiconservative in every cell type, eucaryotic as well as procaryotic.

The replication of eucaryotic DNA occurs when chromatin is dispersed within the nucleus between cell divisions. The chromatin will later condense into compact bodies, the chromosomes, in order to permit an accurate distribution to daughter cells. Each newly formed chromosome will ordinarily contain two identical strands, or sister chromatids, which will be sent to separate daughter nuclei. Experiments conducted at Columbia University in 1958 by J. Herbert Taylor demonstrated that the two chromatids result from semiconservative replication of the single chromatid left after the previous cell division.

Taylor incubated the root cells of a plant (*Bellavalia romana*) in radioactive thymidine (^{3}H-thymidine) between cell divisions, and found that when the chromatin condensed into chromosomes, the

FIGURE 9–7. Cytokinesis in Bacteria. (a) The beginning of division in *Bacillus subtilis*, with septum formation (arrows) at a mesosome (MS). (Note attached DNA.) (b) Nearly completed cytokinesis of *E. coli*, plasmolysed with sucrose to show invagination of both cell wall (CW) and cell membrane (CM). Note separation of daughter DNA molecules. [(a) courtesy of N. Nanninga, *J. Cell Biol.*, **48**:219 (1971); (b) courtesy of M. E. Bayer, *J. Gen. Microbiol.*, **53**:395 (1968).]

two sister chromatids were equally labelled (see Fig. 9–8). When cells that had been labelled in this way were allowed to proceed through a second division cycle, this time without radioactivity, the chromosomes appeared with one labelled chromatid and one chromatid free of label. These experiments suggest that each chromatid might contain only a single long molecule of DNA. If that is the case, the rate of movement of the replication fork must either be very fast or multiple forks must exist.

Autoradiography of nuclei exposed to isotopes for strictly limited periods of time (*pulse labelling*) provides measurements of the rate of movement of replication forks. Measurements by Cairns in 1966 yielded values of about 0.5 μ/min for the rate of DNA replication in cultured human cells. (These were HeLa cells, originally derived from the cervical cancer of a woman named *Henrietta Lacks*.) Although this rate is much slower than the 20–30 μ/min that he had found for *E. coli* chromosomes a few years earlier, apparently the value is not unusual for eucaryotes, as other experiments have shown that the replication rate of eucaryotic DNA ranges from less than a micron per minute to only a couple of microns per minute.

The relatively slow rate of DNA replication in eucaryotes tells us that replication must be proceeding at a number of points at once. Since the human diploid nucleus has a total DNA content of about 175 cm, a complete replication in ten hours (a typical value *in vitro*), even at the relatively rapid rate of 2 μ/min (about

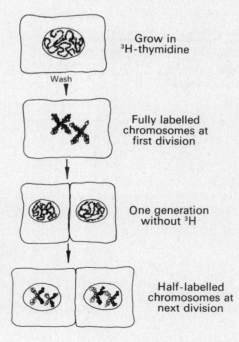

Grow in
³H-thymidine

Wash

Fully labelled
chromosomes at
first division

One generation
without ³H

Half-labelled
chromosomes at
next division

FIGURE 9–8a. Replication of the Eucaryotic Chromosome. Cells are grown in the presence of ³H-thymidine for one replication cycle, resulting in fully labelled chromosomes. After one generation of growth in the absence of tritium, each chromosome has one fully labelled and one unlabelled chromatid—i.e., there is a semiconservative distribution of label.

100 nucleotide pairs per second), would require about 1,500 simultaneously replicating points. Measurements indicate that the actual number of independently replicated units, or *replicons*, is much higher. Autoradiographs place the size of the mammalian replicon at 15–60 μ, with similar lengths for other animals. Replication forks are thought to proceed from either end of each replicon towards its middle. When all replicons in a DNA molecule have been duplicated in this fashion, two complete daughter molecules are formed (see Fig. 9–9).

Several DNA polymerases have been isolated from various eucaryotic cells. Their properties, for the most part, appear to be

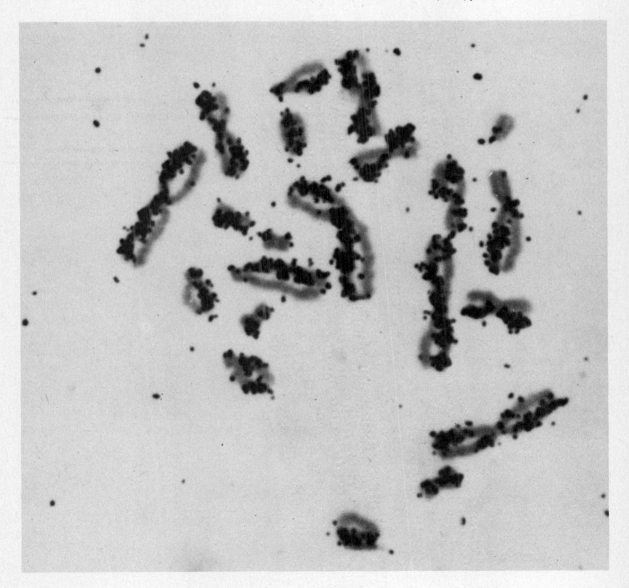

FIGURE 9–8b. Autoradiography of Replicated Chromosomes. Superimposed light micrograph shows distribution of label at the second division. In a few cases, label appears on the sister strand as well, a result of an interchange (crossover) between the strands. [Courtesy of G. Marin and D. M. Prescott, *J. Cell Biol.*, **21**:159 (1964).]

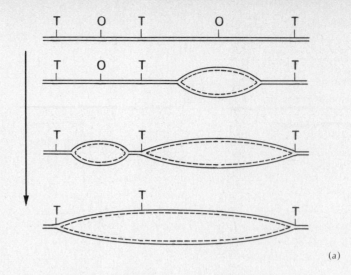

(a)

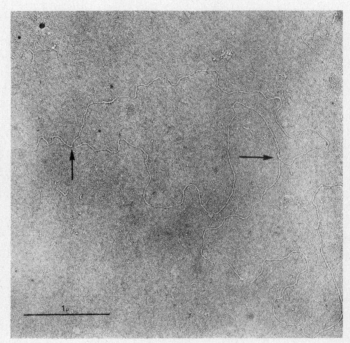

FIGURE 9-9. Replicons. **(a)** Eucaryotic DNA seems to be replicated in multiple discrete units, or replicons, having their own origins (O) and termini (T, shared with the adjacent unit). The sequence shows the replication and joining of two replication units. Dotted line represents newly synthesized DNA. **(b)** Replication of yeast chromosomal DNA, revealing a configuration similar to one of the intermediate steps in part (a). The arrows point to the replication forks. [(a) original description by J. Huberman and A. Riggs, *J. Mol. Biol.*, **32**:327 (1968); (b) courtesy of C. Newlon, T. Petes, L. Hereford, and W. Fangman, *Nature*, **247**:32 (1974).]

(b)

similar to those of the procaryotic replicases. In particular, they support only $5' \rightarrow 3'$ chain growth. The involvement of membrane attachment has also been proposed for eucaryotes, since the euchromatin (dispersed chromatin) of eucaryotic cells is anchored at numerous points to the nuclear membrane. These points correspond to the internal edges of the *annuli*, which are circular thin spots or pores in the nuclear envelope (see Fig. 9–10).

A eucaryotic cell faces the very considerable problem of ensuring the accurate distribution of replicated chromatin to daughter cells.

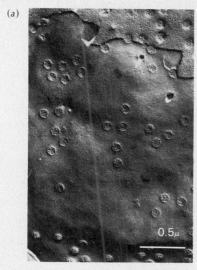

(a)

0.5μ

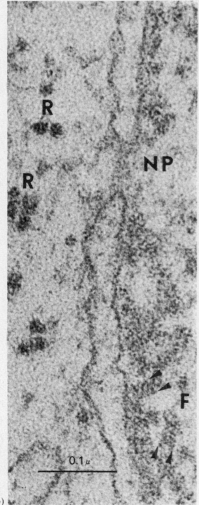

R

R

NP

F

0.1μ

(b)

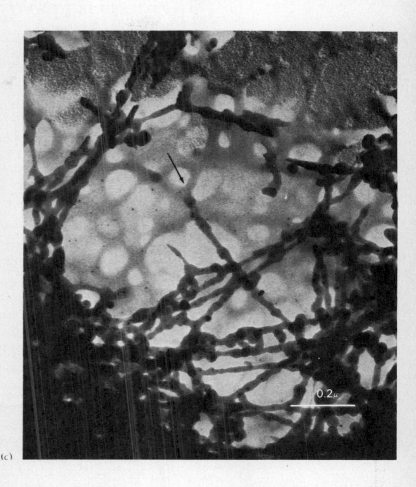

(c)

0.2μ

FIGURE 9–10. Chromatin and the Nuclear Envelope. **(a)** The nuclear envelope of a pig lymphocyte, showing the inner nuclear membrane and fragments of the outer one. Note the pores. **(b)** Cross-section of the nuclear envelope of an immature human lymphocyte. Nuclear pore (NP). Ribosomes (R) are seen in the cytoplasm, chromatin fibers (F and outlined by arrows) in the nucleus. (Also see Fig. 8–24.) **(c)** View from inside a honeybee nucleus, showing an unbroken chromatin fiber attached to the edge of an annulus (arrow). [(a) courtesy of V. Speth and F. Wunderlich, *J. Cell Biol.*, **47**:772 (1970); (b) courtesy of F. Lampert, *Humangenetik*, **13**:285 (1971); (c) courtesy of E. J. DuPraw, *Proc. Nat. Acad. Sci.* (U.S.), **53**:161 (1965).]

Procaryotes distribute their DNA by anchoring it to the cytoplasmic membrane and then allowing growth of the membrane to pull daughter molecules apart. However, all the chromosomes (DNA molecules) in procaryotes are identical, and it is only necessary to ensure that they be distributed more or less evenly through the cytoplasm in order to split their number at cytokinesis. A human cell, on the other hand, with 46 different chromosomes (23 pairs), faces a much more imposing task. Accordingly, a more sophisticated mechanism has evolved to ensure that each daughter cell gets one copy of each chromosome. That mechanism is mitosis. The variation that adds a pairing of homologous chromosomes, so that the two allelic (haploid) sets can be separated into daughter cells, is meiosis. A summary of those processes follows.

MITOSIS. The result of mitosis is ordinarily two new diploid daughter cells, each identical to the diploid parent cell. The process is best described in terms of the cell cycle.

A complete cell cycle usually consists of five stages, labelled G_1, S, G_2, M, and D. The first three of these together were originally believed to be a single *interphase*, or period of rest between cell divisions. That concept arose because visible activity is confined to mitosis (M), during which daughter nuclei form, and to cytokinesis itself, labelled D for division. (Mitosis is sometimes defined to include cytokinesis, eliminating the separate D stage.) However, it is now recognized that most DNA, RNA, protein, and new membrane, is made during interphase, which therefore is the period when metabolic activity is at its greatest.

Although the synthesis of most protein and RNA continues throughout interphase, DNA and histone synthesis is limited to the S period (S for synthesis). Once a cell enters its S stage, it usually proceeds inexorably through the other stages until it is back at G_1 again—i.e., back at the "first gap," or the one preceding DNA synthesis. An animal cell in tissue culture might repeat this sequence every 24 hours, spending approximately 10, 9, and 4 hours in the G_1, S, and G_2 stages, respectively (see Fig. 9–11). Mitosis and cytoplasmic division together may be completed in the remaining hour. Cells not destined for an early repeat of the division cycle are commonly arrested at the G_1 phase or, according to some systems of nomenclature, at a "G_0 stage" between D and G_1.

Mitosis was described by cytologists in the latter part of the nineteenth century. The most detailed reports came from Walter Flemming of the German University of Prague in the 1880s. Since then, the electron microscope and other tools of molecular biology have allowed us a much more intimate look at the process, although the original nomenclature is still retained for most of the structures and events involved. According to this nomenclature,

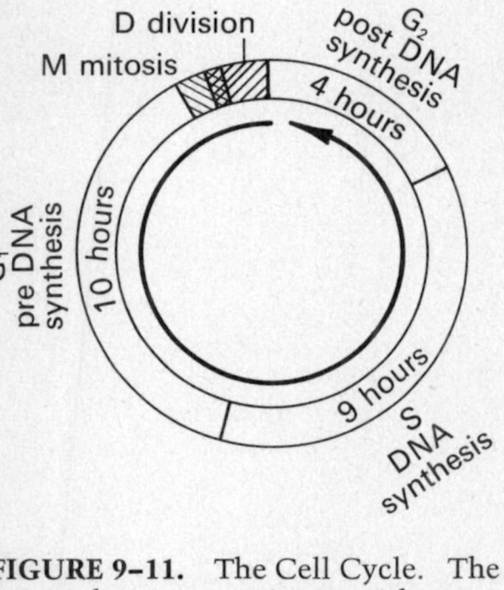

FIGURE 9–11. The Cell Cycle. The time shown is consistent with many lines of cultured mammalian cells. Cell division *in vivo* is generally very much less frequent—in fact, some cells do not divide at all after birth.

TABLE 9.1   A Summary of Mitosis and Meiosis

	MITOSIS	MEIOSIS	
		FIRST DIVISION	SECOND DIVISION
Interphase	(Subdivided into G_1, S, and G_2 stages). Chromosome replication occurs during S phase.	Chromosome replication occurs.	Interkinesis: no replication.
Prophase	Each chromosome is visible as two identical chromatids moving toward center of cell. Spindle apparatus forms. Nuclear envelope fragments.	Each chromosome is visible as two identical chromatids. Homologous pairs synapse to form four-stranded tetrads. A spindle apparatus forms and nuclear envelope fragments.	The chromosomes reappear as double-stranded structures, but their number is haploid. A new spindle apparatus forms and nuclear envelope fragments.
Metaphase	The double-stranded chromosomes align at the center of the nucleus and attach to spindle fibers. Chromatids separate, each attached to spindle fibers but with sister chromatids under the influence of opposite poles.	Tetrads align at center, but without centromere division. Homologous chromosomes—each still consisting of two chromatids—are attached to fibers from opposite poles.	Double-stranded chromosomes align at the center and attach to spindle fibers. This time centromeres divide and chromatids separate, leaving two identical haploid sets in each dividing cell.
Anaphase	Chromosomes, which are now single-stranded, move toward opposite poles. Cleavage furrow (animals) or cell plate (plants) begins to form.	The chromosomes, which are double-stranded, move toward opposite poles. Cleavage begins.	Each haploid set moves toward its own pole, and the second cleavage begins.
Telophase	Spindle dissolves and new nuclear envelopes form. Division of nucleus (karyokinesis) is usually accompanied by a division of the cytoplasm (cytokinesis) to yield two identical, diploid daughter cells.	Spindle dissolves and new nuclear envelopes form. Two nonidentical cells with haploid chromosome number are formed, but each chromosome still consists of two identical (sister) chromatids.	Spindle dissolves and new nuclear envelopes form. Meiosis is complete, leaving a total of four haploid cells of two allelic types.

mitosis is divided rather arbitrarily into four phases, called *prophase, metaphase, anaphase,* and *telophase.* (The prefixes mean, approximately, "before," "between," "toward," and "end," respectively.) Cytokinesis usually takes place in synchrony with the events of telophase. (See Table 9.1, Figs. 9–12 and 9–13.)

During prophase, the chromatin condenses to form visible chromosomes. Meanwhile, the nucleoli usually disperse and disappear. Because of replication during the preceding S phase, a newly condensed chromosome consists of two identical sister chromatids

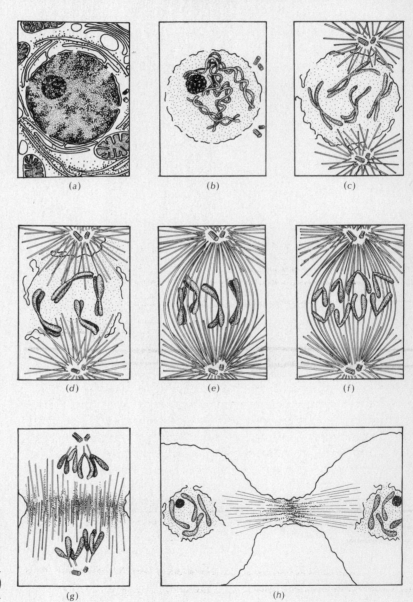

FIGURE 9–12. Mitosis in an Animal Cell. The mechanism of cell division and the presence of centrioles and an aster identify this as an animal cell. (Note the haploid number, $n = 2$.) **(a)** Interphase. **(b)–(d)** Prophase. **(e)** Metaphase. **(f)** Anaphase. **(g)–(h)** Telophase.

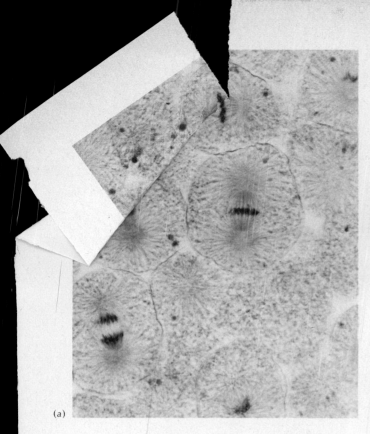

(a)

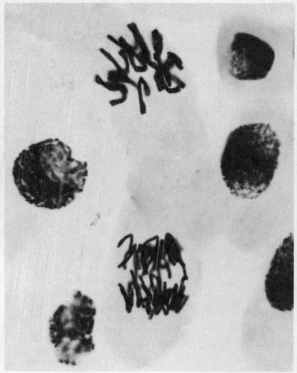

(b)

FIGURE 9–13. Mitotic Figures. **(a)** Mitosis in a whitefish embryo. Note the prominent asters. **(b)** Mitosis in a plant (*Allium* — e.g., garlic) root tip. The chromosomes are more clearly discernible in this preparation, but the spindle apparatus is not stained. (Note whole nuclei with stained chromatin.) [(a) courtesy of W. Etkin, R. M. Devlin, and T. G. Bouffard, *A Biology of Human Concern.* Philadelphia: J. B. Lippincott Co., 1972. By permission. (b) courtesy of the Turtox Collection, General Biologicals, Inc.]

joined together at the centromere. In all of the higher organisms, the nuclear envelope breaks up during this period (see Fig. 9–14).

The nuclear envelope consists of two membranes, each about 100 Å thick, separated by another 100–150 Å space for a total thickness of about 350 Å (see Fig. 9–10). Although it thus represents a substantial structure, its fragmentation is exceedingly rapid once begun, being completed in as little as two minutes. The addition of trypsin can mimic the normal process of dissolution, suggesting that protein hydrolysis may play a role. However, it seems likely that the nuclear membranes come apart into discrete fragments, or subunits — probably indistinguishable from endoplasmic reticulum — rather than all the way to single molecules, thus facilitating later reassembly. In any case, it is clear that dissolution of the envelope removes all barriers to mixing between cytoplasm and nucleoplasm.

Another remarkable prophase event is the assembly of a spindle apparatus, which is a system of microtubules traversing the nuclear region. It is used to separate the two sets of chromosomes so that each can be enclosed in the envelope of a separate daughter nucleus. Formation of the spindle apparatus in animal cells seems to be directed by the centrioles, which are themselves comprised of nine groups of microtubules (usually three to a group, lying in a

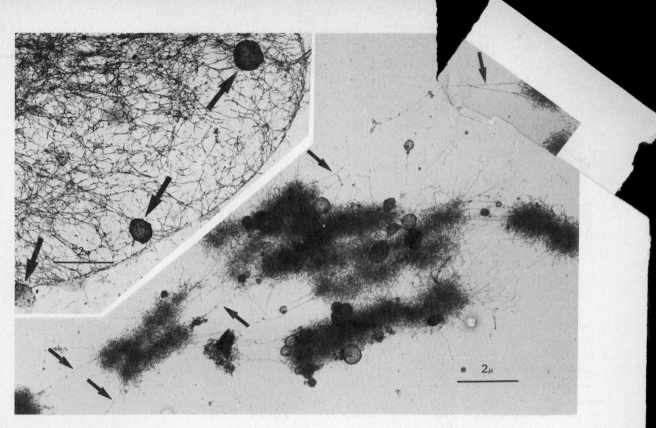

FIGURE 9–14. Chromatin Condensation. Human lymphocyte. The dispersed chromatin characteristic of interphase (inset) is gathered into discrete chromosomes during prophase, accompanied by fragmentation of the nuclear envelope. The large arrows in the inset point to remnants of the nuclear membrane, broken by surface tension forces in the whole-mount preparation technique. The smaller arrows point to single chromatin fibers extending between chromosomes. All chromosomes seem to be interconnected. [Courtesy of F. Lampert, *Humangenetik*, **13**:285 (1971).]

row) arranged about the almost empty core of a cylinder 0.5 μ long by 0.15 μ in diameter.[3]

The interphase animal cell (and some plant cells) usually has a single pair of centrioles lying at right angles to one another near the nuclear membrane. During mitosis, the centrioles migrate to opposite poles of the nuclear region and a new daughter centriole is assembled near each of the original two. A network of fine filaments, presumably the skeleton of the spindle apparatus, can often be seen stretching between migrating centrioles. By the end of prophase, the two sets of centrioles have a well-developed, football-shaped microtubule system (the *spindle*) strung between them and an *aster* (absent in most higher plants) of shorter microtubules radiating outward from each centriole pair.

The assembly of spindle fibers is partly from protein subunits synthesized during the G_2 phase. However, there is ample reason to believe that a major fraction of the proteins involved were

---

3. Centrioles were described in Chap. 1, where they were also implicated, as basal bodies, in the synthesis of cilia and flagella. The mechanism by which they accomplish their task is unknown, although some reports indicate the presence of DNA in them, which suggests that they may be actively engaged in the biosynthesis of some microtubular component.

present in the cell for a longer period of time, and were probably a part of the spindle during the previous mitosis. When one considers the amount of material involved, this is a highly practical mechanism, as the assembled spindle often accounts for 10% or more of the total cellular protein, and in some cases for more than a quarter of the cell's dry weight. The similarity between spindle fibers and other microtubules also suggests that the characteristic rounding of cells prior to mitosis is a consequence of a depolymerization of the microtubule skeleton used to maintain shape, and to the utilization of its subunits in the construction of spindle fibers.

The prophase events just described are those of the higher forms of animal life. Higher plants follow the same pattern, but usually without any obvious participation by centrioles. When one looks at simpler forms of eucaryotic life, however, numerous variations are found. The dinoflagellates, for example, have no spindle, but use a more haphazard scheme for chromosomal distribution. And many protozoans and fungi retain an apparently intact nuclear envelope during mitosis, with the spindle fibers inside it in some cases and outside it in others. We cannot consider all of these exceptions in detail, so the descriptions that follow should be read with that understanding.

During metaphase, spindle fibers become attached to a "granule," called a *kinetochore*, in the centromere region of each chromatid (see Fig. 9–15). After this attachment, the chromosomes line up in the middle of the nuclear region, all in the same plane, sometimes called the *metaphasic plate*. (Alignment may be due to pull on the chromosomes from both poles of the spindle.) Once the chromosomes are aligned at the middle of the spindle, the duplicate chromatids separate by centromere division, with each chromatid still attached to a spindle fiber.

When free of each other, the sister chromatids move apart at a rate somewhere between 0.2 and 4 μ/min. This is anaphase, or the period during which the two sets of chromatids collect at opposite poles of the spindle apparatus. This movement in most cells is accompanied by a shortening of the chromosomes themselves, making them more compact, and by an increase in the pole-to-pole distance of the spindle.

The motive force for chromosome movement, and that required to align the chromosomes at the metaphasic plate, have been the subject of considerable controversy over the years. It has been argued that the force is generated by the chromosomes themselves, with the spindle fibers serving only as guides. However, ultrastructural and chemical studies of chromosomes fail to reveal any contractile elements that might account for a self-propelled movement. On the other hand, microtubules such as those that make up the spindle have been implicated in a number of intracellular transport mechanisms. Therefore, it seems more reasonable to at-

(a)

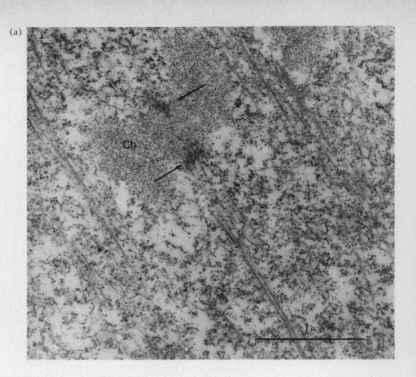

FIGURE 9-15. Spindle Fiber Attachment. (a) Metaphase chromosome (Ch) of a fertilized sea urchin egg. Note the dense kinetochores on both sides, to which spindle fibers are attached (arrows), and the continuous pole-to-pole spindle fibers adjacent to the chromosome. (b) A metaphase myoblast from a chick embryo. Note the spindle fibers (SP), chromosomes (Ch), centrioles (Ce), and mitochondria (M). The adjacent cell is a myotube (MT) in which myofilaments (mf) are developing. Myotubes are multinucleated cells, formed by fusion of myoblasts; they differentiate into mature skeletal muscle cells. [(a) courtesy of Patricia Harris, *J. Cell Biol.*, **25**:73 (1965); (b) courtesy of Y. Shimada, *J. Cell Biol.*, **48**:128 (1971).]

tribute the movement of chromosomes to forces exerted on them by attached spindle fibers.

In one such scheme, microtubules that are attached to chromosomes slide along pole-to-pole microtubules in a mechanism similar to that used to explain the motility of cilia and flagella. Alternately, a non-microtubular "movement factor" might replace the pole-to-pole fibers. Still another theory of chromosome movement attributes the motive force to a disassembly of the microtubules at the poles, coupled with their assembly at the midplane. This latter scheme was used to explain the results of an experiment performed by A. Forer of Duke University in 1965.

When Forer damaged spindle fibers with a finely focused microbeam of ultraviolet light, he found that the damaged region moved at a regular pace toward the nearest pole of the spindle—a pace characteristic of chromosome movements in the same cells. Although the movement, which seemed to occur whether or not a chromosome was also attached to the fiber, could have been initiated by the damage, it gives the impression that spindle fibers act as endless belts, with a constant assembly at the center and disassembly at either pole. This proposal requires a counterflow of tubule subunits toward the center of the nuclear region, a movement for which there is some evidence. In addition, one notes that spindle fibers to which chromosomes are attached are not con-

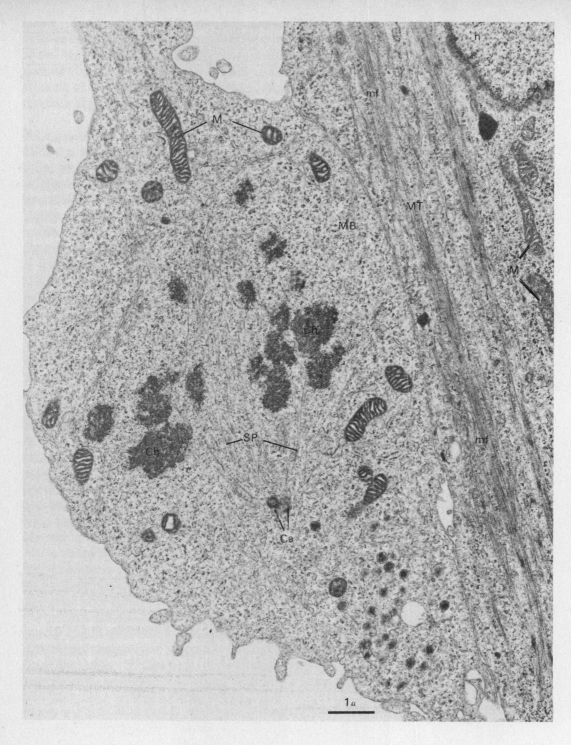

(b)

tinuous pole to pole as are free fibers. Rather, attached fibers seem to "melt" at the poles as they shorten (see Fig. 9–16). In animal cells, the microtubule subunits thus made available noticeably increase the size of the aster.

The movement of chromosomes during anaphase causes them to gather into two diploid groups at opposite ends of the spindle. During the last stage of mitosis, telophase, the spindle apparatus is disassembled, a new nuclear envelope forms about each group of chromosomes to produce two identical daughter nuclei, and the chromatin disperses so that discrete chromosomes are no longer seen. Nucleoli usually reappear at this time (fibrous interior first, granular cortex later). The assembly of the nuclear envelope seems to be from fragments, or subunits, which are quite probably derived in part from the original parent nucleus that broke up during prophase.

Cytokinesis usually occurs in synchrony with the events of telophase, although the first signs of impending division become visible during anaphase. Some of the details of cytokinesis will be considered in a later section.

MEIOSIS. Meiosis produces a total of four haploid cells from each diploid cell with random assortment of the parental chromosome pairs. Superficially, it resembles two successive mitotic divisions; however, there are very important differences and results (see Fig. 9–17 and Table 9.1).

In the first meiotic division, called the *reduction division*, DNA is replicated but centromeres do not divide; a chromatid pair is thus pulled intact into each daughter cell. In the second division, DNA is not replicated but the centromeres do divide, yielding a total of four daughter cells, each with half as many chromosomes as the original parent. Specifically, each daughter cell receives a haploid set consisting of one member of each homologous pair.

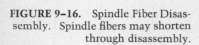

FIGURE 9–16. Spindle Fiber Disassembly. Spindle fibers may shorten through disassembly.

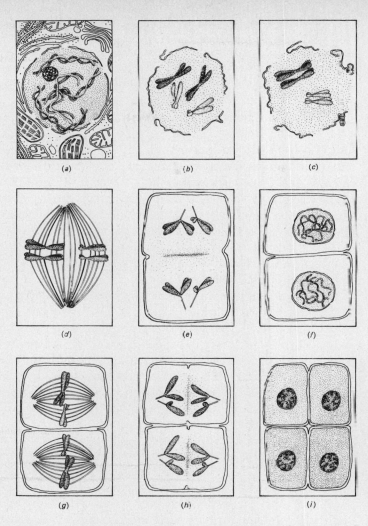

FIGURE 9–17. Meiosis in a Plant Cell. **(a)–(e)** The first meiotic division. Note synapsis at step (c). Completion of the first meiotic division yields two haploid cells containing nonidentical chromosome sets. The chromosomes remain two-stranded—i.e., with two identical chromatids. **(f)** Interphase. **(g)–(i)** The second meiotic division doubles the number of haploid cells to four. (Note that the haploid number of chromosomes is two.)

As most of the events unique to meiosis occur during prophase of the first division, this phase is customarily broken down into five stages:

1. *Leptotene stage* or *leptonema*, characterized by the appearance of single-stranded threads in the light microscope.
2. *Zygotene stage* or *zygonema*, during which homologous chromosomes pair in a process known as *synapsis*. The pairs are referred to as bivalents, since they consist of two chromosomes. It is important to realize that the homologous chromosomes of the bivalent are in register, so that allelic genes are found adjacent to one another. The mechanism of pairing is not understood, but the electron microscope reveals a protein framework between the homologues known as a *synaptinemal* (or *synaptonemal*) *complex*.
3. *Pachytene stage* or *pachynema*, where the chromosomes become more tightly coiled and stain more darkly. Individual chromatids become visible. Since it is now apparent that there are a total of four chro-

matids in each bivalent, the structure may also be called a *tetrad* (Fig.
9-17c).

4. *Diplotene stage* or *diplonema*, identified by a visible separation between the two homologues of each bivalent except at isolated sites of
union known as chiasmata. It is these sites that account for genetic
crossover (interchange of alleles).

5. *Diakinesis*, distinguished from diplotene only by an even tighter coiling
of the chromosomes, which thereby become still shorter and still more
deeply staining. Metaphase follows.

It is the tetrads that align at the center of the spindle during metaphase, with each chromosome falling entirely under the influence of one or the other pole. Hence, the spindle does not separate sister chromatids from each other; rather, it separates the two
homologous sets, one set gathering at each pole. Telophase and
cell division result in the formation of two haploid cells, although
each chromosome in the daughters still consists of two sister
chromatids.

The period between the first and second meiotic divisions is
called *interkinesis* or *interphase*. No DNA replication occurs, so
that chromosomes at the second prophase are the same double-
stranded structures that disappeared at the first telophase. The
second meiotic division proceeds just as normal mitosis would, but
the cell is working with only half as many chromosomes. At the
end of the second meiotic division, then, four haploid cells have
been produced. Two of them carry one allelic set while the other
two cells carry the complementary set.

The products of meiosis in animals are gametes: sperm and eggs
(see Fig. 9–18). For example, in *spermatogenesis* each of the four
haploid products of meiosis differentiates into a spermatozoan (see
Fig. 9–19). The formation of eggs (*oögenesis*), however, produces
three small *polar bodies* for every *ovum*. This unequal division
has the advantage of providing the ovum with more than its share
of stored food and other materials necessary to sustain a new life
during its early stages. (See Fig. 9–20.) The random assortment of
parental chromosomes into haploid gametes means that in a given
human there are 2^{23} or about 8×10^6 possible chromosomal permu-
tations in the gametes—obviously an ample explanation for sibling
diversity.

Oögenesis in the human and some other animals is peculiar in
its timing, for the cell may lie arrested in its first meiotic prophase
for an extended period. In the case of humans, this suspension
starts before birth and lasts until shortly before ovulation of that
particular cell, a period that may be as much as fifty years or even
more. The first polar body is thrown off only during the days immediately preceding ovulation, and the second meiotic division is
not completed unless fertilization occurs. The cell (i.e., the
oöcyte) in arrested prophase is said to be in the *dictyotene* stage
(a variation of diplotene) of the reduction division.

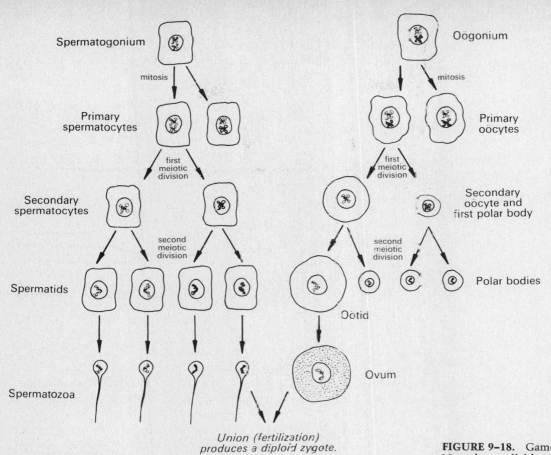

Spermatogonium

mitosis

Primary spermatocytes

first meiotic division

Secondary spermatocytes

second meiotic division

Spermatids

Spermatozoa

Oögonium

mitosis

Primary oöcytes

first meiotic division

Secondary oöcyte and first polar body

second meiotic division

Polar bodies

Oötid

Ovum

Union (fertilization) produces a diploid zygote.

FIGURE 9–18. Gametogenesis. Note the parallel between spermatogenesis and oögenesis. The diagram is drawn with a haploid chromosome number equal to one in both cases.

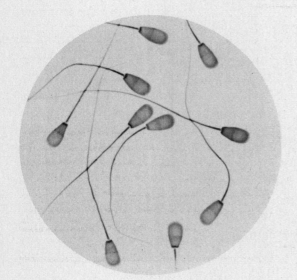

FIGURE 9–19. Bull Sperm [From the Turtox Collection, courtesy of General Biologicals, Inc.]

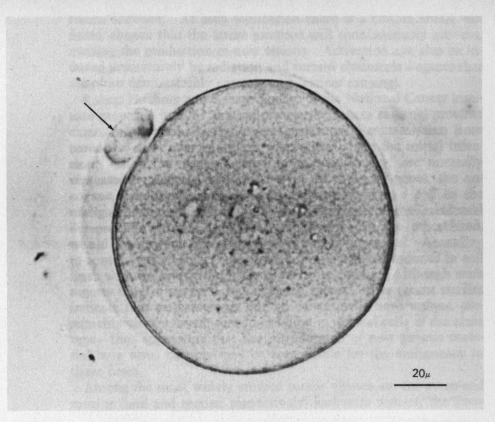

20μ

FIGURE 9-20. A Human Oöcyte and First Polar Body. The oöcyte and its small polar body (arrow) are surrounded by a transparent covering, called the *zona pellucida*. [Courtesy of J. F. Kennedy and R. P. Donahue, *Science*, **164:**1292 (1969). Copyright by the A.A.A.S.]

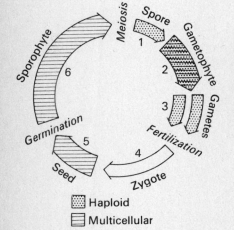

FIGURE 9-21. Alternation of Generations. The higher plants have a life cycle consisting of a mitotically proliferating haploid stage as well as a diploid stage.

Gamete formation in plants also involves meiosis, of course, but the process is complicated in the higher plants by the fact that each has a life cycle consisting of both a multicellular diploid and a multicellular haploid stage (see Fig. 9-21), called its *sporophyte* and its *gametophyte* generation, respectively. In the common garden plants, for example, it is the diploid stage that forms the structure with which we are familiar; the haploid organism is tiny and must be nurtured within specialized organs of the mature diploid plant. In some plants, however (green algae, mosses, and so forth), the multicellular haploid stage is the dominant one. In both cases, the usual pattern is for male and female haploid cells, called *spores*, to be produced by meiosis in the diploid (sporophyte) organism. Spores grow into multicellular male and female structures which, through mitosis, produce more haploid cells corresponding to the actual gametes. (A pollen grain is a male gamete.)

In both animals and plants, male and female haploid gametes unite in a process called *fertilization* to produce a *zygote*, in which the diploid chromosome number is restored. In animals and the simpler plants, the zygote immediately matures to a new diploid organism. In the seed-producing plants, development is arrested at an early multicellular stage as a *seed*, which may remain stable for long periods of time before germination permits a continuation of growth.

9–3 DIVISION OF THE EUCARYOTIC CELL

Mitosis or meiosis divides the nuclear material and encloses the resulting two groups of chromosomes in the envelopes of daughter nuclei. Thereafter, cytokinesis—if it occurs—takes place at a point that distributes the daughter nuclei to different cells.

SYNCHRONY BETWEEN NUCLEAR AND CYTOPLASMIC DIVISION. Cytokinesis is generally, but not always, synchronized with anaphase and telophase in both meiosis and mitosis. A notable exception occurs after fertilization of *Drosophila* eggs, for some 4,000 nuclei are produced by mitoses without any cell division at all. Then, in a remarkable demonstration of synchrony, each nucleus is partitioned into a separate cell by multiple cytokineses. On the other hand, many fungi and some green algae regularly undergo nuclear division without cytokinesis to form a *coenocyte*. In the slime molds, for instance, the coenocyte will become an amoeboid mass, called a *plasmodium*, that can move about and feed on particles of organic matter. (These are the true slime molds, not to be confused with the cellular slime molds.) One also finds multinucleate cells in animals, where they are called *syncytia*. A vertebrate skeletal muscle cell is an example of a syncytial unit, but it is formed by the fusion of immature cells, rather than by nuclear proliferation.

Even in cells where cytokinesis normally accompanies mitosis or meiosis, it can be demonstrated that one is not necessarily linked to the other. This potential independence was first shown by E. Harvey, who in 1936 found that sea urchin eggs are capable of undergoing several rounds of cytokinesis even after their nucleus is removed.

In those cells in which nuclear and cytoplasmic division are normally synchronized, the point where the cells actually divide is clearly influenced by the position of the center of the spindle. For example, in 1919, E. G. Conklin found that when the spindle apparatus of an oocyte was displaced by centrifugation, so was the site of division, altering the relative size of the polar body and ovum. More recently, it has been found that a division furrow can occur anywhere in a typical cell, though its actual position is normally at the center of the spindle. The choice of location is apparently determined early in the mitotic or meiotic cycle, for if the spindle is destroyed at late anaphase, a cleavage furrow forms at the former location of the spindle's center (see Fig. 9–22).

CYTOKINESIS. The division furrow has a dense ring of microfilaments at its leading edge (see Fig. 9–23). It appears that cytokinesis in eucaryotes involves contraction of this ring. This picture emerges from electron microscope studies and from experiments with a drug, *cytochalasin* (cyto-ka-lā'-sin), that causes the ring of microfilaments to disappear and division to stop. When the drug is removed again, the microfilaments are reassembled and division resumes.

The nature of the event that actually triggers formation of a division furrow, and the relationship between division furrow and mitotic spindle, are unclear. There are some studies that provide clues, however. For example, a cell from which the bulk of the cytoplasm has been extracted by glycerol can still form division furrows if ATP is added. And division can be artificially induced by injecting Ca^{2+} into the cell. These properties are reminiscent of

FIGURE 9–22. Cytokinesis without a Spindle. **(a)** The spindle apparatus (clear area in center) of a sea urchin (*Clypeaster*) egg was dissolved by injecting less than a microliter of sucrose solution. **(b)** Cytokinesis occurred on schedule, and in the position that it would have taken had the spindle been left intact. [Courtesy of Y. Hiramoto, *J. Cell Biol.*, **25:**161 (1965).]

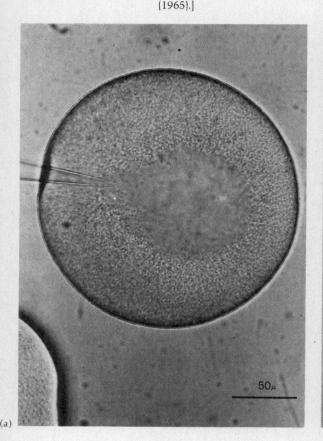

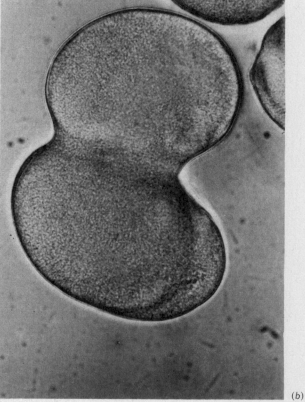

50μ

(a)

(b)

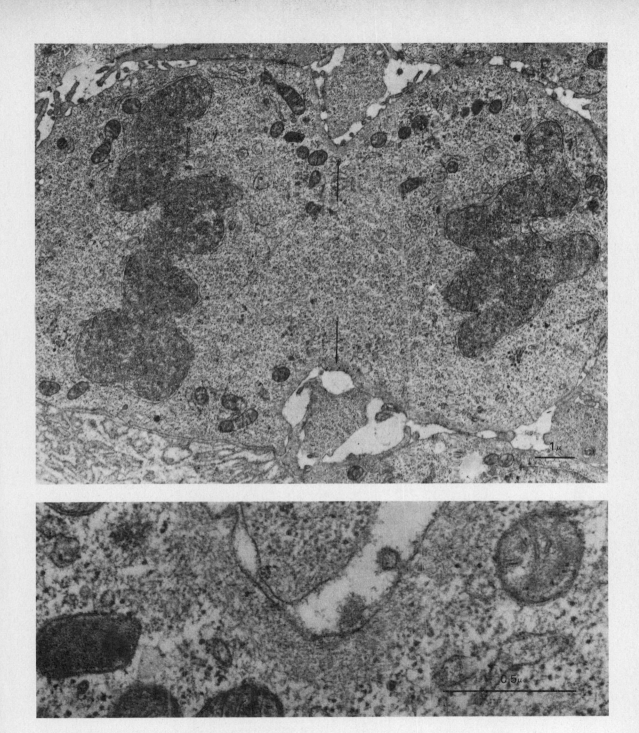

FIGURE 9–23. The Division Furrow. (a) Early telophase in a mouse mammary epithelial cell. Arrows mark the furrow. Note the irregularly shaped nuclei, newly formed around the still partially condensed chromatin. (b) Detail of the upper furrow. The microfilaments are cut in cross-section, but they are packed densely enough to be easily visible. [Courtesy of D. C. Scott and C. W. Daniel, *J. Cell Biol.*, **45**:461 (1970).]

the contractile proteins found in muscle. In fact, as noted earlier, microfilaments are apparently identical to the actin filaments of muscle, and at least some cells contain a myosin as well.

Once the division furrow has contracted to a small diameter, daughter cells pinch off from one another. The behavior of some cells indicates that a new mechanism may be activated at this point. In certain cultured human cells, for example, the cytoplasmic connection between daughter cells persists for some time. This *telophase neck* (see Fig. 9–24) contains fibers from the unassembled spindle apparatus, still running between the two centriole pairs.

Cytokinesis in plants is more like that already described for bacteria, because the cytoplasm is divided by the formation of a new partition, called a *cell plate*, rather than just by the plasma membrane alone (see Fig. 9–25). A cell plate forms at the midplane of the spindle, growing from the center to the edges until the cytoplasm is completely cut in two. It is a membrane that is pieced together from fragments that first line up and then fuse, in a process that reminds us of the assembly of the nuclear envelope. When complete, the cellulose walls of the two daughter cells are then laid down in the center until two cells, each containing its own complete wall, are formed.

9–4 THE REGULATION OF CELL DIVISION

The regulation of cell division is a topic that is getting a great deal of attention from biomedical scientists, in part because of its relationship to the problem of malignancy. A malignant cell (a cancer cell) does not respond to normal controls. In spite of the advances made in treating cancer through the use of surgery, radiation, and drugs that kill rapidly dividing cells, most scientists in the field feel that much more effective therapies could probably be designed if only we knew what makes a cancer cell divide, for then one might find a way of stopping it. Unfortunately, the regulation of cell division is not well understood even in normal cells, and that understanding is a prerequisite to a complete comprehension of how cancer cells are different.

Growth of a cell necessarily precedes its division, for otherwise the cells would get ever smaller at each generation. The rate at which a cell grows, and hence the rate at which an appropriate size for cell division is reached, obviously depends on the rate of metabolism. That, in turn, will be influenced by such things as the surface-to-volume ratio, which limits the capacity to bring in food and get rid of wastes. (The ratio becomes less favorable as the cell gets larger.) Metabolic rate is also dependent on temperature and other environmental factors, including the supply of oxygen and nutrients. The actual rate of growth itself will be a function both of metabolic activity and of the kinds of organic molecules the cell

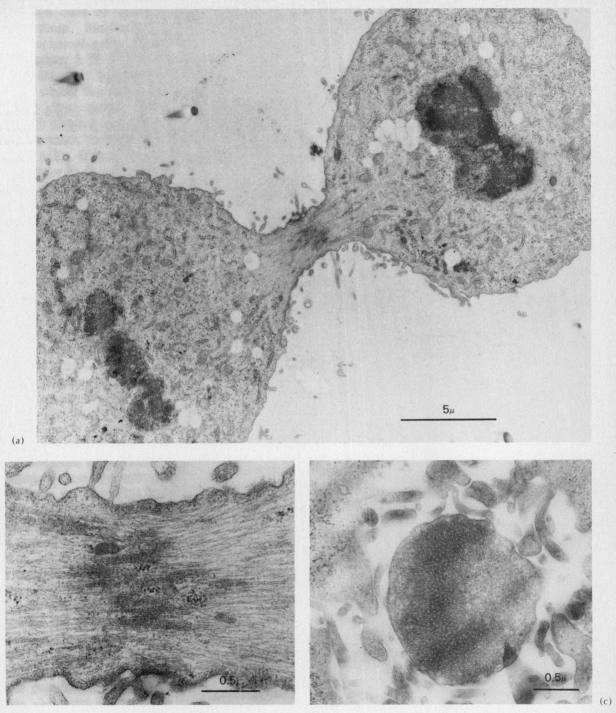

(a)

5μ

(b) 0.5μ

(c) 0.5μ

FIGURE 9–24. The Telophase Neck. Often there seems to be a hesitation just prior to actual separation of daughter cells. That stage is seen here in a cultured animal cell line (WI-38). **(a)** The cell at telophase. The nuclear envelopes have largely reformed, though the chromatin remains condensed. **(b)** Higher magnification of the stem. Note the microtubules, representing remnants of the spindle. **(c)** The stem in cross-section. The microtubules are quite clearly seen. [Courtesy of J. R. McIntosh and S. C. Landis, *J. Cell Biol.*, **49**:468 (1971).]

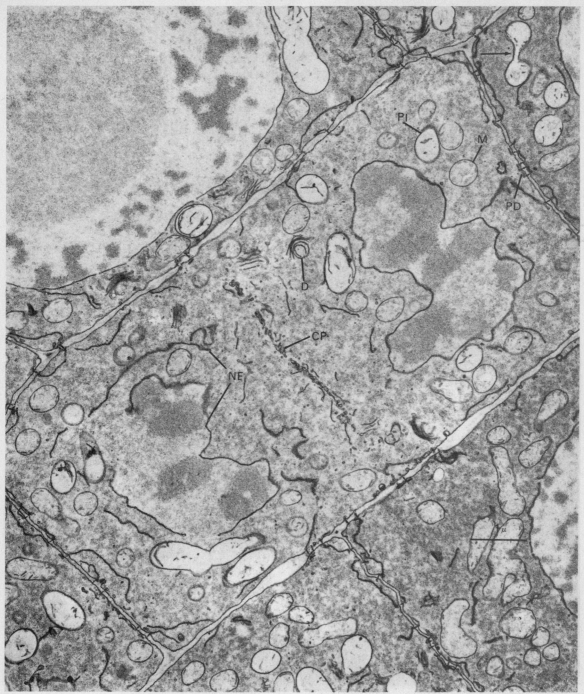

FIGURE 9–25. Cell Plate Formation. A dividing cell of a maize root tip is shown in the process of assembling new nuclear envelopes (NE) around its chromosomes, which are still condensed. A cell plate (CP) is also forming. Both seem to be put together from endoplasmic reticulum. The specimen was fixed with potassium permanganate ($KMnO_4$), which preserves membranes but little else. Mitochondria (M). Plastids (Pl). Note the dividing plastid (arrow at upper right), the plasmodesma (PD, a cytoplasmic channel to the next cell), and a huge nucleus at the upper left, with a prominent nucleolus. Dictysosome (D). [Courtesy of Hilton H. Mollenhauer.]

must make, as opposed to those that are supplied by the growth medium. All these things, then, influence the generation time. In addition, there are specific events and controls associated with the initiation of cell division, some of which will be described. We will start with the simpler, procaryotic systems.

REGULATION OF CELL DIVISION IN PROCARYOTES. A bacterium that is about to divide is not only twice as large as a newly formed cell, but it is growing twice as fast. The larger cell contains twice as much DNA as the smaller cell, and thus it is capable of a growth rate under given conditions that is proportionately faster. This relationship results from newly replicated procaryotic DNA being immediately available for transcription. Most of the genes in eucaryotes, on the other hand, are turned off and on at various times in the cell cycle. Although the DNA content of the nucleus doubles by mid-interphase, the growth rate of a typical eucaryote doubles only after mitotic separation of the chromosomes.

The event that triggers cell division in procaryotes is the completion of a round of replication in any one chromosome. In the case of *E. coli*, replication of the chromosome takes about forty minutes, with another lapse of twenty minutes between the end of a round of replication and cell division. (Since these numbers are fairly constant over a wide range of growth rates, generation times of less than an hour—they may be as short as twenty minutes—obviously require more than one replicating chromosome per cell.) The delay in cell division following replication of the chromosome seems to be influenced by at least one gene, perhaps located near the end of the chromosome and hence replicated last. This model is derived from the discovery of mutants that have lost their ability to synchronize cytokinesis and chromosomal replication. Division sometimes takes place prematurely in such cells, leading to enucleate daughters.

If cell division is triggered by the completion of a round of chromosomal replication, and if the rate of DNA synthesis at a particular replication fork is constant, then generation time must be controlled by the frequency with which replication forks are initiated.

Initiation of bacterial DNA synthesis is determined neither by the presence of existing replication forks nor directly by the completion of a round of replication. It does require the synthesis of protein, although no further synthesis is required to maintain a replication fork once it is formed. A critical ratio of cell mass to DNA content seems to initiate chromosomal replication, implying the need to accumulate an activator or to dilute an inhibitor. Accumulation of an activator that is consumed in the process of initiation seems the more likely option, and there is some experimental evidence to support it. The existence of mutants, defective in initiating replication at moderately high temperatures but not at lower ones, indicates that at least one of the factors involved is a protein capable of exerting positive control over the initiation event. As we shall see, a similar factor is probably present in eucaryotes.

INITIATION OF THE EUCARYOTIC CELL CYCLE. Most cells not destined for immediate division are arrested at the G_1 phase, prior to DNA replication. If such cells can be induced to begin synthesizing DNA, they generally proceed inexorably through the other stages of their cycle until they are back at G_1 again. One might conclude from this behavior that the cell cycle is regulated by DNA synthesis. However, cells arrested at the G_2 phase (after DNA synthesis but before mitosis) have also been identified in a wide variety of plant and animal tissues. DNA replication is not an immediate prerequisite to mitosis in such cells. Obviously, the controls must be more complex.

Several observations lead one to expect the existence of a cytoplasmic component that is capable of triggering cell division. Such a component would explain, for example, why cells with more than one nucleus generally have all their nuclei dividing in synchrony. A cytoplasmic trigger for the initiation of cell division also explains the results of nuclear transplantation experiments: (1) When nuclei are moved from cells at the G_1, G_2, or D phases to cells in the S phase of their cycle, replication of DNA in the transplanted nucleus takes place, followed by mitosis. (2) Conversely, a nucleus transplanted from an S phase cell to one at another stage of the cycle stops synthesizing DNA. And (3) when G_1 and S phase cells are fused, the G_1 nucleus enters the S phase, not the other way around.

There is also reason to believe that this cytoplasmic division factor is a protein, or at least that it requires a protein for its appearance (i.e., that it is not RNA). When *Tetrahymena* (a protozoan) is deprived of an essential amino acid in its G_1 phase, replication never occurs. However, if DNA synthesis is first allowed to begin, removing the amino acid does not arrest the cell cycle. This observation implies that some newly synthesized protein is required in order to start the S phase, for the missing amino acid stops all protein synthesis within a few minutes. Identification of this factor is, unfortunately, still uncertain.

A different approach in the search for a division regulator is to look for growth-correlated changes in known internal regulatory substances. Since cyclic AMP (cAMP) has been associated with so many diverse control functions, an examination of its effects has been made. Ira Pastan of the National Cancer Institute reported in 1971 that the growth rate of fibroblasts in tissue culture (these are the cells that manufacture the connective tissue fiber, collagen) is inversely related to their internal concentration of cAMP—the faster they grow, the lower their cAMP levels. Furthermore, this relationship was found to hold both for normal fibroblasts and for cancerous fibroblasts. (The opposite behavior is seen with cGMP.)

Cyclic AMP could be involved in the regulation of eucaryotic genes, as it is in procaryotic genes, and thus be responsible for producing the protein factor discussed earlier. In eucaryotes, however, cAMP is also known to cause (through protein kinases) the phosphorylation of histones. The lysine-rich histones (i.e., "fraction I"), in particular, seem to undergo phosphorylation at about the time of DNA synthesis. It has been suggested that phosphorylation of this histone fraction initiates chromosomal condensation and hence mitosis or meiosis, while the other major fractions of histones are more concerned with regulating RNA synthesis Reversible phosphorylation of the f1 histone might cause changes in histone-DNA interactions and thus expose sites attractive to DNA polymerase.

CONTACT INHIBITION. One of the first controls over cellular replication to be studied is contact inhibition, which *in vitro* is seen as a cessation of movement and replication once a confluent monolayer is reached in a culture dish. If a strip of cells is removed from such a culture, cells at the border of the "wound" multiply and migrate into the denuded area until the missing cells are replaced. Recent experiments suggest that this phenomenon is actually two processes: the first, a density-dependent inhibition of growth, and the second, a contact inhibition of movement. The density-dependent inhibition of growth appears in fibroblasts to be due to the rate at which substances can diffuse across the thin boundary of liquid just outside the cell membrane (the "diffusion boundary layer"). In the case of cultured epithelial cells (epithelial cells line surfaces), the growth-limiting factor seems to be the availability of surface to which the cells can attach. A wound increases the available surface for those cells at the edges of the wound and provides a greater exposure of cell membrane, fostering both replication and migration.

Experiments to be described in the next chapter indicate that contact inhibition of growth, and conceivably of replication in some cell types, may be due to cytoplasmic bridges between cells that allow the passage of regulatory molecules. At any rate, cytoplasmic connections seem to be common in many cell cultures that exhibit contact inhibition, but uncommon in cultures of tumor cells, which are noted for their insensitivity to contact inhibition.

The loss of contact inhibition exhibited by tumor cells can sometimes be corrected, causing the cells to revert to more normal (i.e., restricted) patterns of growth. The addition of cyclic AMP, dibutyryl cyclic AMP, or theophylline to cultures of tumorous fibroblasts slows their growth and causes some of them to revert to normal physical and replicative characteristics. (Theophyllin, a purine derivative, inhibits the breakdown of cAMP. The butyryl derivative of cAMP is used because it is more lipid soluble and therefore penetrates the cell more easily.)

Conversely, contact inhibition can be prevented in some normal cells by treating them with trypsin or other proteolytic enzymes, an apparent result of altering their surface properties. In addition, some plant glycoproteins called *lectins* (e.g., phytohemagglutinin and concanavalin A) stimulate certain animal cells to undergo mitosis when they would not ordinarily do so. Lectins act by binding to the cell surface at specific carbohydrates. In all cases, however, the changes are reversible: trypsin-treated cells return to normal when they have had a chance to replace the damaged surface features; and simple removal of a lectin is enough to reverse its effect. On the other hand, irreversible changes in growth patterns can be caused by a wide range of chemical and physical stimuli, including some viruses. These are the carcinogens, or cancer-causing agents, that transform a normal cell to the malignant state. Since cancer is a problem in differentiation as well as in simple growth control, it will be treated in more detail in the next chapter.

CHALONES. An experiment reported by B. D. Srinivasan in 1964 suggests the presence of an extracellular substance capable of regulating cell division in damaged tissues. Srinivasan used a needle to make a circular wound in the cornea of a rabbit's eye, and observed a wave of stimulation traveling outward from the center of the wound at a rate of about 17 μ/hr. As the wave passed a given cell, it set in motion a series of events that in 12–14 hours resulted in DNA synthesis and a new cell cycle. Within about 24 hours, the stimulated cells had completed their first new division.

It appears that the destruction of cells establishes some kind of gradient that induces mitotic activity in nearby cells—cells that would not otherwise undergo replication for some time. What is more, motile cells such as fibroblasts can follow this gradient toward the center of the area of destruction. Many attempts have been made to identify a mitotic stimulator in such gradients, but so far without much success. However, a substance discovered in 1960 by W. S. Bullough and E. B. Laurence of London's Birbeck College may be responsible. It is not a mitotic stimulator, but a mitotic depressor. Bullough named the substance *chalone*, which he says comes from a Greek word meaning "to slack off the main sheet of a sloop to slow the vessel down." Chalones are bound to specific receptor sites on the cell membrane. When membranes are damaged, chalones are lost, thereby stimulating cellular proliferation and subsequent repair of the wound.

Chalones have now been found in virtually every tissue that has been examined for them. Most seem to be glycoproteins of moderate size, though some appear to be relatively simple polypeptides. They share the following characteristics:

1. They inhibit mitosis, both *in vivo* and *in vitro*.
2. Although they are tissue-specific, the chalone for a given tissue is remarkably the same from one species to another.
3. The tissue affected by a particular chalone is also the tissue that makes that same kind of chalone.
4. Mitotic suppression by chalones is reversible, apparently without harm to the cell.

Unfortunately, the mode of action of the chalones remains unsolved, but it is known that some chalones require a hormonal cofactor. Included are the epidermal chalones, which require adrenaline.

HORMONAL CONTROL OF CELL GROWTH AND DIVISION. Cellular replication may also be controlled by hormones without involvement by chalones. Cell growth and cell division in plants is controlled by the interaction of two classes of hormones, *auxins* (e.g., indoleacetic acid) and *cytokinins* (e.g., kinetin, a purine derivative). The former group controls enlargement of cells, whereas the cytokinins act synergistically with auxins to bring about cell division. The *in vitro* application of an auxin alone is apt to yield a very large polyploid cell, the result of chromosomal replication without cytokinesis. Such observations have been harnessed in very practical ways, for application of plant hormones can cause even highly differentiated tissue to undergo replication. For example, a third class of plant hormone, the *giberellins* (e.g., gibberellic acid, first isolated from a fungus of the genus *Gibberella*) have been used to increase the rate of growth of celery, producing tenderer stalks.

Animal cells, too, may respond to certain hormones by undergoing mitosis. Thus, the diverse effects of *growth hormone* (or *somatotrophic hormone*, a protein manufactured in the anterior pituitary gland) include stimulating the proliferation of cells of different types. Hormones from the testes and ovaries also cause the growth of certain cells, producing secondary sexual characteristics. And the growth and division of oöcytes mentioned earlier is under the control of *follicle-stimulating hormone* from the pituitary. As with plants, even highly differentiated cells can be induced to undergo mitosis in some cases. One example is the mitotic proliferation of uterine smooth muscle under the influence of the hormone estradiol.

The earlier mention of cAMP and cGMP as possible growth regulators is also relevant here, for we already know that the action of many nonsteroid hormones is mediated through them. Media-

synergistically — working together

steroid — any of a group of cmpds including sterols, bile acids, sex hormones etc. having C-atom ring structure of the sterols

tion in the case of cAMP is apparently accomplished by the presence of hormone receptor sites on the cell surface which, when occupied, activate adenyl cyclase on the cytoplasmic side to produce cAMP from ATP. (Steroid hormones, as noted in Chap. 8, seem to act mostly within the cell, possibly because their greater lipid solubility gives them easier entry.)

There will be some additional discussion of mitotic regulation in the next chapter, but clearly, much work is left to be done in this area. As one might expect, considering its relevance both to a basic understanding of cellular activity and to clinical problems involving tumors, mitotic control is a subject that is getting much attention from scientists throughout the world.

SUMMARY

9-1 Cellular replication is comprised of two main events, division of the nucleus and division of the cytoplasm.

Division of the procaryotic nucleus merely requires the semiconservative replication of a double-stranded molecule of DNA (the procaryotic chromosome) and subsequent separation of the daughter molecules by growth of membrane between them. However, the semiconservative replication of DNA poses some serious problems, stemming from the rapid unwinding of the parent strand and from the fact that known DNA replicases elongate a strand only from its 3' end—i.e., the strand can grow only in a 5' → 3' direction. A discontinuous back-filling of the strand that grows in the 3' → 5' direction and temporary single-strand breaks ahead of the replication fork have been proposed to circumvent these problems.

Some of the more common errors in DNA replication, leading possibly to mutant offspring, are due to tautomeric shifts in the parental base, causing insertion of the wrong base in the daughter strand, and looping, resulting in insertions or deletions.

9-2 The DNA of eucaryotes is replicated semiconservatively, just as it is in procaryotes. In addition, it is distributed semiconservatively to daughter cells, so that each chromatid seems to be composed of half new and half parental DNA—as if it contained a single double-stranded molecule. Replication of chromatin, which appears to involve the activation of many individual replicating units called replicons, occurs during the S stage of the cell cycle when the chromatin is dispersed and anchored to the nuclear envelope at the edge of annuli.

Distribution of replicated DNA is usually via mitosis (Table 9.1), which produces two identical homologous sets.

In contrast, the result of meiosis is a total of four haploid cells, two of which contain identical copies of one haploid set of chromosomes and two of which contain copies of the homologous set. When meiosis is a part of spermatogenesis, the four haploid cells differentiate (grow flagella) to become male gametes, or sperm. When it is a part of oögenesis, one of the four cells will be much larger than the others. The larger cell forms the ovum, or egg, while the other three are left as polar bodies and disappear. In plants, on the other hand, meiosis produces cells (often spores) that commonly develop into multicellular haploid structures from which the true gametes form by mitosis. In all these cases, the male and female gamete will unite at fertilization to produce a new diploid cell, the zygote.

9-3 Cytokinesis (division of the cytoplasm) in animals apparently involves a dense ring of microfilaments just under the plasma membrane, pinching the cell into two daughters. In plants, cytokinesis is accomplished with the aid of a cell plate (a special membrane) that grows across the middle of the cell. In both cases, the location and timing are ordinarily controlled by the position of the spindle, since the division furrow and cell plate form at the position of the metaphasic plate.

Cytokinesis is normally linked with nuclear division, beginning even before the latter is complete. In other words, once a cell cycle is initiated, it usually proceeds at a predictable pace until the original stage of the cell cycle is reached again.

9–4 The completion of a round of DNA replication in *E. coli*, which takes about forty minutes, usually is followed some twenty minutes later by cell division. The existence of mutants in which this coordination is lost implicates a protein in the process. Since DNA polymerization occurs at a relatively constant rate, generation time depends on the rate of initiation of DNA replication, a situation that is probably also true of eucaryotes. In both types of cells, cytokinesis seems to depend on reaching a critical size, so that any condition influencing the rate of metabolism also influences generation time.

Initiating cell division may require a single, rather simple event, such as the induction or activation of a protein. Initiation is probably achieved spontaneously in some cells, limited only by their ability to attain a certain size. However, external controls over cellular replication are common in multicellular organisms. These controls may involve hormones, contact inhibition, chalones, and possibly more specialized kinds of controls not yet discovered.

STUDY GUIDE

9–1 (a) What is the difference between conservative and semiconservative replication of DNA? Describe the experiment that first supported the semiconservative mode. (b) The rapid unwinding of the parent molecule and unidirectional $5' \rightarrow 3'$ chain growth are problems in DNA replication. Explain how they might be circumvented. (c) How might errors occur in replication, and what is their consequence? (d) Describe the process of cytokinesis in procaryotes, including the mechanism for distributing replicated DNA.

9–2 (a) How is the DNA of eucaryotes organized in the interphase nucleus? How is it organized during mitosis and meiosis? (b) What are the five stages of the cell cycle? (c) How is the spindle apparatus formed and how is it thought to function? (d) Summarize the events of mitosis. At what points does meiosis differ, and in what ways? (e) What are the distinguishing features of meiosis as it is found (1) in male animals, (2) in female animals, and (3) in plants?

9–3 (a) To what do we ascribe the formation and activity of a division furrow? (b) How does cytokinesis differ in animal and plant cells?

9–4 (a) What experiments indicate that a new cell cycle is initiated by the synthesis of a specific protein? (Consider both procaryotes and eucaryotes.) (b) How might cyclic AMP be involved in the regulation of cellular replication? (c) What is meant by "contact inhibition?" (d) What is a "chalone?"

REFERENCES

(See Chap. 8 for reprint collections referred to here by their editors' initials.)

THE REPLICATION OF DNA

CAIRNS, JOHN, "The Bacterial Chromosome." *Scientific American*, January 1966. (Offprint 1030.)

CALLAN, H. G., "Replication of DNA in the Chromosomes of Eucaryotes." *Proc. Roy. Soc.* (London) B, **181**:19 (1972).

FILNER, P., "Semi-Conservative Replication of DNA in a Higher Plant Cell." *Exp. Cell Res.*, **39**:33 (1965). Reprinted in LL.

GROSSMAN, L., "Enzymes Involved in the Repair of DNA." *Adv. Radiation Biol.*, **4**:77 (1974).

HOLLAND, I. B., "DNA Replication in Bacteria." *Science Progress* (Oxford), **58**:71 (1970).

JACOB, F., A. RYTER, and F. CUZIN, "On the Association Between DNA and Membrane in Bacteria." *Proc. Roy. Soc.* (London) B, **164**:267 (1966).

KORNBERG, ARTHUR, "The Biological Synthesis of Deoxyribonucleic Acid." *Science*, **131**:1503 (1960) and in *Nobel Lectures, Physiology or Medicine, 1942–1962*. Amsterdam: Elsevier Publ. Co., 1964, p. 665. Nobel Lecture, 1959. Reprinted in JHT.

———, *DNA Synthesis*. San Francisco: W. H. Freeman and Co., 1974.

SMITH, D. M., "DNA Synthesis in Prokaryotes: Replication." *Prog. Biophys. and Mol. Biol.*, **26**:321 (1973).

TAYLOR, J. HERBERT, "The Duplication of Chromosomes." *Scientific American*, June 1958. (Offprint 60.)

_____, "Units of Replication in Chromosomes of Eucaryotes." *Int. Rev. Cytology* **37**:1 (1974).

THE MECHANISM OF CELL DIVISION

BAJER, A., and J. MOLÉ-BAJER, "Architecture and Function of the Mitotic Spindle." *Adv. Cell and Mol. Biol.*, **1**:213 (1971).

_____, "Spindle Dynamics and Chromosome Movements." (*Int. Rev. Cytology*, Suppl. 3). New York: Academic Press, 1972.

BRINKLEY, R. R., and E. STUBBLEFIELD, "Ultrastructure and Interaction of the Kinetochore and Centriole in Mitosis and Meiosis." *Adv. Cell Biol.*, **1**:119 (1970).

DeHARVEN, E., "The Centriole and the Mitotic Spindle." In *The Nucleus* (*Ultrastructure in Biological Systems*, 3), ed. A. J. Dalton and F. Haguenau. New York: Academic Press, 1968, p. 197.

FLEMMING, WALTER, "Contributions to the Knowledge of the Cell and Its Processes." Translated from an 1880 article and reprinted in *J. Cell Biol.*, **25** (**1, pt 2**):3 (1965). An early description of mitosis.

FORD, E. H. R., *Human Chromosomes*. New York: Academic Press, 1973. Structure, morphology, behavior at mitosis and meiosis, abnormalities, and other topics.

FORER, A., "Local Reduction of Spindle Fiber Birefringence in Living *Nephrotoma suturalis* (Loew) Spermatocytes Induced by Ultraviolet Microbeam Irradiation." *J. Cell Biol.*, **25**(**1, pt 2**):95 (1965). Spindle fiber movements.

GRAHAM, J., M. SUMNER, D. CURTIS, and C. PASTERNAK, "Sequence of Events in Plasma Membrane Assembly During the Cell Cycle." *Nature*, **246**:291 (1973). Membrane assembly occurs during interphase, not during cytokinesis.

KAYE, R. R., and I. R. JOHNSTON, "The Nuclear Envelope: Current Problems of Structure and Function." *Sub-Cellular Biochem.*, **2**:127 (1973).

LEDBETTER, MYRON C., "The Disposition of Microtubules in Plant Cells During Interphase and Mitosis." In *Formation and Fate of Cell Organelles* (*Symp. Int. Soc. for Cell Biol.*, **6**), ed. K. B. Warren. New York: Academic Press, 1967, p. 55.

LUYKX, PETER, "Cellular Mechanisms of Chromosome Distribution." (*Int. Rev. Cytology*, Suppl. 2). New York: Academic Press, 1970.

MAZIA, D., "The Cell Cycle." *Scientific American*, January 1974, p. 54.

MOENS, P. B., "Mechanisms of Chromosome Synapsis at Meiotic Prophase." *Int. Rev. Cytology*, **35**:117 (1973). A review of various proposals.

RAPPAPORT, R., "Cytokinesis in Animal Cells." *Int. Rev. Cytology*, **31**:169 (1971).

THE REGULATION OF CELL DIVISION

ABELL, C. and T. MONAHAN, "The Role of Adenosine 3', 5'-Cyclic Monophosphate in the Regulation of Mammalian Cell Division." *J. Cell Biol.*, **59**:549 (1973). A review.

ABERCROMBIE, M., and JOAN E. M. HEAYSMAN, "Observations on the Social Behavior of Cells in Tissue Culture:II. 'Monolayering' of Fibroblasts." *Exp. Cell Res.*, **6**:293 (1954). The first description of contact inhibition.

BECKER, F. F., "Cell Division in Normal Mammalian Tissues." *Annual Rev. of Medicine*, **20**:243 (1969).

BRADBURY, E., R. INGLIS, and H. MATTHEWS, "Control of Cell Division by Very Lysine Rich Histone (Fl) Phosphorylation." *Nature*, **247**:257 (1974).

BULLOUGH, W. S., "The Chalones." *Science J.*, **5**:71 (1969). Natural mitotic inhibitors.

BURGER, MAX M., "Surface Changes in Transformed Cells Detected by Lectins." *Federation Proceedings*, **32**:91 (1973). Surface characteristics of malignant cells are like normal cells undergoing mitosis.

_____, "Proteolytic Enzymes Initiating Cell Division and Escape from Contact Inhibition of Growth." *Nature*, **227**:170 (1970).

deTERRA, NOËL, "Cortical Control of Cell Division." *Science*, **184**:531 (1974). Function of the cell surface in replication.

DULBECCO, R., and J. ELKINGTON, "Conditions Limiting Multiplication of Fibroblastic and Epithelial Cells in Dense Cultures." *Nature*, **246**:197 (1973). Contact inhibition is apparently not what limits replication in tissue culture.

GOSPODAROWICZ, D., "Localization of a Fibroblast Growth Factor and its Effect Alone and with Hydrocortisone on 3T3 Cell Growth." *Nature*, **249**:123 (1974). The growth factor was found in brain and pituitary.

HARTWELL, L., J. CULOTTI, J. PRINGLE, and B. REID, "Genetic Control of the Cell Division Cycle in Yeast." *Science*, **183**:47 (1974). A model.

MARTZ, E., and M. S. STEINBERG, "The Role of Cell–Cell Contact in 'Contact' Inhibition of Cell Division: A Review and New Evidence." *J. Cellular Physiol.*, **79**:189 (1972).

MORLEY, G. D., "Humoral Regulation of Liver Regeneration and Tissue Growth." *Perspectives in Biol. and Med.*, **17**:411 (1974). Tissue and humoral (blood-borne) factors regulating cell division.

ROSS, RUSSELL, "Wound Healing." *Scientific American*, June 1969. (Offprint 1144.)

ROWBURY, R. J., "Bacterial Cell Division: Its Regulation and Relation to DNA Synthesis." *Science Progress* (Oxford), **60**:169 (1972).

SHARON, N., and H. LIS, "Lectins: Cell-Agglutinating and Sugar-Specific Proteins." *Science*, **177**:949 (1972).

SLATER, M., and M. SCHAECHTER, "Control of Cell Division in Bacteria." *Bact. Revs.*, **38**:199 (1974).

SRINIVASAN, B. D., "Chromosome Duplication and the Cell Cycle in Lens Epithelium." *Nature*, **203**:100 (1964).

STOKER, M. G. P., "Role of Diffusion Boundary Layer in Contact Inhibition of Growth." *Nature*, **246**:200 (1973). Contact itself may not be responsible for contact inhibition of growth.

CHAPTER 10

Cellular Differentiation

10-1 GENE–CYTOPLASM INTERACTION IN DEVELOPMENT   428
Cloning of Plants and Animals
Cytoplasmic Influence on Nuclear Function

10-2 INTERCELLULAR COMMUNICATION   433
Evidence for Intercellular Communication
Cytoplasmic Bridges

10-3 STABILITY OF THE DIFFERENTIATED STATE   437
Determination and Transdetermination
Regeneration
Tumors
Characteristics of the Malignant Cell
Tumorigenesis
Differentiation of Tumor Cells

10-4 SENESCENCE   446
Variations in Life Span
Programmed Death

SUMMARY   450
STUDY GUIDE   451
REFERENCES   451

Although the discussions up to now have emphasized features that are common to most cell types, the specialized properties of certain cells have been discussed in previous chapters. We will now take time to further consider some of the mechanisms of *cellular differentiation*, through which specialization is achieved.

Differentiation in the simplest sense refers to the appearance of any new cellular property, whether it involves a change in morphology (form and structure) or in chemical composition. The induction and repression of enzymes, discussed earlier, are simple kinds of differentiation. What interests us most, however, is the more complicated sequence of events by which a cell becomes something quite obviously different from its parents.

10–1 GENE–CYTOPLASM INTERACTION IN DEVELOPMENT

One of the earliest theories of cellular differentiation was proposed in 1892 by August Weismann, who suggested that differentiation is due to a progressive loss of genes. That is, the chromosomes become less complete as the cell becomes more highly specialized, with the particular developmental pathway being determined by which genes are lost. This suggestion seems at first to be quite reasonable, for we know that different genes are expressed in different cell types, resulting in varying enzyme patterns and physical properties. It is now generally accepted, however, that an identical set of genes is found in every nucleated somatic cell of most organisms—that is, in all but the germ cells. Differentiation must therefore be the result of changes in the pattern of gene expression—i.e., *epigenetic changes*—and this, of course, implies the existence of specific gene repressors.

CLONING OF PLANTS AND ANIMALS. That even highly differentiated cells contain a full set of genes has been dramatically demonstrated both for animals and for plants. In 1964, F. C. Steward and his colleagues grew entire carrot plants from cultures containing individual cells and small cell aggregates taken from the phloem tissue of a carrot root. The next year V. Vasil and A. C. Hildebrandt isolated single cells from a tissue culture of tobacco cells and grew complete plants from them. In both cases normal, viable seed was produced by the plants. Apparently, each of the highly differentiated cells from which these plants were started contained all the genes needed for every cell type found in the mature plant.

Though the cloning of plants (that is, the asexual formation of a line of descendants genetically identical to the original) was a spectacular achievement, the parallel experiment carried out with frogs was even more impressive because of the greater complexity of that organism. J. B. Gurdon and his colleagues at Oxford succeeded in cloning frogs in the mid-1960s. They used intestinal cells taken from tadpoles of the South African clawed frog, *Xenopus laevis*, and implanted the cells in unfertilized eggs from which the nucleus had been removed mechanically or destroyed by ultraviolet light. A few of these hybrid cells developed into adult frogs (see Fig. 10–1). To be sure that development was controlled by the transplanted nucleus rather than by the original one, two strains of frogs were used. One strain had the normal complement of two nucleoli per nucleus, while the other strain was a mutant with only one nucleolus per nucleus. The adult product of the experiment always reflected the nucleolar arrangement of the strain that donated the nucleus to the hybrid, not the strain that donated the egg.

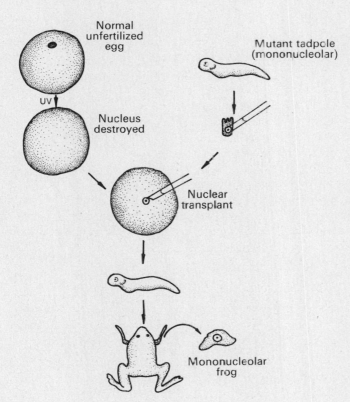

FIGURE 10–1. The Developmental Potential of a Differentiated Nucleus. That differentiation does not necessarily involve the loss of genes was demonstrated by the experiment described here. The nucleus of a normal, unfertilized egg was destroyed and replaced by the nucleus of a differentiated intestinal epithelial cell. The latter was taken from a mutant strain of tadpoles having but one nucleolus per diploid nucleus. Some of the hybrids developed into complete frogs, but always with the mutant nucleolar count. [Described by J. B. Gurdon and V. Uehlinger, *Nature*, **210**:1240 (1966).]

The ability to produce a complete frog from the nucleus of an already differentiated cell created a great deal of interest, both theoretical and practical. On the practical side, embryologists could now foresee the possibility (remote, some would say) of producing clones of genetically identical domestic animals, thus preserving particularly valuable traits instead of taking chances with the genetic mixing inherent in sexual reproduction. (Extended to humans, cloning raises the possibility of a woman giving birth to her own—or to her husband's—identical twin.) On the theoretical side, nuclear transplant experiments have given satisfying evidence that differentiation is, indeed, the result of a particular pattern of gene expression in a cell that carries the full complement characteristic of the species.

skip CYTOPLASMIC INFLUENCE ON NUCLEAR FUNCTION. The nuclear transplant experiments just described suggest the presence of cytoplasmic control over gene expression. The full genetic potential of the transplanted nuclei was realized only when the nuclei were introduced into the cytoplasm of unfertilized eggs.

Other examples of cytoplasmic control over nuclear activity come from work with a single-celled alga of the genus *Acetabularia*, starting with experiments by J. Hämmerling in the 1930s. This graceful tropical seaweed grows to a height of several centimeters, with a funnellike cap on one end of a long slender stalk and an anchoring organ, the *rhizoid*, on the other end (see Figs. 10–2 and 10–3). Each plant is a single cell, with a single nucleus large enough to be visible to the naked eye. At the time of cap formation, the nucleus becomes greatly enlarged; but when the

FIGURE 10–2. *Acetabularia mediterranea.* Each plant is a single cell, up to several centimeters in length. Occasionally one can see a rhizoid (attachment organ, arrow) on the end of the stem. The nucleus is found there. [From the Turtox Collection, courtesy of General Biologicals, Inc.]

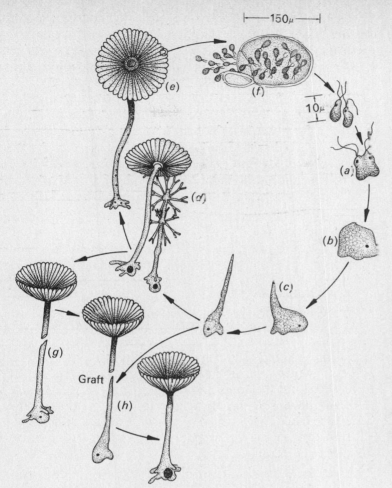

FIGURE 10–3. Cytoplasmic Control of Nuclear Function in *Acetabularia*. The life cycle of *Acetabularia* begins with **(a)** the fusion of two gametes to **(b)** form a zygote, which **(c)** grows a stalk and **(d)** a cap. (From (a) to (d) takes about two months.) **(e)** Maturation is complete when the nucleus undergoes multiple divisions to produce new haploid gametes. The latter migrate into the cap and are packaged and released in cysts **(f)** to start a new cycle. See text for an explanation of the grafting experiment, parts **(g)** and **(h)**. [Described by J. Haemmerling, *Ann. Rev. Plant Physiol.*, **14**:65 (1963).]

cap is cut off a cell at this point, the nucleus immediately shrinks back to its former compact size (Fig. 10–3g). If the mature cap is transplanted to a young cell, the nucleus of the young cell undergoes the meiotic nuclear divisions necessary to create gametes (Fig. 10–3h). In both cases, the cytoplasm clearly affects the activity of the nucleus.

Although cytoplasmic factors are capable of modulating gene expression, one must remember that the components of the cytoplasm are themselves determined by genes: RNAs are the direct products of genes, proteins are produced by translation of RNAs, and other substances not obtained from outside the cell are synthesized via the catalytic activity of enzymes—which are, of course, proteins. Each generation starts with some non-gene material inherited through the cytoplasm of its parental gametes, but these substances would soon become uselessly dilute due to growth if they were not replaced by the activity of genes.

Not all genes reside in the nucleus, however. We have already seen that mitochondria and plastids (chloroplasts) contain genes capable of synthesizing some, but not all, of their own components. There is some evidence for small amounts of genetic material at other cytoplasmic locations in the cells of higher organisms (e.g., in centrioles); there is good evidence for it in protozoa, especially in their *kinetosomes* (basal bodies), *kinetoplasts* (which appear to be a mitochondrial relative), and perhaps at sites in the cortex of the cell other than the kinetosomes. There are also some reports of enzymes in animal cells capable of replicating RNA or making DNA from it. (One or both activities are always found in cells infected by RNA viruses.) Such enzymes would permit cytoplasmic RNA to function in a way normally associated with nuclear genes.

In spite of these exceptions, the bulk of the genetic information in eucaryotic cells clearly resides with its nuclear genes, and so it is these genes that determine most of the properties of a typical eucaryotic cell.

For example, when one removes the stalk of a mature *Acetabularia* cell that is about to undergo spore formation, the cap is able to grow only a new stalk, the isolated stalk itself shows no growth or differentiation at all, but the rhizoid, which contains the nucleus, is able to form a whole new plant (see Fig. 10–4a). In addition, when the nucleus of one species is grafted to the severed stalk of a second species, the resulting cap is always characteristic of the species to which the nucleus belonged. The information necessary to produce a complete new cell is apparently confined largely to the nucleus.

The nucleus probably controls differentiation through the manufacture and release of some diffusible substance, which we might

FIGURE 10–4. Nuclear Control of Cytoplasmic Function in *Acetabularia*. **(a)** A severed stalk has no regenerative capacity, a severed cap only a limited amount. However, the nuclear-containing rhizoid can regenerate a complete new plant. **(b)** Nuclear control is exercised through the synthesis of RNA: If a severed cap is treated with ribonuclease (RNAase), no stalk regeneration occurs; but a stalk that is allowed to remain on the rhizoid for a time after cap removal develops the ability to grow a new cap, apparently through the incorporation of nuclear-derived RNA. [See J. Brachet, *Biochemical Cytology*. New York: Academic Press, 1957.]

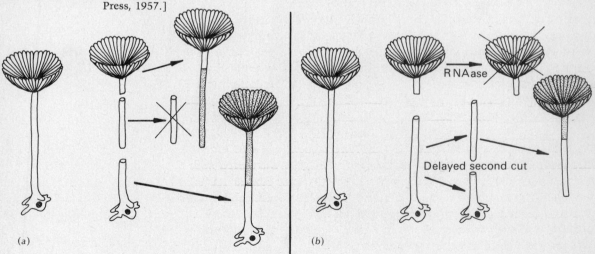

(a)   (b)

expect to be a messenger RNA. There are several kinds of experiments that are consistent with that supposition (see Fig. 10–4b): (1) If the severed cap of a mature *Acetabularia* cell is treated with ribonuclease (which destroys RNA), the expected regeneration of stalk does not occur. (2) If the stalk and rhizoid are left intact for several days after severing the cap, the stalk develops the ability to grow a new cap, independent of the rhizoid. · However, if actinomycin D (which prevents RNA synthesis) is injected into the rhizoid at the time of cap removal, regeneration is prevented. (3) And finally, if the nucleus is removed from a young plant before cap formation, the plant will still form the cap—though, of course, no spores can be produced. Since cap formation is accompanied by the synthesis of several new proteins, the nucleus must produce some long-lived, stable, but inactive messenger RNA's that are translated long after they are made.

The picture that emerges is one in which the nucleus releases to the cytoplasm diffusible substances responsible for altering the differentiated properties of the cell. Some of those substances or their products must be able to reenter the nucleus to turn genes on and off. This view is supported by the nuclear transplant experiments that lead to animal cloning. It is also supported by experiments in which a differentiated cell is fused with an undifferentiated cell of the same species. Though there are numerous pitfalls in the interpretation of cell fusions, there is the clear suggestion of cytoplasmic factors in the undifferentiated cell capable of repressing genes in the more specialized cell. The existence of such factors also explains why the progeny of a cell having specialized functions normally inherits the capacity to carry out those same functions.

10–2 INTERCELLULAR COMMUNICATION

Cellular differentiation in single-celled organisms such as *Acetabularia* appears to involve a timed sequence of changes due to a reciprocal interaction between cytoplasm and nucleus. The environment of the organism might alter the course of events, but the molecular factors involved in the regulation of differentiation are ordinarily provided by the cell itself. This is not necessarily the case in multicellular organisms.

EVIDENCE FOR INTERCELLULAR COMMUNICATION. As the embryonic brain develops in higher animals, two outgrowths called optic vesicles cause the tissue lying over them to form lenses for the new eyes. If an optic vesicle is removed from a salamander embryo and placed under the skin of the embryo at another point, a lens will form there instead. If the optic vesicle of a large, rapidly growing species of salamander is covered by the skin of a smaller, slower growing species at a similar stage of development, the result is an eye that is properly proportioned but intermediate in size. One can show that these changes are due to the intercellular transfer of high molecular weight substances by separating the optic vesicle from the overlying tissue with a cellulose acetate membrane that is impermeable to macromolecules. Such membranes prevent the interactions necessary for differentiation of the overlying tissue.

Some of the simpler animals and plants provide useful models to study the role of this intercellular communication during differentiation. One example is the aggregation and consequent differentiation of cells of the cellular slime mold, *Dictyostelium discoideum.*

The amebas of the cellular slime mold are formed from spores. They replicate and remain free living until they can no longer find food (they normally live mostly on soil bacteria), whereupon some of the cells emit a chemical attractant, called an *acrasin* (identified now as cyclic AMP), that causes other cells to cluster about them. The aggregate, or *pseudoplasmodium* of the cellular slime mold, is a slug-like body with some power of locomotion. However, it eventually settles into a puddle of cells and begins to differentiate, forming a stalk and a fruiting body. The fruiting body is responsible for the formation of a new generation of spores (see Fig. 10–5). (The spores are produced by mitotic nuclear divisions since cellular slime molds remain haploid throughout their life cycle.) Although cells of the slug form appear undifferentiated, their fate is probably determined by the order in which they aggregate. The first cells to come together will eventually be found in the lower stalk of the mature plant, the last cells added are found in the base, and the cells picked up in between form the fruiting body. But when cells are removed from the plant at any stage before the mature fruiting body is completed, the isolated cells revert to the ameboid state. Their ultimate position in a new plant will then be determined by the order in which they are added to the new aggregate.

Thus, in these simple animals and plants, we find evidence for cell recognition and an interaction leading to a mutual differentiation. This transfer of information between cells could be through the release of a substance from one cell and its uptake by the receptor cell, as with the chemotactic aggregation of *Dictyostelium*. Or it could involve direct cytoplasm-to-cytoplasm bridges. Such junctions have long been recognized in plants, where they are called plasmodesmata, and in a few special cases in animals, such as certain steps in egg and sperm formation. But it has recently become clear, largely from the work of W. R. Loewenstein at Columbia University, that direct contact between the cytoplasm of adjacent animal cells may be exceedingly common.

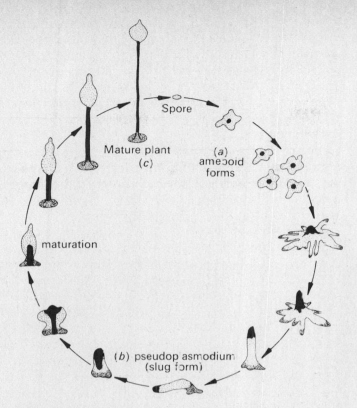

FIGURE 10–5. The Life Cycle of *Dictyostelium discoideum*. The cellular slime mold exists in a free-living, single-celled ameboid form (a), as a motile, multicellular "slug" (pseudoplasmodium) (b), and as a sessile, highly differentiated, spore-producing plant (c). The position and differentiated characteristic of a cell in the mature plant appears to be determined by the order in which it joined the aggregate.

Such bridges form the basis for a highly specific intercellular communication.

CYTOPLASMIC BRIDGES. Low resistance junctions between cells can be demonstrated by using electrodes—in the form of tiny micropipettes filled with a salt solution—to pierce the membranes of cells and thereby measure the flow of electric current (ions) between them (see Fig. 10–6a). The electrical resistance between a single cell and its surrounding solution is very high, reflecting the integrity of the cell membrane and its relative impermeability to ions. Thus, if two cells are impaled by electrodes, the insulation provided by intervening membranes should prevent current from passing between the cells. But when the two cells come together, an electrical pulse introduced to one can sometimes be seen in the second, indicating a low resistance contact between the adjacent cells.

In order to find out more about the nature of low resistance intercellular contacts, Loewenstein injected fluorescein dye into one of the cells engaged in such contact. Within a few minutes, the use of an ultraviolet lamp revealed a telltale green fluorescence in a number of cells grouped together with the injected cell, as in Fig. 10–6b. Thus, the intercellular contacts permit the transfer

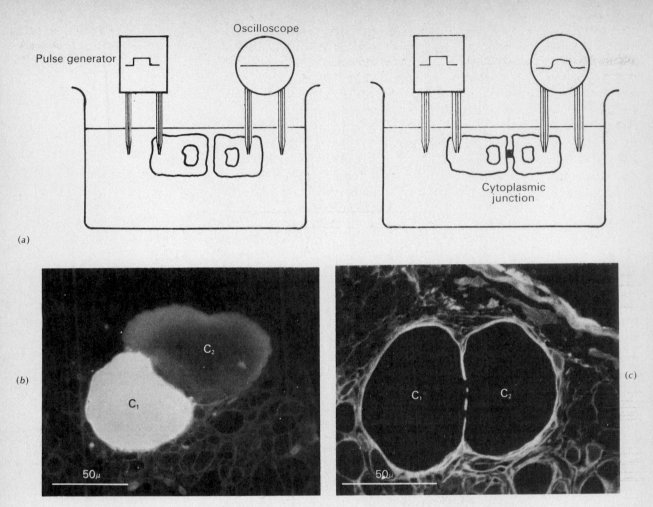

FIGURE 10–6. Intercellular Communication. **(a)** The existence of ion-permeable channels between cells can be demonstrated with a pulse generator (here producing a square wave), using an oscilloscope as a detector. **(b)** Nerve cell processes (axons) of the crayfish (*Procambarus*). The "electrotonic coupling" between these cells is due to channels detectable as described in (a). They can also be detected by intercellular diffusion of a dye, procion yellow M4RS (about 500 amu). The dye is injected into one cell (C₁) and detected by its fluorescence, first in C₁ and later in the adjacent cell (C₂). **(c)** The same dye is added to the extracellular space, where it stains the fatty covering (Schwann sheath) around a similar set of two cells, C₁ and C₂, penetrating neither. (Note the discontinuous nature of the covering between the cells. The cell membranes are intact in these regions, but not stained.) [(b) and (c) courtesy of B. W. Payton, M. V. L. Bennett, and G. D. Pappas, *Science*, **166**:1641 (1969). Copyright by the A.A.A.S.]

not only of simple ions, but of organic molecules at least as large as fluorescein, which has a molecular weight of 332 amu. Subsequent work in Loewenstein's laboratory has shown that the junctions through which fluorescein passes are cytoplasmic channels and that they permit the intercellular movement of a variety of molecules and ions, including at least some of the smaller proteins.

Cytoplasmic bridges of this type have now been found in a wide variety of embryonic and mature tissues and can form even between cells of different species. Their presence would presumably permit a controlled passage of material from one cell to a restricted range of other cells without the uncertainties associated with passage through the surrounding medium. Thus, bridges could account in part for some aspects of embryonic development, for they provide a way for cells to keep track of each other.

Bridges between the cytoplasm of adjacent cells may also explain the phenomenon of contact inhibition of movement, discussed in Chap. 9 as the failure of motile cells to continue to move about once they have reached a confluent monolayer on the bottom of a culture vessel. Contact inhibition *in vitro* is presumably related to the ability of cells to control their movement and possibly their replication *in vivo* during the development of organs and tissues.

10-3 STABILITY OF THE DIFFERENTIATED STATE

Differentiation into various cell lines has been treated as the turning-on of some genes and the turning-off of others. A step called *determination* must precede the actual expression of specialized characteristics. Whatever the nuclear changes that lead to determination and differentiation, those changes are relatively stable and capable of being passed along from one generation to the next as a cell divides.

It is easy to observe that differentiation in a cell is a condition that can be inherited, for usually one can see that daughter cells look like the parent cell. Determination itself, on the other hand, does not ordinarily result in visible changes. Hence, we will consider first a system in which it is demonstrated that determination is inherited in the same way as differentiated functions. That having been established, conditions will be examined that lead to the loss of determination and differentiation, yielding a cell that may be capable of redifferentiation along different lines. Such examples support the view that determination and differentiation result from the programming of gene expression, not from any changes in the genes themselves.

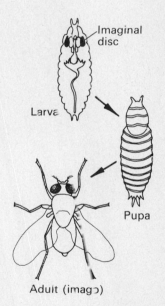

FIGURE 10-7. *Drosophila.* Most of the cells of the larva break down during pupation. Adult structures emerge from determined but undifferentiated larva cells found in clusters called imaginal discs

DETERMINATION AND TRANSDETERMINATION. The fruit fly, *Drosophila melanogaster*, like other insects goes through four distinct phases during its life cycle: embryo, larva, pupa, and adult (also called an *imago*). During pupation, most of the differentiated cells of the larva dissolve, and their components are utilized to form the adult structures (see Fig. 10-7). However, the adult cells themselves are descendants of embryonic tissue carried by the larva, not of the mature cells found therein. This embryonic tissue is present in discrete bundles called *imaginal discs*, although proliferation of disc cells takes place during the larval stage, their differentiation does not begin until it is triggered by hormones produced only during pupation.

One can demonstrate that cells of the imaginal discs are already determined in the larva by removing them from their host and placing them in the abdominal cavity of a second larva. When the new host undergoes pupation, both its own and the transplanted imaginal discs differentiate. The result is a fly with accessory organs. A transplanted leg disc, for example, produces leg structures in the abdomen of the host, and so on. In fact, it can be shown that the cells of a particular disc may be individually determined—that is, certain cells in the leg disc will form hairs, others muscle, and so on.

When imaginal discs are transplanted from a larva to the abdomen of an adult, the cells proliferate, but due to the lack of the proper hormones they do not differentiate. Transplantation of a portion of this tissue from one fly to another can keep the cells viable for many generations, during which time their capacity to differentiate into mature structures is retained, as may be shown by transferring a portion back to a larva about to undergo pupation. Determination is thus seen to be a condition that can be passed from a cell to its progeny.

Although it is normally a stable condition, the determination of proliferating imaginal disc cells does occasionally change. In investigating this phenomenon, Ernst Hadorn found that the change in determination, called *transdetermination*, results in a new pattern of development according to rather rigid rules. For example, when a genital disc was followed through several such changes, the following pattern was observed:

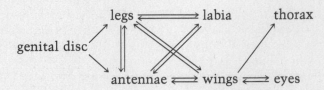

The arrows indicate the direction in which transdetermination can occur. Thus, after a change in determination from presumptive genitalia to antennae, a further change can produce labia, legs, or wings, but not thorax or eye tissue. The latter, however, can arise from cells that have already undergone the transition from presumptive antennae to wings. Hadorn also found that the probability that one of these transitions will take place increases with the number of cell divisions that occur before the test is made.

Transdetermination produces a new genetic programming, changing the capacity of the cells to differentiate. This new condition is as stable as the original condition—that is, the state of determination is normally passed on to progeny as the cells prolif-

erate. However, that transdetermination when it does occur
follows only preset pathways suggests a certain rigidity, or pattern,
to the way in which genes can be turned on and off. This restriction in the observed changes is consistent with our view of development as the sequential activation of genes, with only a restricted
number of possible variations in order.

pick up here

REGENERATION. While transdetermination gives us some idea of
both the stability and capacity for change present in determined
cells, it tells us nothing about the corresponding capacity for
change in already differentiated tissues. We normally think of differentiation as a one-way process, and it is true that the most
highly specialized cells (e.g., nerve and muscle) are also the most
stable. Since DNA replication seems to be a prerequisite to
changes in determination or differentiation, the functional stability
of nerve and muscle cells probably arises from the fact that these
cells rarely, if ever, divide. This is an important consideration, for
a mitotic accident in such highly specialized cells is apt to have
serious consequences for the animal. One way to prevent such accidents is to design the cells so that replication is not necessary.[1]

However, many fully differentiated cells *can* undergo mitosis,
and the loss and regain of differentiated function can and do occur
in certain instances. For example, when the limb of a newt is amputated, the wound is closed in a matter of hours. But underneath
this seal, a mound of embryonic tissue called a *blastema* forms.
The cells of the blastema proliferate and differentiate into a new
limb just like the one that was lost (see Fig. 10–8). The trigger and
control for this process apparently involve the presence of nerve
axons that regrow from the still viable nerve cell bodies in or near
the central nervous system: If innervation is prevented, regeneration does not take place and scar tissue forms instead. (Note, however, that innervation is not essential to the original, embryonic
formation of a limb.)

If a newt is heavily irradiated with X rays before amputation,
regeneration is prevented. Yet if a new limb is grafted to the irradiated animal and then subsequently amputated, a regeneration
blastema may form, depending on where the cut is made. If all of
the grafted tissue is removed during amputation, no new growth
occurs; but if even a small amount of grafted tissue is left behind at
the second amputation, blastema formation and regeneration take
place normally. These and other experiments lead to the conclusion that the embryonic tissue of which the blastema is composed

---

1. The argument is also made that specialized functions and cell division are
antagonistic because both make great demands on the available energy. In other
words, the metabolism of a cell can support one or the other activity but not both.

STABILITY OF THE
DIFFERENTIATED STATE
Section 10–3

439

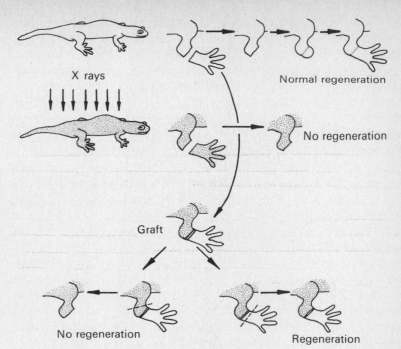

X rays

Normal regeneration

No regeneration

Graft

No regeneration

Regeneration

FIGURE 10–8. Regeneration. When a newt limb is amputated, embryonic cells (a regeneration blastema) appear at the site of injury and, in a few weeks, differentiate to produce a new limb. The cells of the blastema are locally derived, as was demonstrated by grafting a limb to an irradiated newt. Whether one sees regeneration after a second amputation depends on whether unirradiated tissue is left after the second cut. [Described by E. G. Butler, *Anat. Record,* **62:**295 (1935).]

is locally derived from the dedifferentiation of mature structures, including muscle, skin, etc.

Thus, like determination, differentiation itself is not necessarily the result of permanent changes in the cell, but a process that can sometimes be reversed. In addition, we see that a normal pattern of tissue interactions and determinations leading to development can in some animals be initiated in adult tissue.[2]

TUMORS. While the ability of cells to lose their differentiated properties during limb regeneration can be a useful feature in those animals that have it, in other cases a change in the state of differentiation may have drastic consequences, as with the formation of malignant tumors.

A tumor is a *neoplasm*, or abnormal growth of new tissue. Such growths can arise anywhere in the body. It may be quite *benign*, or harmless, such as the common wart, or it may be malignant, in which case it invades and destroys the supporting tissue on which it grows, eventually causing the death of the host. In addition, a malignant tumor, or *cancer*, may slough cells that establish new colonies of malignant cells in other parts of the body. This process

2. A limited amount of limb regeneration can be induced in mammals. No complete regeneration has yet been achieved, however, nor have the experiments been extended to humans, except for the use of small electric currents to promote the healing of broken bones. However, it is not too farfetched to hope that one day a severed human arm or leg can be regrown.

of spreading, called *metastasis*, hastens the time when a vital function will be interrupted by one of the growths and makes surgical removal and other therapies less effective.

Cancer has received a great deal of attention from biological and medical scientists not only because the disease is a major medical problem, but because it represents an aberration in some very basic mechanisms involving the regulation of cellular activity, including growth and differentiation. If we knew exactly how the replication of normal cells is controlled, we would have some important clues as to why this control is not exercised in malignant cells. Conversely, if we knew in detail how cells were converted to the malignant state, it would tell us important facts concerning normal cell control.

Our objective here is to use some of the observations made on cancer cells to further illuminate the problem at hand, namely the mechanisms involved in achieving and maintaining a differentiated state. Cancer cells have failed in one or the other of these tasks. The failure could be due to the loss or gain of one or more genes via mutation or viral infection; or it could be the result of changes in the way normal cellular genes are expressed (i.e. epigenetic changes). That, too, might be caused by viruses.

CHARACTERISTICS OF THE MALIGNANT CELL. Let us start by describing some of the more important characteristics of the malignant cell (see Fig. 10–9). Although these characteristics vary somewhat from one cancer to another and between various stages in the same disease, one of the earliest and most outstanding features is the capacity of the malignant cell for autonomous growth. That is, the cells exist and reproduce independent of normal regulatory mechanisms. The changes include alterations in the cell surface that lead to a decline in the selective adhesiveness of the cells and to a loss of contact inhibition. When normal cells from two different tissues are mixed in a culture, each usually reassociates with its own kind to form structures reminiscent of the original tissue. Mixed aggregates do not form. Cancer cells, however, adhere indiscriminately to all other cell types present, thus interfering with normal reassociation. In addition, while normal cells usually grow to form only a monolayer on the supporting medium and then stop dividing, cancer cells continue to replicate, piling one atop the other to form thick areas called *foci* (see Fig. 10–10). It is this latter property that is most often used to distinguish the malignant cell in tissue culture.

Other characteristic features of cancer cells include their ability to invade the supporting tissues on which they grow and their tendency to metastasize. These features are present to some extent in all cancers, but normally appear later in the disease and may vary in degree. Even the capacity for autonomous growth, which is the

metastasis — the spread of disease from one part of the body to another unrelated to it.

441

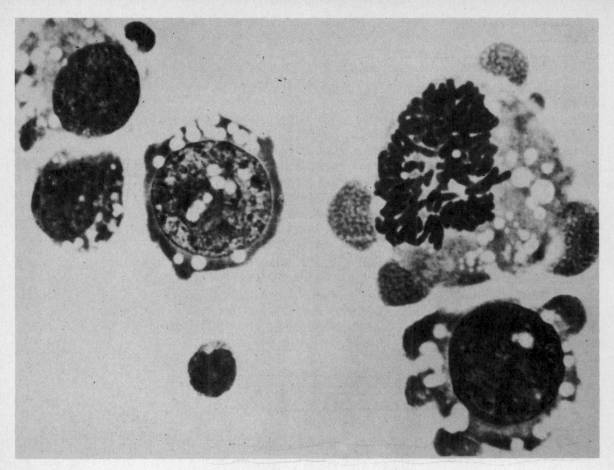

FIGURE 10–9. Cancer Cells. From a mouse mammary tumor grown *in vitro*. Note the large nucleus, the many chromosomes in the mitotic cell, and the outgrowths, or blebs, on the cell surface. [Photo by A. Rivenson. From W. Etkin, R. Devlin, and T. Bouffard, *A Biology of Human Concern*. Philadelphia: Lippincott, Co., 1972.]

most basic feature of any tumor, is sometimes conditional. For example, a young prostate or breast tumor may grow only as long as the proper hormonal stimulus is present, although it often becomes independent of hormonal control in its more advanced stages. In addition to maintaining vestiges of their former sensitivity to normal growth controls, many tumors retain some of the differentiated properties of the normal tissue from which they were derived (i.e., they undergo only a "partial dedifferentiation"). Thus, tumors growing out of an endocrine gland may continue to secrete hormones, leading to gross aberrations in the normal functioning of the body. But tumors may also develop differentiated features of other cell types, such as the secretion of hormones not ordinarily made by the tissue from which the tumor is derived.

TUMORIGENESIS. The search for the causes of all these changes in malignant cells has been complicated by what appears to be a multiplicity of unrelated mechanisms. In the first decade of this century, V. Ellermann and O. Bang of Denmark succeeded in transfer-

442

ring leukemia from one bird to the next with a cell-free filtrate. At about the same time, P. F. Rous demonstrated that cell-free filtrates can also transmit some solid tumors (sarcomas) between birds. The causative agent in both cases is a virus. Although we recognize that malignancies can arise from a wide variety of radiations, chemicals, and physical irritants, there is increasing suspicion that viruses may be a major cause and that some other "causes" may really act by stimulating viral activity.

The genetic material of some viruses can become incorporated into the chromosomes of the host cell, and remain there in a latent (provirus) state through many generations. The genes of the provirus get replicated with the host genes, so that an entire culture of cells, each carrying the provirus, can be grown from a single in-

FIGURE 10-10. Growth of Malignant Cells. Most normal cell lines stop dividing when a confluent monolayer is reached. Malignant cells, on the other hand, pile one atop the other to form foci. One such example is seen here, in a field of normal chick fibroblasts. [Specimen courtesy of G. Beaudreau, Oregon State Univ.]

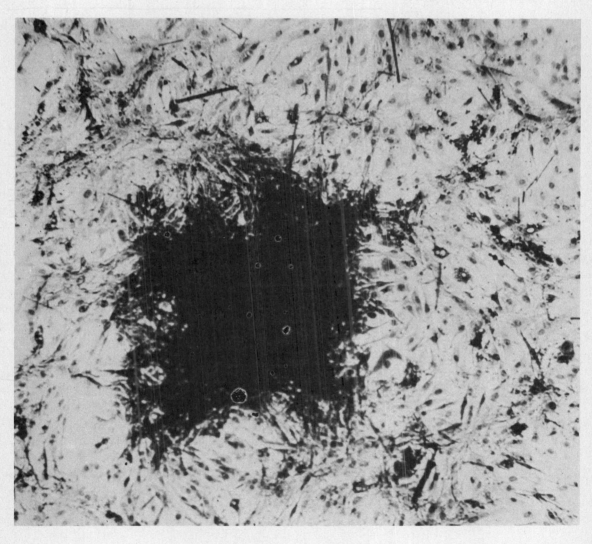

fected ancestor. At each replication there is a certain small, but finite, chance that the latent provirus will spontaneously activate, causing the production of new virions. Activation can also be induced prematurely by radiation and certain chemicals—agents that are often demonstrably carcinogenic (cancer causing).

Robert Huebner and George Todaro of the National Cancer Institute proposed in 1969 that an oncogenic (cancer causing) provirus exists in all normal cells, a result of vertical transmission from parent to child over countless generations since the initial infection. The viral genes, according to this theory, are normally repressed by genetic regulators. One of the viral genes, the *oncogene*, produces a substance that can transform a cell to the malignant state. Cancer would be the result of the accidental derepression of this gene; viral replication, on the other hand, would require that the entire provirus be derepressed. According to some reports, cancer viruses have, in fact, been induced in cell lines with no known prior exposure to the virus. Although such reports tend to support the oncogene theory, other recent studies indicate that at least some human cancer cells have unique, presumably virus-derived, genes not found in normal cells of the same type—thus suggesting that the introduction of new genetic material by a virus particle may be responsible for the malignancy in these cases.

Among the most widely studied tumor viruses are the avian and murine (bird and mouse, respectively) leukemia viruses, the Rous sarcoma virus (RSV) mentioned above, and a pair of very similar viruses that cause solid tumors in animals, SV40 (for Simian Virus 40) and polyoma. These latter two viruses are unusual in their ability to cause tumors in a number of different animals, though not all hosts are equally susceptible to a malignant transformation by them. Polyoma and SV40 contain DNA, although only enough to code for 5–10 proteins, about half of which are essential in the establishment of malignancy. The other cancer viruses just mentioned contain RNA instead of DNA. For several years this was a very puzzling situation, since there was no reason to believe that RNA could incorporate directly into cellular DNA. However, in 1969 Howard Temin reported experimental evidence for the existence of an enzyme, associated with some viruses, that can polymerize DNA complementary to an RNA template. With these RNA-dependent DNA polymerases, or *reverse transcriptases*, one can postulate that all the oncogenic (cancer causing) viruses insert new genetic information into the chromosome of the host. The added genes may result in altered cellular properties, such as a loss of contact inhibition, either directly or by interfering with the normal scheme of gene expression in the host.

DIFFERENTIATION OF TUMOR CELLS. The nuclei of tumor cells may have a full set of normal genes in addition to whatever viral genes they harbor, if any. This genetic completeness has been demonstrated by transplanting nuclei (see Fig. 10-11) and by causing the tumor cells themselves to differentiate. For example, Armin C. Braun of Rockefeller University succeeded in causing complete new plants to develop from single cells taken from a portion of a crown gall tumor, which is a plant tumor. He proceeded by forming a clone of the tumor tissue and then by grafting a portion of it to the cut stem of a tobacco plant tip. Under the influence of the host plant, the grafted tissue regained a portion of its ability to organize mature structures. When some of this still-abnormal growth was grafted to a second tobacco plant, even more function was regained, culminating in the production of seeds from which perfectly normal, healthy plants were grown.

The genetic potential of the malignant cell has also been demonstrated with amphibian regeneration. When a fragment of the Lucké adenocarcinoma (an adenocarcinoma is a cancer of gland tissue) is transplanted from the leopard frog, *Rana pipiens*, to the regeneration blastema of an adult newt amputee, the tumor cells differentiate into muscle, cartilage, and connective tissue as the new limb forms. To establish that the tumor cells themselves are involved in the differentiation, M. Mizell in 1965 amputated a newt tail by cutting through a transplanted tumor already growing in the region. The tumor cells had been heavily labelled with tritium before transplantation, so autoradiographs of the regenerating tissue could be used to determine the fate of the tumor nuclei. In

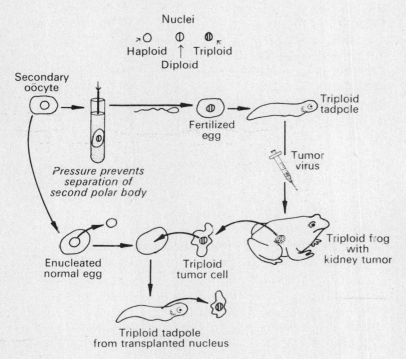

FIGURE 10-11. The Developmental Potential of a Tumor Nucleus. Hydrostatic pressure was used to produce a diploid egg which, after fertilization, became a triploid frog. Nuclear transplantation from a cancer cell of this frog to an enucleated but otherwise normal egg produced a triploid tadpole, thus demonstrating the developmental potential of the tumor nucleus. [Described by R. McKinnel, B. Deggins, and D. Labat, *Science*, 165:394 (1969).]

this way he left no doubt that the tumor cells had been influenced to differentiate into a variety of normal tissue.

It is more difficult to demonstrate the genetic potential of malignant cells in mammals because of their relative lack of regenerative power. However, differentiated (and therefore nonmalignant) cell types are often associated with, and derived from, certain kinds of malignant tumors. To demonstrate the source of these benign tissues, cloned single cells from a human teratocarcinoma, which is a cancer derived from embryonic tissue, were transplanted to the peritoneal cavity of mice. A portion of the cells differentiated into at least a dozen mature cell types, including bone, brain, cartilage, and so forth, though other portions of the tumor remained malignant.

Spontaneous reversion to mature cell types can also be induced *in vitro* in human neuroblastoma cells by manipulating their environment. (Neuroblasts are precursors of mature neurons.) When grown in tissue culture, most such tumors can be made to differentiate to mature neural tissue by the addition of cyclic AMP or other substances. Differentiation has also been observed *in vivo*, where it is associated with occasional spontaneous recovery from the cancer, since the mature neural cells are, of course, benign.

It thus appears that some cancers, at least, are failures in the genetic regulatory mechanisms that govern normal differentiation and not the result of gene gain, loss, or mutation. The frequent reference to a tumor as "dedifferentiated" tissue, or tissue that has reverted to the embryonic state, may be an oversimplified interpretation. However, to the extent that malignancy can be attributed to epigenetic changes rather than to mutation, the description is fairly apt. In support of this interpretation, there have been reports of a similarity between fetal antigens of unknown origin and new antigens (carcinoembryonic antigens or CEA) appearing with certain cancers. (An antigen is any molecule or surface feature that can be detected by immunological tests.) These views have led to expanded research on the possibility of curing cancers not by killing the malignant cells, but by causing them to differentiate to mature, benign cell types. Unfortunately, progress is impeded by an incomplete understanding of the mechanisms by which differentiation is achieved.

10-4 SENESCENCE

The last stage in the development of many cells is senescence, or old age, culminating in death. It is a phenomenon that seems to be limited to the more complicated organisms, however, for there is no obvious counterpart in procaryotes or simple eucaryotes: a bacterial cell, for instance, continues to divide as long as the environment is favorable. It is only cells with a high degree of specialization that seem to have a life cycle that terminates in death.

VARIATIONS IN LIFE SPAN. Human erythrocytes are sometimes used in studies on cellular senescence, for they undergo profound changes as they near the end of their 120-day life span. The activities of various enzymes decrease, the cells become progressively more fragile and more dense, and the ratio of positive to negative ions on their surface changes. (These differences make it possible to separate younger from older erythrocytes.) In addition, the changed surface properties make old erythrocytes susceptible to phagocytosis and destruction by certain scavenger cells, especially in the spleen.

One is tempted to suggest that erythrocytes are a special case, and that the processes of aging cannot be learned by studying such a simple cell. It is more likely, however, that the processes are the same in all cells, at least from the higher animals, even though different cells have differing abilities to withstand the ravages of time. The mammalian erythrocyte is unique in that it does not possess a nucleus and therefore has no biosynthetic activity. It also has a shorter lifetime than most cells, including the erythrocytes of other animals—compare, for example, the 120-day life span of the human erythrocyte with the 1200-day span of the toad red cell. The rapid aging of mammalian erythrocytes suggests that the capacity to maintain vitality is somehow associated with nuclear activity.

The most general assumption concerning the process of aging is that it is due to accumulated insults, such as random "hits" from background radiation. Even if most such damage is repaired, one may expect a gradual loss in ability to accurately synthesize proteins. This theory, which is essentially the "error catastrophe" theory of aging proposed by L. E. Orgel in 1963, predicts a buildup of debris within the cell, some of which may form a part of the age pigment, or *lipofuscin*, that one sees in older cells (see Fig. 10-12).

Accumulated genetic damage cannot be the whole story of aging, however, for different nucleated cell types have vastly different average survival times (see Table 10.1). Yet each cell in a given animal has the same set of genes, and therefore potential access to the same repair mechanisms. (The mammalian erythrocyte and gametes are obvious exceptions.) Of course, exposure to abrasion and other forms of physical or chemical abuse can account for the rapid turnover of epithelial cells, such as the outer layer of skin. But even among cells in a protected environment, turnover rates vary markedly. We must, therefore, consider the possibility that cells have a built-in rate of senescence—i.e., that something "wears out" after a predetermined life span.

PROGRAMMED DEATH. The concept of programmed death certainly has ample precedent in embryology, for it has long been known that the death of certain cells is a necessary part of the embryonic development of form. For example, separation of the

447

(a)

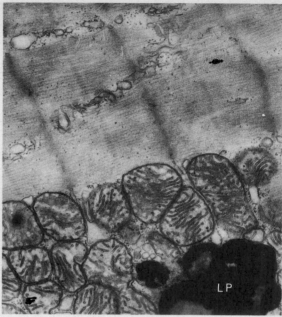

LP

(b)

FIGURE 10–12. The Accumulation of Age Pigment in Rat Cardiac Muscle. **(a)** From a 90-day-old rat. **(b)** From a 1004-day-old rat. The age pigment (lipofuscin, LP) is the dark area. (The contractile filaments seem discontinuous only because they are not parallel to the plane of section.) [Courtesy of N. M. and D. F. Sulkin, *J. Gerontology*, **22:**485 (1967).]

TABLE 10.1   Cellular life spans in the adult mouse[a]

Near or equal to that of the animal itself
   Neurons and associated cells
   Muscle cells of all types
   Brown fat cells
   Osteocytes (bone cells)
   Kidney medullary tubule cells
   Cells of the adrenal medulla
   Stomach zymogen cells
Slow renewal (more than 30 days, but less than the mean life span of the animal)
   Respiratory tract epithelium
   Kidney cortical cells
   Adrenal cortical cells
   Liver parenchymal cells
   Pancreas acinar and islet cells
   Salivary gland cells
   Skin connective tissue cells
   Stomach parietal cells
Fast renewal (less than 30 days)
   Skin (epidermis)
   Cornea
   Epithelium of the mouth and gastrointestinal tract
   Precursors of red and white blood cells

[a] Most data from I. L. Cameron, *Texas Rep. Biol. Med.*, **28:** 203 (1970).

fingers and toes of animals is due to the death of the tissue between them. Similarly, the chick wing is shaped, in part, by the death of selected cells. In an elegant set of experiments with this latter system, J. W. Saunders, Jr., demonstrated that the time of death of these cells is established at a given point during development in much the same way as other characteristics are determined at given times. That is, if cells that are destined to die during the shaping of a wing are transplanted before a certain stage of development, they may either die on schedule or survive as well as other cells, depending on the site of transplantation. After a certain point in the developmental scheme, however, death occurs on schedule no matter where they are moved, or even if they are transferred to tissue culture.

One could form a convincing teleological argument for programmed death, for a species that is immortal is a species that cannot evolve once the upper limit of its population is reached. It is only through successive generations that natural selection can improve a species, or make it better suited to a changing environment. One would expect, therefore, that the most favorable life span would be a period long enough to ensure procreation, but little more, just as a salmon dies after spawning. On this basis, we would predict that animals with long gestation periods, long periods of offspring dependency, or low offspring survival rates, should either live longer or produce huge numbers of descendants. To a reasonable degree, this does seem to be the case.

Further support for the concept of programmed death comes from the work of Leonard Hayflick at Stanford University. He found that cultured human fibroblasts (the cells that produce collagen) obtained from embryonic tissue undergo only about fifty doublings before the culture dies out. When fibroblasts are obtained from persons after birth, the number of potential doublings is less—later established as a loss of 0.2 doublings per year of age, on the average. Furthermore, when cultured cells are frozen, even for a period of years, they "remember" their generation number. That is, when they are thawed and allowed to continue growing, they complete the originally assigned number of generations, and die on schedule.

It does not appear that accumulated radiation damage is the reason for the death of these cultures, for that should continue to some degree even in the frozen state. Hayflick also eliminated the possibility that a buildup of waste products kills the culture: when a male population at the fortieth generation was mixed with a female population at the tenth generation, both ran their normal course, oblivious of the other cells. This, plus careful control of the culture medium and population density, indicates that death is not a result of the gradual poisoning of the culture by a waste product or specific excretion, nor does it come from the depletion of an essential nutrient.

Although the length of survival of cells in tissue culture is quite variable, depending on the source and type of cell being cultured, it is true that most normal human cells cannot be grown indefinitely. When a permanent cell line is established, it is found to contain variants—often with abnormal chromosome numbers—that are somehow immune to normal aging. Tumor cells are particularly easy to culture because the transformation to an "immortal" condition seems already to have taken place in them. This capacity is another consequence of their release from normal controls on growth.

If normal human cells cannot be grown indefinitely in tissue culture while malignant human cells can be, we are tempted to conclude either that examination *in vitro* is not a fair test of life span or that normal human cells are programmed to die. If the latter is true, then we must also conclude that the programming is subject to reversal. The slow decline in the efficiency of bodily functions with age might therefore be traced to a built-in rate of cellular disability and death. If we could but identify the process that appears to result in programmed death, it might be possible to interfere with that process and, for better or worse, dramatically extend the human life span.

SUMMARY

10-1 Differentiation normally results from epigenetic changes rather than from an irreversible loss of genes. This can be shown by cloning experiments in which complete animals and plants are grown from single cells, where either the cell itself (in plants) or the nucleus (in animals) is derived from highly differentiated tissue. However, though nuclear genes are clearly responsible for the emergence of adult features, they are influenced by cytoplasmic factors, the earliest of which, in the case of animals, must be present in the egg before fertilization. Differentiation is viewed as a reciprocal gene/cytoplasm interaction in which regulators that control one stage of cellular development are produced by the activity of genes during the previous stage.

10-2 The development of organized structures requires a mechanism for cell recognition and communication. The cellular slime molds are useful model systems for studying these phenomena, for the developmental paths of the cells of the pseudoplasmodium are determined by the order in which they

were added to the aggregate. Aggregation in this case is due to secreted chemotactic agents, but intercellular communication in this and other tissues may be due in part to cytoplasm-to-cytoplasm bridges capable of passing a variety of small organic molecules and probably at least some smaller proteins. Cytoplasmic bridges provide one mechanism by which cells can keep track of each other in the formation of tissues and organs.

10-3 Though we normally think of them as stable conditions, neither determination nor differentiation is irreversible. Transdetermination in *Drosophila*, for instance, is due to an occasional accidental change in the determination of cells as they proliferate. The pattern of change, however, seems to be highly restricted, an observation that suggests an inherent limitation in the sequence of developmental determinations.

The conditional stability of fully differentiated tissue is revealed by the regeneration of limbs in amphibia and by transformation to malignancy. Limb regeneration involves the dedifferentiation of mature tissue to an embryonic condition and its subsequent proliferation and redifferentiation. This process also indicates the importance of tissue interactions during development, as the capacity to form the proper structures is obviously inherent in the cells and requires nothing that is unique to the embryonic environment.

Tumors also result from changes in differentiated properties, for besides losing their growth restraints (e.g., contact inhibition), transformation to a malignant state may be accompanied either by the loss of specialized functions, by the gain of inappropriate specialized functions, or by both changes.

10-4 The last stage in the development of special-ized cells is old age, or senescence. It is accompanied by profound changes in the properties of the cell, including alterations in both structure and enzymatic activities. It is not clear, however, why these changes should occur, as they are not present in the simplest forms of life, nor in mammalian cells that have been adapted for growth in vitro.

Cell death is not always associated with old age, but is a normal part of the embryonic shaping of hands, feet, and wings. The cells involved seem to be determined for this fate at a particular stage in development. Thus, it is possible that eventual senescence is programmed into a cell in much the same way as any other function.

STUDY GUIDE

10-1 (a) What experiments suggest that the pattern of gene expression in a highly differentiated cell is reversible, and that most such cells still contain a complete set characteristic of the species? (b) What experiments, in animals and plants, indicate that nuclear activity is under the control of cytoplasmic factors? (c) What experiments indicate that the cytoplasmic factors themselves result from an earlier RNA synthesis in the nucleus?

10-2 (a) What features of the cellular slime molds make them useful in studies on cellular differentiation? (b) By what mechanism do the cells of the cellular slime mold find each other? (c) What experiments indicate the presence of intercellular cy-toplasmic bridges in a large number of animal tissues?

10-3 (a) What is transdetermination, and what does it tell us about the stability of determination? (b) What are the origin and ultimate fate of the cells in a regeneration blastema? (c) Why is it believed that at least some malignant cells (plant and animal) still carry a full complement of normal genes?

10-4 (a) Eventual senescence is less common in cells that reproduce themselves as opposed to those that result from the differentiation of a simpler, stem cell. Give examples that support this statement. (b) What role might programmed death play in normal development?

REFERENCES

GENERAL

BELL, EUGENE, ed., Molecular and Cellular Aspects of Development. New York: Harper & Row, 1965. Reprinted papers, with useful introductions.

BRACHET, J., Introduction to Molecular Embryology. New York: Springer-Verlag, 1974. (Paperback.)

COWARD, S. J., ed., Developmental Regulation. Aspects of Cell Differentiation. New York: Academic Press, 1973. A series of papers and reviews.

EBERT, J. D., and I. M. SUSSEX, Interacting Systems in Development (2d ed.). New York: Holt, Rinehart and Winston, 1970. (Paperback.) An excellent introduction to cellular differentiation and morphogenesis.

FANKHAUSER, GERHARD, "Memories of Great Embryologists." American Scientist, 60:46 (1972).

FLICKINGER, R. A., Developmental Biology. Dubuque, Iowa: W. C. Brown Co., 1966. (Paperback.) A collection of original papers.

GABRIEL, M. L., and S. FOGEL, eds., Great Experiments in Biology. Englewood Cliffs, N. J.: Prentice-Hall, 1955. Includes papers by Driesch, Spemann, Mangold, and H. V. Wilson.

GARROD, D. R., "Cellular Development." (Wiley Outlines) New York: John Wiley and Sons, 1973. Patterns of development.

HARRISON, R. G., Organization and Development of the Embryo, ed. Sally Wilens. New Haven, Conn.: Yale Univ. Press, 1969. Reprints of some of Harrison's classical papers, annotated and illustrated.

LOOMIS, W. F., ed., Papers on Regulation of Gene Activity During Development. New York: Harper & Row, 1970.

SUSSMAN, MAURICE, Developmental Biology: Its Cellular and Molecular Foundations. Englewood Cliffs, N. J.: Prentice-Hall, 1973. (Paperback.)

WHITTAKER, J. R., Cellular Differentiation. Belmont, Calif.: Dickenson Publ. Co., 1968. (Paperback.) A book of readings containing excerpts, mostly from modern papers of note.

GENE–CYTOPLASM INTERACTIONS

ASHWORTH, J. M., "Cell Differentiation." (Wiley Outlines) New York: John Wiley and Sons, 1973. Emphasis on biochemistry.

BALLS, M., and F. S. BILLETT, *The Cell Cycle in Development and Differentiation.* London: Cambridge Univ. Press, 1973. See especially the reviews on chalones and nuclear transfers.

COOK, P. R., "On the Inheritance of Differentiated Traits." *Biol. Revs.* (Cambridge), **49:**51 (1974).

DAVIDSON, R. L., "Regulation of Differentiation in Cell Hybrids." *Federation Proceedings,* **30:**926 (1971). Fusion with an undifferentiated cell suppresses expression of specialized characteristics.

DAVIS, F. M., and E. A. ADELBERG, "Use of Somatic Cell Hybrids for Analysis of the Differentiated State." *Bact. Revs.,* **37:**197 (1973).

DEUCHAR, E. M., "Biochemical Aspects of Early Differentiation in Vertebrates." *Adv. in Morphogenesis,* **10:**175 (1973).

FRISTROM, J., and M. YUND, "Genetic Programming for Development in *Drosophila.*" *CRC Crit. Revs. Biochem.* **1:**537 (1973).

GIUDICE, G., "Developmental Biology of the Sea Urchin Embryo." New York: Academic Press, Inc., 1973. Morphogenesis, ultrastructure, biochemistry, and molecular biology.

GURDON, J. B., "Transplanted Nuclei and Cell Differentiation." *Scientific American,* December 1968. (Offprint 1128.)

GURDON, J. B., and C. F. GRAHAM, "Nuclear Changes During Cell Differentiation." *Science Progress* (Oxford), **55:**259 (1967).

HILDEBRANDT, A. C., "Growth and Differentiation of Plant Cell Cultures." In *Control Mechanisms in the Expression of Cellular Phenotypes* (*Symp. of the Int. Soc. for Cell Biol.,* **9**), ed. H. A. Padykula. New York: Academic Press, 1970, p. 147. Whole plants from single cells.

STEWARD, F. C., "From Cultured Cells to Whole Plants: The Induction and Control of their Growth and Morphogenesis." *Proc. Roy. Soc.* (London) *B,* **175:**1 (1970). Croonian Lecture, 1969.

VASIL, V., and A. C. HILDEBRANDT, "Differentiation of Tobacco Plants from Single Isolated Cells in Culture." *Science,* **150:**889 (1965). Reprinted, in part, in Whittaker.

INTERCELLULAR COMMUNICATION

BENNETT, M. V. L., "Function of Electrotonic Junctions in Embryonic and Adult Tissues." *Federation Proceedings,* **32:**65 (1973).

BURTON, ALAN C., "Cellular Communication, Contact Inhibition, Cell Clocks, and Cancer: The Impact of the Work and Ideas of W. R. Loewenstein." *Perspectives in Biol. & Med.,* **14:**301 (1971).

GUSTAFSON, TRYGGVE, "Cell Recognition and Cell Contacts During Sea Urchin Development." In *Cellular Recognition,* eds. R. T. Smith and R. A. Good. New York: Appleton-Century-Crofts, 1969, p. 47.

LOEWENSTEIN, W. R., "Some Reflections on Growth and Differentiation." *Perspectives in Biol. & Med.,* **11:**260 (1968).

———, "Intercellular Communication." *Scientific American,* May 1970. (Offprint 1178.)

———, "Membrane Junctions in Growth and Differentiation." *Federation Proceedings,* **32:**60 (1973).

McNUTT, N. S., and R. S. WEINSTEIN, "Membrane Ultrastructure at Mammalian Intercellular Junctions." *Prog. Biophys. and Mol. Biol.,* **26:**45 (1973).

PAPPAS, G. D., "Junctions Between Cells." *Hospital Practice,* August 1973, p. 39. The basis for ionic coupling.

ROTH, S., "A Molecular Model for Cell Interactions." *Quart. Rev. Biol.,* **48:**541 (1973). Cell surfaces and cellular recognition.

ACETABULARIA AND DICTYOSTELIUM AS MODEL SYSTEMS

BRACHET, J., and S. BONOTTO, *Biology of Acetabularia.* New York: Academic Press, 1970.

GIBOR, AHARON, "Acetabularia: A Useful Giant Cell." *Scientific American,* November 1966. (Offprint 1057.)

HAEMMERLING, J., "The Role of the Nucleus in Differentiation, Especially in Acetabularia." *Symp. of the Soc. for Exp. Biol.,* **17:**127 (1963). Reprinted in Bell.

RAPER, KENNETH B., "The Environment and Morphogenesis in Cellular Slime Molds." In *The Harvey Lectures, 1961–1962* (*Ser. 57*). New York: Academic Press, 1963, p. 111.

REGENERATION

HAY, ELIZABETH D., *Regeneration.* New York: Holt, Rinehart & Winston, 1966.

LAVINE, L. S., I. LUSTRIN, M. H. SHAMOS, R. A. RINALDI, and A. R. LIBOFF, "Electric Enhancement of Bone Healing." *Science,* **175:**1118 (1972).

POLEZHAEV, L. V., "Regeneration of Organs." *Science J.,* **4:**69 (1968). Discusses attempts to extend regeneration to man.

SCHMIDT, ANTHONY J., *Cellular Biology of Vertebrate Regeneration and Repair.* Chicago: The Univ. of Chicago Press, 1968.

THORNTON, CHARLES S., "Amphibian Limb Regeneration." *Adv. Morphogenesis,* **7:**205 (1968).

THE MALIGNANT CELL

BRAUN, ARMIN C., "The Reversal of Tumor Growth." *Scientific American,* November 1965. (Offprint 1024.)

———, "On the Origin of the Cancer Cell." *American Scientist,* **58:**307 (1970).

BURGER, MAX M., "Surface Properties of Neoplastic Cells." *Hospital Practice,* July 1973, p. 55.

BUSCH, HARRIS, ed., *The Molecular Biology of Cancer.* New York: Academic Press, Inc., 1974. A series of reviews of varying difficulty.

CHEDD, GRAHAM, "The Molecular Roots of Cancer." *New Scientist,* **54:**740 (1972). An explanation of the oncogene and alternate hypotheses.

CLOUD, WALLACE, "Stalking the Wild Crown Gall." *The Sciences,* **13:**6 (1973). A readable introduction to the molecular biology of this plant tumor.

DULBECCO, R., "The Induction of Cancer by Viruses." *Scientific American*, April 1967. (Offprint 1069.)

ECKHART, WALTER, "Genetic Modification of Cells by Viruses." *BioScience*, 21:171 (1971).

GULLITON, B. J., "Cancer Virus Theories: Focus of Research Debate." *Science*, 177:44 (1972). Like the preceding reference, an easy account of conflicting theories.

HARRIS, HENRY, "Cell Fusion and the Analysis of Malignancy." *Proc. Roy. Soc.* (London) *B*, 179:21 (1971). The Croonian Lecture, 1971. A condensation appears in *New Scientist*, 51:90 (1971).

HARRIS, R. J., ed., *What We Know About Cancer*. New York: St. Martin's Press, 1972. (Paperback.)

MARX, J. L., "Research News: Cancer." *Science*, 183:1066 and 1279 (1974). The first is on DNA cancer viruses and the second on the cell surface.

MAUGH, T. H. II, "Research News: Cancer." *Science*, 183:147, 940, and 1181 (1974). On carcinoembryonic antigens, chemical carcinogenesis, and RNA cancer viruses, respectively.

PIERCE, G. BARRY, "Differentiation of Normal and Malignant Cells." *Federation Proceedings*, 29:1248 (1970).

ROUS, PEYTON, "The Challenge to Man of the Neoplastic Cell." *Science*, 157:24 (1967). Nobel Lecture, 1966.

RUBIN, H., "The Behavior of Cells Before and After Virus Induced Malignant Transformation." In *The Harvey Lectures, 1965–1966* (Ser. 61). New York: Academic Press, 1967, p. 117.

SPIEGELMAN, S., "DNA and RNA Viruses." *Proc. Roy. Soc.* (London) *B*, 177:87 (1971). The oncogenic RNA viruses

STEVENS, L. C., "Embryonic Potency of Embryoid Bodies Derived from a Transplantable Testicular Teratoma of the Mouse." *Devel. Biol.*, 2:285 (1960). Reprinted in Bell. Embryolike aggregates are present in the tumor mass.

SUSS, R., V. KINZEL, and J. D. SCRIBNER, *Cancer: Experiments and Concepts*. New York: Springer-Verlag, 1973. History and some current approaches to cancer research.

TEMIN, H. M., "Malignant Transformation of Cells by Viruses." *Perspectives in Biol. & Med.*, 14:11 (1970). Reverse transcriptase.

————, "RNA-Directed DNA Synthesis." *Scientific American*, January 1972. (Offprint 1239.)

AGING

HAYFLICK, LEONARD, "Human Cells and Aging." *Scientific American*, March 1968. (Offprint 1103.)

KOHN, ROBERT R., *Principles of Mammalian Aging*. Englewood Cliffs, N. J.: Prentice-Hall, 1971. (Paperback.)

ORGEL, L. E., "The Maintenance of the Accuracy of Protein Synthesis and Its Relevance to Aging." *Proc. Nat. Acad. Sci.* (U. S.), 49:517 (1963). The "error catastrophe" theory.

————, "Aging of Clones of Mammalian Cells." *Nature*, 243:441 (1973). A review and speculation.

SAUNDERS, JOHN W., JR., "Death in Embryonic Systems." *Science*, 154:604 (1966). The role of selective cell death in morphogenesis and development.

TIMIRAS, P. S., "Developmental Physiology and Aging." New York: Macmillan, 1973. A text, emphasizing humans.

WOOLHOUSE, H. W., "Longevity and Senescence in Plants." *Science Prog.* (Oxford), 61:123 (1974). Senescence may be hormone-controlled in plants.

YIELDING, K. L., "A Model for Aging Based on Differential Repair of Somatic Mutational Damage." *Perspectives in Biol. and Med.*, 17:201 (1974).

Index

(Italicized page numbers refer to illustrations or structures)

A

A–band, 286, 287
Abbe's relationship, 48
Acetabularia, 374, 430, 431
  grafting, 431
  nuclear transplants, 432
Acetaldehyde
  in fermentation, 173
Acetate, 100
  citric acid cycle, 195
  titration curve, 86
Acetic acid. *See* Acetate
Acetylcholine, 280
  effect on heart, 284
  neurotransmission, 282
  receptor, 282, 283
Acetyl-SCoA, *196. See also* Acetylcoenzyme A
Acetylcoenzyme A, 195
  energy of hydrolysis, 164
Acetyl glucosamine, 111
Acids, 85
  behavior in water, 85
Aconitase, 197
Aconitate, 197
Acrasin, 434
Actin, 287
  in cell division, 416
  in endocytosis, 262
  evolutionary conservation, 309
  filaments, 287, 289, *290*
  as microfilaments, 309, 311
  in non-muscle cells, 309
  properties, 287
Actinomycin D, 433
Action potential, 267. *See also* Spike potential
Active site, of enzymes, 144
Active transport, 243
Actomyosin, 309
  in non-muscle cells, 309
Adenine, *114*
  nucleosides, 115
  nucleotides, 115
  pairing with inosine, 345
Adenocarcinoma, 445
Adenosine, 115
Adenosine diphosphate, 116

Adenosine triphosphate. *See* ATP
Adenovirus, 43, *45*
Adenylate kinase, 171, 199, 295
  mitochondrial location, 199
  X-ray diffraction of, *134*
Adenylic acid, 115
Adipose cells, 104
ADP, 116
Adrenaline, 280
  and chalones, 423
  effect on heart, 284
Aequorin, 298
Aerobic, 11
Aging, theories, 447
Alanine (ala), *122*
Alcohol (ethanol), 99
  cell permeability, 241, *245*
Alcoholic fermentation, 167
Alcohols, 99, 100
Aldehydes, 100
Aldohexose, 107
Aldolase, 168
Aldose, 107
Aldotriose, 107
Algae, blue-green, 9, 11, *220*
  oxygen evolution by, 11
  photosynthesis, 219
Algae, eucaryotic
  chloroplasts of, *206*
Alicyclic, 98
Aliphatic compounds, 98
Alkanes, 98
Alkenes, 98
Alkynes, 98
Alleles, 319
  crossover, 322
Allelomorphs, 318. *See also* Alleles
Allen, D. W., 144
Allo amino acids, 121
Allolactose, 356
Allosteric regulation, 159
  effectors, 159
Allostery, 145
Altmann staining procedure, 180, 181
Amber mutants, 339, 350
Amebas, 261
  phagocytosis by, 261
Ameboid locomotion, 308, 311, *312*
Amides, 100

Aminoacyl-tRNA, 344
  synthetase, 343
Amine oxidase
  mitochondrial location, 199
Amines, 100
Amino acids, 120, 121
  D and L, 121, *123*
  membrane permeability, 244
  membrane transport, 258
  pK_a, 121, 122
  stereoisomers, 121, *123*
  structures, 122
  as zwitterions, 123
Aminobutyrate, 281
Amoeba, 261
AMP, 115
  in coenzymes, 174
  cyclic, 361. *See also* cAMP
  regulation of glycolysis, 171
Amylopectin, 111
Amylose, 111
Amu (dalton), 94
Anabolic reactions, 166
Anaerobic, 13
Anaphase. *See* Mitosis or Meiosis
Androgens, 106
animal starch, 111. *See also* Glycogen
Annulate lamellae, 240, *243*
Annuli. *See* Nucleus, pores
Anomer, 108
Antibiotics
  ionophorous, 249
  See also specific names
Aperture, 48
  angle, 48
  numerical, 48
Apoenzyme, 143
Arg, *122. See also* Arginine
Arginase, 376
Arginine (arg), *122*
  pK_a, 122
  regulation of biosynthesis, 359
Aromatic compounds, 99
Ascorbic acid (vitamin C), *109*
Ascus, 325
Ashley, C. C., 298
Asn (asparagine), *122*
Asp (aspartic acid), *122*
Asparagine (asn), *122*

Aspartate transcarbamylase, 160
  feedback regulation, 160
  reaction, 160
  subunits of, 161
Aspartic acid (asp), *122*
  pK$_a$, 122
Astbury, W. T., 127
Aster, 404
Astrachan, L., 332
Asymmetric carbons, 96, 107
ATCase. *See* Aspartate Transcarbamylase
ATP, 116, *163*
  coupled reactions, 162
  energy of hydrolysis, 163
  muscle contraction, 298
  photorespiration, 217
  photophosphorylation, 211
  regulation of CTP synthesis, 161
  regulation of glycolysis, 171
  transphosphorylations, 197
  in tRNA charging, 343, 344
ATPase, 190
  See also Sodium pump
  calcium activated, 297
  mitochondrial location, 200
Atropine, 283
Autophagocytosis, 262
Autoradiography, 384, 396, *397*
  pulse labelling, 396
Autosomes, 323
Autotrophism, 11
Auxins, 423
Avery, O. T., 328
Avian leukemia virus, 46, 47
  See Virus, leukemia
Avogadro's number, 68, 208
Axons, 268, *269–273*
  conduction velocity, 279, 280
  perfusion of, 278
  transport, 308
Axoplasm, 273

B

Bacilli, 11
Bacillus, *15, 394*
  iron transport, 251
  phage SP8, 334
Bacteria, *9, 12. See also* Procaryotes and
      specific names
  capsule, 13
    and virulence, 327
  cell wall, 14
  conjugation, 358
  cytokinesis, 393
  discovery, 10
  division, 384, 393
  DNA replication, 384
  F factors, 358
  fimbriae, 13
  flagella, 13
  gender, 358
  gram stain, 11, 14
  growth, 355, 393
  mating, 358
  membrane composition, 233
  mesosome, 15
  operons, 355
    negative control, 359
    positive control, 359

phage receptors, 38
photosynthetic, 219
pili, 13
protoplasts, 393
purple sulfur, 220
sex factors, 358
shapes of, 8, 11, 12
sheath, 13
structure, 13
taxonomy of, 13
wall-membrane adhesions, *14*
Bacteriophage. *See also* Virus
  adsorption to host, 38
  contraction of tail, 308
  discovery, 36
  f2, 350
  lambda, 341
  MS2, 340
  mutants, 392
  R17, 350
  reproduction, 37, *41*
  SP8, 334
  T2, 37, *39*, 329
  T4, 37, *38, 41, 392*
  T6, 37
  virulence, 332
Bang, O., 442
Barnacle, muscle fibers, 298
Barr body, 365, 367
Basal body. *See also* Centrioles
  bacterial, 13
  construction, 32
  eucaryotic, 31, *32*
  kinetosomes, 432
Basal corpuscle. *See* Basal body
Bases, 85
  behavior in water, 86
  organic
    DNA and RNA, 114
    pairing of, 117
Beadle, G. W., 325
Beijerinck, M. W., 35, 43
Benzene, 97
  conjugation of, 97
Benzoic acid, 100
Bernstein, Julius, 275
Beta galactosidase. *See* Galactosidase
Beta rays, 386
Binary fission, 383
Biotin, 325
Blastema, 439, *440*
Blepharisma, 4
Boltzmann, Ludwig, 68
  Boltzmann's law, 68
Bonds
  covalent, 70, 96
  disulfide, 131, *132*
  double, 96
  energy of, 68, 70
  "high energy", 163
  hydrogen, 72, *132*
    in DNA, 119
    in protein, *132*
  hydrophobic, 76, 79, *132*
    in DNA, 119
  ionic, 74
  isopeptide, 131, *132*
  phosphoric acid anhydride, 163
  rotation about, 97
  short-range, 78

stability, 70
strong, 70
triple, 96
van der Waals, 78
  in DNA, 119
weak, 70
Bonner, James, 367
Bound water, 74
Braun, Armin, 445
Britten, R. J., 372, 373
Bronsted acids and bases, 85
Brown, Robert, 7, 18
Brownian motion, 68
Buchner, Eduard, 5, 7
Buffers, 86
Buffering capacity, 86
Bullough, W. S., 422
Burnet, Macfarlane, 43
Butane, 98

C

C3 pathway, 216
C4 pathway, 218, *219*
Ca^{2+}. *See* Calcium
Cairns, John, 386, 396
Calcium
  active transport, 252
  amount in humans, 95
  regulation of muscle, 298
Calcium pump, 297
Calvin, M., 214
Calvin cycle, 215
Calvin-Benson-Bassham cycle, 215
cAMP, 361
  as acrasin, 434
  and cell division, 420
  and hormones, 369
  and tumor cells, 421
Cancer, 440
  foci, 441, *443*
  oncogene theory, 444
  and viruses, 35, 443
Capsule, procaryotic, 13
Carbohydrates, 106
  elementary composition, 95
Carbon
  asymmetric, 96
  bond angles, 96
  prevalence, 95
  properties, 96
  valence, 96
Carbon dioxide
  fixation of, 214
  photosynthetic fixation, *215*, 216, 217,
      219
  structure, 96
Carboxylate, pK$_a$, 103
Carboxylic acids, 99, 100
  pK$_a$, 103
Carcinoembryonic antigens, 446
Cardiolipin, 104, 106
Carotenoids
  absorption spectrum, 209
Catabolic reactions, 166
Catalase, 23, 137
Catalysis, in cells, 143
  enzymatic mechanisms, 145
Cell biology, origins, 6

Cell cycle, 400
control, 416
Cell doctrine, 4
Cellobiose, 110
Cell plate, 416, *418*
Cells
chemical composition, 94
division
eucaryotic, 413
procaryotic, 384
excitable, 267
fractionation, 33, *34*
fusion of, 231, *232, 233*
glycocalyx, 240
growth
factors influencing, 416
membrane
bacterial, 14, *15*
eucaryotic, 16
See also Membranes
molecular composition, 94
osmotic regulation, 256
plant, *17, 27*
plasmolysis, 241
procaryotic, 10
recognition of, 240
size and diversity, 8
somatic, 231, 321
surface features, 240
Cell sap, 26
Cell theory, 4
Cellulose, 110, 112, *113*
in blue-green algae, 13
Cell wall
eucaryotic, 16, *17*
procaryotic, 13, *15, 244*
Centriole, *16,* 31, *407*
in cell division, 402, 403
construction, 32
Centrifugation
in cell fractionation, 34
density gradient, 384
Centromere, 21, 403
division of, 401, 408
Centrosome, 31
Cephalin, 104
cGMP, 368
and cell growth, 421
Chalones, 422
and adrenaline, 423
Chamberlain, C., 35
Changeux, J.-P., 154
Chaos, 261
Chase, Martha, 43, 329
Chemiosmotic theory, 190
Chiasmata, 410
Chitin, 111
Chlamydiae, 13, 46
Chloroamphenicol, 224
Chloride, amount in humans, 95
Chlorophyll, 202
absorption spectrum, 209, 210
light absorption, 208, 209, 210
in quantasomes, 205
relationship to heme, 137
Soret band, 209
structure, 210
Chloroplasts, 26, *27,* 200, *202, 203, 205,*
206, 218, 222. See also Photosynthesis

absorption spectrum, 209
action spectrum, 209
of algae, 206
differentiation, 202
dimorphic, 217, *218*
electron transport, 211, *212*
evolution of, 221
grana, 204
immature, *202*
lamellae, 204, 233
photophosphorylation, 211
quantasomes, 205
replication, *222*
size, 202
stroma, 204
structure, 204
thylakoids, 204
Cholesterol, 106
in membranes, 233
Choline, *106*
Chondriosomes, 180
See also Mitochondria
Chromatids, 21, 323
Chromatin, 20, 363, 364
Barr body, 365, 367
condensation, *404*
euchromatin, 363
heterochromatin, 363
nuclear anchoring, *20,* 21, 398, 399
replication, 395
structure, 363
Chromoplast, 26, 202
Chromosomes, 20, *22,* 364, 365, 366, *406,*
407. See also Chromatin
in cell division, 401
centromere, 21, 403
division of, 401, 408
distribution errors, 323
and genes, 321
giant, 364, 366
puffs, 365, 366
g regions, 373
haploid number, 321
examples, 323
heteromorphic, 323
homologous, 321
kinetochore, 405, *406*
lampbrush, 364, *365*
number, 21, 321, 323
polytene, 364, 366
puffs, 365, 366
replication, 395
sex, 323
structure, 362ff
Chymotrypsin, mechanism of, 148
Cilia, 32, *33,* 308
Cisterna, 23
Cistron, 326
Citrate, 194
regulation of glycolysis, 171
Citrate condensing enzyme, 197
Citric acid cycle, 194
ATP yield, 197
energetics, 197
mitochondrial location, 200
reactions, 196
Cleavage furrow, 414, *415*
Cloning experiments, 423
CMP, 115
Cocaine, *278*

Cocci, 11, *12*
Codon, 335
assignments, 339
nonsense, 339, 350
Coenocyte, 413
Coenzyme, 147, 173
examples, 174
Coenzyme A, 194
Coenzyme Q (ubiquinone), 187, *188*
Cole, K. S., 276
Coliphages, 37
See also Virus, bacterial
Collagen, 129, *130*
and gelatin, 130
Color-blindness, inheritance of, 325, 327
Concanavalin A, 422
Conjugation, in organic chemistry, 97
Conklin, E. G., 413
Contact inhibition, 437, 421
Cooperativity
of enzymes, 151
positive and negative, 152
CoQ (coenzyme Q), 187, *188*
Corepressor, 354
Corey, R. B., 125
Cortisone, 106
Coulomb's law, 75
Covalent bond, 70
Crane, Robert, 256
Crassulacean acid metabolism, 219
Creatine kinase
as M-line protein, 294
reaction, 295
Cristae. *See* Mitochondria
Crossover, 322
Crown gall tumor, 445
Crick, Francis, 117
Cryptic mutants, 249
CsCl gradients, 332, *333,* 384
Cucumber mosaic virus, *44*
Curare, 283
Cyanophyta. See Algae, blue-green
Cyclic AMP. *See* cAMP
Cyclic GMP, 368
and cell growth, 421
Cyclohexane, 98
Cycloheximide, 224
Cysteine (cys), *122*
in papain, 148
pK_a, 122
Cytidine, 115
Cytidine triphosphate,
regulation of synthesis, 160
Cytidylic acid, 115
Cytochalasin, 414
Cytochrome oxidase, 187
Cytochromes
of chloroplasts, 205
electron transport, 181, 187
evolution, 224
heme, 137
ionic charge, 197
photosynthesis, *212*
Cytokinesis, 384
in eucaryotes, 413
control, 420
and microfilaments, 414, *415*
in procaryotes, 393
control, 419
and spindle, 413, *414*

457

Cytokinesis (*cont.*)
  synchrony with nuclear division, 413
Cytokinins, 423
Cyton, of neurons, 268
Cytoplasm, 14, 20
  streaming, 31, 308, 311
Cytoplasmic membrane. *See* Membrane
Cytosine, *115*
  nucleosides and nucleotides, 115
  paired with inosine, 345
Cytosol, 34

D

dADP, 116
Dalton, 94
dAMP, 115
Danielli, J. F., 234
dATP, 116
Davidson, E. H., 373
Davson, Hugh, 235
dCMP, 115
Debye, Peter, 79
deDuve, Christian, 25
Dehydrogenation, 173
Denaturation (protein), 131
Dendrite, 268
Deoxyadenosine, 115
Deoxyadenylic acid, 115
Deoxycitidine, 115
Deoxycitidylic acid, 115
Deoxyglucose, *247*
  membrane transport, 247
Deoxyguanosine, 115
Deoxyguanylic acid, 115
Deoxynucleoside, 114
Deoxyribonucleic acid. *See* DNA
Deoxyribonucleoside, 114
Deoxyribose, 113, *116*
  in DNA, 113, 114, *117*
Deoxythymidine, 115
Depot fat, 104
Depth of field, 53, *54*
DeRobertis, E., 283
Detergent, 77
Determination, 437
Dextrorotation, 96
  by amino acids, 121
Dextrose, 108. *See also* Glucose
dGMP, 115
d'Herelle, Felix, 36, 43
Diakinesis, 410
Diastereo isomers, 121
Diauxic growth, 355
Dibutyryl cyclic AMP, 421
Dictyosome, 21, *418*
  See also Golgi body
Dictyostelium, 434, *435*
Dictyotene, 410
Dielectric constant, 75
  of air, 75
  of vacuum, 75
  of various organic substances, 76
  of water, 75
Diesters, in nucleic acids, 116
Digestive vacuole, formation, 25
Diglyceride, 104
Dihydrouridine, *342*, 367
Dihydroxyacetone phosphate, *168*

in glycolysis, 168
in photosynthesis, *216*
Dimethyl sulfoxide, 298
Dinitrophenol, and electron transport, 188
Dinoflagellates, mitosis, 405
Diphosphoglycerate, 168, *216*
  oxidation of, 173
  in photosynthesis, *216*
Diphosphopyridine nucleotide, 174
  See also NAD+
Diploid, 321
Diplonema, 410
Diplotene, 410
Dipole-dipole interactions, 78
Dipoles, 72, 73
  induced, 78
  interactions in proteins, *132*
Disaccharide, 110
Dispersion forces, 79
Disulfide bond, 131, *132*
  in proteins, *132*
Division, reductive. *See* Meiosis
DMSO, 298
DNA, 113
See also Gene
  amount in tissues, 94
  annealing, 332
  bacterial, 14, 15, 225
  bacteriophage, *39*
  base pairing, 117
  chloroplast, *225*
  denaturation, 332
  density gradient centrifugation, 384, *385*
  dimensions, 119
  endonuclease, 389, *390*
  hybrids with RNA, 332
  hydrogen bonding, 119
  hydrophobic bonding, 119
  infectivity, 330
  ligase, 389, *390*
    temperature sensitive, 389
  melting, 332
  micrographs, *225, 226, 386, 387, 391*
  microscopy of, 58
  mitochondrial, *226, 391*
  mutagens, 393
  mutations, 392
  nickase, 389, *390*
  Okazaki fragments, 389
  polymerases, 388
    eucaryotic, 397
    RNA dependent, 444
  polymerization, 116, 388
  repair, 389
  repetitious, 373
  replicase, 388
  replication, 384, *385, 391*
    conservative, 384, *385*
    errors, 392
    in eucaryotes, 395
    fork, *387*, 388
    initiation control, 419
    initiation point, 388
    rate, 396, 398
    RNA initiation, 390
    semiconservative, 384, *385, 387*
  replicons, 397, *398*
  RNA dependent polymerase, 444
  satellite, 373
  stacking interactions, 119

structure, 117, *118*
synthesis. *See* replication
transcription, 333
  strand selectivity, 333
UV repair, 389
viscosity, 120
X-ray diffraction of, 118, *119*
Dominance, 319, 326
Down's syndrome, 323
DPN+, 174. *See also* NAD+
Drosophila
  chromosomes, 364, *366*
    number, 323
  cytokinesis of egg, 413
  development, *437*
  imaginal discs, 437
  transdetermination, 438
dTMP, 115
Dutrochet, R. J. H., 3, 7
Dynein, 309
Dysentery, 36

E

Ebashi, S., 289
Eccles, J. C., 291
E. coli. See Escherichia coli
Ectoplasm, 28, *55*, 311
EF-1, 346
EF-2, 346
EF-G, 346
EF-T, 346
Einstein, Albert, 207
Einstein (units), 208
Electrochemical equilibrium, 254
Electron, wavelength, 50
Electron microscope. *See* Microscopy,
    electron
Electron transport, 187
  in chloroplasts, 211, *212*
  uncouplers, 188
Electronegativity, 72
Electroplax, 283
Electrostatic fields, 72
  equation, 75
Electrotonic synapse, 281
Ellermann, V., 442
Elongation factors, 346
Embden, G., 167
Emerson, R., 210
Emerson effect, 210
Emulsification, 103
Endergonic reactions, 67
Endocytosis, *258, 259*
  microfilaments in, 262
  selectivity, 261
Endonuclease, 389, *390*
Endoplasmic reticulum, *17, 23, 24*, 351, *353*
  origin from nucleus, 241, *243*
  rough *vs.* smooth, 23
  in protein synthesis, 351
Endosymbiosis, 224
Endothermic reactions, 67
Energy
  of activation, 142, *143*
  barrier, 142
  of bonds, 68
  changes in reactions, 82
  Gibbs, 66

Energy, Gibbs (cont.)
  standard state, 69, 82
Enolase, 168
Enthalpy, 66
Entropy, 66
Enzymes, 5, 112, 142
  active site, 144
  allosteric, 145, 160
    effectors, 159
    regulatory sites, 160
  apoenzyme, 143
  cascade, 369
  catalysis by, 143
    mechanism, 144, 146
  coenzymes, 147
  conformation of, 144
  constitutive, 355
  cooperativity, 151
    mechanisms, 153, 154
    negative, 152
    positive, 152
  denaturation, 143
  feedback regulation, 155
  and genes, 325
  Hill equation, 151, 160
  historical, 5
  holoenzyme, 143
  induced fit theory, 144
  inducible, 354
  inhibition of
    allosteric, 159
    competitive, 156
    non-competitive, 157
  interaction with substrate, 144, 146, 147
  international unit of activity, 177
  kinetics, 149
  lock and key theory, 144
  maximum velocity, 149
    measurement of, 150
  mechanisms of catalysis, 145, 146
  nomenclature, 148
  orbital steering, 147
  phosphorylation of, 369
  product inhibition, 172
  repression, 354
  turnover, 362, 376
    number, 151
Epigenetic changes, 428
Epinephrine. See Adrenaline
Episomes, 358
Equilibrium constant, 69, 84
  and free energy, 69, 84
ER. See Endoplasmic reticulum
Erythrocytes, 180, 336
  aging, 447
  chicken, 232, 233
  hemoglobin, 135
  hemolysis, 255
  membrane, 234, 235, 236
    composition, 233
  metabolism in, 167
  sickle cell, 336
Erythrose 4-phosphate, 216
Escherichia coli, 37, 41, 244, 394
  DNA, 387, 388
  molecular composition, 94
  pyruvate dehydrogenase, 195
  plasmolysis, 244
Eserine, 283
Esotropy, 240

Ester, 100, 104
Estradiol, 423
Estrogens, 106
Ethane, 98
Ethanol. See also Alcohol
  membrane permeability, 241, 245
  from pyruvate, 173
Ethanolamine, 106
Ethers, 100
Exergonic reactions, 67
Exocytosis, 239, 259, 262
  in secretion, 22
Exonuclease, 375
Exothermic reactions, 66
Exotropy, 240
Euchromatin, 363
Eyes
  color, inheritance, 321
  resolution of, 6, 48

F

F_1, 199, 200. See also Mitochondria
F2 phage, 350
F-actin, 287
  helix, 289, 290
Facilitated diffusion, 246
Facultative organisms, 13
FAD, 174, 187
  reduction of, 174
$FADH_2$, 174, 187
  ATP yield, 193
  from citric acid cycle, 195–197
Fat, 104
Fatty acids, 101, 102, 103
  common, 101, 102
  and citric acid cycle, 194, 195
  essential, 103
  mitochondrial degradation, 199
Fermentation, alcoholic, 5, 167
Ferredoxin
  in chloroplasts, 205
  role in photosynthesis, 212
Fertilization, 411, 412
Fibroblasts
  aging of, 449
  and collagen, 129
Fibrocyte, 129. See also Fibroblast
Fields, electrostatic, 72
  equation, 75
Fiers, W., 340
Filaments, 100Å, 28, 29. See also Actin,
  Myosin, Microfilaments, etc.
Filial generations, 318
Fimbriae, 13
First Law of thermodynamics, 67
Fischer, Emil, 144
Fixation (for microscopy), 56
Flagella
  bacterial, 13
    motility of, 303
  eucaryotic, 32, 33
Flavin adenine dinucleotide. See FAD or
  $FADH_2$
Flavin mononucleotide. See FMN
Flavins, electron transport, 187
Flavoproteins, 174
  and ATP synthesis, 193
Flemming, Walter, 400

Flicker fusion, 277
Fluorescein, intercellular transfer, 435
Fluorescence, 208
FMN, 174
  reduction of, 174
Follicle stimulating hormone, 423
Foot and mouth disease, 35
Forer, A., 406
Fraenkel-Conrat, H., 43, 330
Free energy, 81
  of formation, 82
  standard state, 69, 82
Freeze-etching, 59
  of membranes, 236
Freeze-fracture, 59
  of membranes, 236
Fremyella, 220
Frosch, P., 35, 43
Fructofuranose, 108. See also Fructose
Fructose, 108
Fructose diphosphate, 168
  activator of pyruvate kinase, 172
  in glycolysis, 168
Fructose 6-phosphate, 168
  in glycolysis, 168
  in photosynthesis, 216, 219
Fruit fly. See Drosophila
Fumarase, 197
Fumarate, 196, 197
Fungi, mitosis of, 405
Furan, 109
Furanose, 109

G

GABA, 281
G-actin, 287
Galactose, 110
  membrane transport, 247
Galactosemia, 341
β-galactosidase, 354
  induction of, 354
Galactoside acetylase. See Galactoside
  transacetylase
Galactoside permease, 354
Galactoside transacetylase, 354, 355
Gamete, 319, 410, 411, 412
Gametogenesis, 410, 411
Gametophyte, 412
Gamma amino butyric acid, 281
Gamma rays, 207
Gamow, George, 335
Geier, M., 341
Gelatin, 130
Gene
  amplification, 369
  assortment, 319
  autogenous regulation, 359
  of bacteriophage, 329
  catabolite repression, 361
  chemical nature, 327
    in bacteria, 328
    in eucaryotic cells, 330
    in viruses, 329, 330
  and chromosomes, 321
  concept, 318
  constitutive, 355
  corepressors, 354
  crossover, 322

Gene (*cont.*)
  determination, 437
  dominance, 319
    mechanism, 326
  and enzymes, 325
  epigenetic change, 428
  extranuclear, 432
  history, 322
  induction, 353, 354
  linkage, 321, *323*
  loci, 323
  operator, 355
  promoters, *356, 358*
  regulation
    director hypothesis, 367
    eucaryotic, 361
    by histones, 365
    by hormones, 368
    procaryotic, 353
  regulator, 355
  regulons, 359
  repression, 353, 354
  repressor, 354
    isolation, 358
  segregation, 319
  sex linked, 323, 325
  sigma factors, 360
  structural, 355
  theory, 322
  transcription, 333, 334, 360
  translational control, 374
Genetic code, 335, 339, *340*
  codon assignment, 337
  history, 335
  reading of mRNA, 346
  table, 339
  universality, 341
  wobble hypothesis, 339
Genotype, 319
Gerhart, J., 161
Germination, 412
Gibbs free energy, 66, 81
  standard state, 69, 82
Giberellins, 423
Gierer, A., 43
Gilbert, Walter, 358
Glial cells, 270
Gln (glutamine), *122*
Glu. *See* Glutamic acid
Glucokinase, 168
Glucopyranose, 108. *See also* Glucose
Glucosamine, 111
Glucose, *108*
  in glycolysis, 168
  membrane transport, 247
  photosynthesis of, 214
Glucose 1-phosphate, energy of hydrolysis,
    164
Glucose 6-phosphate, 168
  energy of hydrolysis, 164
  in glycolysis, 168
Glutamic acid (glu), *122*
  as neurotransmitter, 281
  pK_a, 122
Glutamine (gln), *122*
Glyceraldehyde, 97, 107
  isomers, 97
Glyceraldehyde phosphate, *168*
  in glycolysis, 168
  in photosynthesis, *216, 219*

Glyceraldehyde phosphate
  dehydrogenase, 168
Glycerides, 102
  table, 104
Glycerol, 102, 104
  esters, 104
  membrane permeability, 245
Glycerol phosphate, energy of hydrolysis,
    164
Glycine (gly), *122*
Glycocalyx, 240
Glycogen, 111, 112, *113*
  amount in tissues, 94
  energy of hydrolysis, 164
Glycogen phosphorylase, 369
Glycolysis, 166
  comparison to aerobic pathways, 197
  energetics, 168, 169, 170
  free energy change, 167
  reactions of, 168
  regulation of, 171
Glycoproteins, 135, 240
  in membranes, 240
Glyoxysomes, 25
GMP, 115
  cyclic, 369
Golgi, Camillo, 21
Golgi apparatus. *See* Golgi body
Golgi body, *17, 21, 23, 351, 352*
  and digestive vacuoles, 262
  glycoprotein synthesis, 240
  membrane structure, 238
  in protein synthesis, 351
Gorter, E., 234
Gram, Christian, 11
Gram's stain, 11
Grana, of chloroplasts, 204, *205*
Green, David E., 191
Grendel, F., 234
Griffith, F., 327
Growth hormone, 423
GTP
  from citric acid cycle, 195–197
  in protein synthesis, 346
  transphosphorylation of, 197
Guanine, *114*
  nucleosides and nucleotides, 115
Guanosine, 115
Guanylic acid, 115
Gurdon, J. B., 429

H

H^+. *See* Hydrogen ion
H_3O^+. *See* Hydronium ion
Hadorn, Ernst, 438
Hair, *127, 128*
  keratin, 128
Hall, B. D., 332
Hämmerling, J., 430
Hanson, J., 287, 291
Haploid, 321
Harvey, E. B., 413
Harvey, E. N., 234
Hatch, M. D., 217
Hatch and Slack pathway, 217
Haurowitz, F., 153
Hayflick, Leonard, 449

Heart
  action potentials, 284, 300
  innervation, 284
Heat
  of fusion, 73
  of reactions, 66
  of vaporization, 73
HeLa cells, 396
Heme, 136
  Soret band, 209
Hemizygotes, 323
Hemoglobin, 135, *136*
  cooperativity, 151
  early studies, 153
  Hb S, 335, 341
  mutants (table), 341
  oxy *vs.* deoxy, *136*
  oxygenation of, 136, 152
  sickle cell, 335, 341
  subunits, 135, *136*
  variants, 341
  X-ray diffraction, 135
Hemolysis, 255
Hemophilia, inheritance, 325, 327
Hemophilus, DNA of, *386*
Henderson-Hasselbalch equation, 86
Henri, V., 149
Heptulose, 107
Hershey, A. D., 43, 329
Heterochromatin, 363
Heterocyclic compounds, 99
Heterokaryon, 231
Heterotrophism, 11
Heterozygote, 319
Hexokinase, 168
Hexose, 107
Hexulose, 107
Hildebrandt, A. C., 428
Hill coefficient, 152
Hill equation, 151, 160
Histidine (his), *122*
  pK_a, 122
  regulation of synthesis, 355
Histones
  in cell division, 421
  classification, 362
  in gene regulation, 365
  mRNA for, 334
  phosphorylation, 421
  synthesis in S phase, 400
HnRNA, 334. *See also* RNA
Hodgkin, A. L., 254, 275, 279, 291
Holley, R. W., 343
Holoenzyme, 143
Homozygote, 319
Hooke, Robert, 2, 7
Hormones, and gene regulation, 368
Huebner, Robert, 444
Humans
  chromosome number, 323
  gender determination, 323, *324*
Huxley, A. F., 279, 291, 300
Huxley, H. E., 287, 291, 294
Hybrid, somatic, 231
Hydration, of ions, 74
Hydride ion, 70
Hydrocarbons, 98
Hydrogen, prevalence, 95
Hydrogenations, 173
Hydrogen bonds, 72

Hydrogen bonds (*cont.*)
  in DNA, 119
  energy of, 72
  in proteins, *132*
Hydrogen ion
  in acid-base reactions, 85
  ionic mobility, 89
Hydrogen peroxide, 25
Hydrolysis, 104
  energy of, 164
Hydronium ion, 85
Hydrophobic bond, 76
  and dispersion forces, 79
  in DNA, 119
  in proteins, *132*
Hydroxyproline, 129, *130*
  in collagen, 129
Hypertonic, 241
H-zone, 287

I

I-band, 286, *287*
Ice, structure, 73
Icosahedral viruses, 43, *45*
Ilu (isoleucine), *122*
Imago, 437
Imino acids, 121
Indoleacetic acid, 423
Induced dipoles, 78
Infrared radiation, 207
Ingram, V. M., 335
Inhibitors (of enzymes)
  allosteric, 159
  competitive, 156
  noncompetitive, 157
Initiation factors, 349
Inosine, 345
Inositol, 106
Insulin, 120
  primary structure, 124, *125*
Intercalation (in DNA), 393
Interphase. *See* Mitosis or Meiosis
Ions, hydration, 74
Ion-dipole interactions, 78
Ionic bonds, 74
Ionic mobilities, 75, 89
Iron, amount in humans, 95
Isocitrate, *196*, 197
Isocitrate dehydrogenase, 197
  crystals, *134*
Isoenzymes (isozymes), 171
Isoleucine (ilu), *122*
Isomers, organic, 96
Iso-osmolar, 241
Isopeptide bond, 131, *132*
  in proteins, *132*
Isoprene, 102
Isotonic, 241
Isozymes (isoenzymes), 171
Ivanovsky, D., 35, 43

J

Jacob, Francois, 331, 337, 355
Janssens, F. A., 322
Janssen, Z., 47
Jenner, Edward, 34, 43

Jobsis, F., 298
Junction
  myoneural, 269
  neuromuscular, 269, 302, *303*, *304*

K

K^+. *See* Potassium ion
K_a, 85
K_{eq}, 68, 85
K_w, 87
Karyokinesis, 401
Katz, Bernard, 275, 283
Kekulé, F., 96
Kendrew, J. C., 131, 135
Keratin, 128, *129*
α-Ketoglutarate, *196*, 197
α-Ketoglutarate dehydrogenase, 197
  mitochondrial location, 199
Ketones, 100
Ketopentose, 107
Ketose, 107
Ketotriose, 107
Keynes, R. D., 254
Khorana, H. G., 337
Kinase, 171
Kinetics
  cooperative, 151
  enzymes, 149
Kinetin, 423
Kinetochore, 405, *406*
Kinetoplast, 432
Kinetosome, 432
Kingsbury, B. F., 180
Kohne, D. E., 372
Kok, Bessel, 210
Kölliker, A., 180, 181
Kornberg, Arthur, 388
Koshland, D. E., Jr., 144, 145, 146, 153
Krebs, H. A., 194
Krebs cycle. *See* Citric acid cycle

L

Lactate, *168*
  in glycolysis, 167, 168
Lactate dehydrogenase, 168
Lactose, 110
  utilization by *E. coli,* 354
Lambda. *See* Bacteriophage
Lamellae, chloroplast, *27*
Langmuir, L., 234
Langmuir trough, 234
Laurence, E. B., 422
Law. *See also* Thermodynamics
  of independent assortment, 319
  of independent segregation, 319
L-dopa, 280
Lecithin, 104, *105*
Lectins, 422
Leder, P., 338
Leeuwenhoek, Anton van, 2, 7, 10, 47
Lenard, P., 207
Leptonema, 409
Leptotene, 409
Leucine (leu), *122*
Leucoplasts, 26, 202
Leukemia viruses, 35, 46, *47*
Leukocytes, 55. *See also* Lymphocyte

Levorotation, 96
  amino acids, 121
Levulose, 108. *See also* Fructose
Lewis, G. N., 70
Li^+. *See* Lithium ion
Liebig, Justus von, 5, 7
Ligase, 389, *390*
  temperature sensitive, 389
Light
  plane-polarized, 96
  speed of, 207
Lignin, 16
Lineweaver-Burk plot, 150
Linoleic acid, 103
Linolenic acid, 103
Lithium ion
  size, 74
  mobility, 75, 89
Lipids, 101
  in membranes, 233
Lipmann, F., 163
Lipofuscin, 447, *448*
Lipoprotein, 135
Liver, molecular composition, 94
Loewenstein, W. R., 434
Loewi, Otto, 284
Löffler, F., 35, 43
Loligo, neurons of, 273
London, F., 79
London dispersion forces, 78
Longitudinal vesicle, 288, *289*, 296
Lucké adenocarcinoma, 445
Lwoff, André, 43, 357
Lymphocyte, 55, 259, *312*
  B cell, *259*
  motility, *312*
  T cell, 55
Lysine (lys), *122*
  pK$_a$, 122
Lysis, by bacteriophage, 36
Lysogeny, 40, *42*
Lysome, 25, 262
  digestive vacuole, 262
  secondary, 262
Lysozyme, 393

M

MacLeod, C. M., 328
Macrophage, *260*
Magnesium, amount in humans, 95
Magnolia, 3
Malaria, and Hb S, 335
Malate, *196*, 197
  in photosynthesis, 219
Malate dehydrogenase, 197
Malignancy, 441
Maltose, 111
Mannose, in membranes, 240
Marker, K., 349
Marsh-Bendall factor, 314
Marsh factor, 314
Marsupials, sex chromosomes, 367
Maternal messages, 372
Matthaei, J. H., 337
McCarty, M., 328
Measles, 35
Meiosis, 321, 401, 408
  anaphase, 401

461

Meiosis (*cont.*)
  bivalents, 410
  centromere division, 401, 408, *409*
  chiasmata, 410
  diakinesis, 410
  dictyotene, 410
  diplonema, 410
  diplotene, 410
  interkinesis, 401, *409*, 410
  interphase, 401, *409*
  leptonema, 409
  leptotene, 409
  metaphase, 401, 410
  pachynema, 409
  pachytene stage, 409
  plants, 409, 412
  prophase, 401, *409*
  synapsis, 409
  synaptinemal complex, 409
  telophase, 401, 410
  tetrads, 410
  zygonema, 409
  zygotene stage, 409
Messenger RNA. *See* RNA, messenger
Membrane, 231ff
  active transport, 248, 251
  annulate lamellae, 240, *243*
  bilayer concept, 234
  carriers, 246
    non-protein, 249
    permeases, 246
    protein, 248
  chloroplast, 204, 205
    structure, 238
  composition, 233, 240
  continuity, 240, *242*, *243*
  Danielli-Davson model, 234
  diversity, 238
  and DNA replication, 389
  electrical potential, 253
  electrochemical equilibrium, 254
  endocytosis, 259
  endoplasmic, 238, 240
  exoplasmic, 238, 240
  facilitated diffusion, 246
  fluid mosaic model, 236, *237*
  fluidity, 231
  fusion, 231, *232*, *233*
  growth, 231
  metabolically coupled transport, 251
  mitochondrial, 183, 200
    structure, 238
  models, 234ff
  neurons, ionic distribution and potential, 273
  origin, 240, *243*
  permeability, 241ff
  permeability coefficient, 245
  permeases, 246
  pores, *237*, 246
  potentials, 253
    permeability effects, 276
  protein crystal model, *237*
  protein-lipid interaction, 237
  sodium pump, 254
  transport, 246
    active, 251
    cryptic mutants, 249
    energetics, 253
    group translocation, 252
  metabolically coupled, 251
    sodium coupled, 256, *257*
  trilaminate appearance, *235*
  unit membrane, *234*, 235
Mendel, Gregor, 318
Mendel's laws, 319
  exceptions, 320
Menten, M. L., 149
Mercaptoethylamine, 194
Mercuric ion, 88
  hydration of, 88
Meromyosin, 290
Merril, Carl, 341
Meselson, M., 384
Mesosome, *13*, *15*
  and cytokinesis, 393
Met. *See* Methionine
Metabolic pathways, principles, 167
Metaphase. *See* Mitosis or Meiosis
Metaphasic plate, 405
Metastasis, 441
Methane, 98
Methanol, membrane permeability, 241, 245
Methionine (met), *122*
Meyer, A., 204
Meyerhof, O., 167
M fibers (myonemes), 308
Micelles, 77, 103
Michaelis constant, 150
  determination of, 150
Michaelis, L., 149
Michaelis-Menten equation, 149
Microbodies, 23. *See also* Peroxisomes
Microfilaments, 28, *29*
  in axons, 269
  and endocytosis, 262
  identity with actin, 309, 311
Micron, 3
Micropinocytosis, 262
Microscopy
  comparison of methods, 52, 55
  depth of field, 53, *54*
  electron, 49
  fixation for, 56
  freeze-etching, *55*, 59, 236
  freeze-fracture, *55*, 59, 236
  light, 48
  metal shadowing, 58
  negative stains, 56
  origins, 47
  phase contrast, 49
  resolution, 48
    light, 48
    scanning electron, 52
    transmission electron, 50
  sample preparation, 54
  scanning, 50, *52*
Microsomes, 23, 331
  preparation of, 34
Microtome, 57
Microtubules, 29, *30*, *31*
  ameboid locomotion, 308
  axonal transport, 308
  in axons, 269
  cell motility, 308
  chromosome movement, 405
  cilia and flagella, 32
  cytoplasmic streaming, 308
  protofilaments of, *30*, 31
spindle, 403
subunits, *31*, 309
Miledi, R., 283
Miscibility, 76
Mitchell, P., 190
Mitochondria, *17*, *18*, 26, 180ff
  cellular distribution, 181
  conformational changes, *192*
  coupling factors, 199, 200
  cristae, 28, 184
  DNA, 224, 226
    transcription, 334
  electron transport, 186, 187
  elementary particles, 184, *186*, 200
  evolution of, 221
  half-life, 262
  history, 180, 181
  inner membrane spheres, *184*, *186*, 200
  intracristal space, 184
  kinetoplasts, 432
  matrix, 184
    composition, 200
  membranes, 183
    composition, 199, 233
    permeability, 200
  outer space, 183
  oxidative phosphorylation, 186
  replication, *223*
  respiratory assemblies, 184, *186*, 199, 200
  respiratory chain, 187
  size, 183
  structure, *182–185*
  structure-function relationship, 199
  turnover, 262
Mitosis, 322, 400, 401
  anaphase, 401, *402*, *403*, 405
  asters, 404
  centrioles, *402*, 403
  chromosome movement, 405
  interphase, 400, 401
  metaphase, 401, *402*, *403*, 405
  metaphasic plate, 405
  prophase, 401, 402
  spindle, 403
  telophase, 401, *402*, 416, *417*
    telophase neck, 416, *417*
Mizell, M., 445
M-line, 287
  protein of, 294
Molecular biology, origins, 6
Mommaerts, W., 305
Monera, kingdom, 10
Mongolism, 323
Monocytes, *258*, *259*
Monod, Jacques, 154, 331, 337, 355
Monoglyceride, 104
Monosaccharides, 106
  D vs. L, 107
  optical activity, 107
Morgan, T. H., 322
Motor end plates, 283
mRNA. *See* RNA, messenger
Mucopolysaccharide, 111
Mucoprotein, 240
Mucus, 240
Mueller, P., 278
Müller-Hill, B., 358
Mumps, 35
Murexide, 298
Muscle, 285ff

462

Muscle (*cont.*)
actin, 287, *290*
atrophy of disuse, 302
Ca²⁺ accumulation, 252
catch muscle, 301
contraction, 291
contraction-relaxation control, 295
cross-bridges, 294
cross-innervation, 307
fiber, 286
graded response, 299
junctional folds, 303
membrane potential, 256
metabolism, 305
mitochondria, *182*
molecular composition, 94
motor units, 302
myofibrils, 286
myosin, 287, *290, 291*
nerve interaction, 302
neuromuscular junction, 302
  smooth, *303*
  striated, *304*
pink, 305
potassium content, 256
red, 303
  metabolism of, 305
refractory period, 300
rigor, 298
sarcoplasmic reticulum, 296
sarcotubular system, 295
sliding filament theory, 291, *293*
smooth, *286*, 300
  and prostaglandins, *103*
sodium content, 256
spike potentials, 284
striated, 285, *286*
synaptic clefts, 303
synaptic trough, 303
syncytium, 413
tetanus, 300
tonus, *300*, 301
transverse tubules, *288*, 296, 300
tropomyosin, 289, *290*
troponin, 289, *290*
twitch, 299
white, 303
  metabolism of, 167, 305
Mutation
frequency, 392
mechanism, 392
mutagens, 393
temperature sensitive, 390
Mycoplasma, 13, 14, 393
Myelin, 270, *272, 273*
composition, 233
and conduction velocity, 279, 280
Myoblasts, *49*, 296
Myofibrils, 286
Myofilaments, 286
Myoglobin, 131, *133*
oxygen binding, 152
X-ray diffraction of, 135
Myo-inositol, 106
Myokinase. *See* Adenylate kinase
Myonemes, 308
Myoneural junction, 269
Myosin
in ameba, 262
in cell division, 416

dimensions, 290
filaments, 287, 290, *291*
in non-muscle cells, 309, 311
in red and white muscle, 307
Myxovirus (Sendai virus), 231, *233*

N

Na⁺. *See* Sodium ion
NAD⁻/NADH, 173, *174*
ATP yield from NADH, 193
in citric acid cycle, 195–197
oxidation-reduction, 173, 174
NADP⁻/NADPH, 173, *174*
in photosynthesis, 205, 208, 212, 213
reduction of, 174
Nass, Margit, 224
Negative staining, 56
Neomycin, 347
resistance to, 347
Neoplasm, 440
Nernst equation, 254
Nerve, *270, 271*. *See also* Neuron, Spike
  potential, Synapse
adrenergic, 280, *303*
autonomic, 280
cholinergic, 280, *303*
excitatory, 284
fiber, 270
impulse. *See* Spike potential
inhibitory, 284
medullated, 270
muscle interaction, 302
myelinated, 270
  conduction velocity, 279, 280
nodes of Ranvier, 273
parasympathetic, 280
saltatory conduction, 279
sympathetic, 280
trunk, 270
vagus, 284
  effect on heart, 284
Nervous system
central, 268
peripheral, 268
Neuraminic acid, 240
Neuroblast, 446
Neuroblastoma, 446
Neurofilaments, 28, 269
and axonal transport, 308
Neuromuscular interaction, 302
Neuromuscular junction, 269, 302, *303*,
  304
transmission, 284
Neuron, *267, 268, 269, 273*. *See also*
  Nerve, Spike potential, Synapse
axonal transport, 308
membrane potentials, 273
metabolism, 167
motor end plate, 283
myelinated, 270
  conduction velocity, 279, 280
refractory period, 275, 277
resting potential, 274, 275
sodium pump, 274
spike potential, 274, *275*
voltage clamp, 276
Neurospora, 325
Neurotransmission, 280
Neurotransmitters, 280

Niacin (nicotinic acid), 100, 173
Nickase, 389, *390*
Nicotinamide adenine dinucleotide.
  See NAD⁺ or NADP⁺
Nicotinamide mononucleotide, 174
Nicotine, 100
Nicotinic acid (niacin), 100, 173
Niedergerke, R., 291
Nirenberg, M., 337, 338
Nitrogen
fixation, 221
prevalence in cells, 95
NMN, 174
NMR, 89
Node of Ranvier, 273
and conduction velocity, 279
Nonsense codons, 339
Noradrenaline, 280
effect on heart, 284
Norepinephrine. *See* Noradrenaline
Nuclear magnetic resonance, 89
Nuclear RNA, 334
Nucleic acids, 112
elementary composition, 95
Nucleolus, *16, 18*, 21, *370, 371*
in cell division, 402, 408
organizer, 371
in protein synthesis, 334
in ribosome synthesis, 348
structure, 370
Nucleosides, 114
nomenclature, 115
Nucleotides, 113
nomenclature, 115
vs. nucleosides, 114
Nucleus, *16, 18, 19*, 364. *See also*
  Nucleolus
annuli, *19, 20, 364*, 399
in cell division, 403
envelope, *17, 18, 19, 20*, 398, *399, 404*
  assembly, 408, *415, 417, 418*
  chromatin attachment, 398, *399*
pores, *19, 20, 364*, 399
transplantation, 420

O

OAA. *See* Oxalacetate
Ochre mutants, 339, 350
O'Conner, M., 298
Okazaki fragments, 389
Olefins, 98
Oleic acid, 103
Oligopeptide, 124
Oligosaccharide, 107, 109
  formation, 109
Oncogene, 444
Oöcyte, 364, 372, 410, *411, 412*
Oögenesis, 410, *411*
Oögonium, *411*
Oötid, *411*
Opal mutants, 339, 350
Opalina, 33
Operon, 355
  control of, 359
Optical activity, 96
Optic vesicles, 433
Organic molecules, classification, 98
Orgel, L. E., 447
Osmotic regulation, 256

Overton, E., 232, 241
Ovum, 410, *411, 412*
Oxalacetate, 194, *196,* 197
Oxaloacetate. *See* Oxalacetate
Oxidation, 172
Oxidative decarboxylations, 195
Oxidative phosphorylation, 188
  ATP yield, 193
  chemical coupling, 189, *193*
  chemiosmotic coupling, 190, *193*
  conformational coupling, 189, *193*
  electrochemical coupling, 189, *193*
  electromechanical coupling, 191
  stoichiometry, 191, *193*
  theories, comparison, 191, *193*
Oxygen
  prevalence in biological systems, 95
  reduction by NADH, 175
Oxyphos. *See* Oxidative phosphorylation

P

P_i, 86
P690, 211
P700, 210
P870, 220
Pachynema, 409
Pachtyene stage, 409
Palade, G. E., 352
Palmitic acid, 103
Palmitoleic acid, 103
Pantothenic acid, 194
Papain, 148
  myosin cleavage, 291
Paramecium, 8
  endosymbiosis, 224
Paramyosin, 290, 301
Pardee, A., 161
Parkinson's disease, 280
Partition coefficients, oil/water, 245
Pastan, Ira, 420
Pasteur, Louis, 5, 7, 10, 35, 43
Pauling, Linus, 97, 125, 335
Peas, chromosome number, 323
Pectin, 111
Penicillin, 393
Pentose, 107
Pentulose, 107
PEP. *See* Phosphoenolpyruvate
Peptide bond, 123, *124*
  conjugation of, 123
  dimensions, 123, *124*
  resonant forms, 123
Peptidyl transferase, 346
Perfusion, of axons, 278
Perikaryon, of neurons, 268
Permeability coefficient, 245
Permeases, 246
  induction, 354
Peroxisome, 23, *24*
Perutz, M. F., 135, 153
Petricciani, J., 341
Pfeffer, W., 241
pH, 85
Phagocytosis, *258,* 259, *260, 261*
  autophagocytosis, 262
  and microfilaments, 262
  selectivity, 261
Phagosome, 262
Phe (phenylalanine), *122*

Phenol, 99
Phenotype, 319
Phenylalanine (phe), *122*
Phosphate
  ionization (pKa), 86
  titration curve, 87
Phosphatides, 104
Phosphatidic acid, 104, 105
Phosphatidyl choline, 104, *105*
Phosphatidyl ethanolamine, 104
Phosphatidyl serine, 104
Phosphocreatine, energy of hydrolysis, 164
Phosphoenolpyruvate, *168*
  energy of hydrolysis, 164
  in glycolysis, 168
  in membrane transport, 252
  in photosynthesis, 217, *219*
Phosphoenolpyruvate synthetase, 252
Phosphofructokinase, 168
  control of, 171
Phosphoglucose isomerase, 168
Phosphoglycerate, *168*
  in glycolysis, 168
  hydrolysis of, 170
  oxidation of, 173
  in photosynthesis, 214, *216*
Phosphoglycerate kinase, 168
Phosphoglyceric acid. *See*
      Phosphoglycerate
Phosphoglyceromutase, 168
Phosphoinositide, 104
Phospholipids, 102, 105
  amount in tissues, 94
  in membranes, 233
  table, 104
Phosphorescence, 208
Phosphoric acid. *See* Phosphate
Phosphorus, prevalence, 95
Phosphorylation
  oxidative, 188
  substrate level, 188
Photoelectric effect, 207
  in chlorophyll, 208
Photons, 207
  energy of, 207
Photophosphorylation, 211
  cyclic, 212
Photorespiration, 217
Photosynthesis
  action spectrum, 210
  C3 pathway, 216
  C4 pathway, 218, *219*
  Crassulacean acid metabolism, 219
  dark reaction, 205, 213
  dual light effect, 210
  dual pigment system, 210
  efficiency, 213
  Emerson effect, 210
  Hatch and Slack pathway, 217, 218, *219*
  history (table), 204
  light reaction, 205, 207
  photophosphorylation, 211
    cyclic, *212*
  pigment systems, 211
  in procaryotes, 219
  quantum efficiency, 211, 213
  stoichiometry, 213
Phycobilosomes, 206
Phycoerythrin, absorption
  spectrum, 209

Phytohemagglutinin, 422
P_1, 86
Picrate, *250*
Pili, 13
Pinocytosis, 259, 262
Pinosome, 262
Pi-pi interactions, 99
pK_a, 85
pK_w, 87
Planck, Max, 207
Planck's constant, 207
Plaques, bacteriophage, 36, *37*
Plasma cell, 353
Plasmalemma. *See* Cell membrane
Plasmodesma, *418,* 434
Plasmodium, 413
Plastids, 26, 202. *See also* Chloroplast
  differentiation of, 202
  division of, *222, 418*
  proplastids, 200, *201*
Plastocyanin, 205
  role in photosynthesis, *212*
Plastoquinone, *188*
  role in photosynthesis, *212*
  similarity to CoQ, 187
Pleated sheet. *See* Proteins, β-structure
Pleuro-pneumonia-like organisms. *See*
  Mycoplasma
Pneumococcus, 327
P/O ratio, 193
pOH, 87
Polar body, 410, *411, 412*
Polar molecules, 73
Polio, virus causation, 35
Pollen, 412
Polynucleotides, energy of hydrolysis, 164
Polyoma, 444
Polypeptide, 124, 131
  configuration in proteins, 125
  energy of hydrolysis, 164
Polyribosome. *See* Polysome
Polysaccharides, 110
Polysome, 346, *347, 349*
Porcelain candle filter, 35
Potassium ion
  amount in humans, 95
  mobility, 75, 89
  size, 74
Potassium permanganate, in microscopy,
  418
PP_i, 116
PPLO. *See* Mycoplasma
Primary structure. *See* Proteins, structure
Pro. *See* Proline
Procaine, 279
Procaryotes, 9ff. *See also* Bacteria
  cell division, 384, 393
  eucaryotic endosymbiosis, 224
Procion yellow, *436*
Proinsulin, *125*
Proline (pro), *122*
  conformation, 123
  as helix-breaker, 130, 131
  hydroxyproline, 129, *130*
    in collagen, 129
  as imino acid, 121
Propane, 98
Prophage, 40
Prophase. *See* Mitosis or Meiosis
Proplastids, 26, 200, *201*

Proplastids (cont.)
  differentiation, 202
  replication, 201
Prostaglandins, 103
Prosthetic groups. See Protein
Protein, 120. See also Enzymes
  amount in tissues, 94
  carbohydrate content, 135
  collagen structure, 129, 130
  conformation, 124
  conformational equilibrium, 137
  conjugated, 135
  crystals, 134
    X-ray diffraction, 133, 134
  denaturation of, 131
  elementary composition, 95
  flavoproteins, 174
  folding in vivo, 137
  glycoprotein, 135
  helices, 125, 126, 128
  lipid content, 135
  lipoprotein, 135
  in membranes, 233
  phosphorylation, 369
  pleated sheets, 127
  prosthetic groups, 135
    flavins, 174
  protomers, 134
  random coils, 131
  structure, 124, 133
    determination in vivo, 137
    primary, 124
    quaternary, 133
  secondary, 125
    α-structure (helices), 126, 128
    β-structure (pleated sheets), 126,
      128, 129
  tertiary, 130
    spontaneous formation, 133
    stability, 131
  solubility, factors, 121
  subunit interactions, 134
  subunits, 133, 134
  synthesis, 342ff
    elongation, 346
    elongation factors, 346
    in eucaryotes, 351
    initiation, 349
    polio virus, 351
    rho factors, 358
    termination, 350
    transcription rate, 360
  turnover, 376
  X-ray diffraction, 133, 134
Protein kinase, 369
Proteus, 12
Protista (classification), 10
Protofilaments (of microtubules), 30, 31
Protomer, 134
Proton. See Hydrogen ion
Protoplasm, 14, 20
Protoplasmic streaming, 308, 311
Protoplasts, 14, 393
Protozoa, mitosis in, 405
Provirus, 40, 42
  induction of, 40
  in malignancy, 443
Pseudoplasmodium, 434
Pseudopod, 28, 311, 312
Ptashne, M., 359

Punnett square, 321
Purine, 99, 114
  nucleosides and nucleotides, 115
Pyran, 109
Pyranose, 109
Pyrenoid, 206
Pyridine, 99
Pyrimidine, 99, 114, 115
  nucleosides and nucleotides, 115
Pyrophosphate (PPᵢ), 116
  in ATP, 163
Pyruvate, 168
  and citric acid cycle, 196
  in glycolysis, 168
  oxidative decarboxylation of, 195
  in photosynthesis, 217, 219
  reduction of, 173
Pyruvate dehydrogenase, 195
  mitochondrial location, 199
Pyruvate kinase, 168
  isozymes, 171
  reaction, 171

Q

Quantasomes, 205
  composition, 205
Quantum (of energy), 208
Quaternary structure. See Protein
Quinones, 188
  in electron transport, 187

R

R17 (bacteriophage), 350
Rabies, 35, 43
Racker, E., 200
Radiation
  electromagnetic, 207
  as mutagen, 393
Random coils (in proteins), 131
Raney nickel, 344
Ranvier, nodes of, 273
Reactions, chemical, 80
  energy changes, 67
  enthalpy driven, 67
  entropy driven, 67
  spontaneous, 67
Recessive traits, 319
Red blood cells. See Erythrocytes
Redox reactions, 172
Reducing sugars, 107
Reductions, 173
Reduction division, 408. See also Meiosis
Refractory period, 275
  of muscle, 300
  of neurons, 277
Regeneration (regrowth), 439
Regulons, 359
Release factors, 350
Replicase, 388
Replication fork, 387, 388
Replicons, 397, 398
Repressor, genetic, 354
  isolation, 358
Resolution, in microscopy, 48
Resonance, 97
Respiratory chain, 187

Respiratory research, history, 181
Rhizoid (of Acetabularia), 430
Riboflavin, 173, 174
  and respiration, 187
Ribofuranose, 108. See also Ribose
Ribonuclease, 337, 432, 433
  in DNA replication, 390
Ribonucleic acid. See RNA
Ribonucleoside, 114
Ribose, 108
  in nucleic acids, 113
  in RNA, 114
Ribosome, 347, 348, 349
  30S, 50S, 70S, 348
  40S (37S), 60S, 80S, 348, 349
  aminoacyl site, 346
  bacterial, 15
  chloroplast, 224
  eucaryotic, 348
    rRNA synthesis, 371
  mitochondrial, 224
  peptidyl site, 346
  physical characteristics, 348
  polysomes (polyribosomes), 346, 347
  procaryotic, 348
    vs. eucaryotic, 224
  rRNA synthesis, 371
  translocation, 346
Ribulose diphosphate, 215, 216
Ribulose diphosphate carboxydismutase,
  215
Ribulose diphosphate carboxylase, 215
Rickettsiae, 13
Ridgway, E. B., 298
RNA, 113, 331
  amount in tissues, 94
  base pairing, 117
  in DNA replication, 390
  and genes, 330
  heterogeneous nuclear (HnRNA), 334
  hybrids with DNA, 332
  messenger (mRNA), 331
    eucaryotic, 334
      turnover, 375
    masked, 372
    maternal messages, 372
    processing, 334, 373
    rate of synthesis, 360
    structure, 375
    synthetic, 338
    translational control, 375
    turnover
      in bacteria, 332
      in eucaryotes, 375
  nuclear (HnRNA), 334
  polymerase, 333
    isolation, 360
    sigma factors, 360
  polymerization, 116
  ribosomal (rRNA), 348
    sizes, 348
    synthesis in eucaryotes, 371
    soluble. See transfer RNA (tRNA)
  transfer (tRNA), 342
    alanyl, 342
    aminoacyl-tRNA, 344
    charged, 345
    in codon assignments, 338
    formylmethionine, 349
    phenylalanyl, 343

RNA, transfer *(cont.)*
  wobble hypothesis, 345
Robertson, J. D., 235
Rootlet, flagellar, 16, *17*
Rous, P. F., 443
Rous sarcoma virus, 444
rRNA. *See* RNA, ribosomal
Rudin, D. O., 278

S

Saccharides, 107
  optical activity, 107
Salmonella, 12
Salt bridge, in proteins, *132*
Saltatory conduction, 279
Sanger, F., 124, 336, 349
Sarcolemma, 286
Sarcomas, 443
Sarcomeres, 286, *287*
Sarcoplasm, 286
Sarcoplasmic reticulum, 296
Sarcosomes, 180, 286. *See also* Mitochondria
Sarcotubular system, 295
Saunders, J. W., Jr., 449
Secondary structure (of proteins), 125
Second law of thermodynamics, 68
Secretory granule, 22, 352
Sedoheptulose phosphate, *216*
Seed, *412*, 413
Sendai virus, 231, *233*
Senescence, 446
Serine (ser), *106*, *122*
  in chymotrypsin and trypsin, 148
Serotonin, 281
Scanning electron microscopy. *See* Microscopy, scanning
Schachman, H. K., 161
Schimke, R. T., 376
Schizokinen, 251
Schizomycetes. See Bacteria
Schleiden, M. J., 3, 7
Schramm, G. S., 43, 330
Schwann cells, 270, *271*
Schwann, Theodor, 3, 7
Scrapies, 45
Sheath, procaryotic, 13
Short-range bonds, 78
Sialic acid, 240
Sickle cell anemia, 335, *336*
Sigma factors, 360
Singer, S. J., 238
Skin, *9*, *127*
Skou, J. C., 254
Slack, C. R., 217
Slater, E. C., 189
Slime molds, 413
  cellular, 434
Smallpox virus, 34
Soap, 77, 102
Sodium ion
  amount in humans, 95
  mobility, 75
  size, 74
Sodium pump, 254, *255*
  electrogenic, 255
  in neurons, 274
  osmotic regulation, 256

Solubility, determining factors, 76
Soma, of neurons, 268
Somatic hybrid, 231
Somatotrophic hormone, 423
Soret band, 209
SP8 (bacteriophage), 334
Sperm. *See* Spermatozoa
Spermatid, *411*
Spermatocyte, *411*
Spermatogenesis, 410, *411*
Spermatogonium, *411*
Spermatozoa, *411*, *412*
Spheroplast, bacterial, 14
Sphingolipids, 102
Sphingosine, 102
Spiegelman, S., 332
Spike potential, 267, 274, *275*. *See also* Nerve, Neuron, and Synapse
conduction
  cable effect, 277, 279
  velocity, 279, 280
  flicker fusion, 277
  initiation of, 277
  in perfused axons, 278
  propagation. *See also* conduction and transmission
    chemical, 277, 278
    local currents, 277, 278
    saltatory, 277, 279
  refractory period, 275, 277
  in synthetic membranes, 278
  transmission, 277
Spindle apparatus, 403
Spirilla, 11
Spirillum, 12
Spontaneous generation theory, 10
Spore, 383, 412
Sporophyte, 412
Squid, neurons of, 273
Srinivasan, B. D., 422
sRNA. *See* RNA, transfer
Stacking interactions, DNA, 119
Stahl, F. W., 384
Standard state free energy, 69, 82
Stanley, Wendell, 42, 43, 330
Staphylococcus, 11, *12*
Starch, 111
  animal. *See* Glycogen
  photosynthesis of, 214–217
Steady state hypothesis and enzymes, 150
Stearic acid, 103
Stereoisomers, 96, 97, 107
  of amino acids, 121, *123*
Steroids, 102, 106
  receptor proteins, 369
Sterols, 106. *See also* Steroids
Steward, F. C., 428
Streptococcus, 11, *12*
Streptomycin, 347
  resistance to, 347
Substrate, 144, 162
Subunits (protein), 133
Succinate, *196*, 197
  in citric acid cycle, 195, *196*, 197
Succinate dehydrogenase, 197
  mitochondrial location, 199
  as mitochondrial marker, *306*
Succinyl CoA, *196*, 197
Succinyl thiokinase, 197
Succulents, 219

Sucrose, *109*, 110
  cell permeability to, 241
  energy of hydrolysis, 164
  formation, 109
Sulfates, pK_a, 103
Sulfonic acids, 100
Sulfur, biological prevalence, 95
Sutherland, E. W., 368
Sutton, Walter, 321
SV40, 444
Synapse, 269, *281*, *282*
  electrotonic, 281
  vesicles, *281*, *282*, *302*, *303*
Synapsis, 409
Synaptic boutons, 282
Synaptic knobs, 269
Synaptic vesicles, *281*, *282*, *302*, *303*
Synaptinemal complex, 409
Syncytia, 413
  muscle, 286

T

T system (of muscle), 296
T2. *See* Bacteriophage
T4. *See* Bacteriophage
Tatum, E. L., 325
Taylor, J. H., 395
Taylor, R. E., 300
TCA cycle. *See* Citric acid cycle
Teichoic acid, 14
Telophase. *See* Mitosis or Meiosis
Temin, Howard, 444
Teratocarcinoma, 446
Terminal cisternae, 296
Terpenes, 102
Tertiary structure. *See* Protein
Tetanus (muscle), 300
Tetrad, 410
Tetrahymena, 4
Theophylline, 421
Thermodynamics, 66
  first law, 67
  second law, 67
  third law, 68
Threonine (thr), *122*
Thylakoids
  of chloroplasts, 204, *205*
Thymidylic acid, 115
Thymine, *115*
  nucleosides and nucleotides, 115
  tautomers, 392
TMV. *See* Virus, tobacco mosaic
Tobacco mosaic virus. *See* Virus, tobacco mosaic
Tobacco rattle virus, *44*
Todaro, George, 444
Tonus (muscle), *300*
TPN$^+$ TPNH, 174. *See also* NADP$^+$
Transcriptase, 333
  reverse, 444
Transcription, 333
  mitochondrial DNA, 334
  rate of synthesis, 360
  strand selectivity, 333
Transformation
  of animal cells, 46
  bacterial, 328
Translation, 334

Translocation (of ribosomes), 346
Transverse tubule, 288, 296, 297, 300
Triad (in muscle), 296
Tricarboxylic acid cycle. *See* Citric acid
 cycle
Triglyceride, 104
Trinitrophenol, *250*
Triose, 107
Triosephosphate isomerase, 168
Triphosphopyridine nucleotide, 174.
 See also NADP⁺
Trisaccharide, 110
Tritium, 386
tRNA. *See* RNA, transfer
Tropocollagen, 129, *130*
Tropomyosin, 289, *290*
 in non-muscle cells, 311
 regulation of contraction, 299
Troponin, 289, *290*
 regulation of contraction, 299
Trypsin mechanism, 148
Tryptophan (try), *122*
Tryptophan pyrrolase, 376
Tryptophan synthetase, 374
Tubocurarine, 283, 284
Tubulin, 309
Tumors, 440
Turaniella, 4
Twort, F. W., 36, 43
Tyrian purple, 298
Tyrosine (tyr), *122*
 pK_a, 122

U

Ubiquinone, *188*
 in electron transport, 187
Ultramicrotome, *57*
Ultraviolet light
 DNA repair, 389
UMP, 115
Uracil, *115*
 nucleosides and nucleotides, 115
 paired with inosine, 345
Urea
 first synthesis, 5
 membrane permeability, *245*
Uridine, 115
Uridylic acid, 115
Uroid, 311
Uronychia, 4

V

Vacuole
 digestive, *260, 261, 262*
 of plants, *17, 26, 27*
 secretory, 22

Vagus nerve, 284
 effect on heart, 284
Valine (val), *122*
Valinomycin, 249, *250*
 as model membrane carrier, 250
van der Waals, J. C., 78
van der Waals forces, 78
 in DNA, 119
van der Waals radius, 80
Vasil, V., 428
Vegetable oil, 102
Velocity (enzymatic reactions), 149
Virchow, Rudolf, 3, 7
Virion. *See* Virus
Viroid, 45
Virus
 adenovirus, *43, 45*
 animal, 35
  host cell penetration, 45
 bacterial. *See* Bacteriophage
 canine hepatitis, *45*
 comparison with cells, 46
 cucumber mosaic, *44*
 discovery, 34
 diseases caused by, 35
 filterable, 35
 leukemia, 35, *46, 47*
 lysogenic, 40, *42*
 membrane, 45
 membrane receptors, 240
 oncogenic, 444
 plant, 35
 polio, translation of, 351
 polyoma, 444
 Rous sarcoma, 444
 Sendai, 231, *233*
 SV40, 444
 tobacco mosaic, 35, 42, *43*
  chemical composition, 330
  crystallization, 330
  infectious RNA, 330, *331*
  subunit structure, 135
 tobacco rattle, *44*
Vitalism, 5
Vitamins
 A, 102
 ascorbic acid (C), 109
 biotin, 325
 D, 102
 E, 102
 K, 102
 niacin, 100
Volkin, E., 332
Voltage clamp, 276

W

Wald, George, quote, 60
Waller, J. P., 350

Warts, 35
Water
 as acid or base, 87
 amount in cells and tissues, 94
 bound, 71, 74
 formation, 142
  enthalpy of, 143
 hard, 102
 intracellular, 89
 ionic mobilities in, 89
 and ions, 74
 membrane permeability, 244, *245*, 246
 molecular diameter, 246
 by oxygen reduction, 175
 reactions of, summary, 89
 solvent properties, 73
 structure, 73
 volume changes with temperature, 73
Watson, J. D., 117
Waxes, 102
Weismann, A., 428
Wilkins, Maurice, 119
Williams, Robley, 43
Wilson, E. B., 321
 quote from, 2
Wobble hypothesis, 339, 345
Wöhler, Friedrich, 5
Wool, 128, *129*
Work, in biological systems, 66
Wyman, J., 144, 154

X

Xenopus
 cloning, 429
 gene amplification, 372
X-rays, 207
X-ray diffraction
 of DNA, 119
 of hemoglobin and myoglobin, 135
 of proteins, *134*
Xylulose 5-phosphate, *216*

Y

Yeast, *383*
 budding, 383
Young, J. Z., 273

Z

Z lines, 286, 287
Zinc, amount in humans, 95
Zona pellucida, *412*
Zwitterions, 123
Zygote, 412

467